Lecture Notes in Computer Science 9709

Commenced Publication in 1973
Founding and Former Series Editors:
Gerhard Goos, Juris Hartmanis, and Jan van Leeuwen

Arnold Beckmann · Laurent Bienvenu
Nataša Jonoska (Eds.)

Pursuit
of the Universal

12th Conference on Computability in Europe, CiE 2016
Paris, France, June 27 – July 1, 2016
Proceedings

 Springer

Editors
Arnold Beckmann
Computer Science
Swansea University
Swansea
UK

Laurent Bienvenu
IRIF
Université Paris-Diderot - Paris 7
Paris
France

Nataša Jonoska
Mathematics and Statistics
University of South Florida
Tampa, FL
USA

ISSN 0302-9743 ISSN 1611-3349 (electronic)
Lecture Notes in Computer Science
ISBN 978-3-319-40188-1 ISBN 978-3-319-40189-8 (eBook)
DOI 10.1007/978-3-319-40189-8

Library of Congress Control Number: 2016940339

LNCS Sublibrary: SL1 – Theoretical Computer Science and General Issues

Printed on acid-free paper

This Springer imprint is published by Springer Nature
The registered company is Springer International Publishing AG Switzerland

Preface

CiE 2016: Pursuit of the Universal
Paris, France, June 27 – July 1, 2016

This year Computability in Europe (CiE) honored the 80th anniversary of A. Turing's paper introducing the Universal Turing Machine. In this context the conference sought better understanding of universal computational frameworks ranging from mathematics, computer science, through various natural sciences such as physics and biology. CiE provides a forum for exchanging ideas on broad aspects of "computability" striving to understand the essence of computation through studies of theoretical models of new paradigms, information processing, encryption, philosophy and history of computing as well as computability in natural and biological systems. This year's CiE conference was held in Paris and through a sequence of tutorials, plenary lectures, and special sessions allowed in-depth discussions and novel approaches in pursuit of the nature of computability. Similarly to previous CiE conferences in this series, CiE 2016 had a broad scope promoting the development of computability-related science.

The conference series is organized under auspices of the Association CiE. The association promotes the development of all areas of mathematics, computer science, as well as natural and engineering sciences that study the notion of "computability," including its philosophical and historical developments. The conference series is a venue where researchers in the field meet and exchange the most novel features of their findings.

CiE 2016 was organized jointly by Université Paris 13 and Université Paris 7, chaired by Paulin de Naurois at Université Paris 13. The previous CiE conferences were held in Amsterdam (The Netherlands) in 2005, Swansea (Wales) in 2006, Siena (Italy) in 2007, Athens (Greece) in 2008, Heidelberg (Germany) in 2009, Ponta Delgada (Portugal) in 2010, Sofia (Bulgaria) in 2011, Cambridge (UK) in 2012, Milan (Italy) in 2013, Budapest (Hungary) in 2014, and Bucharest (Romania) in 2015. The proceedings containing the best submitted papers as well as extended abstracts of invited speakers for all these meetings are published in the Springer series *Lecture*

Notes in Computer Science. The annual CiE conference has risen to be the largest international meeting focused on computability theory issues. CiE 2017 will be held in Turku, Finland. The leadership of the conference series recognizes that there is under-representation of female researchers in the field of computability and therefore incorporates a special session of Women in Computability (WiC) in every CiE conference. WiC was initiated in 2007, and was first funded by the Elsevier Foundation, later taken over by the publisher Elsevier. This year's program, organized by Liesbeth De Mol, besides the regular workshop also provided travel grants for junior female researchers and a mentorship program.

The 39-member Program Committee of CiE 2016 was chaired by Laurent Bienvenu (IRIF, CNRS, and Université Paris 7, France), and Nataša Jonoska (University of South Florida, Tampa, USA). The committee selected the plenary speakers and the special session organizers, and ran the reviewing process of all the regular contributions submitted. We received 40 non-invited contributed paper submissions. Each paper received at least three reviews by the Program Committee and additional reviewers. About 45 % of the submitted papers were accepted for publication in this volume. In addition, this volume contains 19 extended abstracts/papers contributed by plenary speakers and speakers of the invited sessions. The production of the volume would have been impossible without the diligent work of all of the Program Committee members and our expert reviewers. We are very grateful to all the Program Committee members and the reviewers for their excellent work.

All authors who contributed to this conference were encouraged to submit significantly extended versions of their papers with unpublished research content to *Computability*, the journal of the Association CiE.

This year the conference started with a special session honoring the memory of Barry Cooper, one of the initiators and founders of the conference as well as a driving force behind the organization of CiE, including the presidency of the association. The session was organized by Mariya Soskova and the contributors were Theodore Slaman (University of California Berkeley), Andrea Sorbi (University of Siena), Dag Norman (University of Oslo), and Ann Copestake (Cambridge University).

Two tutorials were given by Bernard Chazelle from Princeton University, USA, and Mikolaj Bojanczyk from University of Warsaw, Poland. In addition, the Program Committee invited seven speakers to give plenary lectures: Natasha Alechina (University of Nottingham, UK), Vasco Brattka (Universität der Bundeswehr München, Germany), Delaram Kahrobaei (The City University of New York, USA), Steffen Lempp (University of Wisconsin, USA), André Nies (University of Auckland, New Zealand), Dominique Perrin (Université Paris-Est Marne-la-Vallée, France), and Reed Solomon (University of Connecticut, USA).

Springer generously funded two awards this year, the Best Student Paper Award and Best Paper Award. The winner of the Best Student Paper Award this year was Mikhail Andreev for his contribution "Busy Beavers and Kolmogorov Complexity." The Best Paper Award was given to Olivier Bournez, Nachum Dershowitz and Pierre Neron for their contribution "An Axiomatization of Analog Algorithms."

CiE 2016 has six special sessions: two sessions, Cryptography and Information Theory and Symbolic Dynamics, were organized for the first time in the conference series. The other four special sessions covered new developments in areas previously covered by the conference series: Computable and Constructive Analysis, Computation in Biological Systems, Weak Arithmetic, and History and Philosophy of Computing. Speakers in these special sessions were selected by the special session organizers and were invited to contribute a paper to this volume.

Computable and Constructive Analysis

Organizers. Daniel Graça and Elvira Mayordomo.
Speakers. Mathieu Hoyrup (Inria and University of Lorraine), Arno Pauly (University of Cambridge), Vela Velupillai (New School for Social Research in New York City and University of Trento), Martin Ziegler (KAIST, Daejeon).

Computation in Biological Systems

Organizers. Alessandra Carbone and Ion Petre.
Speakers. Daniela Besozzi (University of Milan-Biccocca), Eugen Czeizler (Åbo Akademi University), Vincent Moulton (University of East Anglia), Eric Tannier (Inria and University of Lyon).

Cryptography and Information Theory

Organizers. Danilo Gligoroski, and Carles Padro.
Speakers. Ludovic Perret (Université Pierre et Marie Curie, France), Ignacio Cascudo (Aarhus University in Denmark), Oriol Farras (Universitat Rovira i Virgili, Spain), Danilo Gligoroski (Norwegian University of Science and Technology - Trondheim).

History and Philosophy of Computing

Organizers. Alberto Naibo and Ksenia Tatarchenko
Speakers. Maël Pégny (IHPST, Paris), Pierre Mounier-Khun (CNRS), Simone Martini (University of Bologna), Walter Dean (University of Warwick).

Symbolic Dynamics

Organizers. Jarkko Kari and Reem Yassawi.
Speakers. Valérie Berthé (University of Paris 7), Emmanuel Jeandel (University of Lorraine), Irène Marcovici (University of Lorraine), Ronnie Pavlov (Denver University).

Weak Arithmetic

Organizers. Lev Beklemishev and Stanislav Speranski.
Speakers. Pavel Pudlák (Academy of Sciences of the Czech Republic), Alexis Bès (University of Paris 12), Leszek Kołodziejczyk (University of Warsaw), Albert Visser (University of Utrecht).

The organizers of CiE 2016 would like to acknowledge and thank the following for their financial support (in alphabetical order): the Association for Symbolic Logic (ASL), the European Association for Theoretical Computer Science (EATCS), and Springer. We would also like to acknowledge the support of our non-financial sponsor, the Association Computability in Europe (CiE).

April 2016

Arnold Beckmann
Laurent Bienvenu
Nataša Jonoska

Organization

Steering Committee

Arnold Beckmann	Swansea University, UK, Chair
Laurent Bienvenu	LIAFA, CNRS and Université de Paris, France
Paola Bonizzoni	Università di Milano-Bicocca, Italy
Alessandra Carbone	Université Pierre et Marie Curie, France
Nataša Jonoska	University of South Florida, USA
Benedikt Löwe	Universiteit van Amsterdam, The Netherlands
Florin Manea	Christian-Albrechts-Universität zu Kiel, Germany
Dag Normann	The University of Oslo, Norway
Mariya Soskova	Sofia University, Bulgaria
Susan Stepney	University of York, UK

Program Committee

Marcella Anselmo	University of Salerno, Italy
Nathalie Aubrun	CNRS, Inria, UCBL, Université de Lyon, France
Georgios Barmpalias	Chinese Academy of Sciences, China
Marie-Pierre Béal	Université Paris-Est, France
Arnold Beckmann	Swansea University, UK
Laurent Bienvenu	LIAFA, CNRS, Université de Paris 7, France
Paola Bonizzoni	Università di Milano-Bicocca, Italy
Alessandra Carbone	Université Pierre et Marie Curie, France
Douglas Cenzer	University of Florida, USA
Liesbeth De-Mol	CNRS, Université de Lille 3, France
Volker Diekert	University of Stuttgart, Germany
David Doty	California Institute of Technology, USA
Jérôme Durand-Lose	LIFO, Université D'Orléans, France
Martin Escardo	University of Birmingham, UK
François Fages	Inria Paris-Rocquencourt, France
Enrico Formenti	Nice Sophia Antipolis University, France
Daniela Genova	University of North Florida, USA
Noam Greenberg	Victoria University of Wellington, New Zealand
Hajime Ishihara	JAIST, Japan
Paulin Jacobé De Naurois	CNRS, LIPN, Université Paris 13, France
Nataša Jonoska	University of South Florida, USA
Jarkko Kari	University of Turku, Finland
Lila Kari	University of Western Ontario, Canada
Margarita Korovina	A.P. Ershov Institute of Informatics Systems, Russia

Marta Kwiatkowska	University of Oxford, UK
Karen Lange	Wellesley College, USA
Benedikt Löwe	Universiteit van Amsterdam, The Netherlands
Florin Manea	Christian-Albrechts-Universität zu Kiel, Germany
Keng Meng Selwyn Ng	Nanyang Technological University, Singapore
Arno Pauly	University of Cambridge, UK
Mario Perez-Jimenez	University of Seville, Spain
Ion Petre	Åbo Akademi University, Finland
Alexis Saurin	PPS and Inria Pi.R2, France
Shinnosuke Seki	University of Electro-Communications, Tokyo, Japan
Paul Shafer	Ghent University, Belgium
Alexander Shen	LIRMM CNRS and University of Montpellier 2, France
Alexandra Soskova	Sofia University, Bulgaria
Mariya Soskova	Sofia University, Bulgaria
Peter van Emde Boas	ILLC-FNWI-Universiteit van Amsterdam, The Netherlands

Additional Reviewers

Asher, Nicholas	Kim, Hwee	Rizzi, Romeo
Barash, Mikhail	Kleinberg, Samantha	Romashchenko, Andrei
Bauwens, Bruno	Kreuzer, Alexander P.	Salo, Ville
Besson, Tom	Kudinov, Oleg	Shafer, Paul
Bodirsky, Manuel	Kuyper, Rutger	Shlapentokh, Alexandra
Calvert, Wesley	Manzoni, Luca	Sibelius, Patrick
Cherubini, Alessandra	Melnikov, Alexander	Sieweck, Philipp
Crabbé, Benoit	Michel, Pascal	Steiner, Rebecca
Cramer, Marcos	Minnes, Mia	Stephan, Frank
Cristescu, Ioana	Monin, Benoît	Szymanik, Jakub
Delacourt, Martin	Nemoto, Takako	Towsner, Henry
Dolce, Francesco	Papazian, Christophe	van Leeuwen, Jan
Ehlers, Thorsten	Paskevich, Andrei	Vatev, Stefan
Fokina, Ekaterina	Petre, Luigia	Westrick, Linda Brown
Hirschfeldt, Denis	Place, Thomas	Yokoyama, Keita
Johannsen, Jan	Porreca, Antonio E.	Zandron, Claudio
Jolivet, Timo	Raimondi, Franco	Ziegler, Martin
Kach, Asher	Rin, Benjamin	

S. Barry Cooper
1943 – 2015

Barry Cooper at the opening of CiE 2009 in Heidelberg.
Photo taken by Peter van Emde Boas, July 2009.

Barry Cooper, founding member and former president of the Association Computability in Europe, died on October 26, 2015, shortly after his 72nd birthday. Born on October 9, 1943, Barry was a leading figure in the UK logic scene all of his academic life, a major figure in computability theory, and in particular degree theory. Most relevant in the context of CiE 2016 is of course that Barry was the driving force of Computability in Europe and without him, our association would not exist. This text is focused on Barry in relation to the Association Computability in Europe, it is based on a short obituary by Benedikt Löwe and Dag Normann [1], and borrows from it with due permission of the authors.

Barry retired from the office of President of the Association CiE in summer 2015, and had the chance to close the association AGM in Bucharest in July 2015 with a speech reminiscing about the history of the association. Barry was very fond of telling the ironic tale of how our association comprising more than a thousand members grew out of a rejected application for European funding.

In order to discuss the negative feedback of the referees, it was decided to have a conference in Amsterdam, which became the first CiE conference. Barry's vision and guidance pushed us along the way, to subsequent CiE conferences and finally to the formal formation of this association in 2008. In 2007 and 2012, he personally co-chaired the Program Committees of the CiE conferences in Siena and Cambridge; the fact that these two events were the two largest CiE conferences to date is a testament to Barry's infectious enthusiasm and inclusive attitude. Barry also realized the potential of the Turing Centenary and made sure that the 100th birthday of Alan Turing was appropriately celebrated during the Alan Turing Year, not just in the UK, but all across the globe; at Turing's alma mater in Cambridge, Barry was one of the organizers of a six-month Turing-related research program at the Isaac Newton Institute for Mathematical Sciences culminating on Turing's 100th birthday, June 23, 2012, on the lawn in front of King's College. In the years after the centenary, Barry renamed the Alan Turing Year to Alan Turing Years. As the media attention to Alan Turing grew, partly due to the Academy-award winning movie *The Imitation Game,* Barry became one of Alan Turing's spokespeople on Twitter and in opinion pieces for *The Guardian.*

A comprehensive account of Barry's impact on Computability in Europe by Benedikt Löwe [2] has been published in the association's journal *Computability.*

Barry was very influential in shaping our thinking about computability in much broader, interdisciplinary terms, which was key to the success of the movement Computability in Europe. His vision will continue to live in us; his stimulating remarks and kindness will be very much missed.

April 2016

Arnold Beckmann
Laurent Bienvenu
Nataša Jonoska

References

[1] Löwe, B., Normann, D.: Barry Cooper (1943–2015). Obituary published on the website of the Association CiE, 28 October 2015
[2] Löwe, B.: Barry Cooper (1943–2015): The Engine of Computability in Europe. Computability **5**(1), 3–11 (2016)

Contents

Contributed Papers

Invited Papers

Verifying Systems of Resource-Bounded Agents

Natasha Alechina$^{(\boxtimes)}$ and Brian Logan

University of Nottingham, Nottingham, UK
{nza,bsl}@cs.nott.ac.uk

Abstract. Approaches to the verification of multi-agent systems are typically based on games or transition systems defined in terms of states and actions. However such approaches often ignore a key aspect of multi-agent systems, namely that the agents' actions require (and sometimes produce) resources. We briefly survey previous work on the verification of multi-agent systems that takes resources into account, and outline some key challenges for future work.

1 Verifying Autonomous Systems

A multi-agent system (MAS) is a system that is composed of multiple interacting agents. An agent is an *autonomous* entity that has the ability to collect information, reason about it, and perform actions based on it in pursuit of its own goals or on behalf of others. Examples of agents are controllers for satellites, non-driver transport systems such as UAVs, health care systems, and even nodes in sensor networks.

Multi-agent systems are ubiquitous. Many distributed software and hardware systems can be naturally modelled as multi-agent systems. Such systems are by the nature of their components extremely complex, and the interaction between components and their environment can lead to undesired behaviours that are difficult to predict in advance. With the increasing use of autonomous agents in safety critical systems, there is a growing need to *verify* that their behaviour conforms to the desired system specification, and over the last decade verification of multi-agent systems has become a thriving research area [24].

A key approach to the verification of MAS is *model checking*. Model checking involves checking whether a model of the system satisfies a temporal logic formula corresponding to some aspect of the system specification. Model checking has the advantage that it is a fully automated technique, which facilitates its use in the MAS development process.[1] A wide range of approaches to model-checking MAS have been proposed in the literature, ranging from the adaptation of standard model-checking tools, e.g., [12,13] to the development of special-purpose model checkers for multi-agent systems, e.g., [22,27].

---

[1] Another strand of work focusses on theorem proving, e.g., [28], but such approaches typically require user interaction to guide the search for a proof.

© Springer International Publishing Switzerland 2016
A. Beckmann et al. (Eds.): CiE 2016, LNCS 9709, pp. 3–12, 2016.
DOI: 10.1007/978-3-319-40189-8_1

2   Resource-Bounded Agents

In many multi-agent systems, agents are *resource-bounded*, in the sense that they require resources in order to act. Actions require time to complete and typically require additional resources depending on the application domain, for example energy or money. For many applications, the availability or otherwise of resources is critical to the properties we want to verify: a multi-agent system will have very different behaviours depending on the resource endowment of the agents that comprise it. For example, an agent with insufficient energy may be unable to complete a task in the time assumed by a team plan, if it has to recharge its battery before performing the task.

However, with a few exceptions which we discuss below, previous work on verification of MAS abstracts away from the fact that many multi-agent systems consist of agents that need resources to operate and that those resources are limited. In particular, current state-of-the-art verification techniques and tools for MAS are unable to verify system properties that depend on the resource production and consumption of the agents comprising the MAS.

In this paper we survey recent work in the emerging field of verification of resource-bounded agents, and highlight a number of challenges that must be overcome to allow practical verification of resource-bounded MAS. We argue that recent work on the complexity of model-checking for logics of strategic ability with resources offers the possibility of significant progress in the field, new verification approaches and tools, and the ability to verify the properties of a large, important class of autonomous system that were previously out of reach.

3   Model-Checking with Resources

In this section we give a brief introduction to model-checking multi-agent systems and explain how standard model checking approaches can be extended with resources.

In a model-checking approach to the verification of multi-agent systems, a MAS is represented by a finite state transition system.[2] A state transition system consists of a set of states and transitions between them. Intuitively, each state of a MAS corresponds to a tuple of states of the agents and of the environment, and each transition corresponds to actions performed by the agents. Each state is labelled with atomic propositions that are true in that state. A standard assumption is that each state in the system has at least one outgoing transition (if a state is a deadlock state in the original MAS, we can model this by adding a transition to itself by some null action and labelling it with a 'deadlock' proposition). Properties of the system to be verified are expressed in an appropriate temporal logic L. The *model-checking problem for L* is, given a state transition system M (and possibly a state s) and an L formula ϕ, check whether ϕ is true in M (at state s).

---

[2] There is work on model-checking infinite state transition systems, see, for example, [11], but in this paper we concentrate on the finite case.

For multi-agent systems, a temporal logic of particular interest is Alternating Time Temporal Logic (ATL) [9]. ATL generalises other temporal logics such as Computation Tree Logic (CTL) [19] (which can be seen as a one-agent ATL) by introducing notions of strategic ability. ATL is interpreted over concurrent game structures (transition systems where edges correspond to a tuple of actions performed simultaneously by all the agents, see the example below). The language of ATL contains atomic propositions, boolean connectives $\neg, \wedge$, etc. and modalities $\langle\!\langle A \rangle\!\rangle\bigcirc$, $\langle\!\langle A \rangle\!\rangle\square$ and $\langle\!\langle A \rangle\!\rangle\mathcal{U}$ for each subset (or coalition in ATL terms) A of the set of all agents, which express the strategic ability of the coalition A. $\langle\!\langle A \rangle\!\rangle\bigcirc\phi$ means that the coalition of agents A has a choice of actions such that, regardless of what the other agents in the system do, ϕ will hold in the next state. $\langle\!\langle A \rangle\!\rangle\square\phi$ means that coalition A has a strategy to keep ϕ true forever, regardless of what the other agents do. A strategy is a choice of actions which either only depends on the current state (memoryless strategy) or on the finite history of the current state (perfect recall strategy). Finally, $\langle\!\langle A \rangle\!\rangle\phi\,\mathcal{U}\,\psi$ means that A has a strategy to ensure that after finitely many steps ψ holds, and in all the states before that, ϕ holds. The model-checking problem for ATL can be solved in time polynomial in the size of the transition system and the property [9], and there exist model-checking tools for ATL, for example, MOCHA [10] and MCMAS [27].

Example. Fig. 1 illustrates a simple ATL model of a system with two agents, 1 and 2, and actions α, β, γ and *idle*. Action tuples on the edges show the actions of each agent, for example, in the transition from state s_I to s, agent 1 performs action α and agent 2 performs *idle*. In this system, in state s_I, agent 1 has a (memoryless) strategy to enforce that p holds eventually in the future no matter what agent 2 does, which can be expressed in ATL as $\langle\!\langle\{1\}\rangle\!\rangle\top\,\mathcal{U}\,p$. Similarly, in s_I agent 1 has a memoryless strategy to keep $\neg p$ true forever, so $\langle\!\langle\{1\}\rangle\!\rangle\square\neg p$ holds in s_I.

3.1 Adding Resources

In order to model multi-agent systems where agents' actions produce and consume resources, it is necessary to modify the approach above in two ways. One is to add resource annotations to the actions in the transition system: for each individual action and each resource type, we need to specify how many units of this resource type the action produces or consumes. For example, suppose that there are two resource types, r_1 and r_2 (e.g., energy and money). Then we can specify that action α in Fig. 1 produces two units of r_1 and consumes one unit of r_2, action β consumes one unit of r_1 and produces one unit of r_2, action γ consumes five units of r_1, and action *idle* does not produce or consume any resources.

The second modification is to extend the temporal logic so that we can express properties related to resources. For example, we may want to express a property that a group of agents A can eventually reach a state satisfying ϕ or can maintain the truth of ψ forever, provided that they have available n_1 units of resource type r_1 and n_2 units of resource type r_2. Such statements about coalitional ability under resource bounds can be expressed in an extension of ATL where coalitional modalities are annotated with a resource bound on the strategies available to the coalition. We call logics where every action is associated with produced and consumed resources and the syntax reflects resource requirements on agents, *resource logics*.

To illustrate the properties resource logics allow us to express, consider the model in Fig. 1 with the production and consumption of resources by actions specified above. In this setting, we can verify if agent 1 can eventually enforce p provided that it has one unit of r_2 in state s_I, or whether the coalition of agents $\{1, 2\}$ can achieve p under this resource bound by working together. There are surprisingly many different ways of measuring costs of strategies and deciding which actions are executable by the agents given the resources available to them, but under at least one possible semantics, the answer to the first question is no and to the second one yes, but the latter requires a perfect recall strategy (the two agents should loop between states s_I and s until they produce a sufficient amount of resource r_1, and then execute actions corresponding to the $\langle \gamma, idle \rangle$ transition from s to s').

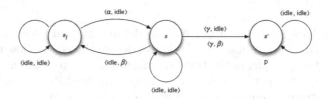

Fig. 1. State transition system.

Clearly, the model-checking problem for temporal logics is a special case of the model-checking problem for the corresponding resource logics. The question is, how much harder does the model-checking problem become when resources are added?

4   A Brief Survey of Resource Logics

In this section, we briefly review recent theoretical work on the development of resource logics. We focus on expressiveness and model-checking complexity, as these features determine the suitability of a particular logic for practical verification.

4.1   Consumption of Resources

Early work on resource logics considered only consumption of resources (no action produced resources), and initial results were encouraging.

One of the first logics capable of expressing resource requirements of agents was a version of Coalition Logic (CL)[3], called Resource-Bounded Coalition Logic (RBCL), where actions only consume (and don't produce) resources. It was introduced in [1] with the primary motivation of modelling systems of resource-bounded reasoners, however the framework is sufficiently general to model any kind of action. The model-checking problem for this logic was shown to be decidable in [5] in polynomial time in the transition system and the property and exponential in the number of resource types.

A resource-bounded version of ATL, RB-ATL, where again actions only consume (and not produce) resources was introduced in [2]. It was also shown that the model-checking problem for this logic is decidable in time polynomial in the size of the transition system and exponential in the number of resource types. (For a single resource type, e.g., energy, the model-checking problem is no harder than for ATL.)

Practical work on model-checking standard computer science transition systems (not multi-agent systems) with resources also falls in the category of consumption-only systems, for example probabilistic model-checking of systems with numerical resources as done using PRISM model-checker [26] assumes costs monotonically increasing with time.

4.2 Adding Production

However, when resource production is considered in addition to consumption, the situation changes. In a separate strand of work, a range of different formalisms for reasoning about resources was introduced in [14,16]. In those formalisms, both consumption and production of resources was considered. In [15] it was shown that the problem of halting on empty input for two-counter automata [25] can be reduced to the model-checking problem for several of their resource logics. Since the halting problem for two-counter automata is undecidable, the model-checking problem for a variety of resource logic with production of resources is undecidable. The reduction uses two resource types (to represent the values of the two counters) and either one or two agents depending on the version of the logic (whether the agents have perfect recall, whether the formula talking about coalition A can also specify resource availability for remaining agents, and whether nested operators 'remember' initial allocation of resources or can be evaluated independently of such initial allocation).

The only decidable cases considered in [14] are an extension of CTL with resources (essentially one-agent ATL) and a version where on every path only a fixed finite amount of resources can be produced. In [14], the models satisfying this property are called bounded, and it is pointed out that RBCL and RB-ATL are logics over a special kind of bounded models (where no resources are produced at all). Other decidability results for bounded resource logics have also been reported in the literature. For example, [20] define a decidable logic, PRB-ATL (Priced Resource-Bounded ATL), where the total amount of resources

---

[3] CL is a fragment of ATL with only the next time $\langle\!\langle A \rangle\!\rangle \bigcirc$ modality.

in the system has a fixed bound. The model-checking algorithm for PRB-ATL requires time polynomial in the size of the model and exponential in the number of resource types and the resource bound on the system. In [21] an EXPTIME lower bound in the number of resource types for the PRB-ATL model-checking problem is shown.

A general logic over systems with numerical constraints called QATL* was introduced in [17]. In that paper, more undecidability results for the model-checking problem of QATL* and its fragments were shown. For example, QATL (Quantitative ATL) is undecidable even if no nestings of cooperation modalities is allowed. The main proposals for restoring decidability to the model-checking problem for QATL in [17] are removing negative payoffs (similar to removing resource production) and also introducing memoryless strategies. Shared resources were considered in [18]; most of the cases considered there have undecidable model-checking (apart from the case of a single shared resource, which has decidable model-checking).

This brief survey of work to date suggest that the main approach until recently to dealing with both resource production and consumption was to bound the amount of produced resources globally in the model. For some systems of resource-bounded agents, this is a reasonable restriction. For example, agents that need energy to function and are able to charge their battery, can never 'produce' more energy than the capacity of their battery. This is a typical bounded system. However, in some cases, although every single application of the agent's actions produces a fixed amount of some resource, repeating this action arbitrarily often will produce arbitrarily large amounts of the resource. This may apply to energy stored in unbounded storage, or to money, or many other natural situations. Recent work suggests that verification of such systems may still be possible.

5   Decidable Unbounded Production

In [6] a version of ATL, RB$\pm$ATL, was introduced where actions both produce and consume resources. The models of the logic do not impose bounds on the overall production of resources, and the agents have perfect recall. The syntax of RB$\pm$ATL is very similar to that of ATL, but coalition modalities have superscripts which represent resource allocation to agents. Instead of stating the existence of *some* strategy, they state the existence of a strategy such that every computation generated by following this strategy consumes at most the given amount of resources. Coming back to the example, the property that agent 1 can eventually enforce p provided that it has one unit of r_2 can be expressed as $\langle\langle\{1\}^{(0,1)}\rangle\rangle\top\,\mathcal{U}\,p$[4] The model-checking problem for RB$\pm$ATL is decidable (RB$\pm$ATL is very similar to one of the resource logics introduced

---

[4] Here, $(0,1)$ is the allocation of 0 units of r_1 and 1 unit of r_2 to agent 1. We only show resource bound for the proponent agents, $\{1\}$ in this case. Versions of resource logic where opponents are also resource-bounded all have an undecidable model-checking problem, see [14].

in [14] for which the decidability of the model-checking problem was left open). The existence of a decidable resource logic with unbounded production was surprising, as it was the first indication that it is possible to automatically verify properties of this important class of resource-bounded multi-agent system.

However, although this result is encouraging, we are not yet at the point of practical verification of such systems. In [6] the lower bound on the complexity of the model-checking problem for $RB \pm ATL$ is shown to be EXPSPACE. The proof of EXPSPACE-hardness is by reduction of the reachability problem for Petri Nets to the model-checking problem for $RB \pm ATL$. Although the Petri Net reachability problem is decidable, the upper bound on its complexity is still unknown; similarly we do not know the upper bound on the $RB \pm ATL$ complexity. The complexity of the model-checking problem $RB \pm ATL$ is thus much higher than that for ATL without resources and the consumption-only resource logics surveyed above. This high complexity makes it difficult to develop practical verification approaches. The only exception is $1\text{-}RB \pm ATL$, $RB \pm ATL$ with a single resource type, where the complexity is PSPACE.

In [3], a new syntactic fragment FRAL of resource logic RAL with decidable model-checking has been identified. It restricts the occurrences of coalitional modalities on the left of the Until formulas; on the other hand, it allows nested modalities to refer to resource allocation at the time of evaluation, rather than always consider a fresh resource allocation, as in $RB \pm ATL$. More precisely, a formula $\langle\!\langle A^b \rangle\!\rangle \phi \mathcal{U} \langle\!\langle A^\downarrow \rangle\!\rangle \psi_1 \mathcal{U} \psi_2$ says that given resource allocation b, coalition A can always reach a state (maintaining ϕ) where with the remaining resources, it can reach ψ_2 while maintaining ψ_1. The complexity of model-checking for this fragment is also open, and is also likely to be high.

Although model-checking of ATL with perfect recall and uniform strategies is undecidable, replacing uniformity with a weaker notion, for example defining it using distributed knowledge, is decidable [23]. Similar results hold for $RB \pm ATL$ with syntactic epistemic knowledge and weaker notions of uniformity, $RB \pm ATSEL$ [4].

Below is a summary of resource logics with decidable model-checking problem. In all of them, the semantics assume that in every state each agent has an available action to do nothing, which produces and consumes no resources (see Table 1).

6   Future Challenges

The $RB \pm ATL$ results offer the possibility of significant progress in the verification of resource-bounded multi-agent systems. However many challenges remain for future research. Below we list three of the most important.

Understanding Sources of Undecidability. Developing a better understanding of the sources of decidability and undecidability (beyond boundedness) will be critical to future progress. As observed in [14], subtle differences in truth conditions for resource logics result in the difference between decidability and

Table 1. Resource logics with decidable model-checking problem

Logic	Resource production	Complexity of model-checking
RBCL	no	EXPTIME (PTIME in model)
RB-ATL	no	EXPTIME (PTIME in model)
PRB-ATL	bounded	EXPTIME
RB ± ATL	yes	EXPSPACE-hard
1-RB ± ATL	yes	PSPACE
FRAL	yes	?
RB ± ATSEL	yes	EXPSPACE-hard

undecidability of the model-checking problem. Preliminary work in this direction is reported in [3].

Lower Complexity. It is useful to discover sources of undecidability and how to construct expressive logics for which the model-checking problem is decidable. However, it is even more important to be able to develop logics, or fragments of existing logics such as RB ± ATL, that are sufficiently expressive for practical problems, but where the model-checking problem has tractable complexity (ideally polynomial in the size of the transition system, as in the case of bounded production logics). Only then would we be able to implement practical model-checking tools for systems of resource-bounded agents.

Practical Tools. Although model checking algorithms have been proposed for several of the logics surveyed, work on implementation is only beginning. We aim to develop practical model-checking tools for verifying resource-bounded MAS by extending the MCMAS model checker [27] to allow the modelling of multi-agent systems in which agents can both consume and produce resources. Work on symbolic encoding of RB-ATL model-checking is reported in [8] and work on symbolic encoding of RB ± ATL model-checking are reported in [7].

Addressing these challenges will allow practical model-checking of resource logics and constitute a major break-through in multi-agent system verification.

References

1. Alechina, N., Logan, B., Nguyen, H.N., Rakib, A.: A logic for coalitions with bounded resources. In: Proceedings of the 21st International Joint Conference on Artificial Intelligence (IJCAI 2009), vol. 2, pp. 659–664. IJCAI/AAAI, AAAI Press (2009)
2. Alechina, N., Logan, B., Nguyen, H.N., Rakib, A.: Resource-bounded alternating-time temporal logic. In: Proceedings of the 9th International Conference on Autonomous Agents and Multiagent Systems (AAMAS 2010), pp. 481–488. IFAAMAS (2010)

3. Alechina, N., Bulling, N., Logan, B., Nguyen, H.N.: On the boundary of(un)decidability: decidable model-checking for a fragment of resource agent logic. In: Yang, Q. (ed.) Proceedings of the 24th International Joint Conference on Artificial Intelligence (IJCAI 2015). AAAI Press, Buenos Aires, July 2015

4. Alechina, N., Dastani, M., Logan, B.: Verifying existence of resource-bounded coalition uniform strategies. In: Rossi, F. (ed.) IJCAI 2016, Proceedings of the 25th International Joint Conference on Artificial Intelligence. IJCAI/AAAI (2016)

5. Alechina, N., Logan, B., Nga, N.H., Rakib, A.: Logic for coalitions with bounded resources. J. Logic Comput. 21(6), 907–937 (2011)

6. Alechina, N., Logan, B., Nguyen, H.N., Raimondi, F.: Decidable model-checking for a resource logic with production of resources. In: Proceedings of the 21st European Conference on Artificial Intelligence (ECAI-2014), pp. 9–14. IOS Press, Prague, August 2014

7. Alechina, N., Logan, B., Nguyen, H.N., Raimondi, F.: Symbolic model-checking for oneresource RB+-ATL. In: Yang, Q. (ed.) Proceedings of the 24th International Joint Conference on Artificial Intelligence (IJCAI 2015). AAAI Press, Buenos Aires, July 2015

8. Alechina, N., Logan, B., Nguyen, H.N., Raimondi, F., Mostarda, L.: Symbolic model-checking for resource-bounded ATL. In: Proceedings of the 2015 International Conference on Autonomous Agents and Multiagent Systems, AAMAS 2015, pp. 1809–1810. ACM (2015)

9. Alur, R., Henzinger, T., Kupferman, O.: Alternating-time temporal logic. J. ACM 49(5), 672–713 (2002)

10. Alur, R., Henzinger, T.A., Mang, F.Y.C., Qadeer, S., Rajamani, S.K., Tasiran, S.: MOCHA: modularity in model checking. In: Hu, A.J., Vardi, M.Y. (eds.) Computer Aided Verification. LNCS, vol. 1247, pp. 521–525. Springer, Heidelberg (1998)

11. Belardinelli, F.: Verification of non-uniform and unbounded artifact-centric systems: decidability through abstraction. In: Bazzan, A.L.C., Huhns, M.N., Lomuscio, A., Scerri, P. (eds.) International conference on Autonomous Agents and Multi-Agent Systems, AAMAS 2014, pp. 717–724. IFAAMAS/ACM (2014)

12. Bordini, R.H., Fisher, M., Visser, W., Wooldridge, M.: Model checking rational agents. IEEE Intell. Syst. 19(5), 46–52 (2004)

13. Bordini, R.H., Fisher, M., Visser, W., Wooldridge, M.: Verifying multi-agent programs by model checking. J. Auton. Agents Multi-Agent Syst. 12(2), 239–256 (2006). http://dro.dur.ac.uk/622/

14. Bulling, N., Farwer, B.: On the (un-)decidability of model checking resource-bounded agents. In: Proceedings of the 19th European Conference on Artificial Intelligence (ECAI 2010). Frontiers in Artificial Intelligence and Applications, vol. 215, pp. 567–572. IOS Press (2010)

15. Bulling, N., Farwer, B.: On the (un-)decidability of model checking resource-bounded agents. Technical report IfI-10-05, Clausthal University of Technology (2010)

16. Bulling, N., Farwer, B.: Expressing Properties of Resource-Bounded Systems: The Logics RTL* and RTL. In: Fisher, M., Novák, P., Dix, J. (eds.) CLIMA X. LNCS, vol. 6214, pp. 22–45. Springer, Heidelberg (2010)

17. Bulling, N., Goranko, V.: How to be both rich and happy: combining quantitative and qualitative strategic reasoning about multi-player games (extended abstract). In: Mogavero, F., Murano, A., Vardi, M.Y. (eds.) Proceedings 1st International Workshop on Strategic Reasoning, SR 2013. EPTCS, vol. 112, pp. 33–41 (2013)

18. Bulling, N., Nguyen, H.N.: Model checking resource bounded systems with shared resources via alternating büchi pushdown systems. In: Chen, Q., Torroni, P., Villata, S., Hsu, J., Omicini, A. (eds.) PRIMA 2015: Principles and Practice of Multi-Agent Systems. LNCS, vol. 9387, pp. 640–649. Springer, Heidelberg (2015)
19. Clarke, E.M., Emerson, E.A., Sistla, A.P.: Automatic verification of finite-state concurrent systems using temporal logic specifications. ACM Trans. Program. Lang. Syst. **8**(2), 244–263 (1986)
20. Monica, D.D., Napoli, M., Parente, M.: On a logic for coalitional games with priced-resource agents. Electr. Notes. Theor. Comput. Sci. **278**, 215–228 (2011)
21. Della Monica, D., Napoli, M., Parente, M.: Model checking coalitional games in shortage resource scenarios. In: Proceedings of the 4th International Symposium on Games, Automata, Logics and Formal Verification (GandALF 2013. EPTCS, vol. 119, pp. 240–255 (2013)
22. Dennis, L.A., Fisher, M., Webster, M.P., Bordini, R.H.: Model checking agent programming languages. Autom. Softw. Eng. **19**(1), 5–63 (2012)
23. Dima, C., Tiplea, F.L.: Model-checking ATL under imperfect information and perfect recall semantics is undecidable. CoRR abs/1102.4225 (2011). http://arxiv.org/abs/1102.4225
24. Fisher, M., Dennis, L.A., Webster, M.P.: Verifying autonomous systems. Commun. ACM **56**(9), 84–93 (2013)
25. Hopcroft, J.E., Ullman, J.D.: Introduction to Automata Theory. Addison-Wesley, Languages and Computation (1979)
26. Kwiatkowska, M., Norman, G., Parker, D.: PRISM 4.0: verification of probabilistic real-time systems. In: Gopalakrishnan, G., Qadeer, S. (eds.) CAV 2011. LNCS, vol. 6806, pp. 585–591. Springer, Heidelberg (2011)
27. Lomuscio, A., Qu, H., Raimondi, F.: MCMAS: a model checker for the verification of multi-agent systems. In: Bouajjani, A., Maler, O. (eds.) CAV 2009. LNCS, vol. 5643, pp. 682–688. Springer, Heidelberg (2009)
28. Shapiro, S., Lespérance, Y., Levesque, H.J.: The cognitive agents specification language and verification environment for multiagent systems. In: Proceedings of the First International Joint Conference on Autonomous Agents and Multiagent Systems (AAMAS 2002), pp. 19–26. ACM, New York (2002)

Effective S-adic Symbolic Dynamical Systems

Valérie Berthé[1]([⊠]), Thomas Fernique[2], and Mathieu Sablik[3]

[1] IRIF, CNRS UMR 8243, Univ. Paris Diderot, Paris, France
`berthe@liafa.univ-paris-diderot.fr`
[2] LIPN, CNRS UMR 7030, Univ. Paris 13, Villetaneuse, France
`fernique@lipn.fr`
[3] I2M UMR 7373, Aix Marseille Univ., Marseille, France
`mathieu.sablik@univ-amu.fr`

Abstract. We focus in this survey on effectiveness issues for S-adic subshifts and tilings. An S-adic subshift or tiling space is a dynamical system obtained by iterating an infinite composition of substitutions, where a substitution is a rule that replaces a letter by a word (that might be multi-dimensional), or a tile by a finite union of tiles. Several notions of effectiveness exist concerning S-adic subshifts and tiling spaces, such as the computability of the sequence of iterated substitutions, or the effectiveness of the language. We compare these notions and discuss effectiveness issues concerning classical properties of the associated subshifts and tiling spaces, such as the computability of shift-invariant measures and the existence of local rules (soficity). We also focus on planar tilings.

Keywords: Symbolic dynamics · Adic map · Substitution · S-adic system · Planar tiling · Local rules · Sofic subshift · Subshift of finite type · Computable invariant measure · Effective language

1 Introduction

Decidability in symbolic dynamics and ergodic theory has already a long history. Let us quote as an illustration the undecidability of the emptiness problem (the *domino problem*) for multi-dimensional subshifts of finite type (SFT) [8,40], or else the connections between effective ergodic theory, computable analysis and effective randomness (see for instance [14,35,44]). Computability is a notion that has also appeared as a major understanding tool in the study of multi-dimensional subshifts of finite type with the breakthrough characterization by M. Hochman and T. Meyerovitch of the entropies of multi-dimensional subshifts of finite type as the non-negative right recursively enumerable numbers [34] (see also [30] in the one-dimensional case). Let us mention also the realization of effective subshifts (with factor and projective subaction operation) from higher-dimensional subshifts of finite type [3,19,33]. It is now clear that

This work was supported by ANR DynA3S and QuasiCool. We would like to thank warmly the anonymous referees, M. Rigo and F. Durand for their useful comments.

© Springer International Publishing Switzerland 2016
A. Beckmann et al. (Eds.): CiE 2016, LNCS 9709, pp. 13–23, 2016.
DOI: 10.1007/978-3-319-40189-8_2

sofic and effective subshifts are closely related, in particular for substitutive subshifts and tilings. Indeed, contrarily to the one-dimensional case, substitution subshifts are known to have (colored) local rules (they are SFT or sofic) in the higher-dimensional framework [26,31,38].

We focus here on effectiveness issues for S-adic subshifts (and tilings). An S-adic expansion is a way to represent (or to generate) words (one-dimensional and multi-dimensional ones), or tilings, by composing infinitely many substitutions. A (word) substitution is a morphism of the free monoid: it replaces letters by words. Substitutions can also be defined in the higher-dimensional framework: they replace letters by multi-dimensional patterns, and act on multi-dimensional words (configurations) in $\mathbb{Z}^d$; more generally, substitutions can also generate and act on tilings, by replacing a tile by a finite union of tiles. An infinite word u (or a d-dimensional configuration, or a tiling) admits an S-adic expansion if

$$u = \lim_{n \to \infty} \sigma_0 \sigma_1 \cdots \sigma_{n-1}(a_n),$$

where $(\sigma_n)_{n \in \mathbb{N}}$ is a sequence of substitutions, and $(a_n)_{n \in \mathbb{N}}$ a sequence of letters. For more on substitutions, see e.g. [28], and for more on S-adic words and tilings, see [9,29]. There is a deep parallelism between subshifts associated with such expansions (under natural assumptions like primitivity) and Bratteli–Vershik systems endowed with adic transformations, hence the terminology 'adic', with the letter S referring to 'substitution'. This connection between adic models and substitutions has been widely investigated; see e.g. [24], or [10] and the references therein. Recall also that any Cantor minimal system admits a Bratteli–Vershik representation [32], which illustrates the representation power of this notion.

Without any further assumption on the S-adic representation, every infinite word admits an S-adic expansion (according to Cassaigne's construction, see e.g. [9, Remark 3]). One thus needs to introduce suitable assumptions on these S-adic representations in order to find a good balance between the expressive power of such representations and the information provided by their existence. Let us illustrate this with [2] where it is proved that multi-dimensional S-adic subshifts, obtained by applying an effective sequence of substitutions chosen among a finite set of substitutions, are sofic subshifts.

Basic notions and definitions on substitutions and S-adic subshifts and tilings are recalled in Sect. 2. We discuss some decidability results in the one-dimensional setting for substitutive words in Sect. 3. Section 4 focuses on effectiveness for S-adic subshifts. Lastly, multi-dimensional Sturmian words and planar tilings are considered in Sect. 5.

2   Definitions

2.1   Subshifts

Let $\mathcal{A}$ be finite *alphabet* and $d \geq 1$. A *configuration* u is an element of $\mathcal{A}^{\mathbb{Z}^d}$. A pattern p is an element of $\mathcal{A}^D$, where $D \subset \mathbb{Z}^d$ is a finite set, called its *support*.

Denote $\mathcal{A}^*$ the set of patterns. A *translate* of the pattern p by $\mathbf{m} \in \mathbb{Z}^d$ is denoted $p+\mathbf{m}$ and has $D+\mathbf{m}$ for support. A pattern $p \in \mathcal{A}^D$ is a *factor* of a configuration $u = (u_\mathbf{n})_{\mathbf{n} \in \mathbb{Z}^d}$ if there exists $\mathbf{m} \in \mathbb{Z}^d$ such that the restriction of u to $D + \mathbf{m}$ coincides with $p + \mathbf{m}$. The set of factors (up to translation) of u is called its *language*.

The set $\mathcal{A}^{\mathbb{Z}^d}$ endowed with the product topology is a compact metric space. A d-dimensional *subshift* $X \subset \mathcal{A}^{\mathbb{Z}^d}$ is a closed and shift-invariant set of configurations in $\mathcal{A}^{\mathbb{Z}^d}$, where the *shifts* $\sigma_\mathbf{m}$ with $\mathbf{m} \in \mathbb{Z}^d$ are defined as $\sigma_\mathbf{m} : \mathcal{A}^{\mathbb{Z}^d} \to \mathcal{A}^{\mathbb{Z}^d}$, $(u_\mathbf{n})_{\mathbf{n} \in \mathbb{Z}^d} \mapsto (u_{\mathbf{n}+\mathbf{m}})_{\mathbf{n} \in \mathbb{Z}^d}$. The shifts provide a natural action of $\mathbb{Z}^d$.

A subshift X can be defined by providing its *language*, that is, the set of patterns (up to translation) that occur in configurations in X. It can be defined equivalently by providing the set of forbidden patterns. Subshifts of *finite type* (also called SFT) are the subshifts such that the set of their forbidden patterns is finite. *Sofic subshifts* are images of SFT under a factor map, where a factor map $\pi : X \to Y$ between two subshifts X and Y is a continuous, surjective map such that $\pi \circ \sigma_\mathbf{m} = \sigma_\mathbf{m} \circ \pi$, for all $\mathbf{m} \in \mathbb{Z}^d$.

Definition 1 (Computable subshift). *A subshift is said to be*

- Π_1-**computable** *or* **effective** *if its language is co-recursively enumerable;*
- Σ_1-**computable** *if its language is recursively enumerable;*
- Δ_1-**computable** *or* **decidable** *if its language is recursive.*

A subshift is said to *be linearly recurrent* if there exists $C > 0$ such that every pattern whose support is a translate of $[-Cn, Cn]^d$ contains every factor whose support is a translate of $[-n, n]^d$. The *frequency* $f(p)$ of a pattern in a d-dimensional configuration u is defined as $\limsup_n |x_n|_p/(2n+1)^d$, where x_n is the restriction of u to $[-n, n]^d$, and $|x_n|_p$ stands for the number of occurrences of p in x_n. If the lim sup is in fact a limit, then the frequency is said to exist. A subshift is said to be *uniquely ergodic* if it admits a unique shift-invariant measure; in this case, pattern frequencies do exist. A subshift is said to be *minimal* if every non-empty closed shift-invariant subset is equal to the whole set. A minimal and uniquely ergodic subshift is said *strictly ergodic*. Any pattern which appears in a strictly ergodic subshift has a positive frequency. For more on multidimensional subshifts, see e.g. [11, Chap. 8, 9].

2.2 Substitutions and S-adic Subshifts

A *substitution* s over the alphabet $\mathcal{A}$ is a map $s : \mathcal{A} \longrightarrow \mathcal{A}^*$. Let $\mathscr{S}$ be a finite set of substitutions; we want to define how a pattern of substitutions $\mathbf{s} \in \mathscr{S}^D$ acts on a pattern $p \in \mathcal{A}^D$, with $D \subset \mathbb{Z}^d$ finite. This general definition allows us to apply simultaneously different substitutions; we are in the non-deterministic case of [38]. We thus introduce *concatenation rules* which specify how the respective images of two adjacent tiles must be glued. A pattern of substitutions $\mathbf{s} \in \mathscr{S}^D$ is said to be *compatible* with a pattern $p \in \mathcal{A}^D$ (made of cells) if it is *consistent* (the image of a cell does not depend on the sequence of concatenation rules that are used, patterns have a unique image) and *non-overlapping* (the images of two cells do not overlap).

When $\mathbf{s}$ and p are compatible, the (unique) image of p by $\mathbf{s}$ is denoted as $\mathbf{s}(p)$. If all letters of $\mathbf{s}$ are equal to the same substitution $s \in \mathscr{S}$, the $\mathscr{S}$-pattern is said to be *s-constant*. This corresponds to the classical case of the action of one substitution (the deterministic case in [38]). An $\mathscr{S}$-*super-tile of order* n corresponds to n iterations of (compatible) patterns of substitutions applied to a letter. We define the $\mathscr{S}$-adic subshift $X_{\mathscr{S}}$ as the set of the configurations for which every pattern appears in an $\mathscr{S}$-super-tile. Here we can compose all substitutions in $\mathscr{S}$. This notion plays a role for the description of planar tilings introduced in Sect. 5.

We now introduce the usual S-adic setting by applying only constant patterns of substitutions. We take a sequence of substitutions $S = (s_n)_{n \in \mathbb{N}} \in \mathscr{S}^{\mathbb{N}}$; the shift acting on S is denoted as σ ($\sigma(S) = (s_{n+1})_{n \in \mathbb{N}} \in \mathscr{S}^{\mathbb{N}}$). The S-*super-tile of order* 0 *and type* $a \in \mathcal{A}$ is defined as the letter a, whereas the S-super-tile of order $n + 1$ and type a is the image of the $\sigma(S)$-super-tile of order n and type a by a s_0-constant $\mathscr{S}$-pattern. A super-tile of order n can thus be defined by a word of size n in $\mathscr{S}^*$ together with a letter. The sequence S is said to be a *directive sequence*. We then define the S-adic subshift X_S as the set of the configurations for which every pattern appears in an S-super-tile. For a closed subset $\mathbf{S} \subset \mathscr{S}^{\mathbb{N}}$, we also define the $\mathbf{S}$-adic subshift $X_{\mathbf{S}} = \bigcup_{S \in \mathbf{S}} X_S$.

For $d = 1$, a natural way to define concatenation rules is to consider that the image of two consecutive letters is obtained as the concatenation of the two image words. Thus a substitution can be viewed as a non-erasing endomorphism of the free monoid $\mathcal{A}^*$. For example the Fibonacci substitution on the alphabet $\{a, b\}$ is defined by $\sigma \colon a \mapsto ab$, $b \mapsto a$. For $d = 2$, if all the supports of the images by an element of $\mathscr{S}$ are rectangular, concatenation rules of two adjacent letters consist in the concatenation of the two image patterns as long as the two glued edges have the same size. Rectangular substitutions are considered e.g. in [38] (see also [16] for the notion of shape-symmetric rectangular substitutions).

It is possible to extend the notion of substitution to geometric tilings. A *tiling* of $\mathbb{R}^d$ is a collection of compact sets which cover topologically $\mathbb{R}^d$, that is, with the interiors of the tiles being pairwise disjoint. In general, a tile-substitution in $\mathbb{R}^d$ is given by a set of prototiles $T_1, \ldots, T_m \subset \mathbb{R}^d$, an expanding map and a rule how to dissect each expanded prototile into translated copies of some prototiles T_i. These *geometric tiling substitutions* are considered e.g. in [31]. It is also possible to define S-adic tilings in this context (see [29]).

There are further strategies for defining substitutions such as described in [28]. For instance, one can also use a global information; see the formalism introduced in [1] that allows the generation of multi-dimensional Sturmian words considered in Sect. 5; this formalism also provides concatenation rules [25].

3 Some Decisions Problems for Substitutions

In the one-dimensional case, numerous decidability results exist for fixed points of substitutions (D0L words), and their images by general morphisms (HD0L words). More precisely, let $\mathcal{A}$, $\mathcal{B}$, be finite alphabets. We consider two morphisms $\sigma \colon \mathcal{A}^* \to \mathcal{A}^*$, $\phi \colon \mathcal{A}^* \to \mathcal{B}^*$; an infinite word of the form $\lim_n \sigma^n(u)$ (respectively $\phi(\lim_n \sigma^n(u)))$ is a D0L (respectively an HD0L word or morphic word), for u finite word.

We focus here on some decision problems that can be solved using the notion of return word and derived sequence (see e.g. [20]). Let σ be a primitive substitution. It generates a minimal subshift X_σ. A return word to a word u of its language is a word w of the language such that uw admits exactly two occurrences of u, with the second occurrence of u being a suffix of uw. One can recode sequences of the subshift via return words, obtaining *derived sequence* (see e.g. [20]). Note that even if analogous notions exist in the higher-dimensional case and for tilings [39], this is not sufficient to yield a direct generalization of the results described below.

The HD0L ω-equivalence problem (which has been open for more than 30 years) is solved in [21] for primitive morphisms: it is decidable to know whether two HD0L words are equal (see the references in [21] for the D0L case). The decidability of the ultimate periodicity of HD0L infinite sequences has also been a long-standing problem: it is decidable to know whether an HD0L word is ultimately periodic. See [21] for the primitive case, and [22] for the general case. See also the references in [22] for the D0L case. This problem is closely related to the decidability of the ultimate periodicity of recognizable sets of integers in some abstract numeration systems [7]. It is also proved in [23] that the uniform recurrence of morphic sequences is decidable.

The particular case of constant-length substitutions (automatic sequences) has also been widely studied; see e.g. [17,42] where decision procedures are produced based on the connections between first-order logic and automata such as developed in [15] where the equivalence between being p-recognizable and p-definability is developed. For more references, see also the book [41].

4   Effectiveness for S-adic Subshifts and Local Rules

We discuss here several effectivity notions for S-adic subshifts concerning their directive sequences, pattern frequencies, or else their language. We also focus on the existence of local rules. We only consider here iterations by constant $\mathscr{S}$-patterns. We recall that $\mathscr{S}$ is finite.

A closed subset $\mathbf{S} \subset \mathscr{S}^{\mathbb{N}}$ is *effectively closed* if the set of (finite) words which do not appear as prefixes of elements of $\mathbf{S}$ is recursively enumerable (one enumerates forbidden prefixes). An effectively closed set is not necessarily a subshift.

A set of substitutions $\mathscr{S}$ has *a good growing property* if there are finitely many ways of gluing super-tiles, and if the size of the super-tiles of order n grows with n: there exists a finite set of patterns $\mathcal{P} \subset \mathcal{A}^*$ such if a pattern formed by a super-tile of order n surrounded by super-tiles of order n is in the language of $X_{\mathscr{S}^{\mathbb{N}}}$, then it appears as the n-iteration by a constant $\mathscr{S}$-pattern of a pattern of $\mathcal{P}$, and, moreover, if for every ball of radius R, there exist $n \in \mathbb{N}$ such a translate of this ball is contained in all the supports of super-tiles of order n. Clearly non-trivial rectangular substitutions or geometrical substitutions (such as defined in [31]) verify this property.

Proposition 1. *Let* $\mathbf{S} \subset \mathscr{S}^{\mathbb{N}}$ *be a closed subset. If* $X_{\mathbf{S}}$ *is effective, then there exists an effective closed subset* $\mathbf{S}' \subset \mathscr{S}^{\mathbb{N}}$ *such that* $X_{\mathbf{S}} = X_{\mathbf{S}'}$. *The reciprocal is true if* $\mathscr{S}$ *has the good growing property.*

Proof. Assume that $X_{\mathbf{S}}$ is effective. The complement of its language is recursively enumerable. Let $\mathbf{S}'$ be the effective closed set such that a word in $\mathscr{S}^*$ is a forbidden prefix if the associated super-tiles are in the complement of the language of $X_{\mathbf{S}}$. Clearly $X_{\mathbf{S}'} = X_{\mathbf{S}}$.

Conversely, consider the $\mathbf{S}$-adic subshift $X_{\mathbf{S}}$ where $\mathbf{S}$ is effectively closed and let $\mathcal{P}$ be the set of patterns given by the good growing property. Let $\mathcal{P}'$ be the set of patterns in $\mathcal{P}$ that occur in $X_{\sigma^n(\mathbf{S})}$ for infinitely many n. A pattern p is in the language of $X_{\mathbf{S}}$ if it appears in the image by an n-iteration of a pattern of $\mathcal{P}'$, where n is the first order where the support of p is included in all super-tiles of order n. Since $\mathcal{P}'$ is finite and the prefixes of $\mathbf{S}$ are co-recursively enumerable, the same holds for the language of $X_{\mathbf{S}}$.

Definition 2 (Computable frequencies and measure). *Let X be a subshift. X is said to have* computable frequencies *if the frequencies of patterns exist and are uniformly computable. A shift-invariant measure is said to be* computable *if the measure of any cylinder is uniformly computable.*

Remark 1. Computability of letter frequencies does not say much on the algorithmic complexity of a subshift: take a subshift $X \subset \{0,1\}^{\mathbb{Z}}$ and consider the subshift Y obtained by applying to each configuration of X the substitution $0 \mapsto 01, 1 \mapsto 10$. The subshift Y admits letter frequencies (they are both equal to $1/2$), and it has the same algorithmic complexity as X.

Proposition 2. *Let X be a subshift. If X is effective and uniquely ergodic, then its invariant measure is computable and it is decidable. If X is minimal and its frequencies are computable, then its language is recursively enumerable. If X is minimal and effective, then it is decidable.*

Proof. Let X be a d-dimensional subshift. We assume X effective and uniquely ergodic. Let us prove that the frequency of any pattern is computable. We use the following algorithm that takes as an argument the parameter e that stands for the precision. We consider a finite pattern p. At step n, one produces all 'square' patterns of size n with support being a translate of $[-n, n]^d$ that do not contain the n first forbidden patterns (they do not need to belong to the language of X). For each of these square patterns of size n, one computes the number of occurrences of p in it, divided by $(2n + 1)^d$. We continue until these quantities belong to an interval of length e. This algorithm then stops, and taking an element of the interval provides an approximation of the frequency of p up to precision e. Indeed, the square patterns of size n contain the square patterns of size n of X. It remains to prove that the algorithm stops. Suppose it does not, then, for all n, one can find two patterns of size n, x_n and x'_n, that do not contain the n first forbidden patterns and such that $||x_n|_p/(2n+1)^d - |x'_n|_p/(2n+1)^d| > e$. By compactness, we can extract two configurations x and x' that do not contain forbidden patterns (they thus belong to the subshift X) such that the frequency of p in x is distinct from the frequency of p in x'. This contradicts the unique ergodicity of X.

We now assume X minimal with computable pattern frequencies (frequencies are positive). One can decide whether the frequency of a pattern is larger than a given value. This thus implies that the language is recursively enumerable.

Assume that X is minimal and effective. Consider the square patterns of size n that do not contain the n first forbidden patterns. If a pattern belongs to all these patterns for some n, then it belongs to the language. Otherwise, consider size $n + 1$. The algorithm stops if a pattern is in the language by minimality.

Corollary 1. *Let X_S be a strictly ergodic S-adic subshift defined with respect to a directive sequence $S \in \mathscr{S}^{\mathbb{N}}$ such that $\mathscr{S}$ satisfies the good growing property. The following conditions are equivalent:*

1. *There exists a computable sequence S' such that $X_S = X_{S'}$;*
2. *The unique invariant measure of X_S is computable;*
3. *The subshift X_S is decidable.*

Proof. We first assume (1). By Proposition 1, X_S is effective and Proposition 2 yields that its unique measure is computable and that X_S is decidable.

We now assume (2). Let d stand for the cardinality of the alphabet of the substitutions in $\mathscr{S}$. The letter frequency vector is in the cone defined by the product of the incidence matrices of the directive sequence. The incidence matrix of a substitution s is a square matrix whose entry of index (i, j) counts the number occurrences of the letter i in $s(j)$. Let M_n stand for the incidence matrix of the substitution s_n. The letter frequency belongs to the cone $\bigcap_n M_1 \cdots M_n \mathbb{R}_+^d$, which is one-dimensional by unique ergodicity. Given a precision e, one can compute n such that the columns of $M_1 \cdots M_n$ are expected to be at a distance less than e from the letter frequency vector. We fix a cylinder around the direction provided by the letter frequency vector with precision e. Now we test finite products of n substitutions in $\mathscr{S}$. We consider the cone obtained by taking the product of the incidence matrices, and check whether it intersects the cylinder. If it does not intersect the cylinder, one gets a forbidden product of substitutions, which proves that $\{S\}$ is effectively closed.

It remains to prove that (3) implies (1). There exists a closed effective set such that $X_S = X_{\mathbf{S}}$, by Proposition 1. For every $S' \in \mathbf{S}$, one has $X_{S'} = X_S$, by minimality. We then exhibit a computable S' in $\mathbf{S}$ as follows: for any n, take the first prefix for the lexicographic order among the prefixes of elements in $\mathbf{S}$ such that $s_0 s_1 \cdots s_n(a)$ is in the language of X_S.

Existence of Local Rules. A natural question in tiling theory is to find local rules which only produce aperiodic tilings. The first examples of aperiodic subshifts of finite type were based on hierarchical structures [8,40]: substitutive structures are known to be able to force aperiodicity. Note that a non-trivial sub-stitutive subshift cannot be sofic in dimension 1: it has zero topological entropy whereas non-trivial sofic subshifts have positive entropy. In dimension $d \geq 2$, under natural assumptions, it is known for different types of substitutions that substitutive tilings can be enforced with (colored) local rules. The ideas is always to force a hierarchical structure, as in Robison's tiling, where each change of level is marked by the type of the super-tile of this level, and the rule used is trans-mitted for super-tiles of lower order. For rectangular substitutions, the result is proved in [38] (with the result being more general since the substitutions are non-deterministic). The case of geometrical substitutions is handled in [31] and the result is also true in a more combinatorial way [26].

In the case of rectangular substitutions it is shown in [2] that the **S**-adic subshift $X_{\mathbf{S}}$ is sofic if and only if it can be defined by a set of directive sequences **S** which is effectively closed. A similar result for more general substitutions is expected; the difficulty relies in the ability to exhibit a rectangular grid to use the simulation (see [3,19]) of a one-dimensional effective subshift by a two-dimensional sofic subshift. One can also ask whether linearly recurrent effective subshifts are sofic. Note that such a statement cannot hold for computability reasons (there are uncountably many linearly recurrent subshifts) without any effectivity assumption. Note also that in the one-dimensional case, linearly recur-rent subshifts are primitive S-adic [20].

5   An Application: Planar Tilings

As an example of S-adic configurations, we consider *multi-dimensional Sturmian words*. The associated tilings belong to the more general class of *planar tilings*. A (canonical) planar tiling is an approximation of an affine d-plane E in $\mathbb{R}^n$, via the cut-and-project method (see e.g. [4]). Such a tiling can be lifted into the tube $E + [0,t]^n$: the space E is called the *slope* and the smallest possible t the *thickness*. Planar tilings are closely related to discrete planes in discrete geometry and provide models of quasicrystals. The case $t = 1$ and $d = n - 1$ corresponds to the multi-dimensional Sturmian case. In terms of configurations, a multidimensional Sturmian word is defined as the coding of a $\mathbb{Z}^d$-action by d rotations $R_{\alpha_i} : \mathbb{R}/\mathbb{Z} \to \mathbb{R}/\mathbb{Z}$, $x \mapsto x + \alpha_i$ $(1 \leq i \leq d)$, where the α_i are positive real numbers. We assume $1, \alpha_1, \cdots, \alpha_d$ rationally independent and $\sum_i \alpha_i < 1$. A *multidimensional Sturmian word* $u \in \{1, 2, \cdots, d+1\}^{\mathbb{Z}^d}$ is defined as follows: there exists ρ, a partition of $\mathbb{R}/\mathbb{Z}$ into $d + 1$ semi-open intervals, d of lengths α_i, and one of length $1 - \sum \alpha_i$, such that $u_{\mathbf{n}} = i$ if and only if $n_1\alpha_1 + \cdots + n_d\alpha_d + \rho \in I_i$ [12]. The cut-and-project framework is larger than the S-adic framework but multidimensional Sturmian words are S-adic [13] (via the formalism of [1]).

The study of the connections between the existence of local rules for a planar tiling and the parameters of its slope started with [18,36,37,43]. In particular, it was proven in [36] that a slope enforced by undecorated local rules is necessarily algebraic (this is however not sufficient, see e.g. [5,6]). However, computability comes into play when the tiles can be decorated. Decorations indeed allow the transfer of information through the tiling, and this was used in [27] to prove that a slope can be enforced by such rules if and only if it is computable.

References

1. Arnoux, P., Ito, S.: Pisot substitutions and Rauzy fractals. Bull. Belg. Math. Soc. Simon Stevin **8**, 181–207 (2001)
2. Aubrun, N., Sablik, M.: Multidimensional effective S-adic systems are sofic. Distrib. Theor. **9**, 7–29 (2014)
3. Aubrun, N., Sablik, M.: Simulation of effective subshifts by two-dimensional subshifts of finite type. Acta Applicandae Mathematicae **126**, 35–63 (2013)
4. Baake, M., Grimm, U.: Aperiodic Order: A Mathematical Invitation, vol. 1. Cambridge University Press,Cambridge (2013)
5. Bédaride, N., Fernique, T.: No weak local rules for the $4p$-fold tilings. Disc. Comput. Geom. **54**, 980–992 (2015)
6. Bédaride, N., Fernique, T.: When periodicities enforce aperiodicity. Commun. Math. Phys. **335**, 1099–1120 (2015)
7. Bell, J.P., Charlier, E., Fraenkel, A.S., Rigo, M.: A decision problem for ultimately periodic sets in non-standard numeration systems. Int. J. Algebra Comput. **9**, 809–839 (2009)
8. Berger, R.: The Undecidability of the Domino Problem, vol. 66. Memoirs of the American Mathematical Society, Providence (1966)
9. Berthé, V., Delecroix, V.: Beyond substitutive dynamical systems: S-adic expansions, RIMS Lecture note 'Kokyuroku Bessatu' B46, pp. 81–123 (2014)
10. Berthé, V., Rigo, M. (eds.): Combinatorics, Automata and Number Theory, Encyclopedia of Mathematics and its Applications, vol. 135. Cambridge University Press, Cambridge (2010)
11. Berthé, V., Rigo, M. (eds.): Combinatorics, Words and Symbolic dynamics, Encyclopedia of Mathematics and its Applications, vol. 159. Cambridge University Press, Cambridge (2016)
12. Berthé, V., Vuillon, L.: Tilings and rotations on the torus: a two-dimensional generalization of Sturmian sequences. Discrete Math. **223**, 27–53 (2000)
13. Berthé, V., Bourdon, J., Jolivet, T., Siegel, A.: Generating discrete planes with substitutions. In: Karhumäki, J., Lepistö, A., Zamboni, L. (eds.) WORDS 2013. LNCS, vol. 8079, pp. 58–70. Springer, Heidelberg (2013)
14. Bienvenu, L., Day, A.R., Hoyrup, M., Mezhirov, I., Shen, A.: A constructive version of Birkhoff's ergodic theorem for Martin-Löf random points. Inf. Comput. **210**, 21–30 (2012)
15. Bruyère, V., Hansel, G., Michaux, C., Villemaire, R.: Logic and p-recognizable sets of integers. Bull. Belg. Math. Soc. Simon Stevin **12**, 191–238. Correction to: "Logic and p-recognizable sets of integers". Bull. Belg. Math. Soc. Simon Stevin **14**, 577 (1994)
16. Charlier, E., Kärki, T., Rigo, M.: Multidimensional generalized automatic sequences and shape-symmetric morphic words. Discrete Math. **310**, 1238–1252 (2010)
17. Charlier, E., Rampersad, N., Shallit, J.: Enumeration and decidable properties of automatic sequences. Int. J. Found. Comput. Sci. **23**, 1035–1066 (2012)
18. de Bruijn, N.G.: Algebraic theory of Penrose's nonperiodic tilings of the plane. Nederl. Akad. Wetensch. Indag. Math. **43**, 39–66 (1981)

19. Durand, B., Romashchenko, A.E., Shen, A.: Effective closed subshifts in 1D can be implemented in 2D. In: Blass, A., Dershowitz, N., Reisig, W. (eds.) Fields of Logic and Computation. LNCS, vol. 6300, pp. 208–226. Springer, Heidelberg (2010)

20. Durand, F.: Linearly recurrent subshifts have a finite number of non-periodic subshift factors, Ergodic Theor. Dynam. Syst. **20**, 1061–1078. Corrigendum and addendum, Ergodic Theor. Dynam. Syst. **23**, 663–669 (2003)

21. Durand, F.: HD0L ω-equivalence and periodicity problems in the primitive case. Unif. Distrib. Theor. **7**(1), 199–215 (2012)

22. Durand, F.: Decidability of the HD0L ultimate periodicity problem. RAIRO - Theor. Inf. Appl. **47**, 201–214 (2013)

23. Durand, F.: Decidability of uniform recurrence of morphic sequences. Int. J. Found. Comput. Sci. **24**, 123–146 (2013)

24. Durand, F., Host, B., Skau, C.: Substitutional dynamical systems, Bratteli diagrams and dimension groups. Ergodic Theor. Dyn. Syst. **19**(4), 953–993 (1999)

25. Fernique, T.: Local rule substitutions and stepped surfaces. Theor. Comput. Sci. **380**, 317–329 (2007)

26. Fernique, T., Ollinger, N.: Combinatorial substitutions and sofic tilings. In: JAC (2010)

27. Fernique, T., Sablik, M.: Local rules for computable planar tilings automata. In: JAC (2012)

28. Frank, N.P.: A primer of substitution tilings of the Euclidean plane. Expo. Math. **26**, 295–326 (2008)

29. Frank, N.P., Sadun, L.: Fusion: a general framework for hierarchical tilings of Rd. Geom. Dedicata. **171**, 149–186 (2014)

30. Gangloff, S., de Menibus, B.H.: Computing the entropy of one-dimensional decidable subshifts. arXiv:1602.06166

31. Goodman-Strauss, C.: Matching rules and substitution tilings. Ann. Math. **147**, 181–223 (1998)

32. Herman, R.H., Putnam, I.F., Skau, C.F.: Ordered Bratteli diagrams, dimension groups and topological dynamics. Int. J. Math. **3**, 827–864 (1992)

33. Hochman, M.: On the dynamics and recursive properties of multidimensional symbolic systems. Inventiones Mathematicae **176**, 131–167 (2009)

34. Hochman, M., Meyerovitch, T.: A characterization of the entropies of multidimensional shifts of finite type. Ann. Math. **171**, 2011–2038 (2010)

35. Galatolo, S., Hoyrup, M., Rojas, C.: Effective symbolic dynamics, random points, statistical behavior, complexity and entropy. Inf. Comput. **208**, 23–41 (2010)

36. Le, T.Q.T.: Local rules for quasiperiodic tilings. In: The Mathematics of Long-Range Aperiodic Order, Waterloo, ON, 1995. NATO Advanced Science Institutes Series C: Mathematical and Physical Sciences, vol. 489, pp. 331–366. Kluwer Academic Publisher, Dordrecht (1997)

37. Levitov, L.S.: Local rules for quasicrystals. Commun. Math. Phys. **119**, 627–666 (1988)

38. Mozes, S.: Tilings, substitution systems and dynamical systems generated by them. J. d'analyse mathématique **53**, 139–186 (1989)

39. Priebe, N.M.: Towards a characterization of self-similar tilings in terms of derived Vorono tessellations. Geom. Dedicata. **79**, 239–265 (2000)

40. Robinson, R.M.: Undecidability and nonperiodicity for tilings of the plane. Invent. Math. **12**, 177–209 (1971)

41. Rigo, M.: Formal Languages, Automata and Numeration Systems. Wiley, Hoboken (2014)

42. Shallit, J.: Decidability and enumeration for automatic sequences: a survey. In: Bulatov, A.A., Shur, A.M. (eds.) CSR 2013. LNCS, vol. 7913, pp. 49–63. Springer, Heidelberg (2013)
43. Socolar, J.E.S.: Weak matching rules for quasicrystals. Commun. Math. Phys. **129**, 599–619 (1990)
44. V'yugin, V.V.: Effective convergence in probability, and an ergodic theorem for individual random sequences. Theor. Probab. Appl. **42**, 42–50 (1997)

Reaction-Based Models of Biochemical Networks

Daniela Besozzi[1,2(✉)]

[1] Department of Informatics, Systems and Communication,
University of Milano-Bicocca, Viale Sarca 336, 20126 Milano, Italy
daniela.besozzi@unimib.it
[2] SYSBIO.IT, Centre of Systems Biology, Milano, Italy

Abstract. Mathematical modeling and computational analyses of biological systems generally pose to modelers questions like: "Which modeling approach is suitable to describe the system we are interested in? Which computational tools do we need to simulate and analyze this system? What kind of predictions the model is expected to give?". To answer these questions, some general tips are here suggested to choose the proper modeling approach according to the size of the system, the desired level of detail for the system description, the availability of experimental data and the computational costs of the analyses that the model will require. The attention is then focused on the numerous advantages of reaction-based modeling, such as its high level of detail and easy understandability, or the possibility to run both deterministic and stochastic simulations exploiting the same model. Some notes on the computational methods required to analyze reaction-based models, as well as their parallelization on Graphics Processing Units, are finally provided.

1 Introduction

In many fields of life sciences, researchers are taking more and more frequently advantage of mathematical modeling and computational analysis as complementary tools to experimental laboratory methods. In the recent years, plenty of multidisciplinary works largely proved the increased capability of the synergistic integration between computational and experimental research, to achieve faster and better comprehension of biological systems. In this context, many frameworks and methodologies that were originally developed in computer science and engineering have been applied to study a variety of biological systems [1,2]. Anyway, the choice of the most adequate modeling and computational approach is not a straightforward process, which should always be guided by the nature of the system of interest, by the experimental data in hand and, above all, by the *scientific question* that motivates the development of the model.

Many times, laboratory experiments simply aim at identifying the existence of some sort of "difference" between what is happening in the biological system before/after a perturbation, or with/without a key component of that system (for instance, the response to a stress state *vs.* the physiological state, the phenotype of a mutant organism *vs.* the wild type, etc.). In these cases, statistical methods can be used to establish the significance of such a difference by using experimental

© Springer International Publishing Switzerland 2016
A. Beckmann et al. (Eds.): CiE 2016, LNCS 9709, pp. 24–34, 2016.
DOI: 10.1007/978-3-319-40189-8_3

data that are generally easy to measure [3]. When the scope of the modeling is to make predictions about the *emergent behaviours* of the system in conditions that were not previously analyzed, then also the complexity and the technical challenges of laboratory experiments increase up to the measurement of quantitative time-series, which most of the time require time-consuming and expensive procedures [4,5]. In these situations, the available biotechnological equipments or experimental methodologies could represent a bottleneck to the measurement of biological data at the level of precision or sampling frequency that are beneficial for mathematical modeling, therefore limiting or hampering the effective application of many computational approaches. Albeit the drawbacks that characterize these complex experiments, the availability of reliable time-series data is an indispensable mean to calibrate *mechanistic parametric models*, which are usually assumed to be the most likely candidate to achieve an in-depth comprehension of biological systems. Indeed, these models can lead to quantitative predictions of the system dynamics: by considering in details the mutual regulations among the system components, they allow to infer the possible temporal evolution of the system in both physiological and perturbed conditions.

Although some general guidelines can be suggested to facilitate the choice of the modeling approach, the most widespread strategies—that is, *interaction-based*, *constraint-based*, *mechanism-based* and *logic-based* models—are characterized by peculiar limits and strengths, as discussed hereby. One of the most challenging problem to date, anyway, is how to define *integrated models* having a unified mathematical formalism and able to bridge different layers of the cell functioning (gene expression, signal transduction, metabolism, etc.) [6,7]. Although some striking examples of this integration exist [8], the problem is still far from a general-purpose solution.

The aim of this paper is to provide a brief (non-exhaustive) overview of existing modeling approaches, together with some general tips for choosing the proper formalism for any given biological system under investigation. In particular, the attention is here focused on the advantages of reaction-based modeling, a mechanistic and parametric approach especially useful to describe and analyze networks of biochemical reactions [9]. This modeling approach was successfully used to investigate various signal transduction pathways and to predict their emergent behaviour under different conditions (see, e.g., [10–16]). The analysis of reaction-based models typically involves the execution of stochastic or deterministic simulations, and most often requires to face challenging computational problems necessary for model definition and calibration. The benefits of using high-performance computing solutions to parallelize the simulations and reduce the computational burden of these tasks are concisely sketched as conclusive remarks to this work.

2   A Few Tips for the Modeling of Biological Systems

The starting point for the definition of the model of a biological system is, in general, a diagram or a conceptual map, graphically representing the current biological knowledge about that system. In general, these diagrams show

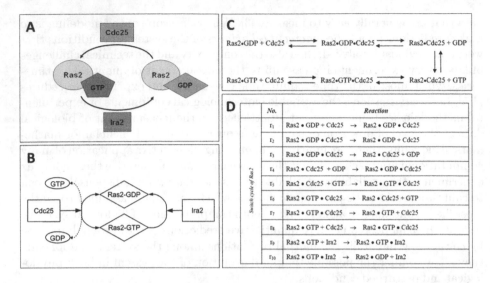

Fig. 1. The switch cycle of protein Ras2: from diagrams to reaction-based model. Panels **A** and **B** represent the interactions among proteins Ras2, Cdc25, Ira2, and nucleotides GTP and GDP. Briefly, Ras2 is a GTPase—that is, an enzyme that catalyzes the hydrolysis of guanosine triphosphate (GTP) into guanosine diphosphate (GDP)—which cycles between an active state, when bound to GTP, and an inactive state, when bound to GDP. The active state of Ras2 is positively regulated by Cdc25, a protein that stimulates the GDP to GTP exchange, and negatively regulated by Ira2, a protein that stimulates the GTPase activity of Ras2. The intracellular ratio of GTP and GDP is also involved in the regulation of Ras2 activity, since Cdc25 stimulates the exchange of these nucleotides according to their relative concentration. Panels **C** and **D** formally describe the cascade of biochemical reactions and the formation of molecular complexes among proteins and nucleotides.

the molecular components involved in the system and their mutual relations, usually specifying also the formation of molecular complexes, and they possibly highlight the activation/inhibition of some component or the presence of positive/negative feedback regulations. For instance, panels A and B in Fig. 1 show two similar diagrammatic representations of the switch cycle of Ras2 proteins (see caption for more details). Although both diagrams in Fig. 1 represent the same biochemical process, panel B better characterizes the roles of GTP and GDP in their interaction with proteins Ras2 and Cdc25, which is instead not inferable by panel A. This example emphasizes that diagrams of cellular processes could be subject to ambiguous interpretation, or might not explicitly represent all biological knowledge about that process.

In order to avoid any possible misinterpretations, mathematical modeling should be used to precisely and unambiguously describe what is known to occur within the biological system of interest, and possibly include any hypothesis that should then be tested by further laboratory experiments. In the case of the switch cycle of Ras2 protein, an example of precise formalization of these diagrams is given in panel C, which shows a typical representation of reactions scheme used in biochemistry, while panel D shows the *reaction-based model* that explicitly describes all direct and inverse biochemical reactions occurring in Ras2 switch cycle (the interested reader is referred to [12,15] for more details about the modeling of this process in the context of the Ras/cAMP/PKA signal transduction pathway in yeast). Reaction-based models belong to the family of mechanism-based modeling approach, which is hereby discussed in the context of the choice of the most adequate formalism for the biological system of interest.

Considering the available knowledge about the biochemical, physical and regulatory properties of all system components and their mutual interactions, the very first step in the definition of a mathematical model consists in the establishment of the *purpose* (or scientific questions) of the model. This step is fundamental in order to recognize what kind of novel insights the modeling process should add to the experimental investigation. Although some general guidelines can be suggested to facilitate the choice of the modeling approach, the four features schematized in Fig. 2 emphasize the limits and the strengths of the most widespread strategies, that is, *interaction-based, constraint-based, mechanism-based* and *logic-based* models (see figure caption for more details).

Unfortunately, there exists no biunivocal correspondence between each of these modeling approaches and any specific biological system, that allows to define a crisp separation among the variety of mathematical formalisms and to select the most suitable strategy for the scientific question under examination. Actually, this is a game with no winners: Fig. 2 shows that mechanism-based models are characterized by the highest level of detail in the formalization of the system, so that they can achieve quantitative predictions of the system dynamics, but they require a lot of data measurements and usually have the highest computational costs. On the other side of the spectrum, interaction-based models can be analyzed with generally low execution times, but they are characterized by the lowest level of detail and hence, in general, they are only used to investigate topological properties of biological networks. Constraint-based models are used to investigate flux distributions in metabolic networks, up to genome-wide size systems, while Boolean/fuzzy logic models have been exploited to study gene regulatory networks or signal transduction pathways having different system sizes. Because of space limits, the interested reader is referred to [17–23] for more information on these modeling approaches and examples of their application.

Despite their highest predictive power with respect to other modeling approaches, mechanism-based models are generally subject to a relevant criticism: usually, they are not exploited to formally describe systems with a large number of components and interactions, so that they cannot be used to explain the functioning of large-scale systems or whole cells. Anyway, even a small-scale quantitative model can be useful to better understand a complex cellular process, together with its feedback/feedforward regulations and non-linear behaviour, and predict its dynamics in physiological and perturbed conditions. To this aim, the model should be as detailed as possible, but should also retain a simple formalization, so that also experimentalists can easily intervene in the discussion and in the development of the model itself. Reaction-based models satisfy these conditions: some of their peculiar strengths are described in the following section.

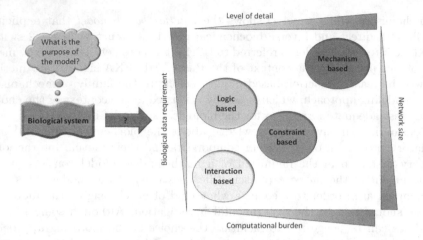

Fig. 2. Given a biological system and the purpose of the model, the modeling approach should be chosen by taking into account four aspects. First, the *size of the system*, defined in terms of the number of components and interactions, which can be in the order of a few units or tens for small-scale models, up to hundreds and thousands for large-scale models. Second, the necessary *level of detail* has to be chosen to fully describe the components of the system and their mutual interactions. Third, the type and the quality of *experimental data* that are already available or could be measured with proper protocols often constrain the model definition according to a specific approach. Fourth, the *computational burden* that the chosen approach will bring about in the simulation and analysis of the model, and that sometimes might represent the trade-off in the choice between quantitative or qualitative models. These features usually (but not strictly) lead to model signal transduction or small metabolic pathways with the mechanism-based approach; protein-protein interaction networks with the interaction-based approach; gene regulation networks with the interaction/logic-based approach; whole-cell metabolic networks with the constraint-based approach. The modeling approaches are graphically represented with an increasing intensity scale, according to their *quantitative* feature: for instance, interaction-based models are characterized by a simplified representation of the system and can only derive qualitative (topological) outcomes from computational analyses, while mechanism-based models are highly detailed and fully parameterized, and have the highest quantitative predictive capability. In order to be *predictive*, models should be defined in such a way that both the expected and the unknown dynamical patterns are not directly "hard-wired" into the mathematical description, but they globally appear as the consequence of the chemico-physical interactions between the species included in the model.

3   The Advantages of Reaction-Based Models

Given a biological system Ω, a reaction-based model of Ω formally consists in a set of molecular species $\mathcal{S} = \{S_1, \ldots, S_N\}$ and a set of biochemical reactions $\mathcal{R} = \{r_1, \ldots, r_M\}$. A generic biochemical reaction is given in the form r_j: $\sum_{i=1}^{N} \alpha_{ji} S_i \rightarrow \sum_{i=1}^{N} \beta_{ji} S_i$, where $\alpha_{ji}, \beta_{ji} \in \mathbb{N}$ are the stoichiometric coefficients of the i-th reactant and the i-th product of the j-th reaction, respectively, which specify how many molecules of each species are involved in the reaction.

It is worth highlighting that reaction-based models are *quantitative parametric* models, which means that each species S_i is characterized by an amount X_i (given either as an integer number of molecules or a real-value concentration), and each reaction r_j is characterized by a kinetic constant $k_j \in \mathbb{R}^+$, encompassing its physical and chemical properties. The temporal evolution of Ω can be simulated by updating the state of the system at consecutive time steps, where the state at some time t is a vector $\mathbf{X}(t) = (X_1(t), \ldots, X_N(t))$, having as components the amounts of all species at time t. Usually, the initial state of the system is given by specifying the amount of each species at some fixed $t = 0$, chosen as the beginning time point of both experimental measurement and computational analysis.

Since it relies on syntactic and semantic features similar to the language of biochemistry, reaction-based modeling represents for experimentalists the most easily comprehensible formalization of a biological system. Indeed, it does not require any expertise in mathematical or computational modeling. This is a key strength of reaction-based models, since their correctness and biological plausibility can be easily assessed by experimentalists, without the need for modelers to translate the formal (mathematical) system into a "narrative" (and else possibly ambiguous) description of the model itself. Therefore, reaction-based models can largely facilitate the communication and the crosstalk between modelers and experimentalists, and possibly step up interdisciplinary analysis of biological systems.

Reaction-based modeling does not rely on the use of approximate kinetic functions, such as Michaelis-Menten rate law for enzymatic processes or Hill functions for cooperative binding, that are usually exploited in systems of differential equations. On the contrary, it explicitly provides a detailed and accurate description of the molecular interactions and control mechanisms (including feedback or feedforward regulation) that take place in cellular processes. In addition, since all molecular species and their mutual biochemical reactions appear as atomic entities in the model, they can be analyzed either independently from each other or in combination with other components, in order to determine the corresponding influence on the system behaviour. This feature allows modelers to measure all system components and, above all, to precisely determine the role of each species and reaction in the overall functioning of the system.

The reaction-based formalization is general enough to describe any kind of process determined by interacting components, by simply assigning the appropriate semantics to the set of species and to the set of reactions. For instance, it can be easily used to model ecological systems: in this case, the "species" correspond to the individuals of different populations, while the "reactions" describe the type of relationships occurring between these populations. Assuming, e.g., prey-predator interactions between two populations, as in the Lotka-Volterra equations, a reaction-based model can be specified by considering (1) the set $\mathcal{S} = \{A, X, Y\}$, where A, X, Y represent, respectively, the food resource, the prey and the predator individuals, and (2) the set of reactions $\mathcal{R} = \{r_1 : A + X \to X + X; r_2 : X + Y \to Y + Y; r_3 : Y \to \lambda\}$, where r_1 describes the growth of preys, according to available resources, r_2 describes the interaction between preys and predators and the consequent growth of predators, while r_3 describes the natural death of predators (symbol λ here denotes the die-off of individual Y).

In contrast to other formalisms, any reaction-based model can be easily refined or extended without any labored adjustment in the formalization of the former model (see also [24]). Adding new species or reactions to a model formalized as a system of differential equations would instead require, in general, the modification of many of its differential equations. Thanks to this feature, reaction-based modeling is suitable to modular construction of larger and larger models, whereby an initial core of species and reactions can be extended to take into account other cellular processes (e.g., to introduce new species or new reactions after the former model has been validated with laboratory experiments).

Reaction-based models can be exploited to run both stochastic and deterministic simulations of the temporal evolution of the system. On the one hand, according to the stochastic formulation of chemical kinetics [25], the formalization of a biochemical network into a set of species and a set of reactions can be straightforwardly used to run stochastic simulation algorithms [26]. On the other hand, any reaction-based model can be automatically converted into a corresponding system of ordinary differential equations, and then simulated by means of some numerical integration algorithm [27]. This automatic conversion can be done by considering the law of mass action [28], an empirical chemical law stating that the rate of a reaction is proportional to the product of the concentration of its reactants, each one raised to the power of the corresponding stoichiometric coefficient. This allows to derive a rate equation for each species appearing in the reaction-based model, formally describing how its concentration changes in time according to all reactions where it appears either as reactant or product. Therefore, any biochemical system defined by means of a reaction-based model can be also formalized as a set of coupled (non-linear) first order ordinary differential equations. The duality of the stochastic and deterministic interpretation of a reaction-based model concerns also reaction constants and molecular amounts, which can be bidirectionally converted into appropriate numerical values—i.e., number of molecules into concentration, stochastic constant into kinetic rate, or viceversa—assuming that the dimension of the volume where reactions take place is known [25,29].

4   Conclusive Remarks

In order to gain novel insights into the functioning of a biological system, various computational methods—such as parameter sweep analysis, sensitivity analysis, parameter estimation, reverse engineering (see [30,31] and references therein)—can be exploited to make predictions on the way the system behaves both in physiological and perturbed conditions. To this aim, the response of the system is usually analyzed using distinct model parameterizations, that is, different initializations of species amounts and/or reaction constants. These methods require the execution of a large number of simulations, whose computational burden can rapidly overtake the capabilities of Central Processing Units (CPUs), especially for mechanism-based models. In these cases, high-performance computing solutions can be exploited, such as Graphics Processing Units (GPUs), which recently gained ground in many fields related to life sciences [32,33].

Despite better computational performances, the use of GPUs in scientific applications is weakened by the striking difference between GPU-based and CPU-based computing, which demands specific programming skills to the purpose of fully exploiting the peculiar GPU's architecture. To overcome these drawbacks and to provide user-friendly tools to analyze the dynamics of reaction-based models, coarse-grain GPU implementations of both stochastic and deterministic simulation algorithms were presented in [34,35]. These tools allow to simultaneously execute a massive number of simulations, each one characterized by a different model parameterization, and were proven to largely speed up burdensome computational analyses of various cellular processes [13,34,35].

An additional benefit is that these parallel simulators can be used as core engines to apply other computational methods, like parameter estimation (PE) or reverse engineering (RE). PE consists in the inference of the unknown values for species amounts or reaction constants, and it is often tackled by means of global optimization techniques [36,37], while RE consists in devising a plausible cascade of spatio-temporal interactions in a network of biochemical reactions. Both problems require the availability of time-series experimental data (e.g., quantitative amount of some molecular species), needed as target to assess the goodness of the inferred model and/or its parameterization.

In the context of reaction-based models, the RE problem was faced in [38], while a PE method was initially introduced in [39]. The latter is based on a Swarm Intelligence heuristics, called Particle Swarm Optimization, which exploits a set of random candidate model parameterizations (each one codified by a particle in the swarm) to the purpose of converging to an optimal solution. In particular, the method in [39] exploits a multi-swarm topology, whereby each swarm is associated to time-series data measured in some experimental condition. The method is inspired by the quite common scenario of experimental research, where experiments are executed to observe the biological system in different conditions, and data are collected in both physiological and perturbed states. In the multi-swarm approach, all swarms cooperate to estimate a *unique* model parameterization, which is able to generate the system dynamics that better fits *all* the data that were measured in all tested experimental conditions. The quality of the model parameterization is evaluated by a fitness function defined, e.g., as the distance between the experimental data at each sampled time point, and the corresponding species amounts obtained (at the same time points) by means of dynamic simulations.

This PE method was further refined to fully exploit the parallel acceleration granted by GPUs [40], as well as to take advantage of fuzzy rule-based systems [41] or to test different strategies for the initialization of candidate solutions (i.e., particles' position within the search space) [42]. These refinements were shown to improve both the computational and effectiveness performance of the PE method.

Hopefully, the dissemination of the advantages of reaction-based models and of GPU-accelerated tools for their analysis will provoke a widespread adoption of this formalism by modelers. This could allow a more easy definition and validation of mathematical models, the prediction of unknown behaviours and the identification of critical factors of biochemical networks in reduced times, as well as a faster design of focused laboratory experiments to better comprehend the malfunctioning of cellular processes.

References

1. Bartocci, E., Lió, P.: Computational modeling, formal analysis, and tools for systems biology. PLoS Comput. Biol. **12**(1), e1004591 (2016)
2. Wellstead, P., Bullinger, E., Kalamatianos, D., Mason, O., Verwoerd, M.: The rôle of control and system theory in systems biology. Annu. Rev. Control **32**(1), 33–47 (2008)
3. Wolkenhauer, O.: Why model? Front. Physiol. **5**(21), 1–5 (2014)
4. Akhtar, A., Fuchs, E., Mitchison, T., Shaw, R., St Johnston, D., Strasser, A., Taylor, S., Walczak, C., Zerial, M.: A decade of molecular cell biology: achievements and challenges. Nat. Rev. Mol. Cell Biol. **12**(10), 669–674 (2011)
5. Welch, C., Elliott, H., Danuser, G., Hahn, K.: Imaging the coordination of multiple signalling activities in living cells. Nat. Rev. Mol. Cell Biol. **12**(11), 749–756 (2011)
6. Cvijovic, M., Almquist, J., Hagmar, J., Hohmann, S., Kaltenbach, H.M., Klipp, E., Krantz, M., Mendes, P., Nelander, S., Nielsen, J., Pagnani, A., Przulj, N., Raue, A., Stelling, J., Stoma, S., Tobin, F., Wodke, J.A.H., Zecchina, R., Jirstrand, M.: Bridging the gaps in systems biology. Mol. Genet. Genomics **289**(5), 727–734 (2014)
7. Gonçalves, E., Bucher, J., Ryll, A., Niklas, J., Mauch, K., Klamt, S., Rocha, M., Saez-Rodriguez, J.: Bridging the layers: towards integration of signal transduction, regulation and metabolism into mathematical models. Mol. Biosyst. **9**(7), 1576–1583 (2013)
8. Karr, J.R., Sanghvi, J.C., Macklin, D.N., Gutschow, M.V., Jacobs, J.M., Bolival, B., Assad-Garcia, N., Glass, J.I., Covert, M.: A whole-cell computational model predicts phenotype from genotype. Cell **150**(2), 389–401 (2012)
9. Besozzi, D.: Computational methods in systems biology: case studies and biological insights. In: Petre, I. (ed.) Proceedings of 4th International Workshop on Computational Models for Cell Processes. EPTCS, vol. 116, pp. 3–10 (2013)
10. Amara, F., Colombo, R., Cazzaniga, P., Pescini, D., Csikász-Nagy, A., Muzi Falconi, M., Besozzi, D., Plevani, P.: *In vivo* and *in silico* analysis of PCNA ubiquitylation in the activation of the Post Replication Repair pathway in *S. cerevisiae*. BMC Syst. Biol. **7**(24) (2013)
11. Besozzi, D., Cazzaniga, P., Dugo, M., Pescini, D., Mauri, G.: A study on the combined interplay between stochastic fluctuations and the number of flagella in bacterial chemotaxis. EPTCS **6**, 47–62 (2009)
12. Besozzi, D., Cazzaniga, P., Pescini, D., Mauri, G., Colombo, S., Martegani, E.: The role of feedback control mechanisms on the establishment of oscillatory regimes in the Ras/cAMP/PKA pathway in *S. cerevisiae*. EURASIP J. Bioinform. Syst. Biol. **1**, 10 (2012)

13. Cazzaniga, P., Nobile, M.S., Besozzi, D., Bellini, M., Mauri, G.: Massive exploration of perturbed conditions of the blood coagulation cascade through GPU parallelization. BioMed Res. Int. (2014). Article ID 863298
14. Intosalmi, J., Manninen, T., Ruohonen, K., Linne, M.L.: Computational study of noise in a large signal transduction network. BMC Bioinformatics **12**(1), 1–12 (2011)
15. Pescini, D., Cazzaniga, P., Besozzi, D., Mauri, G., Amigoni, L., Colombo, S., Martegani, E.: Simulation of the Ras/cAMP/PKA pathway in budding yeast highlights the establishment of stable oscillatory states. Biotechnol. Adv. **30**, 99–107 (2012)
16. Petre, I., Mizera, A., Hyder, C.L., Meinander, A., Mikhailov, A., Morimoto, R.I., Sistonen, L., Eriksson, J.E., Back, R.J.: A simple mass-action model for the eukaryotic heat shock response and its mathematical validation. Nat. Comput. **10**(1), 595–612 (2011)
17. Barabási, A.L., Oltvai, Z.N.: Network biology: understanding the cell's functional organization. Nat. Rev. Genet. **5**(2), 101–113 (2004)
18. Bordbar, A., Monk, J.M., King, Z.A., Palsson, B.Ø.: Constraint-based models predict metabolic and associated cellular functions. Nat. Rev. Genet. **15**(2), 107–120 (2014)
19. Cazzaniga, P., Damiani, C., Besozzi, D., Colombo, R., Nobile, M.S., Gaglio, D., Pescini, D., Molinari, S., Mauri, G., Alberghina, L., Vanoni, M.: Computational strategies for a system-level understanding of metabolism. Metabolites **4**(4), 1034–1087 (2014)
20. Novère, N.L.: Quantitative and logic modelling of molecular and gene networks. Nat. Rev. Genet. **16**(3), 146–158 (2015)
21. Morris, M.K., Saez-Rodriguez, J., Sorger, P.K., Lauffenburger, D.A.: Logic-based models for the analysis of cell signaling networks. Biochemistry **49**(15), 3216–3224 (2010)
22. Stelling, J.: Mathematical models in microbial systems biology. Curr. Opin. Microbiol. **7**(5), 513–518 (2004)
23. Wilkinson, D.: Stochastic modelling for quantitative description of heterogeneous biological systems. Nat. Rev. Genet. **10**(2), 122–133 (2009)
24. Iancu, B., Czeizler, E., Czeizler, E., Petre, I.: Quantitative refinement of reaction models. Int. J. Unconv. Comput. **8**(5/6), 529–550 (2012)
25. Gillespie, D.T.: Exact stochastic simulation of coupled chemical reactions. J. Comput. Phys. **81**, 2340–2361 (1977)
26. Gillespie, D.T.: Stochastic simulation of chemical kinetics. Annu. Rev. Phys. Chem. **58**, 35–55 (2007)
27. Butcher, J.C.: Numerical Methods for Ordinary Differential Equations. John Wiley & Sons, Ltd., Chichester (2003)
28. Voit, E.O., Martens, H.A., Omholt, S.W.: 150 years of the mass action law. PLoS Comput. Biol. **11**(1), e1004012 (2015)
29. Wolkenhauer, O., Ullah, M., Kolch, W., Cho, K.H.: Modeling and simulation of intracellular dynamics: choosing an appropriate framework. IEEE Trans. Nanobiosci. **3**(3), 200–207 (2004)
30. Aldridge, B.B., Burke, J.M., Lauffenburger, D.A., Sorger, P.K.: Physicochemical modelling of cell signalling pathways. Nat. Cell Biol. **8**, 1195–1203 (2006)
31. Chou, I.C., Voit, E.O.: Recent developments in parameter estimation and structure identification of biochemical and genomic systems. Math. Biosci. **219**(2), 57–83 (2009)
32. Demattè, L., Prandi, D.: GPU computing for systems biology. Brief Bioinform. **11**(3), 323–333 (2010)

33. Harvey, M.J., De Fabritiis, G.: A survey of computational molecular science using graphics processing units. WIREs Comput. Mol. Sci. **2**(5), 734–742 (2012)
34. Nobile, M.S., Cazzaniga, P., Besozzi, D., Pescini, D., Mauri, G.: cuTauLeaping: a GPU-powered tau-leaping stochastic simulator for massive parallel analyses of biological systems. PLoS ONE **9**(3), e91963 (2014)
35. Nobile, M.S., Besozzi, D., Cazzaniga, P., Mauri, G.: GPU-accelerated simulations of mass-action kinetics models with cupSODA. J. Supercomput. **69**(1), 17–24 (2014)
36. Dräger, A., Kronfeld, M., Ziller, M.J., Supper, J., Planatscher, H., Magnus, J.B.: Modeling metabolic networks in *C. glutamicum*: a comparison of rate laws in combination with various parameter optimization strategies. BMC Syst. Biol. **3**(5) (2009)
37. Moles, C.G., Mendes, P., Banga, J.R.: Parameter estimation in biochemical pathways: a comparison of global optimization methods. Genome Res. **13**(11), 2467–2474 (2003)
38. Nobile, M.S., Besozzi, D., Cazzaniga, P., Pescini, D., Mauri, G.: Reverse engineering of kinetic reaction networks by means of Cartesian genetic programming and particle swarm optimization. In: 2013 IEEE Congress on Evolutionary Computation, vol. 1, pp. 1594–1601. IEEE (2013)
39. Nobile, M.S., Besozzi, D., Cazzaniga, P., Mauri, G., Pescini, D.: A GPU-based multi-swarm PSO method for parameter estimation in stochastic biological systems exploiting discrete-time target series. In: Giacobini, M., Vanneschi, L., Bush, W.S. (eds.) EvoBIO 2012. LNCS, vol. 7246, pp. 74–85. Springer, Heidelberg (2012)
40. Nobile, M.S., Besozzi, D., Cazzaniga, P., Mauri, G., Pescini, D.: Estimating reaction constants in stochastic biological systems with a multi-swarm PSO running on GPUs. In: Soule, T. (ed.) Proceedings of 14th International Conference on Genetic and Evolutionary Computation Conference Companion, GECCO Companion 2012, pp. 1421–1422. ACM (2012)
41. Nobile, M.S., Pasi, G., Cazzaniga, P., Besozzi, D., Colombo, R., Mauri, G.: Proactive particles in swarm optimization: a self-tuning algorithm based on fuzzy logic. In: Proceedings of IEEE International Conference on Fuzzy Systems (FUZZ-IEEE), pp. 1–8 (2015)
42. Cazzaniga, P., Nobile, M.S., Besozzi, D.: The impact of particles initialization in PSO: parameter estimation as a case in point. In: IEEE Conference on Computational Intelligence in Bioinformatics and Computational Biology (CIBCB), pp. 1–8 (2015)

Comparative Genomics on Artificial Life

Priscila Biller[1,2], Carole Knibbe[2,3], Guillaume Beslon[2,3],
and Eric Tannier[2,4(✉)]

[1] University of Campinas, São Paulo, Brazil
[2] INRIA Grenoble Rhône-Alpes, 38334 Montbonnot, France
eric.tannier@inria.fr
[3] Université Lyon 1, LIRIS, UMR5205, 69622 Villeurbanne, France
[4] Université Lyon 1, LBBE, UMR5558, 69622 Villeurbanne, France

Abstract. Molecular evolutionary methods and tools are difficult to validate as we have almost no direct access to ancient molecules. Inference methods may be tested with simulated data, producing full scenarios they can be compared with. But often simulations design is concomitant with the design of a particular method, developed by a same team, based on the same assumptions, when both should be blind to each other. *In silico* experimental evolution consists in evolving digital organisms with the aim of testing or discovering complex evolutionary processes. Models were not designed with a particular inference method in mind, only with basic biological principles. As such they provide a unique opportunity to blind test the behavior of inference methods. We give a proof of this concept on a comparative genomics problem: inferring the number of inversions separating two genomes. We use Aevol, an *in silico* experimental evolution platform, to produce benchmarks, and show that most combinatorial or statistical estimators of the number of inversions fail on this dataset while they were behaving perfectly on ad-hoc simulations. We argue that biological data is probably closer to the difficult situation.

Keywords: Comparative genomics · In silico experimental evolution · Benchmark · Rearrangements

1 Validation of Evolutionary Inferences

The comparative method in evolutionary biology consists in detecting similarities and differences between extant organisms, and, based on more or less formalized hypotheses on the evolutionary processes, infer ancestral states explaining the similarities and an evolutionary history explaining the differences.

A common concern in all evolutionary studies is the validity of the methods and results. Results concern events that were supposed to occur in a deep past (up to 4 billion years) and they have no other trace today than the present molecules used by the comparative method.

As we cannot travel back in time to verify the results, there are several ways to assess the validity of molecular evolution studies: theoretical considerations about the models and methods (realism, consistency, computational complexity,

© Springer International Publishing Switzerland 2016
A. Beckmann et al. (Eds.): CiE 2016, LNCS 9709, pp. 35–44, 2016.
DOI: 10.1007/978-3-319-40189-8_4

model testing, ability to generate a statistical support or a variety of the solutions) [24], coherence with fossil records [26], or ancient DNA [11], or empirical tests when the solution is known, on experimental evolution [17] or simulations. Each method has its caveats. Models for inference have to adopt a compromise between realism, consistency and complexity. Ancient DNA is rarely available, usually not in an assembled shape. Fossils are also rare and provide a biased sampling of ancient diversity. Experimental evolution is expensive, time-consuming and limited in the number of generations it can provide.

Simulation is the most popular validation tool. Genome evolution can be simulated *in silico* for a much higher number of generations than in experimental evolution, at a lower cost. All the history can be recorded in details, and compared with the inference results. A problem with simulations, however, is that they necessarily oversimplify genome evolution processes. Moreover, very often, even if they are designed to be used by another team for inference [4,10,14,15,23], they encode the same simplifications as the inference methods. For example, only fixed mutations are generated because only these are visible by inference methods, selection is tuned to fit what is visible by the inference methods; genes are evolutionary units in simulations because they are the units taken for inference. Everything is designed thinking of the possibilities of the inference methods, leading to easy unrealistic instances.

This mode of *ad-hoc* simulation has been widely applied to test estimators of rearrangement distances, and in particular inversion distances [5,7,9,12,22]. The problem consists in comparing two genomes and estimating the number of inversions (a rearrangement that reverses the reading direction of a genomic segment) that have occurred in the evolutionary lineages separating them. To construct a solution, conserved genes or synteny blocks are detected in the two genomes, and a number of inversions explaining the differences in gene orders is estimated. A lot of work has consisted in finding shortest scenarios [13]. Statistical estimations need a model. The standard and most used model depicts genomes as permutations of genes and assumes that an inversion reverses a segment of the permutation, taken uniformly at random over all segments. When simulators are designed to validate the estimators, they also use permutations as models of gene orders, and inversions on segments of this permutations, chosen uniformly at random. Estimators show good performances on such simulations, but transforming a genome into a permutation of genes is such a simplification from both parts that it means nothing about any ability to estimate a rearrangement distance in biological data [8].

We propose to use simulations that were not designed for validation purposes. It is the case, in artificial life, of *in silico* experimental evolution [18], and in particular of the *Aevol* platform [3,19]. Aevol contains, among many other features, all what is needed to test rearrangement inference methods. The genomes have gene sequences and non coding sequences organized in a chromosome, and evolve with inversions, in addition to substitutions, indels, duplications, losses, translocations. Rearrangements are chosen with a uniform random model on the genome, which should fit the goals of the statistical estimators, but is different from a uniform random model on permutations [8]. We tested 10 different estimators of inversion distance on 18 different datasets generated by Aevol. The difference with ad-hoc simulations is striking. Most estimators completely fail to give a close estimate in a vast majority of conditions. We argue that the reason for this failure lies in realistic features in artificial genomes that are very likely to reproduce the failure on real data.

We first describe the principle of the estimators, then the principles of the simulator, with its goals and its functioning. We will show how to process its results to test statistical estimators of rearrangement distances.

2 Comparative Genomics: Estimating an Inversion Distance

We tested 10 estimators of the number of inversions separating two genomes, called ID (the inversion distance) [16], CL for Caprara and Lancia [9], EH for Eriksen and Hultman [12], Badger [20], BD for Berestycki and Durrett [5], LM for Lin and Moret [22], BGT for Biller, Guéguen and Tannier [7], AA for Alexeev and Alekseyev [2], ER1 and ER2 for Erdös-Renyi 1 and 2 [8].

For 8 of them (ID, LM, BGT, Badger, EH, BD, CL, AA), a genome is defined as a signed permutation, π over $\{1, \ldots, n\}$, that is, an ordering of the elements of $\{1, \ldots, n\}$ where each element is given a sign, $+$ or $-$ ($+$ usually omitted), representing the reading direction of an element. The elements of the permutation are *genes*, or *solid regions*, the ones that are never cut by inversions. All inversions have the same probability. For the two remaining estimators (ER1 and ER2), a genome is made up of two components: the same signed permutation, and in addition a vector p of $n + 1$ breakage probabilities, $p_i > 0$, $0 \leq i \leq n$, with $\sum_i p_i = 1$. An inversion of the segment $[\pi_i, \ldots, \pi_j]$ has probability $p_{i-1} p_j$.

Suppose A and B are two signed permutations. We define the *breakpoint graph* of A and B as the graph with $2n + 2$ vertices and $2n + 2$ edges: for each element $i \in \{1, \ldots, n\}$, define two vertices i_t and i_h, plus two additional vertices 0_h and $n+1_t$; then for any two consecutive numbers ab of A, join two extremities by an A-edge: first is a_h if a is positive, a_t otherwise, second is b_t if b is positive, b_h otherwise. Additionally, if a is the first element of the permutation, join 0_h and a_t if a is positive, a_h otherwise, and if b is the last element of the permutation, join $n + 1_t$ and b_h if b is positive, b_t otherwise. Do the same for B, and call the edges B-edges.

An *adjacency* of a genome A is an A-edge in the breakpoint graph. It is a *common adjacency* with a genome B if it is also a B-edge, otherwise it is a *breakpoint*. Breakpoint graphs have a uniform degree of 2 on all vertices, thus they are sets of disjoint cycles alternating between A-edges and B-edges. We note b the number of breakpoints, c the number of cycles of the breakpoint graph, and c_2 the number of cycles with 4 edges.

The parsimony estimator (ID) is the minimum number of inversions necessary to transform A into B, which is close to $n + 1 - c$ [16]. Badger is a Bayesian sampler of inversion scenarios and computes an *a posteriori* probable distance.

The others all work with the method of moments. This consists in computing an expected value for one or two observable parameters of the breakpoint graph (b, c_2, c or a combination of two of them) if A and B are separated by k inversions. It is a function of k and n: $f_n(k)$. It is never computed exactly, approximate formulas or computation principles are given. Then k is estimated as $\hat{k} = f_n^{-1}(p)$ for the observed value p of the parameter. LM, CL, BGT, ER1 are based on the expected value of b. EH and BD are based on the expected value of c. ER2 is based on expected values for b and c_2, and AA uses expected values for b and c. The two latter use two values because they also consider n as unknown and estimate it as well as k.

3   Artificial Life: *In Silico* Experimental Evolution and the Aevol Platform

Unlike many simulators used to validate phylogenetic inference methods [4,10, 14,15,23], Aevol does not represent a species by a single lineage undergoing fixed mutations. Like forward-in-time simulators used in population genetics, it explicitly represents all genotypes present in the population and simulates spontaneous mutations, which can be deleterious, neutral or beneficial. An important difference, however, is that the selection coefficients of mutations are not predefined for each locus nor drawn from a random distribution. Instead, an artificial chemistry is used to decode each genome present in the population and compute its phenotype, which is its ability to perform a computational task (see details below). Point mutations or small indels can alter gene sequences and non coding sequences. A local mutation in a gene can have a different effect on phenotype and fitness, depending on the genomic background (other genes). Chromosomal rearrangements like duplications, deletions, translocations or inversions can occur anywhere in the chromosome sequence. They can alter gene number and gene order and disrupt genes.

Figure 1 summarizes the functioning of Aevol. We give a high level description here, and emphasize that the tool has many other possibilities than being used as a bench mark. For a complete description and some of its possibilities, see [3,19]. Genomes are circular sequences on a binary alphabet. A population of typically 1000 genomes lives at a given generation. Genes are segments situated on a transcribed sequence (*i.e.*, a sequence starting after a promoter and ending at a terminator sequence) starting after a Ribosome Binding Start and a Start codon and ended by a Stop codon on the same reading frame. Inside a gene, a coding sequence is translated into a protein sequence using a genetic code on size three codons. This protein sequence encodes the parameters of a piecewise linear function that indicates the contribution (in $[-1, 1]$) of the protein to each abstract "phenotypic trait" in $[0, 1]$. All proteins encoded in a genome are summed to produce the phenotype, which is thus a piece-wise linear function indicating the level of each phenotypic trait in $[0, 1]$.

This phenotype is then compared with a target function indirectly representing the environment of the individual. The difference between the two is

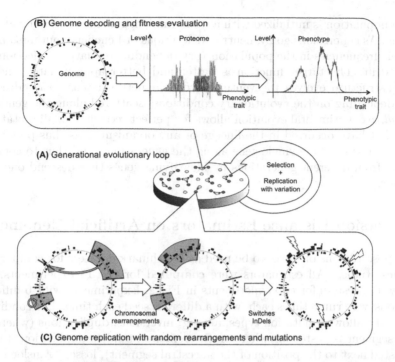

Fig. 1. Overview of the Aevol model. **(A)** A population of genomes is simulated. At each generation, all genomes are evaluated and the fittest ones are replicated to produce the next generation. The replication process includes variation operators. The joint actions of selection, genetic drift and variation make the population evolve. **(B)** Overview of the genome decoding process. Left: Each individual own a circular double-strand genome with scattered genes. Right: The individual's phenotype is the level of each abstract phenotypic trait in [0, 1]. It is compared a target representing the optimal phenotype given the environment. Middle: Each gene is decoded into a protein that contributes to a small subset of phenotypic traits. More precisely, the sequence of the gene is decoded into three reals that specify the mean, width and height of a triangular kernel function. All the proteins are then summed up to compute the phenotype. The individual displayed here was obtained after 460.000 generations of evolution in Aevol under a mutational pressure of 10^{-6} mutations/bp/generation for local events and 10^{-5} mutations/bp/generation for chromosomal rearrangements (see below). Its genome is 6898 bp long. It encodes 113 genes and 35 RNAs (not shown). 28.4 % of the genome is non-coding. **(C)** Overview of the replication process. During its replication each genome may undergo chromosomal rearrangements affecting large DNA segments (here an inversion and a translocation) and local mutations (point mutations or small InDels).

used to compute the *fitness* of the genome. To produce the next generation, genomes with high fitness are replicated in the following generation with higher probability than genomes with low fitness. During replication, local mutations and chromosomal rearrangements are performed on the genomes, at a spontaneous rate fixed by simulation parameters.

The population is initialized with a same random genome containing at least one gene. As generations go by, neutral, deleterious or beneficial mutants appear and their frequencies in the population vary depending on natural selection and genetic drift. The target function is better and better approximated and the genome structure evolves to eventually contain between tens to hundreds of genes (depending on the evolutionary conditions) scattered along the genome.

In silico experimental evolution allows for perfect recording of all mutational events that have occurred in the lineage of any organism. It is thus possible to trace the evolution of a single gene along the generations, and thus to compare genomes from different generations by identifying genes that descend one from the other.

4   Inversion Distance Estimators on Artificial Genomes

We propose 18 runs of Aevol to be used as benchmark datasets for comparative genomics studies. All estimators were computed for the 18 experiments, and we show the results for two experiments in Fig. 2. Experiments with 6 different conditions were run 3 times each, with a different seed each time. The conditions concern the allowed mutation types, among: inversions, duplications (where the copied segment is pasted anywhere on the genome), tandem duplications (where it is pasted next to the position of the ancestral segment), losses, translocations, point mutations and small InDels. Mutation rates were set to 5.10^{-6} mutation per base per generation for point mutations and InDels, and 10^{-5} for larger allowed rearrangements. All runs were stopped after 15000 generations with a genome containing approximately 100 genes. We make accessible, for each of the 18 runs, the input parameters, and for each generation, the list of genes, their coordinates on the genome, and their genealogy (how they relate to each other across generations). Material can be uploaded here: http://aevol.inrialpes. fr/resources/benchmark/cie_2016.

From Aevol output we compute signed permutations of genes without duplicates which model the relative order of genes compared with the last generation. We keep, in each generation, only the genes that have a unique descendant in the last generation, with no duplication event in its history between this generation and the last. Only the last generations can contain such genes, so permutations are only computed for a few hundred generations.

The results for two different runs out of 18 are shown in Fig. 2. The two were chosen for extreme but informative behaviors. The first run allowed for inversions, duplications and small mutations (A), while the second one allowed for translocations and tandem duplications in addition (B). At each generation we keep the genome in the ancestry lineage of the fittest genome at generation 15 000. The true number of inversions is compared with the estimated one, according to 7 estimators (we removed 3 of them because the curves were indistinguishable from others). The results highly depend on the conditions. On the (A) part, all estimators except AA are estimating a rather good number of inversions up to 50 events. On the (B) part, we cut the graph after 100 generations

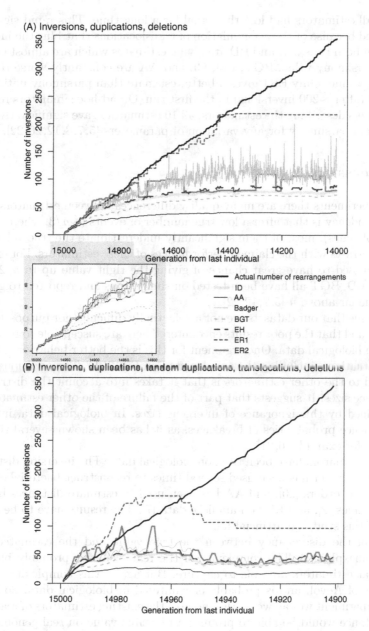

Fig. 2. The results of 7 estimators: ID, Badger, EH, BGT, ER1, ER2, AA. The other ones gave indistinguishable curves (BD from EH, LM and CL from BGT). x axis is the generation number. y axis is the number of inversions. All generations are compared with the last one, number 15000. The true number is the black solid line, and the others are estimated numbers. These Aevol runs includes (A) inversions, duplications and deletions (B) inversions, duplications, tandem duplication, translocations and deletions. The number of compared genes is from 119 to 109 (A), and from 92 to 49 (B).

because all estimators had lost the signal for a long time. This rapid signal loss is expected because of the accumulation of translocations that blur the inversion signal. On both runs, EH and BD are giving estimates which are almost equal to ID, the parsimony value. BGT, LM, CL and AA are constantly worse than the parsimony value. Only ER1 gives a better estimate than parsimony until generation 600 (after ~200 inversions) in the first run. On ad-hoc simulators reported in the papers describing the estimators, all 10 estimators gave significantly better results than parsimony for any variation of parameters [5,7,8,12,20,22].

5   Discussion

In our experiments there are many quality differences between estimators. But a general tendency is that after a low true number of events ($\sim n/3$, where n is the number of genes), most of them significantly underestimate the true value. This highly contrasts with the claimed performances of these estimators. For example ID is supposed to have great chance of giving the right value up to $n/2$, while LM, EH, BD, BGT all have been tested on simulations and reported to give the right value far above n [5,7,12,22].

We argue that our datasets are not artefactually difficult (nor purposely *made* difficult), and that the poor results encountered here are susceptible to reflect real results on biological data. One argument for this is the better behavior of ER1 in several situations, including the one depicted on Fig. 2(A). The addition of ER1 compared to the other estimators is that it takes into account the distribution of intergene sizes. It suggests that part of the failure of the other estimators can be explained by this ignorance of intergene sizes. In biological data, intergene sizes influence probabilities of breakages, as it has been shown several times on mammals for example [6,21].

Some estimators have been tried on biological data. The inversion distance is often used. Badger has been used several times to reconstruct bacterial or mitocondrial gene orders [20], and AA has been used to estimate distances between Yeast genomes [2], and ER2 on amniote data [7]. The results have to be read in regard of this study on artificial life.

Part of the discrepancy between the true value and the estimated value remains unexplained. The complexity of the real scenarios probably blurs the signal that estimators are able to capture. But again, this complexity is not a specificity of Aevol, and is probably encountered in biological data. So by this simple experiment we can worry that none of the existing estimators of rearrangement distance would be able to produce a plausible value on real genomes.

Future Work. We tested only the estimation of the number of inversions. But only with the runs we have already computed, a lot more can be done: estimation of the proportion of translocations as in [1], or estimating both inversions and duplications as in [25]. Artificial genomes could in principle not only be used by comparative genomics inference methods, but by a larger set of molecular evolution studies. For the moment the sequences are made of 0s and 1s, which is not a problem to study gene order, but can be disturbing for sequence analyses. This way of coding sequences is on another hand a good sign that Aevol was not developed for benchmarking purposes. In a close future nucleotidic and proteic sequences with the biological alphabet will be added to extend the benchmarking possibilities of the model.

Also we work with only one lineage, and compare only two genomes here, because Aevol evolves a single population. A useful addition will be speciation processes, in order to be able to compare several genomes.

On the Blind Multidisciplinarity. This study experiments a singular kind of interdisciplinarity. Obviously communities from comparative genomics and artificial life have to work together in order to make such results possible. But, on the opposite, these results are only possible because both communities first work in relative isolation. If they had defined their working plans together, spoke to each other too often or influenced each other's way of thinking evolutionary biology, the work would have lost some value. Indeed, what makes the difficulty here for comparative genomicists is that they have to infer histories on data for which they have no stranglehold on the processes, just as for biological data, but on which they also have the correct answer, just not as for biological data.

Acknowledgement. This work was funded by FAPESP grant 2013/25084-2 to PB, ANR-10-BINF-01-01 Ancestrome to ET and ICT FP7 European programme EVOEVO to CK and GB.

References

1. Alexeev, N., Aidagulov, R., Alekseyev, M.A.: A computational method for the rate estimation of evolutionary transpositions. In: Ortuño, F., Rojas, I. (eds.) IWBBIO 2015, Part I. LNCS, vol. 9043, pp. 471–480. Springer, Heidelberg (2015)
2. Alexeev, N., Alekseyev, M.A.: Estimation of the true evolutionary distance under the fragile breakage model. Arxiv (2015). http://arxiv.org/abs/1510.08002
3. Batut, B., Parsons, D.P., Fischer, S., Beslon, G., Knibbe, C.: In silico experimental evolution: a tool to test evolutionary scenarios. BMC Bioinformatics **14**(S15), S11 (2013)
4. Beiko, R.G., Charlebois, R.L.: A simulation test bed for hypotheses of genome evolution. Bioinformatics **23**(7), 825–831 (2007)
5. Berestycki, N., Durrett, R.: A phase transition in the random transposition random walk. Probab. Theory Relat. Fields **136**, 203–233 (2006)
6. Berthelot, C., Muffato, M., Abecassis, J., Crollius, H.R.: The 3d organization of chromatin explains evolutionary fragile genomic regions. Cell Rep. **10**(11), 1913–1924 (2015)
7. Biller, P., Guéguen, L., Tannier, E.: Moments of genome evolution by double cut-and-join. BMC Bioinform. **16**(Suppl 14), S7 (2015)
8. Biller, P., Knibbe, C., Guéguen, L., Tannier, E.: Breaking good: accounting for the diversity of fragile regions for estimating rearrangement distances. Genome Biol. Evol. (2016, in press)

9. Caprara, A., Lancia, G.: Experimental and statistical analysis of sorting by reversals. In: Sankoff, D., Nadeau, J.H. (eds.) Comparative Genomics, pp. 171–183. Springer, Amsterdam (2000)

10. Dalquen, D.A., Anisimova, M., Gonnet, G.H., Dessimoz, C.: ALF-a simulation framework for genome evolution. Mol. Biol. Evol. **29**(4), 1115–1123 (2012)

11. Duchemin, W., Daubin, V., Tannier, E.: Reconstruction of an ancestral yersinia pestis genome and comparison with an ancient sequence. BMC Genom. **16**(Suppl 10), S9 (2015)

12. Eriksen, N., Hultman, A.: Estimating the expected reversal distance after a fixed number of reversals. Adv. Appl. Math. **32**, 439–453 (2004)

13. Fertin, G., Labarre, A., Rusu, I., Tannier, E., Vialette, S.: Combinatorics of Genome Rearrangements. MIT Press, London (2009)

14. Fletcher, W., Yang, Z.: Indelible: a flexible simulator of biological sequence evolution. Mol. Biol. Evol. **26**(8), 1879–1888 (2009)

15. Hall, B.G.: Simulating DNA coding sequence evolution with EvolveAGene 3. Mol. Biol. Evol. **25**(4), 688–695 (2008)

16. Hannenhalli, S., Pevzner, P.A.: Transforming men into mice (polynomial algorithm for genomic distance problem). In: Proceedings of 36th Annual Symposium on Foundations of Computer Science (1995)

17. Hillis, D.M., Bull, J.J., White, M.E., Badgett, M.R., Molineux, I.J.: Experimental phylogenetics: generation of a known phylogeny. Science **255**(5044), 589–592 (1992)

18. Hindré, T., Knibbe, C., Beslon, G., Schneider, D.: New insights into bacterial adaptation through in vivo and in silico experimental evolution. Nat. Rev. Microbiol. **10**, 352–365 (2012)

19. Knibbe, C., Coulon, A., Mazet, O., Fayard, J.-M., Beslon, G.: A long-term evolutionary pressure on the amount of noncoding DNA. Mol. Biol. Evol. **24**(10), 2344–2353 (2007)

20. Larget, B., Simon, D.L., Kadane, J.B.: On a Bayesian approach to phylogenetic inference from animal mitochondrial genome arrangements (with discussion). J. Roy. Stat. Soc. B **64**, 681–693 (2002)

21. Lemaitre, C., Zaghloul, L., Sagot, M.-F., Gautier, C., Arneodo, A., Tannier, E., Audit, B.: Analysis of fine-scale mammalian evolutionary breakpoints provides new insight into their relation to genome organisation. BMC Genom. **10**, 335 (2009)

22. Lin, Y., Moret, M.E.: Estimating true evolutionary distances under the DCJ model. Bioinformatics **24**(13), i114–i122 (2008)

23. Mallo, D., De Oliveira Martins, L., Posada, D.: Simphy: phylogenomic simulation of gene, locus, and species trees. Syst Biol. **65**, 334–344 (2016)

24. Steel, M., Penny, D.: Parsimony, likelihood, and the role of models in molecular phylogenetics. Mol. Biol. Evol. **17**(6), 839–850 (2000)

25. Swenson, K.M., Marron, M., Earnest-DeYoung, J.V., Moret, B.M.E.: Approximating the true evolutionary distance between two genomes. J. Exp. Algorithmics **12**, 3.5 (2008)

26. Szőllősi, G.J., Boussau, B., Abby, S.S., Tannier, E., Daubin, V.: Phylogenetic modeling of lateral gene transfer reconstructs the pattern and relative timing of speciations. Proc. Natl. Acad. Sci. U. S. A. **109**(43), 17513–17518 (2012)

Computability and Analysis, a Historical Approach

Vasco Brattka[1,2](✉)

[1] Faculty of Computer Science, Universität der Bundeswehr München,
München, Germany
[2] Deptartment of Mathematics and Applied Mathematics,
University of Cape Town, Cape Town, South Africa
Vasco.Brattka@cca-net.de

Abstract. The history of computability theory and the history of analysis are surprisingly intertwined since the beginning of the twentieth century. For one, Émil Borel discussed his ideas on computable real number functions in his introduction to measure theory. On the other hand, Alan Turing had computable real numbers in mind when he introduced his now famous machine model. Here we want to focus on a particular aspect of computability and analysis, namely on computability properties of theorems from analysis. This is a topic that emerged already in early work of Turing, Specker and other pioneers of computable analysis and eventually leads us to the very recent project of classifying the computational content of theorems in the Weihrauch lattice.

1 Introduction

Probably Émil Borel was the first mathematician who had an intuitive understanding of computable real number functions and he anticipated some basic ideas of computable analysis as early as at the beginning of the 20th century. It was in his introduction to measure theory where he felt the need to discuss such concepts and we can find for instance the following crucial observation in [5,6].

Theorem 1 (Borel 1912). *Every computable real number function $f : \mathbb{R}^n \to \mathbb{R}$ is continuous.*

Strictly speaking, Borel's definition of a computable real number function was a slight variant of the modern definition (see [2] for details and translations) and his definition was informal in the sense that no rigorous notion of computability or of an algorithm was available at Borel's time.

It was only Alan Turing who introduced such a notion with the help of his now famous machine model [43,44]. Interestingly, also Turing was primarily interested in computable real numbers (hence the title of his paper!) and not so much in functions and sets on natural numbers that are the main objects

V. Brattka—Supported by the National Research Foundation of South Africa. This article uses some historical insights that were established in [2].

A. Beckmann et al. (Eds.): CiE 2016, LNCS 9709, pp. 45–57, 2016.
DOI: 10.1007/978-3-319-40189-8_5

of study in modern computability theory. Turing's definition of a computable real number function is also a slight variant of the modern definition (see [2] for details).

We conclude that computability theory was intertwined with analysis since its early years and here we want to focus on a particular aspect of this story that is related to computability properties of theorems in analysis, which are one subject of interest in modern computable analysis [17,27,38,45].

2   Some Theorems from Real Analysis

In his early work [43] Turing already implicitly discussed the computational content of some classical theorems from analysis. Some of his rather informal observations have been made precise later by Specker and others [2]. Ernst Specker was probably the first one who actually gave a definition of computable real number functions that is equivalent to the modern one [41,42]. The following theorem is one of those theorems that are implicitly discussed by Turing in [43].

Theorem 2 (Monotone Convergence Theorem).   *Every monotone increasing and bounded sequence of real numbers* $(x_n)_n$ *has a least upper bound* $\sup_{n \in \mathbb{N}} x_n$.

What Turing observed (without proof) is that for a computable sequence $(x_n)_n$ of this type the least upper bound is not necessarily computable. A rigorous proof of this result was presented only ten years later by Specker [41].

Proposition 3 (Turing 1937, Specker 1949).   *There is a computable monotone increasing and bounded sequence* $(x_n)_n$ *of real numbers such that* $x = \sup_{n \in \mathbb{N}} x_n$ *is not computable.*

Specker used (an enumeration of) the halting problem $K \subseteq \mathbb{N}$ to construct a corresponding sequence $(x_n)_n$ and such sequences are nowadays called *Specker sequences*. Then the corresponding non-computable least upper bound is $x = \sum_{i \in K} 2^{-i}$. While the Monotone Convergence Theorem is an example of a non-computable theorem, Turing also discusses a (special case) of the Intermediate Value Theorem [43], which is somewhat better behaved.

Theorem 4 (Intermediate Value Theorem).   *Every continuous function* $f : [0, 1] \to \mathbb{R}$ *with* $f(0) \cdot f(1) < 0$ *has a zero* $x \in [0, 1]$.

And then Turing's observation, which was stated for the general case by Specker [42] could be phrased in modern terms as follows.

Proposition 5 (Turing 1937, Specker 1959).   *Every computable function* $f : [0, 1] \to \mathbb{R}$ *with* $f(0) \cdot f(1) < 0$ *has a computable zero* $x \in [0, 1]$.

A rigorous proof could utilize the trisection method, a constructive variant of the well-known bisection method and can be found in [45]. Hence, while the Monotone Convergence Theorem does not hold computably, the Intermediate Value Theorem does hold computably, at least in a *non uniform* sense. It was claimed by Specker [42] (without proof) and later proved by Pour-El and Richards [38] that this situation changes if one considers a sequential version of the Intermediate Value Theorem.

Proposition 6 (Specker 1959, Pour-El and Richards 1989). *There exists a computable sequence $(f_n)_n$ of computable functions $f_n : [0,1] \to \mathbb{R}$ with $f_n(0) \cdot f_n(1) < 0$ for all $n \in \mathbb{N}$ and such that there is no computable sequence $(x_n)_n$ of real numbers $x_n \in [0,1]$ with $f_n(x_n) = 0$.*

For their proof Pour-El and Richards used two c.e. sets that are computably inseparable. Their result indicates that the Intermediate Value Theorem does not hold computably in a *uniform* sense. In fact, it is known that the Intermediate Value Theorem is not computable in the following fully uniform sense: namely, there is no algorithm that given a program for $f : [0,1] \to \mathbb{R}$ with $f(0) \cdot f(1) < 0$ produces a zero of f. Nowadays, we can express this as follows with a partial multi-valued map [45].

Proposition 7 (Weihrauch 2000). $\mathsf{IVT} :\subseteq \mathcal{C}[0,1] \rightrightarrows [0,1], f \mapsto f^{-1}\{0\}$ *with* $\mathrm{dom}(\mathsf{IVT}) = \{f \in \mathcal{C}[0,1] : f(0) \cdot f(1) < 0\}$ *is not computable.*

For general represented spaces X, Y we denote by $\mathcal{C}(X, Y)$ the space of continuous functions $f : X \to Y$ endowed with a suitable representation [45] (and the compact open topology) and we use the abbreviation $\mathcal{C}(X) := \mathcal{C}(X, \mathbb{R})$. In fact, IVT is not even continuous and this observation is related to the fact that the Intermediate Value Theorem has no constructive proof [4].

Another theorem discussed by Specker [42] is the Theorem of the Maximum.

Theorem 8 (Theorem of the Maximum). *For every continuous function $f : [0,1] \to \mathbb{R}$ there exists a point $x \in [0,1]$ such that $f(x) = \max f([0,1])$.*

Grzegorczyk [22] raised the question whether every computable function $f : [0,1] \to \mathbb{R}$ attains its maximum at a computable point. This question was answered in the negative by Lacombe [29] (without proof) and later independent proofs were provided by Lacombe [30, Theorems VI and VII] and Specker [42].

Proposition 9 (Lacombe 1957, Specker 1959). *There exists a computable function $f : [0,1] \to \mathbb{R}$ such that there is no computable $x \in [0,1]$ with $f(x) = \max f([0,1])$.*

Similar results have also been derived by Zaslavskiĭ [46]. Specker used a Kleene tree for his construction of a counterexample. A Kleene tree is a computable counterexample to Weak Kőnig's Lemma.

Theorem 10 (Weak Kőnig's Lemma). *Every infinite binary tree has an infinite path.*

Kleene [26] has proved that such counterexamples exist and (like Proposition 6) this can be easily achieved using two computably inseparable c.e. sets.

Proposition 11 (Kleene 1952). *There exists a computable infinite binary tree without computable paths.*

It is interesting to note that even though computable infinite binary trees do not need to have computable infinite paths, they do at least have paths that are *low*, which means that the halting problem relative to this path is not more difficult than the ordinary halting problem. In this sense low paths are "almost computable". The existence of such solutions has been proved by Jockusch and Soare in their now famous Low Basis Theorem [25].

Theorem 12 (Low Basis Theorem of Jockusch and Soare 1972). *Every computable infinite binary tree has a low path.*

Such low solutions also exist in case of the Theorem of the Maximum 8 for computable instances. The Monotone Convergence Theorem 2 is an example of a theorem where not even low solutions exist in general, e.g., Specker's sequence is already an example of a computable monotone and bounded sequence with a least upper bound that is equivalent to the halting problem and hence not low.

Another case similar to the Theorem of the Maximum is the Brouwer Fixed Point Theorem.

Theorem 13 (Brouwer Fixed Point Theorem). *For every continuous function $f : [0,1]^k \to [0,1]^k$ there exists a point $x \in [0,1]^k$ such that $f(x) = x$.*

It is an ironic coincidence that Brouwer, who was a strong proponent of intuitionistic mathematics is most famous for his Fixed Point Theorem that does not admit a constructive proof. It was Orevkov [34] who proved in the sense of Markov's school that there is a computable counterexample and Baigger [3] later proved this result in terms of modern computable analysis.

Proposition 14 (Orevkov 1963, Baigger 1985). *There is a computable function $f : [0,1]^2 \to [0,1]^2$ without a computable $x \in [0,1]^2$ with $f(x) = x$.*

One can conclude from Proposition 5 that such a counterexample cannot exist in dimension $k = 1$. Baigger also used two c.e. sets that are computably inseparable for his construction.

Yet another theorem with interesting computability properties is the Theorem of Bolzano-Weierstraß.

Theorem 15 (Bolzano-Weierstraß). *Every sequence $(x_n)_n$ in the unit cube $[0,1]^k$ has a cluster point.*

It was Rice [39] who proved that a straightforward computable version of the Bolzano-Weierstraß Theorem does not hold and Kreisel pointed out in his review of this article for the Mathematical Reviews of the American Mathematical Society that he already proved a more general result [28].

Proposition 16 (Kreisel 1952, Rice 1954). *There exists a computable sequence $(x_n)_n$ in $[0, 1]$ without a computable cluster point.*

In fact, this result is not all too surprising and a simple consequence of Proposition 3. What is more interesting is that in case of the Bolzano-Weierstraß Theorem there are even computable bounded sequences without *limit computable* cluster point. Here a point is called *limit computable* if it is the limit of a computable sequence. However, this was established only much later by Le Roux and Ziegler [31] (answering a question posed by Giovanni Lagnese on the email list *Foundations of Mathematics* [fom] in 2006).

Proposition 17 (Le Roux and Ziegler 2008). *There exists a computable sequence $(x_n)_n$ in $[0, 1]$ without a limit computable cluster point.*

This is in notable contrast to all aforementioned results that all admit a limit computable solution for computable instances. For instance, the Monotone Convergence Theorem 2 itself implies that every monotone bounded sequence is convergent and hence every computable monotone bounded sequence automatically has a limit computable supremum. Hence, in a certain sense the Bolzano-Weierstraß Theorem is even less computable than all the other results mentioned in this section.

In this section we have only discussed a selection of theorems that can illustrate a certain variety of possibilities that occur. The computational content of several other theorems from real analysis has been studied. For instance Aberth [1] constructed a computable counterexample in the Russian sense for the Peano Existence Theorem for solutions of ordinary differential equations. Later Pour-El and Richards [37] constructed another counterexample in the modern sense of computable analysis for this theorem. The Riemann Mapping Theorem is an interesting example of a theorem from complex analysis that was studied by Hertling [23]. Now we turn to functional analysis.

3   Some Theorems from Functional Analysis

Starting with the work of Metakides, Nerode and Shore [32,33] theorems from functional analysis were studied from the perspective of computability theory. In particular the aforementioned authors studied the Hahn-Banach Theorem.

Theorem 18 (Hahn-Banach Theorem). *Let X be a normed space over the field $\mathbb{R}$ with a linear subspace $Y \subseteq X$. Then every linear bounded functional $f : Y \to \mathbb{R}$ has a linear bounded extension $g : X \to \mathbb{R}$ with $||g|| = ||f||$.*

Here $||f|| := \sup_{||x|| \leq 1} |f(x)|$ denotes the operator norm. The result holds analogously over the field $\mathbb{C}$. Here and in the following a *computable metric space* X is just a metric space together with a dense sequence such that the distances can be computed on that sequence (as a double sequence of real numbers). If the space has additional properties or ingredients, such as a norm that generates the metric, then it is called a *computable normed space* or in case of completeness also a *computable Banach space*. If, additionally, the norm is generated by an inner product, then the space is called a *a computable Hilbert space*. A subspace $Y \subseteq X$ is called *c.e. closed* if there is a computable sequence $(x_n)_n$ in X such that one obtains $\overline{\{x_n : n \in \mathbb{N}\}} = Y$ (where the $\overline{A}$ denotes the closure of A). Metakides, Nerode and Shore [33] constructed a computable counterexample to the Hahn-Banach Theorem.

Proposition 19 (Metakides, Nerode and Shore 1985). *There exists a computable Banach space X over $\mathbb{R}$ with a c.e. closed linear subspace $Y \subseteq X$ and a computable linear functional $f : Y \to \mathbb{R}$ with a computable norm $\|f\|$ such that every linear bounded extension $g : X \to \mathbb{R}$ with $\|g\| = \|f\|$ is non-computable.*

Similarly, as the computability status of the Brouwer Fixed Point Theorem 13 was dependent on the dimension k of the underlying space $[0, 1]^k$, the computability status of the Hahn-Banach Theorem 18 is dependent on the dimension and other aspects of the space X [10]. Nerode and Metakides [32] observed that for finite-dimensional X no counterexample as in Proposition 19 exists. However, even in this case the theorem is not uniformly computable [10]. Under all conditions that guarantee that the extension is uniquely determined, the Hahn-Banach Theorem is fully computable; this includes for instance all computable Hilbert spaces [10].

A number of further theorems from functional analysis were analyzed by the author of this article and these include the Open Mapping Theorem, the Closed Graph Theorem and Banach's Inverse Mapping Theorem [7,11]. Another theorem that falls into this category is the Uniform Boundedness Theorem [9]. These examples are interesting, since they behave differently from all aforementioned examples. We illustrate the situation using Banach's Inverse Mapping Theorem.

Theorem 20 (Banach's Inverse Mapping Theorem). *If $T : X \to Y$ is a bijective, linear and bounded operator on Banach spaces X, Y, then its inverse $T^{-1} : Y \to X$ is bounded too.*

Here we obtain the following computable version [11].

Proposition 21 (B. 2009). *If $T : X \to Y$ is a computable, bijective and linear operator on computable Banach spaces X, Y, then its inverse $T^{-1} : Y \to X$ is computable too.*

That is, every bijective and linear operator T with a program admits also a program for its inverse T^{-1}, but there is not general method to compute such a program for T^{-1} from a program for T in general as the following result shows [11].

Proposition 22 (B. 2009). *Inversion BIM $:\subseteq \mathcal{C}(\ell_2, \ell_2) \to \mathcal{C}(\ell_2, \ell_2), T \mapsto T^{-1}$ restricted to bijective, linear and bounded $T : \ell_2 \to \ell_2$ is not computable (and not even continuous).*

Analogously, there is also a sequential counterexample [11].

Proposition 23 (B. 2009). *There exists a computable sequence $(T_n)_n$ of computable, bijective and linear operators $T_n : \ell_2 \to \ell_2$ such that the sequence $(T_n^{-1})_n$ of their inverses is not computable.*

Hence, in several respects the Banach Inverse Mapping Theorem behaves similarly to the Intermediate Value Theorem: it is non-uniformly computable, but not uniformly computable. Yet we will see that the uniform content of both theorems is different.

With our final example of a theorem from functional analysis we want to close the circle and mention a result that behaves similarly to the Monotone Convergence Theorem 2, namely the Fréchet-Riesz Representation Theorem.

Theorem 24 (Fréchet-Riesz Representation Theorem). *For every linear bounded functional $f : H \to \mathbb{R}$ on a Hilbert space H there exists a unique $y \in H$ such that $f = f_y$ and $\|f\| = \|y\|$, where $f_y : H \to \mathbb{R}, x \mapsto \langle x, y \rangle$.*

Here $\langle\,,\,\rangle$ denotes the inner product of the Hilbert space H. For every computable $y \in H$ the functional f_y is computable with norm $\|f_y\| = \|y\|$. Since the norm $\|y\|$ is always computable for a computable $y \in H$, it is immediately clear that it suffices to construct a computable functional $f : H \to \mathbb{R}$ without a computable norm $\|f\|$ in order to show that this theorem cannot hold computably. In fact, a Specker-like counterexample suffices in this case [20].

Proposition 25 (B. and Yoshikawa 2006). *There exists a computable functional $f : \ell_2 \to \mathbb{R}$ such that $\|f\|$ is not computable and hence there cannot be a computable $y \in H$ with $f = f_y$ and $\|f\| = \|y\|$.*

4   A Classification Scheme for Theorems

If we look at the different examples of theorems that we have presented in the preceding sections then it becomes clear that theorems can behave quite differently with respect to computability. For one, the uniform and the non-uniform behavior can differ and the levels of computability can be of different complexities (computable, low, limit computable, etc.). On the other hand, certain theorems seem to be quite similar to each other, for instance the Monotone Convergence Theorem is similar in its behavior to the Fréchet-Riesz Theorem.

This naturally leads us to the question whether there is a classification scheme that allows to derive all sorts of computability properties of a theorem once it has been classified according to the corresponding scheme. The best known classification scheme for theorems in logic is Reverse Mathematics, i.e., the project to classify theorems in second order arithmetic according to certain axioms that are required to prove the corresponding theorem [40]. It turns out that this classification scheme is not fine enough for our purposes, because it only captures theorems in a non-uniform sense. In order to preserve computability properties such as lowness that are not closed under product, we also need a classification scheme that is more resource sensitive than reverse mathematics.

Such a classification scheme has been developed over the previous eight years using the concept of Weihrauch reducibility [12–16,18,21,35,36]. If X, Y, Z, W are represented spaces, then $f :\subseteq X \rightrightarrows Y$ is *Weihrauch reducible* to $g :\subseteq Z \rightrightarrows W$, if there are computable multi-valued functions H, K such that $\emptyset \neq H(x, gK(x)) \subseteq f(x)$ for all $x \in \text{dom}(f)$. In symbols we write $f \leq_W g$ in this situation. If the reduction works in both directions, then we write $f \equiv_W g$. It can be shown that this reducibility induces a lattice structure [14,36].

Now a theorem of logical form $(\forall x \in X)(x \in D \implies (\exists y \in Y) \, P(x, y))$ can be interpreted as a multi-valued function $f :\subseteq X \rightrightarrows Y, x \mapsto \{y \in Y : P(x, y)\}$ with $\text{dom}(f) = D$. For instance, we obtain the following multi-valued functions for the theorems that we have considered in the previous sections (some of which are formulated in greater generality here). Here Tr denotes the set of binary trees and $[T]$ the set of infinite paths of a tree T.

- MCT $:\subseteq \mathbb{R}^{\mathbb{N}} \to \mathbb{R}, (x_n)_n \mapsto \sup_{n \in \mathbb{N}} x_n$ restricted to monotone bounded sequences.
- IVT $:\subseteq \mathcal{C}[0, 1] \rightrightarrows [0, 1], f \mapsto f^{-1}\{0\}$ with $\text{dom}(\text{IVT}) := \{f : f(0) \cdot f(1) < 0\}$.
- MAX$_X$ $:\subseteq \mathcal{C}(X) \rightrightarrows \mathbb{R}, f \mapsto \{x \in X : f(x) = \max f(X)\}$ for computably compact[1] computable metric spaces X and, in particular, for $X = [0, 1]$.
- WKL $:\subseteq \text{Tr} \rightrightarrows 2^{\mathbb{N}}, T \mapsto [T]$ restricted to infinite binary trees.
- BFT$_n$ $: \mathcal{C}([0, 1]^n, [0, 1]^n) \rightrightarrows [0, 1]^n, f \mapsto \{x : f(x) = x\}$ for $n \geq 1$.
- BWT$_X$ $:\subseteq X^{\mathbb{N}} \rightrightarrows X, (x_n)_n \mapsto \{x : x$ cluster point of $(x_n)_n\}$, restricted to sequences that are in a compact subset of X.
- BIM$_{X,Y}$ $:\subseteq \mathcal{C}(X, Y) \to \mathcal{C}(Y, X), T \mapsto T^{-1}$, restricted to bijective, linear, bounded T and for computable Banach spaces X, Y.
- FRR$_H$ $:\subseteq \mathcal{C}(H) \to H, f_y \mapsto y$ for computable Hilbert spaces H.
- Z$_X$ $:\subseteq \mathcal{C}(X) \to \mathbb{R}, f \mapsto f^{-1}\{0\}$ for computable metric spaces X.

We have not formalized the Hahn-Banach Theorem here and point the reader to [21]. The last mentioned problem Z_X is the *zero problem*, which is the problem to find a zero of a continuous function that admits at least one zero. By [19, Theorem 3.10] we obtain $\text{Z}_X \equiv_W \text{C}_X$ for the *choice problem* of every computable metric space. We are not going to define C_X here, but whenever we use it we will actually take Z_X as a substitute for it. The following equivalences were proved in [13]:

Theorem 26 (Choice on the natural numbers). *The following are all Weihrauch equivalent to each other and complete among functions that are computable with finitely many mind changes* [12]:

1. *Choice on natural numbers* $\text{C}_{\mathbb{N}}$.
2. *The Baire Category Theorem (in appropriate formulation).*
3. *Banach's Inverse Mapping Theorem* $\text{BIM}_{\ell_2, \ell_2}$.
4. *The Open Mapping Theorem for* ℓ_2.

---

[1] See [19] for a definition of computably compact.

5. *The Closed Graph Theorem for ℓ_2.*
6. *The Uniform Boundedness Theorem for ℓ_2.*

Hence the equivalence class of choice $C_\mathbb{N}$ on the natural numbers contains many theorems that are typically proved with and closely related to the Baire Category Theorem.

We prove that the Theorem of the Maximum is equivalent to the zero problem of $[0, 1]$.

Theorem 27. $\text{MAX}_X \equiv_W Z_X$ *for every computably compact computable metric space X.*

Proof. We prove $Z_X \leq_W \text{MAX}_X$. Given a continuous function $f : X \to \mathbb{R}$ with $A = f^{-1}\{0\} \neq \emptyset$, we can compute the function $g : X \to \mathbb{R}$ with $g := -|f|$. Then $\text{MAX}(g) = f^{-1}\{0\} = A$. This proves the claim. We now prove $\text{MAX}_X \leq_W Z_X$. Given a continuous function $f : X \to \mathbb{R}$ with $\text{MAX}(f) = A \neq \emptyset$, we can compute $g : X \to \mathbb{R}$ with $g := f - \max f(X)$, since X is computably compact. Now we obtain $g^{-1}\{0\} = \text{MAX}(f) = A$. This proves the claim.   □

We now arrive at the following result that is compiled from different sources. It shows that the equivalence class of choice on Cantor space contains several problems whose non-computably was proved with the help of Weak König's Lemma or with the help of two c.e. sets that are computably inseparable. We point out that the sequential version of the Intermediate Valued Theorem formulated in Proposition 6 can be modeled by *parallelization*. For $f :\subseteq X \rightrightarrows Y$ we define its *parallelization* $\widehat{f} :\subseteq X^\mathbb{N} \rightrightarrows Y^\mathbb{N}, (x_n)_n \mapsto \times_{n=0}^\infty f(x_n)$, which lifts f to sequences. Parallelization is a closure operation in the Weihrauch lattice [14].

Theorem 28 (Choice on Cantor space). *The following are all Weihrauch equivalent to each other and complete among non-deterministically computable functions with a binary sequence as advice [12]:*

1. *Choice on Cantor Space $C_{2^\mathbb{N}}$.*
2. *Weak König's Lemma WKL [14,21].*
3. *The Theorem of the Maximum $\text{MAX}_{[0,1]}$ (Theorem 27).*
4. *The Hahn-Banach Theorem (Gherardi and Marcone 2009) [21].*
5. *The parallelization $\widehat{\text{IVT}}$ of the Intermediate Value Theorem [13].*
6. *The Brouwer Fixed Point Theorem BFT_n for dimension $n \geq 2$ [18].*

We note that [18] contains the proof for the Brouwer Fixed Point Theorem only for dimension $n \geq 3$ and the results for $n = 2$ is due to Joseph Miller. It is easy to see that $\text{IVT} \equiv_W \text{BFT}_1$ [18]. We mention that this result implies Propositions 6, 9, 11, 14 and 19 and constitutes a more general uniform classification. In some cases the proofs can easily be derived from known techniques and results, in other cases (for instance for the Hahn-Banach Theorem and the Brouwer Fixed Point Theorem) completely new techniques are required. One can prove that the equivalence classes appearing in Theorems 26 and 28 are incomparable [13]. The next equivalence class that we are going to discuss is an upper bound of both.

We first prove that the Monotone Convergence Theorem MCT is equivalent to the Fréchet-Riesz Representation Theorem FRR_H.

Theorem 29. $\mathsf{FRR}_H \equiv_W \mathsf{MCT}$ *for every computable infinite-dimensional Hilbert space H.*

Proof. Since every infinite-dimensional computable Hilbert space is computably isometrically isomorphic to ℓ_2 by [20, Corollary 3.7], if suffices to consider $H = \ell_2$. We first prove $\mathsf{FRR}_{\ell_2} \leq_W \mathsf{MCT}$. Given a functional $f : \ell_2 \to \mathbb{R}$ we need to find a $y \in \ell_2$ such that $f_y = f$ and $||y|| = ||f||$. There is a computable sequence $(x_n)_n$ in ℓ_2 such that $\overline{\{x_n : n \in \mathbb{N}\}}$ is dense in $\{x \in \ell_2 : ||x|| \leq 1\}$. Hence $||f|| = \sup_{||x|| \leq 1} |f(x)| = \sup_{n \in \mathbb{N}} |f(x_n)| = \sup_{n \in \mathbb{N}} \max_{i \leq n} |f(x_i)|$ and hence we can compute $||f||$ with the help of MCT. Now given $f = f_y$ and $||f||$ we can easily compute y be evaluating $f_y(e_n)$ on the unit vectors e_n. We still need to prove $\mathsf{MCT} \leq_W \mathsf{FRR}_{\ell_2}$. By [8, Proposition 9.1] it suffices to show that we can utilize FRR_{ℓ_2} to translate enumerations g of sets $A \subseteq \mathbb{N}$ into their characteristic functions. Let us assume that $A = \{n : n + 1 \in \mathrm{range}(g)\}$. Without loss of generality we can assume that no value different from zero appears twice in $(g(n))_n$. Using the idea of [20, Example 4.6] we choose $a_k := 2^{g(k)-1}$ if $g(k) \neq 0$ and $a_k := 0$ otherwise. Then $a = (a_k)_k \in \ell_2$ and we can compute $f \in \mathcal{C}(\ell_2)$ with $f(x) := \sum_{k=0}^{\infty} x_k a_k = \langle x, a \rangle$. Now, with the help of FRR_{ℓ_2} we obtain a $y \in \ell_2$ with $||y|| = ||f|| = ||a|| = \sqrt{\sum_{k=0}^{\infty} |a_k|^2}$. But using the number $||y||^2$ we can decide A, since its binary representation has in the even positions the characteristic function of A. □

We note that FRR_H for finite-dimensional spaces H is computable. Altogether we obtain the following result for this equivalence class.

Theorem 30 (The limit). *The following are all Weihrauch equivalent to each other and complete for limit computable functions:*

1. *The limit map* $\lim$ *on Baire space* [8].
2. *The parallelization* $\widehat{\mathsf{C}_{\mathbb{N}}}$ *of choice on the natural numbers* [13].
3. *The parallelization* $\widehat{\mathsf{BIM}}$ *of Banach's Inverse Mapping Theorem* [13].
4. *The Monotone Convergence Theorem* MCT [16].
5. *The Fréchet-Riesz Representation Theorem* FRR *for* ℓ_2 *(Theorem 29).*
6. *The Radon-Nikodym Theorem (Hoyrup, Rojas, Weihrauch 2012)* [24].

This theorem implies Propositions 3, 23 and 25. Finally, we mention that we also have a concept of a *jump* $f' :\subseteq X \rightrightarrows Y$ for every $f :\subseteq X \rightrightarrows Y$, which essentially replaces the input representation of X in such a way that a name of $x \in X$ for f' is a sequence that converges to a name in the sense of f. This makes problems potentially more complicated since less input information is available. It allows us to phrase results as the following [16].

Theorem 31 (B., Gherardi and Marcone 2012). $\mathsf{WKL}' \equiv_W \mathsf{BWT}_{\mathbb{R}}$.

This result does not only imply Proposition 17, but also the following [16].

Corollary 32 (B., Gherardi and Marcone 2012). *Every computable sequence $(x_n)_n$ in the unit cube $[0,1]^n$ has a cluster point $x \in [0,1]^n$ that is low relative to the halting problem.*

Here x is *low relative to the halting problem* if $x' \leq_T \emptyset''$ (some authors would only call this a partial relativization of lowness). In light of Proposition 17 this is one of the strongest positive properties that one can expect for a cluster point. These examples demonstrate that a classification of the Weihrauch degree of a theorem yields a large variety of computability properties of the theorem, uniform and non-uniform ones on the one hand, and positive and negative ones on the other hand.

References

1. Aberth, O.: The failure in computable analysis of a classical existence theorem for differential equations. Proc. Am. Math. Soc. **30**, 151–156 (1971)
2. Avigad, J., Brattka, V.: Computability and analysis: the legacy of Alan Turing. In: Downey, R. (ed.) Turing's Legacy: Developments from Turing's Ideas in Logic, LNL, vol. 42, pp. 1–47. Cambridge University Press, Cambridge, UK (2014)
3. Baigger, G.: Die Nichtkonstruktivität des Brouwerschen Fixpunktsatzes. Arch. Math. Logic **25**, 183–188 (1985)
4. Beeson, M.J.: Foundations of Constructive Mathematics. A Series of Modern Surveys in Mathematics. Springer, Berlin (1985)
5. Borel, É.: Le calcul des intégral définies. Journal de Mathematiques pures et appliquées **8**(2), 159–210 (1912)
6. Borel, É.: La théorie de la mesure et al théorie de l'integration. In: Leçons sur la théorie des fonctions, pp. 214–256. Gauthier-Villars, Paris, 4 edn. (1950)
7. Brattka, V.: Computability of Banach space principles. Informatik Berichte 286, FernUniversität Hagen, Fachbereich Informatik, Hagen, June 2001
8. Brattka, V.: Effective Borel measurability and reducibility of functions. Math. Logic Q. **51**(1), 19–44 (2005)
9. Brattka, V.: Computable versions of the uniform boundedness theorem. In: Chatzidakis, Z., et al. (eds.) Logic Colloquium 2002. LNL, vol. 27, pp. 130–151. ASL, Urbana (2006)
10. Brattka, V.: Borel complexity and computability of the Hahn-Banach Theorem. Arch. Math. Logic **46**(7–8), 547–564 (2008)
11. Brattka, V.: A computable version of Banach's inverse mapping theorem. Ann. Pure Appl. Logic **157**, 85–96 (2009)
12. Brattka, V., de Brecht, M., Pauly, A.: Closed choice and a uniform low basis theorem. Ann. Pure Appl. Logic **163**, 986–1008 (2012)
13. Brattka, V., Gherardi, G.: Effective choice and boundedness principles in computable analysis. Bull. Symbolic Logic **17**(1), 73–117 (2011)
14. Brattka, V., Gherardi, G.: Weihrauch degrees, omniscience principles and weak computability. J. Symbolic Logic **76**(1), 143–176 (2011)
15. Brattka, V., Gherardi, G., Hölzl, R.: Probabilistic computability and choice. Inf. Comput. **242**, 249–286 (2015)
16. Brattka, V., Gherardi, G., Marcone, A.: The Bolzano-Weierstrass theorem is the jump of weak Kőnig's lemma. Ann. Pure Appl. Logic **163**, 623–655 (2012)
17. Brattka, V., Hertling, P., Weihrauch, K.: A tutorial on computable analysis. In: Cooper, S.B., et al. (eds.) New Computational Paradigms: Changing Conceptions of What is Computable, pp. 425–491. Springer, New York (2008)
18. Brattka, V., Le Roux, S., Pauly, A.: Connected choice and the Brouwer fixed point theorem. arXiv 1206.4809. http://arxiv.org/abs/1206.4809

19. Brattka, V., Presser, G.: Computability on subsets of metric spaces. Theor. Comput. Sci. **305**, 43–76 (2003)
20. Brattka, V., Yoshikawa, A.: Towards computability of elliptic boundary value problems in variational formulation. J. Complex. **22**(6), 858–880 (2006)
21. Gherardi, G., Marcone, A.: How incomputable is the separable Hahn-Banach theorem? Notre Dame Journal of Formal Logic **50**(4), 393–425 (2009)
22. Grzegorczyk, A.: Computable functionals. Fundamenta Mathematicae **42**, 168–202 (1955)
23. Hertling, P.: An effective Riemann mapping theorem. Theor. Comput. Sci. **219**, 225–265 (1999)
24. Hoyrup, M., Rojas, C., Weihrauch, K.: Computability of the Radon-Nikodym derivative. Computability **1**(1), 3–13 (2012)
25. Jockusch, C.G., Soare, R.I.: Π_1^0 classes and degrees of theories. Trans. Am. Math. Soc. **173**, 33–56 (1972)
26. Kleene, S.C.: Recursive functions and intuitionistic mathematics. In: Proceedings of the International Congress of Mathematicians, Cambridge, 1950, vol. 1, pp. 679–685. AMS, Providence (1952)
27. Ko, K.I.: Complexity Theory of Real Functions. Progress in Theoretical Computer Science. Birkhäuser, Boston (1991)
28. Kreisel, G.: On the interpretation of non-finitist proofs, II: interpretation of number theory, applications. J. Symbolic Logic **17**, 43–58 (1952)
29. Lacombe, D.: Remarques sur les opérateurs récursifs et sur les fonctions récursives d'une variable réelle. C. R. Acad. Paris 241, 1250–1252, théorie des fonctions, November 1955
30. Lacombe, D.: Les ensembles récursivement ouverts ou fermés, et leurs applications à l'Analyse récursive. C. R. Acad. Paris 245, 1040–1043, logique (1957)
31. Le Roux, S., Ziegler, M.: Singular coverings and non-uniform notions of closed set computability. Math. Logic Q. **54**(5), 545–560 (2008)
32. Metakides, G., Nerode, A.: The introduction of non-recursive methods into mathematics. In: Troelstra, A., van Dalen, D. (eds.) The L.E.J. Brouwer Centenary Symposium, pp. 319–335. North-Holland, Amsterdam (1982)
33. Metakides, G., Nerode, A., Shore, R.A.: Recursive limits on the Hahn-Banach theorem. In: Rosenblatt, M. (ed.) Errett Bishop: Reflections on Him and His Research. AMS, Providence (1985). Contemp. Math. vol. 39, pp. 85–91
34. Orevkov, V.P.: A constructive mapping of the square onto itself displacing every constructive point (Russian). Doklady Akademii Nauk **152**, 55–58 (1963). Translated in: Soviet Math. - Dokl. vol. 4, pp. 1253–1256 (1963)
35. Pauly, A.: How incomputable is finding Nash equilibria? J. Univers. Comput. Sci. **16**(18), 2686–2710 (2010)
36. Pauly, A.: On the (semi)lattices induced by continuous reducibilities. Math. Logic Q. **56**(5), 488–502 (2010)
37. Pour-El, M.B., Richards, J.I.: A computable ordinary differential equation which possesses no computable solution. Ann. Math. Logic **17**, 61–90 (1979)
38. Pour-El, M.B., Richards, J.I.: Computability in Analysis and Physics. Springer, Berlin (1989)
39. Rice, H.G.: Recursive real numbers. Proc. Am. Math. Soc. **5**, 784–791 (1954)
40. Simpson, S.G.: Subsystems of Second Order Arithmetic, 2nd edn. Cambridge University Press, Poughkeepsie (2009)
41. Specker, E.: Nicht konstruktiv beweisbare Sätze der Analysis. J. Symbolic Logic **14**(3), 145–158 (1949)

42. Specker, E.: Der Satz vom Maximum in der rekursiven Analysis. In: Heyting, A. (ed.) Constructivity in Mathematics, pp. 254–265. North-Holland, Amsterdam (1959)
43. Turing, A.M.: On computable numbers, with an application to the "Entscheidungsproblem". Proc. London Math. Soc. **42**(2), 230–265 (1937)
44. Turing, A.M.: On computable numbers, with an application to the "Entscheidungsproblem". Correction Proc. London Math. Soc. **43**(2), 544–546 (1938)
45. Weihrauch, K.: Computable Analysis. Springer, Berlin (2000)
46. Zaslavskiĭ, I.D.: Disproof of some theorems of classical analysis in constructive analysis (Russian). Usp. Mat. Nauk **10**(4), 209–210 (1955)

The Brouwer Fixed Point Theorem Revisited

Vasco Brattka[1,2], Stéphane Le Roux[3], Joseph S. Miller[4], and Arno Pauly[3(✉)]

[1] Faculty of Computer Science, University of the Armed Forces Munich,
Neubiberg, Germany
[2] Department of Mathematics and Applied Mathematics,
University of Cape Town, Cape Town, South Africa
Vasco.Brattka@cca-net.de
[3] Département d'Informatique, Université libre de Bruxelles, Brussels, Belgium
Stephane.Le.Roux@ulb.ac.be, Arno.Pauly@cl.cam.ac.uk
[4] Department of Mathematics, University of Wisconsin–Madison, Madison, USA
jmiller@math.wisc.edu

Abstract. We revisit the investigation of the computational content of the Brouwer Fixed Point Theorem in [7], and answer the two open questions from that work. First, we show that the computational hardness is independent of the dimension, as long as it is greater than 1 (in [7] this was only established for dimension greater than 2). Second, we show that restricting the Brouwer Fixed Point Theorem to L-Lipschitz functions for any $L > 1$ also does not change the computational strength, which together with prior results establishes a trichotomy for $L > 1$, $L = 1$ and $L < 1$.

1 Introduction

In this paper we continue with the programme to classify the computational content of mathematical theorems in the Weihrauch lattice (see [3–5,8,11,13, 17,18]). This lattice is induced by Weihrauch reducibility, which is a reducibility for partial multi-valued functions $f :\subseteq X \rightrightarrows Y$ on represented spaces X, Y. Intuitively, $f \leq_W g$ reflects the fact that the function f can be realized with a single application of the function g as an oracle. Hence, if two functions are equivalent in the sense that they are mutually reducible to each other, then they are equivalent as computational resources, as far as computability is concerned.

Many theorems in mathematics are actually of the logical form

$$(\forall x \in X)(\exists y \in Y)\ P(x, y)$$

and such theorems can straightforwardly be represented by a multi-valued function $f : X \rightrightarrows Y$ with $f(x) := \{y \in Y : P(x, y)\}$ (sometimes partial f are needed,

The majority of this work was done while Le Roux was at the Department of Mathematics, Technische Universität Darmstadt, Germany and Pauly was at the Computer Laboratory, University of Cambridge, United Kingdom.

This project has been supported by the National Research Foundation of South Africa (NRF) and by the German Research Foundation (DFG) through the German-South African project (DFG, 445 SUA-1 13/20/0).

© Springer International Publishing Switzerland 2016
A. Beckmann et al. (Eds.): CiE 2016, LNCS 9709, pp. 58–67, 2016.
DOI: 10.1007/978-3-319-40189-8_6

where the domain captures additional requirements that this input x has to satisfy). In some sense the multi-valued function f directly reflects the computational task of the theorem to find some suitable y for any x. Hence, in a very natural way the classification of a theorem can be achieved via a classification of the corresponding multi-valued function that represents the theorem. In this paper we attempt to classify the Brouwer Fixed Point Theorem.

Theorem 1 (Brouwer Fixed Point Theorem 1911). *Every continuous function $f : [0,1]^n \to_\bullet [0,1]^n$ has a fixed point $x \in [0,1]^n$.*

The fact that Brouwer's Fixed Point Theorem cannot be proved constructively has been confirmed in many different ways; most relevant for us is the counterexample in Russian constructive analysis by Orevkov [16], which was transferred into computable analysis by Baigger [1].

Constructions similar to those used for the above counterexamples have been utilized in order to prove that the Brouwer Fixed Point Theorem is equivalent to Weak König's Lemma in reverse mathematics [21,22] and to analyze computability properties of fixable sets [14], but a careful analysis of these reductions reveals that none of them can be straightforwardly transferred into a *uniform* reduction in the sense that we are seeking here. The results cited above essentially characterize the complexity of fixed points themselves, whereas we want to characterize the complexity of finding the fixed point, given the function. This requires full uniformity.

In the Weihrauch lattice the Brouwer Fixed Point Theorem of dimension n is represented by the multi-valued function $\mathsf{BFT}_n : \mathcal{C}([0,1]^n, [0,1]^n) \rightrightarrows [0,1]^n$ that maps any continuous function $f : [0,1]^n \to [0,1]^n$ to the set of its fixed points $\mathsf{BFT}_n(f) \subseteq [0,1]^n$. The question now is where BFT_n is located in the Weihrauch lattice?

In order to approach this question, we introduce a choice principle CC_n that we call *connected choice* and which is just the closed choice operation restricted to connected subsets. That is, in the sense discussed above, CC_n is the multi-valued function that represents the following mathematical statement: every non-empty connected closed set $A \subseteq [0,1]^n$ has a point $x \in A$. Since closed sets are represented by negative information (i.e. by an enumeration of open balls that exhaust the complement), the computational task of CC_n consists in finding a point in a closed set $A \subseteq [0,1]^n$ that is promised to be non-empty and connected and that is given by negative information.

One of our main results, presented in Sect. 3, is that the Brouwer Fixed Point Theorem is equivalent to connected choice for each fixed dimension n, i.e. $\mathsf{BFT}_n \equiv_W \mathsf{CC}_n$. This result allows us to study the Brouwer Fixed Point Theorem in terms of the function CC_n that is easier to handle since it involves neither function spaces nor fixed points. This is also another instance of the observation that several important theorems are equivalent to certain choice principles (see [3]) and many important classes of computable functions can be calibrated in terms of choice (see [2]). For instance, closed choice on Cantor space $C_{\{0,1\}^{\mathbb{N}}}$ and on the unit cube $C_{[0,1]^n}$ are both easily seen to be equivalent to Weak

Kőnig's Lemma WKL, i.e. WKL $\equiv_W C_{\{0,1\}^{\mathbb{N}}} \equiv_W C_{[0,1]^n}$ for any $n \geq 1$. Studying the Brouwer Fixed Point Theorem in the form of CC_n now amounts to comparing $C_{[0,1]^n}$ with its restriction CC_n.

Our second main result, given in Sect. 5, is that from dimension two onwards connected choice is equivalent to Weak Kőnig's Lemma, i.e. $CC_n \equiv_W C_{[0,1]}$ for $n \geq 2$.

This refutes an earlier conjecture [7] by some of the authors that connected choice in dimension two be computationally simpler than connected choice in dimension three. We then also consider the restriction of Brouwer's Fixed Point theorem to Lipschitz functions in Sect. 4. In the following Sect. 2 we start with a short summary of relevant definitions and results regarding the Weihrauch lattice.

This extended abstract does not contain any proofs. Sections 1, 2 and 3 are taken mostly from [7]. An extended version including the omitted proofs can be found as [6].

2   The Weihrauch Lattice

In this section we briefly recall some basic results and definitions regarding the Weihrauch lattice. The original definition of Weihrauch reducibility is due to Weihrauch and has been studied for many years (see [9,23–25]). Only recently it has been noticed that a certain variant of this reducibility yields a lattice that is very suitable for the classification of mathematical theorems (see [2–5,8,10,17,18]). The basic reference for all notions from computable analysis is [26], alternatively see [19].

The Weihrauch lattice is a lattice of multi-valued functions on represented spaces. A *representation* δ of a set X is just a surjective partial map $\delta :\subseteq \mathbb{N}^{\mathbb{N}} \to X$. In this situation we call (X, δ) a *represented space*. In general we use the symbol "$\subseteq$" in order to indicate that a function is potentially partial. Using represented spaces we can define the concept of a realizer. We denote the composition of two (multi-valued) functions f and g either by $f \circ g$ or by fg.

Definition 1 (Realizer). *Let* $f :\subseteq (X, \delta_X) \rightrightarrows (Y, \delta_Y)$ *be a multi-valued function on represented spaces. A function* $F :\subseteq \mathbb{N}^{\mathbb{N}} \to \mathbb{N}^{\mathbb{N}}$ *is called a* realizer *of* f, *in symbols* $F \vdash f$, *if* $\delta_Y F(p) \in f\delta_X(p)$ *for all* $p \in \mathrm{dom}(f\delta_X)$.

Realizers allow us to transfer the notions of computability and continuity and other notions available for Baire space to any represented space; a function between represented spaces will be called *computable* if it has a computable realizer, etc. Now we can define Weihrauch reducibility.

Definition 2 (Weihrauch reducibility). *Let* f, g *be multi-valued functions on represented spaces. Then* f *is said to be* Weihrauch reducible *to* g, *in symbols* $f \leq_W g$, *if there are computable functions* $K, H :\subseteq \mathbb{N}^{\mathbb{N}} \to \mathbb{N}^{\mathbb{N}}$ *such that* $K\langle \mathrm{id}, GH \rangle \vdash f$ *for all* $G \vdash g$. *Moreover,* f *is said to be* strongly Weihrauch reducible *to* g, *in symbols* $f \leq_{sW} g$, *if there are computable functions* K, H *such that* $KGH \vdash f$ *for all* $G \vdash g$.

Here $\langle,\rangle$ denotes some standard pairing on Baire space. We note that the relations $\leq_W$, $\leq_{sW}$ and $\vdash$ implicitly refer to the underlying representations, which we mention explicitly only when necessary. It is known that these relations only depend on the underlying equivalence classes of representations, but not on the specific representatives (see Lemma 2.11 in [4]). We use $\equiv_W$ and $\equiv_{sW}$ to denote the respective equivalences regarding $\leq_W$ and $\leq_{sW}$, and by $<_W$ and $<_{sW}$ we denote strict reducibility.

A particularly useful multi-valued function in the Weihrauch lattice is closed choice (see [2–4,8]) and it is known that many notions of computability can be calibrated using the right version of choice. We will focus on closed choice for computable metric spaces, which are separable metric spaces such that the distance function is computable on the given dense subset. We assume that computable metric spaces are represented via their Cauchy representation (see [26] for details).

By $\mathcal{A}_-(X)$ we denote the set of closed subsets of a metric space X, where the index "$-$" indicates that we work with negative information. This information is given by a representation $\psi_- : \mathbb{N}^\mathbb{N} \to \mathcal{A}_-(X)$, defined by $\psi_-(p) := X \setminus \bigcup_{i=0}^\infty B_{p(i)}$, where B_n is some standard enumeration of the open balls of X with center in the dense subset and rational radius. The computable points in $\mathcal{A}_-(X)$ are called *co-c.e. closed sets*. We now define closed choice for the case of computable metric spaces.

Definition 3 (Closed Choice). *Let* $\mathbf{X}$ *be a computable metric space. Then the* closed choice *operation* $\mathsf{C_X} :\subseteq \mathcal{A}_-(X) \rightrightarrows X$ *of this space is defined by* $\mathrm{dom}(\mathsf{C}_X) := \{A \in \mathcal{A}_-(X) : A \neq \emptyset\}$ *and* $x \in \mathsf{C_X}(A)$ *iff* $x \in A$.

Intuitively, C_X takes as input a non-empty closed set in negative representation (i.e. given by ψ_-) and it produces an arbitrary point of this set as output. For short we use the notation $\mathcal{A}_n := \{A \in \mathcal{A}_-([0,1]^n) : A \neq \emptyset\}$ for the space of non-empty closed subsets with representation ψ_- in the following.

3   Brouwer's Fixed Point Theorem and Connected Choice

In this section we want to show that the Brouwer Fixed Point Theorem is computably equivalent to connected choice for any fixed dimension. We first define these two operations. By $\mathcal{C}(X,Y)$ we denote the *set of continuous functions* $f : X \to Y$ and for short we write $\mathcal{C}_n := \mathcal{C}([0,1]^n, [0,1]^n)$.

Definition 4 (Brouwer Fixed Point Theorem). *By* $\mathsf{BFT}_n : \mathcal{C}_n \rightrightarrows [0,1]^n$ *we denote the operation defined by* $\mathsf{BFT}_n(f) := \{x \in [0,1]^n : f(x) = x\}$ *for* $n \in \mathbb{N}$.

We note that BFT_n is well-defined, i.e. $\mathsf{BFT}_n(f)$ is non-empty for all f, since by the Brouwer Fixed Point Theorem every $f \in \mathcal{C}_n$ admits a fixed point x, i.e. with $f(x) = x$. We now define connected choice.

Definition 5 (Connected choice). *By* $\mathsf{CC}_n :\subseteq \mathcal{A}_n \rightrightarrows [0,1]^n$ *we denote the operation defined by* $\mathsf{CC}_n(A) := A$ *for all non-empty connected closed* $A \subseteq [0,1]^n$ *and* $n \in \mathbb{N}$. *We call* CC_n connected choice (of dimension n).

Hence, connected choice is just the restriction of closed choice $C_{[0,1]^n}$ to connected sets. We also use the following notation for the set of fixed points of a function $f \in C_n$.

Definition 6 (Set of fixed points). By $\mathsf{Fix}_n : C_n \to A_n$ we denote the function with $\mathsf{Fix}_n(f) := \{x \in [0,1]^n : f(x) = x\}$.

It is easy to see that Fix_n is computable, since $\mathsf{Fix}_n(f) := (f - \mathrm{id})^{-1}\{0\}$ and it is well-known that closed sets in A_n can also be represented as zero sets of continuous functions (see [26]).

Definition 7 (Connectedness components). By $\mathsf{Con}_n : A_n \rightrightarrows A_n$ we denote the map with $\mathsf{Con}_n(A) := \{C : C \text{ is a connectedness component of } A\}$ for every $n \geq 1$.

Theorem 2 (Connectedness components). $\mathsf{Con}_n \equiv_{sW} \mathsf{WKL}$ for $n \geq 1$.

We note that the Brouwer Fixed Point Theorem can be decomposed to $\mathsf{BFT}_n = \mathsf{CC}_n \circ \mathsf{Con}_n \circ \mathsf{Fix}_n$.

The main result of this section will be that the Brouwer Fixed Point Theorem and connected choice are (strongly) equivalent for any fixed dimension n (see Theorem 3 below).

The direction $\mathsf{CC}_n \leq_{sW} \mathsf{BFT}_n$ can be seen as a uniformization of an earlier construction of Baigger [1] that is in turn built on results of Orevkov [16]. This part of the construction was explained in some detail by Potgieter in [20].

For the other direction $\mathsf{BFT}_n \leq_{sW} \mathsf{CC}_n$ of the reduction we uniformize ideas from the third author's PhD thesis [14]. A central technique is topological degree theory. For the uniform aspects of both directions, a representation of closed sets via trees of rational complexes is employed.

The first observation is that the map $\mathsf{Con}_n \circ \mathsf{Fix}_n$ is computable (which might be surprising in light of Theorem 2).

Proposition 1. $\mathsf{Con}_n \circ \mathsf{Fix}_n : C_n \rightrightarrows A_n$ is computable for all $n \in \mathbb{N}$.

Since $\mathsf{BFT}_n \supseteq \mathsf{CC}_n \circ \mathsf{Con}_n \circ \mathsf{Fix}_n$ we can directly conclude $\mathsf{BFT}_n \leq_{sW} \mathsf{CC}_n$ for all n. Together with $\mathsf{CC}_n \leq_{sW} \mathsf{BFT}_n$ we obtain the following theorem.

Theorem 3 (Brouwer Fixed Point Theorem). $\mathsf{BFT}_n \equiv_{sW} \mathsf{CC}_n$ for all n.

It is easy to see that in general the Brouwer Fixed Point Theorem and connected choice are not independent of the dimension. In case of $n = 0$ the space $[0,1]^n$ is the one-point space $\{0\}$ and hence $\mathsf{BFT}_0 \equiv_{sW} \mathsf{CC}_0$ are both computable. In case of $n = 1$ connected choice was already studied in [3] and it was proved that it is equivalent to the Intermediate Value Theorem IVT (see Definition 6.1 and Theorem 6.2 in [3]).

Corollary 1 (Intermediate Value Theorem). $\mathsf{IVT} \equiv_{\mathrm{sW}} \mathsf{BFT}_1 \equiv_{\mathrm{sW}} \mathsf{CC}_1$.

It is also easy to see that the Brouwer Fixed Point Theorem BFT_2 in dimension two is more complicated than in dimension one. For instance, it is known that the Intermediate Value Theorem IVT always offers a computable function value for a computable input, whereas this is not the case for the Brouwer Fixed Point Theorem BFT_2 by Baigger's counterexample [1]. We continue to discuss this topic in Sect. 5.

Here we point out that Proposition 1 implies that the fixed point set $\mathsf{Fix}_n(f)$ of every computable function $f : [0,1]^n \to [0,1]^n$ has a co-c.e. closed connectedness component. The converse direction is true, too, and in a uniform way: We denote by $(f,g) :\subseteq X \rightrightarrows Y \times Z$ the *juxtaposition* of two functions $f :\subseteq X \rightrightarrows Y$ and $g :\subseteq X \rightrightarrows Z$, defined by $(f,g)(x) = (f(x),g(x))$.

Theorem 4 (Fixability). $(\mathsf{Fix}_n, \mathsf{Con}_n \circ \mathsf{Fix}_n)$ *is computable and has a multivalued computable right inverse for all* $n \in \mathbb{N}$.

Roughly speaking a closed set $A \in \mathcal{A}_n$ together with one of its connectedness components is as good as a continuous function $f \in \mathcal{C}_n$ with A as set of fixed points. As a non-uniform corollary we obtain immediately Miller's original result.

Corollary 2 (Fixable sets, Miller 2002). *A set* $A \subseteq [0,1]^n$ *is the set of fixed points of a computable function* $f : [0,1]^n \to [0,1]^n$ *if and only if it is non-empty and co-c.e. closed and contains a co-c.e. closed connectedness component.*

4   The Lipschitz Trichotomy

It seems to be a natural question[1] to what extent finding fixed points becomes easier if the class of functions to be considered is further restricted. In particular we will denote by $L-\mathsf{LBFT}_n$ the restriction of BFT_n to L-Lipschitz functions.

Proposition 2. *For* $L_1, L_2 > 1$ *we find that* $L_1-\mathsf{LBFT}_n \equiv_{\mathrm{W}} L_2-\mathsf{LBFT}_n$.

Proof. If f is L_1-Lipschitz and $L_2 > 1$, then $\mathrm{id} + \frac{L_2-1}{L_1+1}(f - \mathrm{id})$ is L_2-Lipschitz and has the same fixed points as f.

With some additional constructions and a careful analysis, the proof of Theorem 3 can be adapted to yield:

Theorem 5. $2-\mathsf{LBFT}_n \equiv_{\mathrm{W}} \mathsf{BFT}_n \equiv_{\mathrm{W}} \mathsf{CC}_n$.

Being L-Lipschitz for $L < 1$ implies the uniqueness of the fixed point, which in turn implies the computability of $L-\mathsf{LBFT}_n$ for $L < 1$. The remaining $L = 1$ case is also a special (since finite-dimensional) case of the Browder-Goehde-Kirk Fixed Point theorem. Its Weihrauch degree was studied by Neumann in [15], and shown to be equivalent to XC_n – closed choice for convex sets in $[0,1]^n$.

---

[1] Which was put to the authors by Kohlenbach.

Theorem 6 (Le Roux and Pauly [12]).

$$CC_1 \equiv_W XC_1 <_W XC_2 <_W XC_3 <_W \ldots <_W C_{[0,1]}$$

Corollary 3 (Lipschitz dichotomy in dimension 1).

- $L-\mathsf{LBFT}_1 \equiv_W \mathrm{id}$, *iff* $L < 1$
- $L-\mathsf{LBFT}_1 \equiv_W CC_1$, *iff* $L \geq 1$

Corollary 4 (Lipschitz trichotomy). *Let* $n > 1$.

- $L-\mathsf{LBFT}_n \equiv_W \mathrm{id}$, *iff* $L < 1$
- $L-\mathsf{LBFT}_n \equiv_W XC_n$, *iff* $L = 1$
- $L-\mathsf{LBFT}_n \equiv_W C_{[0,1]}$, *iff* $L > 1$

5   Classifying Connected Choice

In this section we want to discuss the degree of connected choice, in particular in relation to the dimension of the ambient space. We will consider three geometric constructions: The one employed in the original proof by Orevkov/Baigger – this construction is insufficient for the uniform aspects. Then a simple construction showing that connected choice is computably complete from dimension three onwards in the sense that it is strongly equivalent to Weak Kőnig's Lemma. Finally, a significantly more involved construction shows even connected choice in two dimensions to be computably complete, too.

A superficial reading of the results of Orevkov [16] and Baigger [1] can lead to the wrong conclusion that they actually provide a reduction of Weak Kőnig's Lemma to the Brouwer Fixed Point Theorem BFT_n of any dimension $n \geq 2$. However, this is only correct in a non-uniform way and the corresponding uniform result does not follow from the known constructions. The Orevkov-Baigger result is built on the following fact.

Proposition 3 (Mixed cube). *The function* $M :\subseteq \mathcal{A}_-[0,1] \to \mathcal{A}_2$ *with* $M(A) = (A \times [0,1]) \cup ([0,1] \times A)$ *is computable and maps non-empty closed sets* $A \subseteq [0,1]$ *to non-empty connected closed sets* $M(A) \subseteq [0,1]^2$.

It follows straightforwardly from the definition that the pairs $(x, y) \in M(A)$ are such that one out of two components x, y is actually in A. In order to express the uniform content of this fact, we introduce the concept of a fraction.

Definition 8 (Fractions). *Let* $f :\subseteq X \rightrightarrows Y$ *be a multi-valued function and* $0 < n \leq m \in \mathbb{N}$. *We define the fraction* $\frac{n}{m}f :\subseteq X \rightrightarrows Y^m$ *such that* $\frac{n}{m}f(x)$ *is the set of all* $(y_1, ..., y_m) \in \mathrm{range}(f)^m$ *with* $|\{i : y_i \in f(x)\}| \geq n$ *for all* $x \in \mathrm{dom}(\frac{n}{m}f) := \mathrm{dom}(f)$.

The idea of a fraction $\frac{n}{m}f$ is that it provides m potential answers for f, at least $n \leq m$ of which have to be correct. The uniform content of the Orevkov-Baigger construction is then summarized in the following result.

Proposition 4 (Dimension two). $\frac{1}{2}\mathsf{C}_{[0,1]} \leq_{sW} \mathsf{CC}_2$.

However, the following results shows that the uniform content of the preceding proposition is very weak, as it cannot even solve closed choice on the two-point space **2** (which is equivalent to LLPO):

Proposition 5. $\mathsf{C}_2 \not\leq_W \frac{1}{2}\mathsf{C}_{[0,1]}$

That is, given a closed set $A \subseteq [0,1]$ we can utilize connected choice CC_2 of dimension 2 in order to find a pair of points (x,y) one of which is in A. This result directly implies the counterexample of Baigger [1] because the fact that there are non-empty co-c.e. closed sets $A \subseteq [0,1]$ without computable points immediately implies that $\frac{1}{2}\mathsf{C}_{[0,1]}$ is not non-uniformly computable (i.e. there are computable inputs without computable outputs) and hence CC_2 is also not non-uniformly computable.

Corollary 5 (Orevkov 1963, Baigger 1985). *There exists a computable function $f : [0,1]^2 \to [0,1]^2$ that has no computable fixed point $x \in [0,1]^2$. There exists a non-empty connected co-c.e. closed subset $A \subseteq [0,1]^2$ without computable point.*

Instead, we shall use a different construction to classify connected choice from three dimensions upwards:

Proposition 6 (Twisted cube). *The function $T :\subseteq \mathcal{A}_{-}[0,1] \to \mathcal{A}_3$ with $T(A) = (A \times [0,1] \times \{0\}) \cup (A \times A \times [0,1]) \cup ([0,1] \times A \times \{1\})$ is computable and maps non-empty closed sets $A \subseteq [0,1]$ to non-empty connected closed sets $T(A) \subseteq [0,1]^3$.*

Here tuples $(x_1, x_2, x_3) \in T(A)$ have the property that at least one of the first two components provide a solution $x_i \in A$, but the third component provides the additional information which one surely does. If x_3 is close to 1, then surely $x_2 \in A$ and if x_3 is close to 0, then surely $x_1 \in A$. If x_3 is neither close to 0 nor 1, then both $x_1, x_2 \in A$. Hence, there is a computable function H such that $\mathsf{C}_{[0,1]} = H \circ \mathsf{CC}_3 \circ T$, which proves $\mathsf{C}_{[0,1]} \leq_{sW} \mathsf{CC}_3$. Together with Theorem 3 we obtain the following conclusion.

Theorem 7 (Completeness of three dimensions). *For $n \geq 3$ we obtain $\mathsf{CC}_n \equiv_{sW} \mathsf{BFT}_n \equiv_{sW} \mathsf{WKL} \equiv_{sW} \mathsf{C}_{[0,1]}$.*

We note that the reduction $\mathsf{CC}_n \leq_{sW} \mathsf{C}_{[0,1]^n}$ holds for all $n \in \mathbb{N}$, since connected choice is just a restriction of closed choice and $\mathsf{C}_{[0,1]^n} \equiv_{sW} \mathsf{C}_{[0,1]} \equiv_{sW} \mathsf{WKL}$ is known for all $n \geq 1$ (see [2]).

Originally, three of the authors had conjectured in [7] that $\mathsf{CC}_2 <_W \mathsf{C}_{[0,1]}$. However, a more involved construction actually establishes that:

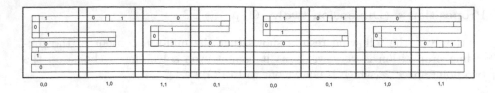

Fig. 1. The geometric pattern after the third round

Theorem 8 (Completeness of two dimensions). $CC_2 \equiv_W C_{[0,1]}$

The proof of Theorem 8 exhibits a reduction $\widehat{C_2} \leq_W CC_2$ instead, using the equivalence $\widehat{C_2} \equiv_W C_{\{0,1\}^{\mathbb{N}}}$ from [4]. The geometric pattern constructed produces an infinitely long line which is then subdivided based on both the information obtained about the input to $\widehat{C_2}$, as well as the order in which this information is found. A glimpse of the construction might be gained from Fig. 1.

6   Two Versus Three Dimensions

A noticeable difference between the construction from the proof of Theorem 8 and Proposition 6 is that the latter yields even a path-connected set, whereas the former does not. Thus, *path-connected choice* is computably-complete from dimension three onwards, but might be simpler in dimension two.

While the status of path-connected choice in dimension two remains open, we can exhibit a related choice principle distinguishing two from three dimensions.

Definition 9. *We say that $A \in \mathcal{A}_2$ has a straight cross, if there are $x, y \in [0,1]$, $\delta > 0$ s.t $\forall \varepsilon \in (-\delta, \delta)$ $(x + \varepsilon, y) \in A \wedge (x, y + \varepsilon) \in A$. Let $\dagger C_{[0,1]^2}$ be choice for sets having a straight cross.*

Proposition 7. $\dagger C_{[0,1]^2} \leq_W \frac{1}{2} C_{[0,1]} \star C_{\mathbb{N}}$.

Corollary 6. $\dagger C_{[0,1]^2} <_W CC_2$

Proof. Combine Proposition 7 with the Fractal Absorption Theorem from [12].

An analogous argument would not succeed in dimension 3, as $\frac{2}{3} C_{[0,1]} \equiv_W C_{[0,1]}$ by a majority-voting argument.

References

1. Baigger, G.: Die Nichtkonstruktivität des Brouwerschen Fixpunktsatzes. Archiv für mathematische Logik und Grundlagenforschung **25**, 183–188 (1985)
2. Brattka, V., de Brecht, M., Pauly, A.: Closed choice and a uniform low basis theorem. Ann. Pure Appl. Logic **163**(8), 968–1008 (2012)
3. Brattka, V., Gherardi, G.: Effective choice and boundedness principles in computable analysis. Bull. Symbolic Logic **1**, 73–117 (2011). arXiv:0905.4685

4. Brattka, V., Gherardi, G.: Weihrauch degrees, omniscience principles and weak computability. J. Symbolic Logic **76**, 143–176 (2011). arXiv:0905.4679
5. Brattka, V., Gherardi, G., Marcone, A.: The Bolzano-Weierstrass theorem is the jump of weak König's lemma. Ann. Pure Appl. Logic **163**(6), 623–625 (2012). arXiv:1101.0792
6. Brattka, V., Le Roux, S., Pauly, A.: Connected choice and Brouwer's fixed point theorem (2012). http://arxiv.org/abs/1206.4809
7. Brattka, V., Le Roux, S., Pauly, A.: On the computational content of the brouwer fixed point theorem. In: Cooper, S.B., Dawar, A., Löwe, B. (eds.) CiE 2012. LNCS, vol. 7318, pp. 56–67. Springer, Heidelberg (2012)
8. Gherardi, G., Marcone, A.: How incomputable is the separable Hahn-Banach theorem? Notre Dame J. Formal Logic **50**(4), 393–425 (2009)
9. Hertling, P.: Unstetigkeitsgrade von Funktionen in der effektiven Analysis. Ph.D. thesis, Fernuniversität, Gesamthochschule in Hagen, Oktober 1996
10. Higuchi, K., Pauly, A.: The degree-structure of Weihrauch-reducibility. Logical Meth. Comput. Sci. **9**(2), 1–17 (2013)
11. Hoyrup, M., Rojas, C., Weihrauch, K.: Computability of the Radon-Nikodym derivative. Computability **1**(1), 3–13 (2012)
12. Le Roux, S., Pauly, A.: Finite choice, convex choice and finding roots. Logical Methods in Computer Science (2015). http://arxiv.org/abs/1302.0380
13. Le Roux, S., Pauly, A.: Weihrauch degrees of finding equilibria in sequential games. In: Beckmann, A., Mitrana, V., Soskova, M. (eds.) CiE 2015. LNCS, vol. 9136, pp. 246–257. Springer, Heidelberg (2015). http://dx.doi.org/10.1007/978-3-319-20028-625
14. Miller, J.S.: Π^0_1 Classes in Computable Analysis and Topology. Ph.D. thesis. Cornell University (2002)
15. Neumann, E.: Computational problems in metric fixed point theory and their Weihrauch degrees. Logic. Methods Comput. Sci. **11**(4) (2015)
16. Orevkov, V.: A constructive mapping of a square onto itself displacing every constructive point. Sov. Math. IV Trans. Doklady Akademie Nauk SSSR. **152**(1), 55 (1963). Published by the American Mathematical Society
17. Pauly, A.: How incomputable is finding Nash equilibria? J. Univers. Comput. Sci. **16**(18), 2686–2710 (2010)
18. Pauly, A.: On the (semi) lattices induced by continuous reducibilities. Math. Logic Q. **56**(5), 488–502 (2010)
19. Pauly, A.: On the topological aspects of the theory of represented spaces. Computability (2016). http://arxiv.org/abs/1204.3763
20. Potgieter, P.H.: Computable counter-examples to the Brouwer fixed-point theorem (2008). http://arxiv.org/abs/0804.3199
21. Shioji, N., Tanaka, K.: Fixed point theory in weak second-order arithmetic. Ann. Pure Appl. Logic **47**, 167–188 (1990)
22. Simpson, S.: Subsystems of Second Order Arithmetic. Springer, Berlin (1999)
23. von Stein, T.: Vergleich nicht konstruktiv lösbarer Probleme in der Analysis. Diplomarbeit, Fachbereich Informatik. FernUniversität Hagen (1989)
24. Weihrauch, K.: The degrees of discontinuity of some translators between representations of the real numbers. Informatik Berichte 129, FernUniversität Hagen, Hagen, July 1992
25. Weihrauch, K.: The TTE-interpretation of three hierarchies of omniscience principles. Informatik Berichte 130, FernUniversität Hagen, Hagen, September 1992
26. Weihrauch, K.: Computable Analysis. Springer-Verlag, New York (2000)

Secret Sharing Schemes with Algebraic Properties and Applications

Ignacio Cascudo[✉]

Department of Mathematics, Aalborg University, Aalborg, Denmark
ignacio@math.aau.dk

Abstract. Secret sharing concerns the distribution of some secret information among a number of parties and is among the most well known tools in cryptography. Secret sharing schemes with certain additional algebraic properties, known as linearity and multiplicativity, have important applications in the area of secure multiparty computation and other areas such as zero knowledge proofs. Secret sharing also has a strong relationship with coding theory and motivates new problems in that field. I will survey several of the recent results in the area and some of their applications.

1 Introduction

Secret sharing, introduced in [26], is among the most well known tools in cryptography. Secret sharing schemes allow to distribute some secret information among a number of parties by sending some *share* (some other piece of information) to each of them, in such a way that a small number of shares give no information about the secret which was shared, but a large enough number of shares allow to reconstruct the secret entirely. Secret sharing schemes are useful as a stand-alone primitive (for distributed information storage), but also play an important role as a building block of many cryptographic protocols.

Before defining secret sharing formally, the following notation is fixed. A *vector of random variables* is a vector $\mathbf{X} = (X_i)_{i \in \mathcal{I}}$ such that the *index-set* $\mathcal{I}$ is finite and non-empty and the X_i's are random variables defined on the same *finite* probability space. If $\mathbf{X}$ is a vector of random variables with index-set $\mathcal{I}$ and $A \subset \mathcal{I}$ with $A \neq \emptyset$, then $\mathbf{X}_A$ denotes the vector of random variables $(X_i)_{i \in A}$. For each $i \in \mathcal{I}$, we denote by the caligraphic $\mathcal{X}_i$ the finite *alphabet* where X_i takes its values. Let $H(X)$ denote the Shannon entropy of a random variable X.

Definition 1 (Secret Sharing Scheme). *A secret sharing scheme is a vector of random variables* $\mathbf{S}$ *with index-set* $\mathcal{I} = \{0, \dots, n\}$ *and such that:*

– Uniformity of the secret: *The set* $\mathcal{S}_0$ *satisfies* $|\mathcal{S}_0| > 1$ *and the variable* S_0 *satisfies* $H(S_0) = \log_2 |\mathcal{S}_0|$, *i.e.,* S_0 *has a-priori the uniform distribution on* $\mathcal{S}_0$.

A major part of this work was written while the author was working at the Department of Computer Science, Aarhus University, Denmark.

A. Beckmann et al. (Eds.): CiE 2016, LNCS 9709, pp. 68–77, 2016.
DOI: 10.1007/978-3-319-40189-8_7

- Joint reconstruction: $H(S_0|\mathbf{S}_{\mathcal{I}\setminus\{0\}}) = 0.$, *i.e.*, S_0 *is completely determined by the rest of the variables in* $\mathbf{S}$.

We abbreviate $\mathcal{I} \setminus \{0\} = \{1, \ldots, n\}$ by $\mathcal{I}^*$ and its cardinality is denoted by $n(\mathbf{S})$. The variable S_0 is the secret, while the remaining variables S_i, $i \in \mathcal{I} \setminus \{0\}$ are the shares.

As mentioned before, secret sharing schemes can be used to distribute secret information. If some party (a dealer) wants to distribute certain information $s \in \mathcal{S}_0$ among a set of $n(\mathbf{S})$ parties (that we assumed indexed by $\mathcal{I}^*$), this can be done in the following way: the dealer samples a vector $\mathbf{a}' = (s, a_1, \ldots, a_n)$ according to the distribution of the conditioned variable $\mathbf{S}|S_0 = s$ and then, for each $i \in \mathcal{I}^*$, he sends the value $a_i \in \mathcal{S}_i$, taken by the variable S_i, to the i-th party. The vector $\mathbf{a} = (a_1, \ldots, a_n)$ is then called a sharing of s.

Definition 2 (Reconstruction and privacy sets). *Let* $\mathbf{S}$ *be a secret sharing scheme. Let* $A \subset \mathcal{I}^*$ *with* $A \neq \emptyset$. *Then:*

- A *is a* reconstructing set *if* $H(S_0|\mathbf{S}_A) = 0$, *i.e., the shares of the set* A *jointly determine the secret with probability 1.*
  $\mathbf{S}$ *has* r-reconstruction *if each subset of* $\mathcal{I}^*$ *of cardinality at least* r *is a reconstructing set. The* reconstruction threshold, *denoted by* $r(\mathbf{S})$, *is the smallest* r *such that* $\mathbf{S}$ *has* r-reconstruction.
- *On the other hand,* A *is a* privacy set *if* $H(S_i|\mathbf{S}_A) = H(S_i)$, *i.e., the a posteriori uncertainty about the secret when given the shares for* A, *equals the a priori uncertainty about the secret (or equivalently,* S_i *and* $\mathbf{S}_A$ *are independent). By definition,* $\emptyset$ *is a privacy set.*
  $\mathbf{S}$ *has* t-privacy *if each subset of* $\mathcal{I}^*$ *of cardinality at most* t *is a privacy set. The* privacy threshold, *denoted by* $t(\mathbf{S})$, *is the largest* t *such that* $\mathbf{S}$ *has* t-privacy.

Obviously if a secret sharing scheme has t-privacy and r-reconstruction, it must hold that $r > t$. Typically, for applications it is interesting to have t-privacy for as large t as possible, while having r-reconstruction for as small r as possible.

2 Linear Secret Sharing

Linear secret sharing schemes are the most well known class of secret sharing schemes. They are described next.

Definition 3. *Let* $\mathbb{F}_q$ *be a finite field. A* linear secret sharing scheme (LSSS) *over* $\mathbb{F}_q$ *is a secret sharing scheme* Σ *where the secret and share spaces* $\mathcal{S}_j$ *are* $\mathbb{F}_q$-vector spaces and $\mathbf{S}$ has the uniform distribution on a $\mathbb{F}_q$-linear subspace $V \leq \times_{j \in \mathcal{I}} \mathcal{S}_j$.

From now on we focus on LSSSs where all the share spaces are the field $\mathbb{F}_q$ and the secret space is $\mathbb{F}_q^k$, for some $k \geq 1$. In that case, it is useful to describe such secret sharing schemes from the point of view of linear codes. Indeed, the support of the variable $\mathbf{S}$ is a linear code C over $\mathbb{F}_q$ of length $k + n$, i.e., a linear subspace of $\mathbb{F}_q^{k+n}$ (we denote this by $C \leq \mathbb{F}_q^{k+n}$). Conversely, we have the following construction:

Definition 4. *Let* $1 \leq k < m$ *be integers. Given a linear code* $C \leq \mathbb{F}_q^m$, *we define the vector of random variables* $\mathbf{S}^{(k)}(C) = (S_i)_{i=0,\ldots,n}$ *(where* $n = m - k$*), whose distribution is given by selecting* $\mathbf{c}$ *uniformly at random in* C *(we index its coordinates as* $\mathbf{c} = (c_{0,1}, c_{0,2}, \ldots, c_{0,k}, c_1, c_2 \ldots, c_n)$*) and defining the secret* S_0 *to be* $(c_{0,1}, c_{0,2}, \ldots, c_{0,k})$ *and the* i-*th share* S_i *to be* c_i *for* $i = 1, \ldots, n$.

Proposition 1. $\mathbf{S}^{(k)}(C)$ *is a secret sharing scheme as long as* C *satisfies two conditions, where* $\mathbf{c}_0$ *will denote the vector* $(c_{0,1}, c_{0,2}, \ldots, c_{0,k})$:

1. *for all* $\mathbf{x} \in \mathbb{F}_q^k$, *there is a word* $\mathbf{c} \in C$ *with* $\mathbf{c}_0 = \mathbf{x}$ *and*
2. *there is no vector in* C *of the form* $\mathbf{c} = (\mathbf{x}, 0, \ldots, 0)$ *where* $\mathbf{x} \in \mathbb{F}_q^k \setminus \{\mathbf{0}\}$.

The advantage of this representation is that we can obtain valuable information about the reconstructing and privacy sets of $\mathbf{S}^{(k)}(C)$ from the supports of the words in C. For $A \subseteq \mathcal{I}^*$, $\mathbf{c} \in C$, let $\mathbf{c}_A$ denote the vector of coordinates $(c_i)_{i \in A}$. Then one can show the following results.

Proposition 2 (based on [11,25]). *Let* $A \subseteq \mathcal{I}^*$.

- *Reconstruction: A is a reconstructing set in the scheme* $\mathbf{S}^{(k)}(C)$ *if and only if for all* $\mathbf{c} \in C$ *such that* $\mathbf{c}_A = \mathbf{0}$, *we have* $\mathbf{c}_0 = \mathbf{0}$.
- *Privacy: A is a privacy set in the scheme* $\mathbf{S}^{(k)}(C)$ *if and only if for all* $\mathbf{s} \in \mathbb{F}_q^k$, *there exists* $\mathbf{c} \in C$ *such that* $\mathbf{c}_0 = \mathbf{s}$ *and* $\mathbf{c}_A = \mathbf{0}$.

Another characterization makes use of the dual code $C^\perp$ of C. Remember that for a linear code $C \leq \mathbb{F}_q^m$, $C^\perp$ is the set of all $\mathbf{x} \in \mathbb{F}_q^m$ such that, for every $\mathbf{c} \in C$, the inner product of $\mathbf{x}$ and $\mathbf{c}$ is 0. $C^\perp$ is a linear code of the same length as C.

Proposition 3 (based on [11,25]). *Let* $A \subseteq \mathcal{I}^*$ *and* $B := \mathcal{I}^* \setminus A$.

- *Reconstruction: A is a reconstructing set in the scheme* $\mathbf{S}^{(k)}(C)$ *if and only if for all* $\mathbf{s} \in \mathbb{F}_q^k$ *there exists* $\mathbf{c}^* \in C^\perp$ *with* $\mathbf{c}_0^* = \mathbf{s}$ *and* $\mathbf{c}_B^* = \mathbf{0}$.
- *Privacy: A is a privacy set in the scheme* $\mathbf{S}^{(k)}(C)$ *if and only if for all* $\mathbf{c}^* \in C^\perp$ *with* $\mathbf{c}_B^* = \mathbf{0}$, *we have* $\mathbf{c}_0^* = \mathbf{0}$.

Using any of these two propositions we can observe that, in the case $k = 1$, every set A of shares is either a privacy or a reconstructing set. However, when $k > 1$, there are sets which are neither privacy nor reconstructing sets, i.e., the knowledge of that set of shares gives some partial information about the secret, but does not determine it completely. In fact, it is always satisfied that

Proposition 4. *If* $A \subseteq B \subseteq \mathcal{I}^*$ *are such that* A *is a privacy set and* B *is a reconstruction set of* $\mathbf{S}^{(k)}(C)$, *then* $|B| - |A| \geq k$. *Consequently,* $r(\mathbf{S}^{(k)}(C)) - t(\mathbf{S}^{(k)}(C)) \geq k$.

Furthermore, from the characterizations above we can bound the reconstruction and privacy thresholds of $\mathbf{S}^{(k)}(C)$ in terms of the minimum distances of the codes C and $C^\perp$. Let $d(C)$ denote the minimum distance of C, i.e., the minimum Hamming distance between two distinct codewords in C, which, since C is linear, is also the minimum Hamming weight of a nonzero codeword in C.

Proposition 5. *The reconstruction and privacy thresholds of* $\mathbf{S}^{(k)}(C)$ *satisfy* $r(\mathbf{S}^{(k)}(C)) \leq n - d(C) + k + 1$ *and* $t(\mathbf{S}^{(k)}(C)) \geq d(C^{\perp}) - k - 1$.

The most famous example of a secret sharing scheme is Shamir scheme, which is linear. Its original description was only for the case $k = 1$, but here a version of it is described that admits secrets in $\mathbb{F}_q^k$ for larger k. Let k, t, n satisfy $1 \leq t + k - 1 \leq n$ and $n + k \leq q$. Fix $\alpha_1, \ldots, \alpha_n, \beta_1, \ldots, \beta_k$ pairwise distinct elements in $\mathbb{F}_q$. Denote by $\mathbb{F}_q[X]_{\leq t+k-1}$ the set of univariate polynomials of degree less than $t + k - 1$ and coefficients in $\mathbb{F}_q$. In order to share a secret $\mathbf{s} \in \mathbb{F}_q^k$ among n parties with this scheme, a polynomial f is chosen in $\mathbb{F}_q[X]_{\leq t+k-1}$ uniformly at random under the additional condition that $f(\beta_j) = s_j$ for $j = 1, \ldots, k$. Then, the i-th share is defined as the evaluation $f(\alpha_i) \in \mathbb{F}_q$. This secret sharing scheme is hence constructed as $\mathbf{S}^{(k)}(C)$ from the Reed Solomon code

$$C = \{(f(\beta_1), \ldots, f(\beta_k), f(\alpha_1), \ldots, f(\alpha_n)) : f \in \mathbb{F}_q[X]_{\leq t+k-1}\}.$$

Shamir's secret sharing scheme has t-privacy and $(t + k)$-reconstruction, and it attains both the bounds given by Propositions 4 and 5.

Linear secret sharing schemes have a property which is of key importance for applications, as it will be explained in Sect. 5: suppose several secrets $\mathbf{s}^{(1)}, \mathbf{s}^{(2)}, \ldots, \mathbf{s}^{(\ell)}$ are shared with $\mathbf{S}$, where the sharing for $\mathbf{s}^{(j)}$ is $(\mathbf{a}_1^{(j)}, \ldots, \mathbf{a}_n^{(j)})$. Then given $\lambda^{(1)}, \ldots, \lambda^{(\ell)} \in \mathbb{F}_q$, the linear combinations $(\sum_{j=1}^{\ell} \lambda^{(j)} \mathbf{a}_i^{(j)})_{i=1,\ldots,n}$ constitute a sharing of $\sum_{j=1}^{\ell} \lambda^{(j)} \mathbf{s}^{(j)}$ in the same scheme $\mathbf{S}$.

3 Multiplicativity

For applications, it would be useful if a property similar to what we just mentioned held also for multiplications. Ideally, we would like that given sharings of $\mathbf{s}$ and $\mathbf{s}'$, the local coordinatewise products of the shares constituted a sharing (in the same scheme) of the coordinatewise product $\mathbf{s} * \mathbf{s}'$ of the secrets. Unfortunately, it is easy to see that LSSSs cannot satisfy this property unless in trivial cases. The notion of multiplicative linear secret sharing schemes, introduced in [12], is a relaxation of this property.

Definition 5. *An ideal LSSS* $\mathbf{S}^{(k)}(C)$ *has* $\hat{r}$*-product reconstruction if for every set* $B \subseteq \mathcal{I}$ *of size at least* $\hat{r}$*, there exists some linear function* $\rho_B : \mathbb{F}_q^{|B|} \to \mathbb{F}_q^k$ *such that for every* $(\mathbf{c}_0, c_1, \ldots, c_n), (\mathbf{d}_0, d_1, \ldots, d_n) \in C$*, it holds that* $\rho_B(\mathbf{c}_B * \mathbf{d}_B) = \mathbf{c}_0 * \mathbf{d}_0$.

This means that the product of two secrets can be reconstructed from the set of the products of the individual shares of the players in B.

Definition 6 ([12]). *A multiplicative LSSS is a LSSS with* n*-product reconstruction.*

A t*-strongly multiplicative LSSS is a LSSS with* t*-privacy and* $(n-t)$*-product reconstruction.*

In order to study the multiplicativity of a LSSS, it is useful to introduce the following notion.

Definition 7 (m-th power of a linear code). *Let $C \subseteq \mathbb{F}_q^n$ be a linear code over $\mathbb{F}_q$, $d > 0$ an integer. Let*

$$C^{*m} := \mathbb{F}_q \langle \{ c^{(1)} * c^{(2)} \ldots * c^{(m)} : (c^{(1)}, c^{(2)}, \ldots, c^{(m)}) \in C^m \} \rangle$$

Remark 1. By definition, $\mathbf{S}^{(k)}(C)$ has $\hat{r}$-product reconstruction if and only if $\mathbf{S}^{(k)}(C^{*2})$ has $\hat{r}$-reconstruction. Consequently, $\mathbf{S}^{(k)}(C)$ has $\hat{r}$-product reconstruction for $\hat{r} = n - d(C^{*2}) + k + 1$ and it is t-strongly multiplicative if $d(C^{\perp}), d(C^{*2}) \geq t + k + 1$.

Proposition 6. *Shamir's scheme (with threshold t) has $(2t + 2k - 1)$-product reconstruction, as long as $2t + 2k - 1 \leq n$. Hence, Shamir's scheme is t-strongly multiplicative as long as $3t + 2k - 1 \leq n$.*

This stems from the fact that if C is a Reed-Solomon code given by evaluations of the polynomials in $\mathbb{F}_q[X]_{\leq t+k-1}$, then C^{*2} is another Reed-Solomon code given by the evaluations, in the same points, of the polynomials in $\mathbb{F}_q[X]_{\leq 2t+2k-2}$.

On the other hand, the following limitation is easy to argue in view of the characterization from Proposition 2.

Proposition 7. *Any LSSS with t-privacy and $\hat{r}$-product reconstruction has $(\tilde{r} - t)$-reconstruction. Consequently it needs to satisfy $\hat{r} \geq 2t + k$. As a consequence of this, any t-strongly multiplicative LSSS satisfies $3t + k \leq n$.*

In the case $k = 1$, the bounds are tight: Shamir's scheme attains the maximum value of t (with respect to the number of players n) for which there is a t-strongly multiplicative LSSS. As far as the author knows, the bound above is not known to be tight for $k > 1$:

Conjecture 1. Any LSSS with t-privacy and $\hat{r}$-product reconstruction satisfies $\hat{r} \geq 2t + 2k - 1$ and any t-strongly multiplicative LSSS satisfies $3t + 2k - 1 \leq n$.

4   Asymptotics

Shamir's scheme achieves several good properties from the point of view of the applications. It matches the bound for $r(\mathbf{S}^{(k)}(C)) - t(\mathbf{S}^{(k)}(C))$ given in Proposition 4. In particular, for $k = 1$, one has the property that any set of t players has no information about the secret, while any set of $t + 1$ can reconstruct it. Furthermore, possibly up to an additive factor k, Shamir's scheme also attains, for a given n, the largest possible t for which there can be t-strong multiplication.

However, Shamir's scheme has the limitation that the number of players n is upper bounded by the size of the field. Indeed, in the definition we have given it is required that $k + n \leq q$ (one can slightly weaken this by adding, as an evaluation point, the coefficient of order $t + k - 1$ of the evaluation polynomial, so that this condition is relaxed as $k + n \leq q - 1$, and all the properties are preserved).[1]

This raises the question whether similar properties can be attained by LSSS **S** if we fix the size of the field q, but want the number of players $n(\mathbf{S})$ to be arbitrarily large. However, the following result sets a restriction in what we can hope to achieve.

Theorem 1. [6] *For any LSSS* $\mathbf{S} = \mathbf{S}^{(1)}(C)$ *we have* $r(\mathbf{S}) - t(\mathbf{S}) \geq \frac{n(\mathbf{S})+2}{2q-1}$. *Consequently, if* **S** *has* $\hat{r}$-*product reconstruction, then* $\hat{r} - 2t(\mathbf{S}) \geq \frac{n(\mathbf{S})+2}{2q-1}$ *and if* **S** *is t-strongly multiplicative, then* $3t \leq (1 - \frac{1}{2q-1})n(\mathbf{S})$.

In fact, by applying somewhat more elaborate arguments, [6] shows a stronger bound for t-strongly multiplicative schemes, namely $3t \leq (1 - \frac{3q-2}{3q^2-3q+1})n$. Extensions for the case where the space of secrets is $\mathbb{F}_q^k$ are also shown in [6].

These results imply that, when n is much larger than q, we will lose a factor $c(q) \cdot n$ in the maximum value of t for which we can have t-strong multiplicativity with respect to what Shamir's scheme attains. The question then is if we can achieve $t = \Omega(n)$ (i.e. $t \geq c'(q)n$ for some other constant $c'(q) > 0$) at all. This is answered by the following result.

Theorem 2 ([2,4,7,10]). *For any finite field* $\mathbb{F}_q$, *there exists an infinite family of secret sharing schemes* $\mathbf{S}^{(k_n)}(C_n)$, *indexed by* $n \in \mathbb{N}$, *such that* $\mathbf{S}^{(k_n)}(C_n)$ *has* n *players and is* t_n-*strongly multiplicative, where* $k_n, t_n = \Omega(n)$ *as* $n \to \infty$.

The result makes use of algebraic geometric codes defined on asymptotically good families of algebraic function fields [19]. It was originally proved for all fields of large enough cardinality in [10]. This in particular included all fields $\mathbb{F}_q$ with $q \geq 49$ and square. Later, the result was extended to every finite field $\mathbb{F}_q$ in [2], by using certain dedicated concatenation of codes, where the outer code is defined to be a code from the family in [10], defined on an appropriate constant-degree extension field of $\mathbb{F}_q$ and the inner code is a Reed-Solomon code over $\mathbb{F}_q$. In [4], it was shown that, by choosing the parameters carefully, and applying more involved algebraic-geometric arguments which, in particular, include bounding the size of the 2-torsion subgroup of the divisor class group of the function fields involved, one can use the algebraic geometric construction in [10] (without the concatenation technique in [2]) directly to show the result for all fields $\mathbb{F}_q$ with $q \geq 16$, and also $q = 8, 9$. Furthermore, the actual values of t_n promised by the construction are larger for smaller fields. The next result is an example of which values of t_n, k_n (with respect to n) can be attained in some cases. For the full results see [2,4,7,10].

---

[1] Furthermore, in the case $k > 1$, one can replace the k evaluation points for the secret by a primitive element of the extension field $\mathbb{F}_{q^k}$, whereby one only needs $n \leq q - 1$, and the privacy and reconstruction thresholds are preserved. The multiplicativity properties hold now with respect to the product in $\mathbb{F}_{q^k}$ (for the secrets) instead of the coordinate-wise product in $\mathbb{F}_q^k$.

Theorem 3. *Let* $\mathbb{F}_q$ *be a finite field. There exists an infinite family of secret sharing schemes* $\mathbf{S}^{(k_n)}(C_n)$, *indexed by* $n \in \mathbb{N}$ *such that* $\mathbf{S}^{(k_n)}(C_n)$ *has* n *players,* t_n-*privacy and* $\hat{r}_n$-*product reconstruction, where* $k_n = \lfloor \kappa n \rfloor$, $t_n = \lfloor \tau n \rfloor$, $\hat{r}_n = \lfloor \hat{\rho} n \rfloor$ *in the following cases:*

- *q square with* $q \geq 49$, *as long as* $1 \geq \hat{\rho} \geq 2\kappa + 2\tau + \frac{4}{\sqrt{q}-1}$.
- *q square, even, with* $q \geq 16$, *as long as* $\hat{\rho} \leq 1$ *and*

$$\hat{\rho} - \frac{h_2(\hat{\rho})}{\log q} - 2\tau - 2\frac{h_2(\tau)}{\log q} \geq \left(2 + \frac{(\sqrt{q}+1)\log q + 1}{(q-1)\log q}\right)\kappa + \frac{(\sqrt{q}+1)\log q + 1}{(q-1)\log q}.$$

- $q = 2$, *as long as* $\hat{\rho} \leq 1$ *and*

$$\hat{\rho} - \frac{h_2(\hat{\rho})}{4} - 30\tau - \frac{h_2(15\tau)}{2} \geq \left(2 + \frac{21}{60}\right)\kappa + \frac{21}{60},$$

where in all cases $h_2(\cdot)$ *denotes the binary entropy function.*

A drawback of these constructions is the high complexity of constructing the generator matrices of algebraic geometric codes defined on the asymptotically good families of function fields from [19]. Unfortunately, so far there is no known way to attain the result of Theorem 2 using "elementary" techniques, for example randomized constructions or polynomial evaluations. This holds even if we drop the condition t_n-strongly multiplicative in Theorem 2 and we require instead t_n-privacy and n-product reconstruction, still with $t_n, k_n = \Omega(n)$. However, if we weaken our requirements further and set $k_n = 1$, then there exists a much more elementary construction using random self-dual codes [11].

Nevertheless, randomized constructions do not seem to have good multiplicative properties in general. In fact, for random codes C, squaring seems to be a "destructive" operation, in the sense that the dimension of C^{*2} grows as the square of the dimension of C. Indeed, [3] showed that

Theorem 4. *Let* $\ell \leq N \leq \frac{\ell(\ell+1)}{2}$. *Write* $u = \frac{\ell(\ell+1)}{2} - N$. *Let* C *be a code chosen uniformly at random among all codes of length* N *and dimension* ℓ. *Then*

$$\Pr\left[C^{*2} = \mathbb{F}^N\right] = 1 - 2^{-\Theta(\ell)} - 2^{-\Theta(u)}.$$

This means that if, in particular $\ell = \Theta(N)$, then C^{*2} will be the full space $\mathbb{F}^N$ with overwhelming probability.

5   Applications

Secure Computation

Secure computation is concerned with the following situation: n parties, each holding some private input x_i, want to correctly compute $f(x_1, x_2, \ldots, x_n)$ for some agreed upon function f in such a way that the intended output is the only new information released. This guarantee should be fulfilled even if a small number t of the players are corrupted and cheat. Corruption is passive if corrupted players follow the specified protocol (but compare their views of the protocol to try to gain additional information) and active if they can behave arbitrarily. For formal security definitions and more information about the topic, see [13].

For the case of a *computationally unbounded* adversary, it is known [1,9] that secure computation of any function is possible as long as less than $n/2$ players are passively corrupted, or less than $n/3$ players are actively corrupted[2]. The results make use of the linearity and multiplicativity of Shamir's scheme. Later in [12], it was shown that, in fact, secure multiparty computation protocol secure against a passive (resp. active) adversary corrupting t players can be constructed from any multiplicative LSSS with t-privacy (resp. t-strong multiplicative LSSS).

The idea of the protocol in the passive case is as follows: assume for the moment $f : \mathbb{F}^n \to \mathbb{F}$ is a linear function. Suppose each input is in $\mathbb{F}$, $\mathbf{S} := \mathbf{S}^{(1)}(C)$ is a LSSS over $\mathbb{F}$ with t-privacy. At the beginning of the protocol, each player can secret share her input x_i with $\mathbf{S}$. Then each player applies f to her received shares. By linearity, the resulting set of values is a sharing of $f(x_1, \ldots, x_n)$ as mentioned at the end of Sect. 2. So now players can broadcast the resulting shares, and reconstruct the output. Ideally, we would want to extend this idea to any function f. Since every function can be computed as an arithmetic circuit over a finite field (for example a Boolean circuit), it would be enough if a sharing of ss' can be created from the sharings of s and s'. However, we mentioned at the beginning of Sect. 3, if s and s' are secret shared with a LSSS $\mathbf{S} := \mathbf{S}^{(1)}(C)$, and each player multiplies her shares, even if the original scheme was multiplicative, the resulting set of values is not a sharing of ss' in the same scheme $\mathbf{S}^{(1)}(C)$, but in $\mathbf{S}^{(1)}(C^{*2})$.

A better alternative is obtained by allowing interaction among the players. Again let $\mathbf{S}^{(1)}(C)$ be multiplicative. By definition there is a linear function ρ such that $ss' = \rho(c_1 c_1', \ldots, c_n c_n')$ where c_i, c_i' are the shares of s and s' respectively. The following protocol allows to create a sharing ss' in the scheme $\mathbf{S}^{(1)}(C)$: Each player i shares the product $c_i c_i'$ of his shares with $\mathbf{S}^{(1)}(C)$. Now, players have shares of $c_1 c_1', c_2 c_2', \ldots, c_n c_n'$ and they can locally compute shares of $ss' = \rho(c_1 c_1', \ldots, c_n c_n')$ since ρ is linear. This can be shown to be secure against a passive adversary corrupting at most t parties if the scheme has t-privacy and it is enough to argue that any function can be securely computed by n players in this situation.

The case of an active adversary requires additional techniques. The problem is that corrupted players (which can deviate from the protocol) may choose to deal inconsistent shares to the honest players or to share wrong values. The solution in [12] involves using *verifiable secret sharing*, in which the dealer sends additional information to each player which can be used to verify that the shares have been well constructed. If the underlying LSSS is t-strongly multiplicative,

---

[2] Here we suppose each pair of players is connected by a secure point-to-point channel, but we do not assume the existence of a broadcast channel.

this leads to a secure protocol to compute any function in the presence of an adversary corrupting at most t players.

Two-Party Cryptography and Other Applications

In recent years, secure multi-party computation protocols have surprisingly been applied on cryptographic problems involving only two players. This is due to the MPC-in-the-head technique, which was introduced in [22] for the purpose of zero knowledge proofs with high communication efficiency, where a virtual multiparty computation protocol with a large number of parties is run by of the two players at some point of the protocol. The ideas from [22] have inspired other applications in areas such as multiparty computation with dishonest majority [17,24], OT combiners [20], OT from noisy channels [21], correlation extraction [23] and UC homomorphic commitment schemes [8,16,18]. In many of these applications, the best results in terms of efficiency are attained by setting up multiparty computation protocols with a very large number of players, while the size of the field should be small. This has been a motivation for the results on asymptotics of t-strongly multiplicative LSSS discussed in Sect. 4.

Finally, one can generalize the concept of multiplicative secret sharing schemes in a number of ways. A generalization is given by the notion of arithmetic codices introduced in [5]: this notion allows to address also other applications in two-party cryptography (zero knowledge proofs of algebraic relations [14]), algebraic complexity (see [7]) and more recently has been applied back in the realm of error correcting, more specifically in the problem of local decodability of Reed-Muller codes (see [15]).

References

1. Ben-Or, M., Goldwasser, S., Wigderson, A.: Completeness theorems for non-cryptographic fault-tolerant distributed computation. In: Proceedings of STOC 1988, pp. 1–10. ACM Press (1988)
2. Cascudo, I., Chen, H., Cramer, R., Xing, C.: Asymptotically good ideal linear secret sharing with strong multiplication over *any* fixed finite field. In: Halevi, S. (ed.) CRYPTO 2009. LNCS, vol. 5677, pp. 466–486. Springer, Heidelberg (2009)
3. Cascudo, I., Cramer, R., Mirandola, D., Zemor, G.: Squares of random linear codes. IEEE Trans. Inf. Theor. **61**(3), 1159–1173 (2015)
4. Cascudo, I., Cramer, R., Xing, C.: The torsion-limit for algebraic function fields and its application to arithmetic secret sharing. In: Rogaway, P. (ed.) CRYPTO 2011. LNCS, vol. 6841, pp. 685–705. Springer, Heidelberg (2011)
5. Cascudo, I., Cramer, R., Xing, C.: The arithmetic codex. In: Proceedings of IEEE Information Theory Workshop (ITW 2012), pp. 75–79 (2012)
6. Cascudo, I., Cramer, R., Xing, C.: Bounds on the threshold gap in secret sharing and its applications. IEEE Trans. Inf. Theor. **59**(9), 5600–5612 (2013)
7. Cascudo, I., Cramer, R., Xing, C.: Torsion limits and Riemann-Roch systems for function fields and applications. IEEE Trans. Inf. Theor. **60**(7), 3871–3888 (2014)
8. Cascudo, I., Damgård, I., David, B., Giacomelli, I., Nielsen, J.B., Trifiletti, R.: Additively homomorphic UC commitments with optimal amortized overhead. In: Katz, J. (ed.) PKC 2015. LNCS, vol. 9020, pp. 495–515. Springer, Heidelberg (2015)

9. Chaum, D., Crépeau, C., Damgård, I.: Multi-party unconditionally secure protocols. In: Proceedings of STOC 1988, pp. 11–19. ACM Press (1988)
10. Chen, H., Cramer, R.: Algebraic geometric secret sharing schemes and secure multiparty computations over small fields. In: Dwork, C. (ed.) CRYPTO 2006. LNCS, vol. 4117, pp. 521–536. Springer, Heidelberg (2006)
11. Chen, H., Cramer, R., Goldwasser, S., de Haan, R., Vaikuntanathan, V.: Secure computation from random error correcting codes. In: Naor, M. (ed.) EUROCRYPT 2007. LNCS, vol. 4515, pp. 291–310. Springer, Heidelberg (2007)
12. Cramer, R., Damgård, I.B., Maurer, U.M.: General secure multi-party computation from any linear secret-sharing scheme. In: Preneel, B. (ed.) EUROCRYPT 2000. LNCS, vol. 1807, pp. 316–334. Springer, Heidelberg (2000)
13. Cramer, R., Damgård, I., Nielsen, J.B.: Secure Multiparty Computation and Secret Sharing - An Information Theoretic Approach. Cambridge University Press
14. Cramer, R., Damgård, I., Pastro, V.: On the amortized complexity of zero knowledge protocols for multiplicative relations. In: Smith, A. (ed.) ICITS 2012. LNCS, vol. 7412, pp. 62–79. Springer, Heidelberg (2012)
15. Cramer, R., Xing, C., Yuan, C.: On Multi-point Local Decoding of Reed-Muller Codes. Manuscript (2016). http://arxiv.org/abs/1604.01925
16. Damgård, I., David, B., Giacomelli, I., Nielsen, J.B.: Compact VSS and efficient homomorphic UC commitments. In: Sarkar, P., Iwata, T. (eds.) ASIACRYPT 2014, Part II. LNCS, vol. 8874, pp. 213–232. Springer, Heidelberg (2014)
17. Damgård, I., Zakarias, S.: Constant-overhead secure computation of Boolean circuits using preprocessing. In: Sahai, A. (ed.) TCC 2013. LNCS, vol. 7785, pp. 621–641. Springer, Heidelberg (2013)
18. Frederiksen, T.K., Jakobsen, T.P., Nielsen, J.B., Trifiletti, R.: On the complexity of additively homomorphic UC commitments. In: Kushilevitz, E., et al. (eds.) TCC 2016-A. LNCS, vol. 9562, pp. 542–565. Springer, Heidelberg (2016). doi:10.1007/978-3-662-49096-9_23
19. Garcia, A., Stichtenoth, H.: A tower of Artin-Schreier extensions of function fields attaining the Drinfeld-Vlăduţ bound. Inventiones Math. 121, 211–222 (1995)
20. Harnik, D., Ishai, Y., Kushilevitz, E., Nielsen, J.B.: OT-combiners via secure computation. In: Canetti, R. (ed.) TCC 2008. LNCS, vol. 4948, pp. 393–411. Springer, Heidelberg (2008)
21. Ishai, Y., Kushilevitz, E., Ostrovsky, R., Prabhakaran, M., Sahai, A., Wullschleger, J.: Constant-rate oblivious transfer from noisy channels. In: Rogaway, P. (ed.) CRYPTO 2011. LNCS, vol. 6841, pp. 667–684. Springer, Heidelberg (2011)
22. Ishai, Y., Kushilevitz, E., Ostrovsky, R., Sahai, A.: Zero-knowledge from secure multiparty computation. In: Proceedings of 39th STOC, San Diego, CA, USA, pp. 21–30 (2007)
23. Ishai, Y., Kushilevitz, E., Ostrovsky, R., Sahai, A.: Extracting correlations. In: Proceedings of 50th IEEE FOCS, pp. 261–270 (2009)
24. Ishai, Y., Prabhakaran, M., Sahai, A.: Founding cryptography on oblivious transfer – efficiently. In: Wagner, D. (ed.) CRYPTO 2008. LNCS, vol. 5157, pp. 572–591. Springer, Heidelberg (2008)
25. Massey., J.: Minimal codewords and secret sharing. In: Proceedings of the 6th Joint Swedish-Russian International Workshop on Information Theory (1993)
26. Shamir, A.: How to share a secret. Commun. ACM 22(11), 612–613 (1979)

Squeezing Feasibility

Walter Dean[✉]

Department of Philosophy, University of Warwick, Coventry CV4 7AL, UK
W.H.Dean@warwick.ac.uk
http://go.warwick.ac.uk/whdean

Abstract. This note explores an often overlooked question about the characterization of the notion *model of computation* which was originally identified by Cobham [5]. A simple formulation is as follows: what primitive operations are allowable in the definition of a model such that its time and space complexity measures provide accurate gauges of practical computational difficulty? After exploring the significance of this question in the context of subsequent work on machine models and simulations, an adaptation of Kreisel's *squeezing argument* [17] for Church's Thesis involving Gandy machines [11] is sketched which potentially bears on this question.

1 Cobham's Problem

The goal of this note is to highlight a historically significant but often over-looked question about the characterization of what is commonly called a *model of computation*. A simple formulation is as follows: what primitive operations are allowable in the definition of such a model such that its time and space complexity measures provide accurate gauges of practical computational difficulty?

A version of this problem was posed in the paper "The intrinsic computational difficulty of functions" [5] in which Alan Cobham first defined the class **FP** of functions computable in polynomial time and conjectured that it coincides with the informal notion of a *feasibly computable function*:

[W]e find that it makes a definite difference what class of computational methods and devices we consider in our attempt to formalize the definition [of a feasibly computable function]. The problem is reminiscent of, and obviously closely related to, that of the formalization of the notion of effectiveness. But the emphasis is different in that the physical aspects of the computation process are here of predominant concern. The question of what may legitimately be considered to constitute a step of a computation is quite unlike that of what constitutes an effective operation ... If we are to make fine distinctions, say between [**TIME**(n)] and [**TIME**(n^2)], then we must have an equally fine analysis of all phases of the computational process ... We must be prepared to argue that we haven't taken

© Springer International Publishing Switzerland 2016
A. Beckmann et al. (Eds.): CiE 2016, LNCS 9709, pp. 78–88, 2016.
DOI: 10.1007/978-3-319-40189-8_8

too broad a class for [**TIME**(n)] and thus admitted to it functions not in actuality computable in a number of steps linearly bounded by the lengths of its arguments.[1]   [5], pp. 29–30

Cobham illustrated the significance of the choice of basic operations by posing the following question: is there a model-independent – or "absolute" – sense in which integer multiplication is intrinsically harder to compute than integer addition?[2] A theory of computational complexity based on a model of computation which treats both addition and multiplication as operations which can be computed in a single primitive step is clearly incapable of providing a principled answer to this question. But since any model of computation must treat *some* operations as primitive, how are we to make a principled choice? Cobham suggested that not only was this one of the "fundamental problems" facing the development of such a theory but that it "may well call for considerable patience and discrimination" ([5] p. 30).

2   Machine Models and the Invariance Thesis

My present goal will not be to attempt to resolve Cobham's problem directly. For in hindsight it is clear that many of the issues he identified are likely related to fundamental unresolved questions about the separation of complexity classes. I merely wish to argue that some of the surrounding issues may be clarified by reflecting further on the developments which led to the adoption of the multi-tape Turing machine $\mathfrak{T}$ as the standard model employed in complexity theory and its use in formulating what is now often called the *Cobham-Edmonds Thesis*:

> A function $f(x)$ is *feasibly computable* just in case it is computed by a machine $T \in \mathfrak{T}$ such that the running time complexity of T is asymptotically bounded by $|x|^k$ for some fixed $k \in \mathbb{N}$ (i.e. such that $\text{time}_T(|x|) \in O(|x|^k)$ where $|x|$ denotes the size of the input).

This proposal is often presented in textbooks as playing the same role for complexity theory which Church's Thesis plays for computability theory – i.e. that of providing a mathematically precise analysis of the pre-theoretical notion of *feasibility* (i.e. computability *in practice*) in the same way which Church's Thesis is

---

[1] Cobham characterized the class $\mathcal{L}$ of feasibly computable functions as those computable by an informal algorithm in "a number of steps ... bounded by polynomials in the lengths of the numbers involved". He proposed that $\mathcal{L}$ coincided with the functions definable using a restricted recursion scheme known as *limited recursion on notation*. He then conjectured that the class of functions definable in this manner coincided with **FP** under its contemporary definition – i.e. the class of functions computable in polynomial time by a deterministic Turing machine. This was later demonstrated by Rose [20].

[2] "Is it harder to multiply than to add? ... I grant I have put [this] question loosely; nevertheless, I think the answer ought to be *yes*" [5], p. 24. By the mid-1960s it was already possible to marshall a good deal of evidence in favor of this hypothesis – e.g., [7].

generally taken to provide an analysis of the pre-theoretical notion of *effectivity* (i.e. computability *in principle*). This in turn suggests that the complexity class **FP** enjoys a status similar to that of the class of Turing computable functions as an "intrinsic" or "absolute" standard of computational difficulty. But although it remains controversial whether this is indeed the case, it is not the identification of feasibility with polynomial time computability to which I wish to draw attention. Rather, it is the rationale for using $\mathfrak{T}$ in the definition of "polynomial time" and "polynomial space" provided that these quantities are intended to serve as useful metrics of computational difficulty.

An early precedent for this was provided by Hartmanis and Stearns [13] who demonstrated the first hierarchy theorems showing that more functions become computable relative to a sufficiently large increase in the number of allowable steps consumed or tape cells accessed during the course of a Turing machine computation. Soon after this, simulation results were obtained linking the time or space bounds for $\mathfrak{T}$ to a variety of other tape-based models which might have initially seemed more or less powerful. For instance if $f(x)$ can be computed in time $O(t(n))$ using a multi-tape machine from $\mathfrak{T}$, then it can be computed in time $O(t(n)^2)$ using a *single*-tape machine. And if $f(x)$ can be computed in time $O(t(n))$ using a machine with a 2-*dimensional tape* then it can be computed in time $O(t(n)^{3/2})$ by a machine from $\mathfrak{T}$ using multiple one-dimensional tapes.

Similar results were also obtained for the basic RAM model $\mathfrak{R}$ introduced by Cook and Reckhow [8]. Recall that this model consists of a finite program, one or more accumulator registers, an instruction counter, and an infinite collection of memory registers $R_0, R_1, \ldots$ which may contain integers of unbounded size. The operations performable by $\mathfrak{R}$ include reading inputs and printing outputs, conditional jump operations based on comparisons between the accumulator registers or other registers, instructions for transferring values between registers and the accumulator (inclusive of indirect addressing), and the following arithmetical operations: *addition* and *subtraction*. In calculating the running time complexity of a RAM machine, it is often assumed that these operations can be carried out in *unit time* – i.e. as a single unmediated step.

Although this latter assumption is evidently false for tape-based models, a basic simulation result about the relative efficiency of $\mathfrak{R}$ and $\mathfrak{T}$ is as follows: if $f(x)$ can be computed in time $O(t(n))$ using a machine from $\mathfrak{R}$, then it can be computed in time $O(t(n)^3)$ using a machine from $\mathfrak{T}$. Similar results apply to some (although as we will see not *all*) register-based models which are obtained by modifying $\mathfrak{R}$ in various straightforward ways – e.g. allowing that values stored in registers may be treated as binary numerals on which it is possible to apply bit-wise Boolean operations in parallel.[3]

The significance of these results is borne out by the observation that $\mathfrak{R}$ is sufficiently flexible to allow for implementations of many algorithms which arise in mathematical practice – say in number theory, linear algebra or graph theory – which are not only reasonably direct, but also have running time proportional

---

[3] See [27] for full definitions of $\mathfrak{T}$ and $\mathfrak{R}$ and references for the simulation results just summarized.

(often by scalar factors) to their informal counterparts.[4] This provides inductive confirmation that measuring computational complexity relative to $\mathfrak{R}$ provides an accurate gauge of the practical difficulty of solving computational problems of the sort which we confront in practice.

It has become conventional to call a model of computation with this property (such as $\mathfrak{R}$) *reasonable*. On the basis of simulation results of the sort mentioned above, van Emde Boas [27] has proposed that the notion of "reasonability" should be extended to other models as follows:

INVARIANCE THESIS: Suppose that $\mathfrak{N}_0$ is a reasonable model of computation. Then another model $\mathfrak{N}_1$ is also reasonable just in case every machine in $\mathfrak{N}_0$ can be simulated by a machine $\mathfrak{N}_1$ with polynomially bounded overhead in time and a constant-factor overhead in space and conversely with respect to $\mathfrak{N}_1$ and $\mathfrak{N}_0$.[5]

It thus follows from the identification of $\mathfrak{R}$ as a reasonable model coupled with the simulation results mentioned above that the Invariance Thesis predicts that $\mathfrak{T}$ and its variants (e.g. single-tape and multi-dimensional tape machines) are also reasonable models. Some complications remain in the formulation of a version of the Invariance Thesis which is sufficiently general to simultaneously cover all of the apparently reasonable register models which have been proposed.[6] But as we have seen, the original class of tape-based models which satisfy the Invariance Thesis with respect to $\mathfrak{T}$ is quite robust. This demonstrates that we would not change the extensional characterization of the class of polynomial time computable functions if we had elected to measure complexity with respect to single tape Turing machines, machines with two dimensional tapes, machines with stacks or counters instead of conventional tapes, etc.

Basing the definition of **FP** on $\mathfrak{R}$ would also yield an extensionally equivalent characterization. However, this depends more sensitively on the details of the register-based model considered. Consider, for instance, the so-called *MRAM* model $\mathfrak{M}$ obtained by supplementing the instructions available in $\mathfrak{R}$ with unit time multiplication. It follows by results of Hartmanis and Simon [12] and Bertoni et al. [2] that what can be computed in polynomial *time* relative to $\mathfrak{M}$ coincides with what can be computed in polynomial *space* relative to $\mathfrak{R}$ (and hence also $\mathfrak{T}$).[7] To put this point more precisely, let $\mathbf{C}_{\mathfrak{N}}$ denote the class of languages decidable using the time or space resource $\mathbf{C}$ relative to the model $\mathfrak{N}$.

---

[4] Considerable inductive support for this claim is supplied by Knuth's [15] detailed implementations of a wide range of algorithms using a variant of $\mathfrak{R}$ which satisfies the Invariance Thesis when its multiplication and division operations are removed.

[5] I.e. for all $N_0 \in \mathfrak{N}_0$ there exists $N_1 \in \mathfrak{N}_1$ computing the same function such that $\text{time}_{N_1}(|x|) \in O(\text{time}_{N_0}(|x|)^k)$ and $\text{space}_{N_1}(|x|) \in O(\text{space}_{N_0}(|x|))$ and conversely.

[6] See [27] for discussion of caveats pertaining both to how the relevant time and space measures must be formulated and the status of the requirement that the relevant overheads be achieved with respect to the same simulation.

[7] Hartmanis and Simon showed this originally with respect to a variant of the MRAM model which allows parallel bit-wise Boolean operations in the manner described above. Bertoni et al. showed that such operations were not necessary in the presence of both addition and multiplication.

Then the results just described establish that the class $\mathbf{P}_{\mathfrak{M}}$ of problems decidable in polynomial time relative to the MRAM model contains the class $\mathbf{PSPACE}_{\mathfrak{T}}$ of problems decidable in polynomial space by a multi-tape Turing machine.

Although it is presently unknown whether $\mathbf{P}_{\mathfrak{T}} \neq \mathbf{PSPACE}_{\mathfrak{T}}$, it is widely believed that the former class is a proper subset of the latter. Relative to this conjecture, $\mathfrak{M}$ is *not* a reasonable model of computation.[8] And despite the fact that $\mathfrak{M}$ is obtained from $\mathfrak{R}$ by a seemingly innocent extension, there is good evidence for such a classification. For it is not hard to see that polynomial time and polynomial space *coincide* for $\mathfrak{M}$ – i.e. $\mathbf{P}_{\mathfrak{M}} = \mathbf{PSPACE}_{\mathfrak{M}}$. This has traditionally been taken to be the touchstone of models which allow for parallelism involving exponential growth in the number of processing units – e.g. the k-PRAM model $\mathfrak{P}$ of Savitch and Stimson [21] which in a single step allows a RAM-like device to create $k \geq 2$ copies of itself (inclusive of all its registers). It is easy to see, however, that there exist concrete instances of problems which we cannot currently decide in practice but which are in $\mathbf{P}_{\mathfrak{P}}$ (and hence also in $\mathbf{P}_{\mathfrak{M}}$). van Emde Boas [27] refers to such models as inhabiting what he refers to as the *second machine class*. Of such models he writes that "it is not clear at all that they can be considered to be reasonable; it seems that the marvelous speed-ups obtained by the parallel models of the second machine class require severe violations of basic laws of nature." (p. 14).

The foregoing observations illustrate that not only is the choice of basic operations central to developing an account of intrinsic computational difficulty, but that Cobham was prescient in using the contrast in the apparent difficulty of addition and multiplication to illustrate one of the fundamental challenges of providing such an account. For although it is easy to see that there is an $O(n)$ algorithm for computing integer addition relative to $\mathfrak{T}$ (where n is the maximum of the lengths of the inputs), it is currently unknown whether this is the case for multiplication with respect to both $\mathfrak{T}$ and $\mathfrak{R}$.[9] In choosing a benchmark for defining complexity classes, there are, of course, clear practical reasons to move from the single-tape Turing machine model (which does not allow for a linear time implementation of addition) to the multi-tape model $\mathfrak{T}$, and again from $\mathfrak{T}$ to a register-based model such as $\mathfrak{R}$. But the apparent sensitivity of "polynomial time" with respect to the inclusion or exclusion of arithmetical operations from this class suggests that we have yet to realize Cobham's hope of finding an intrinsic rationale for using this model (together with the Invariance Thesis) as a means of characterizing feasible computability.

---

[8] For if $\mathfrak{M}$ satisfied the Invariance Thesis with respect to $\mathfrak{R}$, this would also entail that there was a time and space efficient simulation of MRAM machines by multi-tape Turing machines from which $\mathbf{P}_{\mathfrak{T}} = \mathbf{PSPACE}_{\mathfrak{T}}$ would follow.

[9] Schönhage and Strassen [23] showed that the $O(n^2)$ time complexity of the "naive" carry multiplication algorithm can be improved to $O(n \log(n) \log(\log n))$, while also conjecturing a lower bound of $\Omega(n \log(n))$. The Schönhage-Strassen bound has recently been improved by Fürer [10] who presented an $n \log n 2^{O(\log^*(n))}$ algorithm.

3 Arguing for the Cobham-Edmonds Thesis

Drawing on the analogy with Church's Thesis, there are at least three ways one might go about arguing for the Cobham-Edmonds Thesis. First one might attempt to show inductively that **FP** accurately circumscribes the class of functions which we can compute in practice. Second, one might attempt to argue that this class of functions is "intrinsic" in virtue of being definable by extensionally coincident but independently motivated definitions. And third, one might attempt to argue directly that at least one of these definitions serves as a correct analysis of the pre-theoretical notion of a function computable in practice.

These strategies have been explored in as much detail in the complexity theoretic setting as have the analogous arguments in favor of Church's Thesis. Nonetheless, it is evident that they all face complications which their computability-theoretic counterparts do not have to confront. There are, for instance, well known objections even to the claim that **FP** extensionally coincides with feasible computability both on the basis that the former class admits functions computable only by Turing machines whose running time complexity involves "infeasible" constant and scalar factors and also because of the potential vagueness or interest relativity in our background notion of feasibility itself.[10]

With respect to the second argument, recall that there are several well-known alternative characterizations of effective computability in terms of (e.g.) general recursiveness or unrestricted grammars which yield extensionally equivalent definitions of the class of Turing computable functions. We now also know of alternative characterizations of **FP** – e.g. in terms of Cobham's own characterization in terms of limited recursion on notation, the provably recursive functions of certain systems of bounded arithmetic, or logical definability as studied within descriptive complexity theory.

There is, however, at least one salient difference between these cases. For in the original instances, the alternative characterization of the Turing computable functions are arrived at by generalizing definitions to achieve a maximal characterization of what can potentially be computed relative to a certain class of formalisms which meet only very general finiteness requirements – e.g. by eliminating restrictions on the allowable forms of recursion schemes or production rules. This, however, is not the case for the sort of "machine independent" characterizations of **FP** just alluded to. In particular, by changing obvious parameters in the relevant definitions we can also arrive at characterizations of well-known complexity classes which are both narrower and broader than **FP**.[11] Thus while

---

[10] Although I will put aside such issues here, rejoinders to these objections can be found in textbooks such as [1]. See also [9].

[11] For instance, while **FP** coincides with the provably recursive functions of Buss's system S_2^1, similar characterizations of the Linear Time Hierarchy or Polynomial Hierarchy arise as the provably recursive functions of similar theories [3]. And while $P_{\mathcal{I}}$ is characterizable as the set of languages describable in first-order logic with a least fixed point operator in the presence of a linear order, similar (and in many instances simpler) characterizations of narrower classes (like AC^0) or broader classes (like $NP_{\mathcal{I}}$) are also available [14]. See [4] for function algebra descriptions of additional complexity classes such as $PSPACE_{\mathcal{I}}$.

this class is robust in the sense of admitting independent definitions, the alternative definitions which are presently available lack the sort of canonicity which is traditionally thought to accompany the various characterizations of Turing computability whose coincidence is often cited in favor of Church's Thesis.

4   The Squeezing Argument

Let us now consider the status of the third potential sort of argument for the Cobham-Edmonds Thesis – i.e. that of attempting to show that one of the available characterizations of **FP** provides a conceptual analysis of the pre-theoretical notion of a feasibly computable function. Presuming that mathematical practice affords us a sufficiently robust understanding of this notion, nothing appears to preclude the possibility of such an argument. But if we proceed in this manner, it would seem that we run directly into Cobham's problem – i.e. if we lack a prior mathematical characterization of feasibility, how are we to make a principled decision as to what basic operations may legitimately be included in a model of computation on which to base the analysis?

This might at first appear to leave us at an impasse. But I now wish to suggest that it may be possible to modify an indirect argument which has sometimes been put forward in favor of Church's Thesis to argue for the Cobham-Edmonds Thesis. The method in question is a variant of the so-called *squeezing argument* due to Kreisel. Kreisel [17] originally used this method to argue that the *pre-theoretically* identified class INFVAL of statements which are valid in virtue of their logical form extensionally coincides with the class FORMVAL of formal validities of first-order logic. In order to reach this conclusion he argued that INFVAL could be "squeezed" between FORMVAL and the class PROV of formulas provable in a given proof system which is sound and complete for first-order logic.

First note that since we regard the rules of (say) the sequent calculus as individually sound in the informal sense, a simple induction yields that any statement derivable using these rules from no premises is informally valid. Since provability is hence a *sufficient* condition for informal validity, we have that

(1) PROV ⊆ INFVAL

On the other hand, note that even if we do not start out with an entirely settled notion of what it means for a statement to be valid in virtue of its logical form, we presumably acknowledge that no statement which possesses a traditional Tarskian countermodel (i.e. a model with a set-sized domain) can be valid in this sense. Since Tarskian validity is thus a *necessary* condition for informal validity, we thus also have that

(2) INFVAL ⊆ FORMVAL

Note, however, that it is a consequence of the Completeness Theorem for first-order logic that FORMVAL ⊆ PROV. And from this it follows from (1) and (2) that INFVAL = FORMVAL as desired.

On at least two occasions [18,19] Kreisel hinted that the squeezing argument might be adapted to Church's Thesis. The evident strategy here would be to attempt to "squeeze" the pre-theoretically identified class EFFCOMP of effectively computable functions between the class of functions $\mathfrak{N}_0$-COMP and $\mathfrak{N}_1$-COMP computable relative to models of computation $\mathfrak{N}_0$ and $\mathfrak{N}_1$. If it could be established that membership in $\mathfrak{N}_0$-COMP was a sufficient condition for effectivity and that membership in $\mathfrak{N}_1$-COMP was a necessary condition, then we could similarly conclude that EFFCOMP $= \mathfrak{N}_0$-COMP provided that we could demonstrate that $\mathfrak{N}_0$-COMP $= \mathfrak{N}_1$-COMP.

Although Kreisel was not explicit about the forms which such models might take, Turing [25] may plausibly be regarded as providing an argument that there is good justification for taking $\mathfrak{N}_0 = \mathfrak{T}_1$ (i.e. the single tape Turing machine). For in the case of each defining property of this model – e.g. the finiteness of the number of distinguishable symbols and "states of mind", the discreteness and one-dimensionality of the tape, the fact that only a single cell can be scanned at a given step, etc. – he gives an argument that a computational method satisfying these properties jointly is effective. And for this reason Turing computability is traditionally regarded as *sufficient* for effective computability.

But as Turing himself admits (p. 249), it is less clear that these specific conditions are also necessary properties of an effective method. To take his own example, for instance, it is evident that some effective methods (e.g. carry addition and multiplication) take advantage of a two-dimensional computing medium. And we now also typically allow that certain forms of parallel processing – e.g. of the sort employed in conventional parallel algorithms for sorting and matrix multiplication – are also be effective. It is thus not immediately clear how to choose a model $\mathfrak{N}_1$ which simultaneously meets the conditions that $\mathfrak{T}_1$-COMP $= \mathfrak{N}_1$-COMP and is also such that if a function *failed* to be $\mathfrak{N}_1$-computable, then we could be reasonably assured that was not effective computability.

A potential example of such a model was provided by Gandy [11]. His model $\mathfrak{G}$ formalizes a notion of computation on hereditarily finite sets which which is intended to generalize Turing's requirements as far as possible, compatible with the assumption that the described computations can be carried "mechanistically" – i.e. by a finite, discrete, deterministic device which can be physically constructed.[12] The constraints on such devices take the form of what Sieg [24] calls *boundedness* and *locality* conditions – i.e. i) there is a finite bound on the number of distinct symbol configurations which can be recognized in a single step and also on the number of distinct internal states; ii) only observed symbol configurations within a bounded distance of the previously observed configurations may be modified in a single step. These conditions are sufficiently liberal to allow for direct formalizations of algorithms which arise in mathematical practice in something like the manner of $\mathfrak{R}$, thus providing some evidence that EFFCOMP $\subseteq \mathfrak{G}$-COMP. But despite

---

[12] The graph-theoretic models of Kolomogorv and Uspensky [16] and Schönhage [22] are inspired by similar considerations. However Gandy's model differs not only defining states in defining states more abstractly – i.e. as hereditarily finite sets (which may contain unboundedly many labels which are treated as *urelements*) of bounded rank – but also in that it allows for parallelism in the sense that machines may locally transform discontiguous parts of states in a single step.

the abstractness of the Gandy-Sieg conditions, it is still possible to show that the $\mathfrak{T}$-Comp $=$ $\mathfrak{G}$-Comp – i.e. the class of functions computable by Gandy machines is coincident with the class of Turing computable functions.

If it is accepted that the Gandy-Sieg conditions provide a reasonable upper bound on the class of symbolic operations which can be performed in a single step, then the result just cited can be invoked to complete the squeezing argument for Church's Thesis in a manner which avoids the need to provide a prior characterization of what might be meant by an *effective primitive operation* – i.e. one of the sort illustrated but not fully characterized by Turing's analysis. This feature of the argument might also inspire hope that it could also be adapted to provide an argument for the Cobham-Edmonds Thesis which circumvents the need to directly confront Cobham's problem - i.e. that of giving a definition of a *feasible primitive operation* prior to an analysis of *feasibly computable*.

The relevant adaptation would attempt to squeeze the informally identified class of feasibly computable functions FEASCOMP between the classes of $\mathbf{FP}_{\mathfrak{Q}_0}$ and $\mathbf{FP}_{\mathfrak{Q}_1}$ which are computable in polynomial time relative to models $\mathfrak{Q}_0$ and $\mathfrak{Q}_1$. In this case, we wish $\mathfrak{Q}_0$ to be what might be called a *minimally reasonable model* – i.e. one which satisfies the Invariance Thesis and also is such that the inclusion of $f(x)$ in $\mathbf{FP}_{\mathfrak{Q}_0}$ is sufficient to ensure that $f(x)$ is intuitively feasible. On the other hand, we would desire to show that $\mathfrak{Q}_1$ is *maximally reasonable* – i.e. provides maximally direct implementations of feasible algorithms subject to the requirement of satisfying the Invariance Thesis.

Although we must ultimately demonstrate that $\mathbf{FP}_{\mathfrak{Q}_0} = \mathbf{FP}_{\mathfrak{Q}_1}$, nothing prohibits us from choosing $\mathfrak{Q}_0 = \mathfrak{T}$ or even $\mathfrak{T}_1$. On the basis of the foregoing, one might additionally hope that it is also admissible to choose $\mathfrak{Q}_1 = \mathfrak{G}$. However Gandy's model was explicitly designed to allow for the sort parallel evolution exhibited by cellular automata such as Conway's Game of Life. It is thus not immediately clear whether $\mathfrak{G}$ is a reasonable model in the sense discussed above or even whether it fails to belongs to the second machine class.[13]

Per van Emde Boas's prior admonishment, it is hence unclear whether $\mathfrak{G}$ may be directly employed in a squeezing argument for feasibility in the envisioned manner. As such, we have yet to find an argument which puts the Cobham-Edmonds Thesis on the same footing as Church's Thesis. We are, however, left with a potentially promising question to explore in this regard – i.e. is it possible to find natural modifications to the Gandy-Sieg boundness and locality conditions which yield a model of computation which is still Turing complete but which may also be shown to satisfy the Invariance Thesis with respect to the reference model $\mathfrak{R}$?

---

[13] In fact it seems likely that $\mathfrak{G}$ is a *not* a reasonable model. For it is easy to see that Gandy machines can simulate Schönage's storage modification machines with constant factor time overhead. As the latter model admits a *linear time* (i.e. $O(n)$) multiplication algorithm, Gandy machines do as well. And as Cook [6] (p. 403) observes, "we are forced to conclude that either multiplication is easier than we thought or that Schönhage's machines cheat". See Schöngage [22] (pp. 506–507) and van Emde Boas [26] (p. 110) for similar observations.

References

1. Arora, S., Barak, B.: Computational Complexity: A Modern Approach. University Press, Cambridge (2009)
2. Bertoni, A., Mauri, G., Sabadini, N.: Simulations among classes of random access machines and equivalence among numbers succinctly represented. In: Annals of Discrete Mathematics. North-Holland Mathematics Studies, vol. 109, pp. 65–89. Elsevier (1985)
3. Buss, R.: Bounded Arithmetic. Bibliopolis, Naples (1986)
4. Clote, P.: Boolean Functions and Computation Models. Springer, Heidelberg (2002)
5. Cobham, A.: The intrinsic computational difficulty of functions. In: Proceedings of the Third International Congress for Logic, Methodology and Philosophy of Science, Amsterdam, pp. 24–30. North-Holland (1965)
6. Cook, S.: An overview of computational complexity. Commun. ACM **26**, 401–408 (1983)
7. Cook, S., Aanderaa, S.: On the minimum computation time of functions. Trans. Am. Math. Soc. **142**, 291–314 (1969)
8. Cook, S., Reckhow, R.: Time bounded random access machines. J. Comput. Syst. Sci. **7**(4), 354–375 (1973)
9. Dean, W.: Computational complexity theory. In: Zalta, E.N. (ed.) The Stanford Encyclopedia of Philosophy. Fall 2015 Ed. (2015)
10. Fürer, M.: Faster integer multiplication. SIAM J. Comput. **39**(3), 979–1005 (2009)
11. Gandy, R.: Church's thesis and principles for mechanisms. In: Barwise, H.K.J., Kunen, K. (eds.) The Kleene Symposium, vol. 101, pp. 123–148. North Holland, Amsterdam (1980)
12. Hartmanis, J., Simon, J.: On the power of multiplication in random access machines. In: 1974 IEEE Conference Record of 15th Annual Symposium on Switching and Automata Theory, pp. 13–23. IEEE (1974)
13. Hartmanis, J., Stearns, R.: On the computational complexity of algorithms. Trans. Am. Math. Soc. **117**(5), 285–306 (1965)
14. Immerman, N.: Descriptive Complexity. Springer, Heidelberg (1999)
15. Knuth, D.: The Art of Computer Programming, vol. I–III. Addison Wesley, Boston (1973)
16. Kolmogorov, A., Uspensky, V.: To the definition of algorithms. Uspekhi Mat. Nauk **13**(4), 3–28 (1958)
17. Kreisel, G.: Informal rigour and completeness proofs. In: Problems in the Philosophy of Mathematics, pp. 138–186 (1967)
18. Kreisel, G.: Which number theoretic problems can be solved in recursive progressions on Π_1^1-paths through $\mathcal{O}$? J. Symbolic Logic **37**(2), 311–334 (1972)
19. Kreisel, G.: Church's thesis and the ideal of informal rigour. Notre Dame J. Formal Logic **28**(4), 499–519 (1987)
20. Rose, H.: Subrecursion: Functions and Hierarchies. Clarendon Press, Oxford (1984)
21. Savitch, W., Stimson, M.: Time bounded random access machines with parallel processing. J. ACM **26**(1), 103–118 (1979)
22. Schönhage, A.: Storage modification machines. SIAM J. Comput. **9**(3), 490–508 (1980)
23. Schönhage, A., Strassen, V.: Schnelle Multiplikation großer Zahlen. Computing **7**(3–4), 281–292 (1971)
24. Sieg, W.: On computability. In: Irvine, A. (ed.) Philosophy of Mathematics, Handbook of the Philosophy of Science, vol. 4, pp. 549–630. North Holland, Amsterdam (2009)

25. Turing, A.: On computable numbers, with an application to the Entscheidungsproblem. Proc. Lond. Math. Soc. **42**(2), 230–265 (1936)
26. van Emde Boas, P.: Space measures for storage modification machines. Inf. Process. Lett. **30**(2), 103–110 (1989)
27. van Emde Boas, P.: Machine models and simulations. In: Van Leeuwen, J. (ed.) Handbook of Theoretical Computer Science (vol. A): Algorithms and Complexity. MIT Press, Cambridge (1990)

Recent Advances in Non-perfect Secret Sharing Schemes

Oriol Farràs[(✉)]

Universitat Rovira i Virgili, Tarragona, Catalonia, Spain
oriol.farras@urv.cat

Abstract. A secret sharing scheme is non-perfect if some subsets of players that cannot recover the secret have partial information about it. This paper is a survey of the recent advances in non-perfect secret sharing schemes. We provide an overview of the techniques for constructing efficient non-perfect secret sharing schemes, bounds on the efficiency of these schemes, and results on the characterization of the ideal ones. We put special emphasis on the connections between non-perfect secret sharing schemes and polymatroids, matroids, information theory, and coding theory.

Keywords: Secret sharing · Non-perfect secret sharing · Ideal secret sharing scheme · Information ratio · Polymatroid · Matroid

1 Introduction

A *secret sharing scheme* is a cryptographic primitive that is used to protect a *secret value* by dividing it into *shares*. Secret sharing is used to prevent both the disclosure and the loss of the secret value. In the typical scenario, each share is sent privately to a different *player*. Then a subset of players is *qualified* if their shares determine the secret value, and *forbidden* if their shares do not contain any information on the secret value. In this work we just consider schemes that are *information-theoretically secure*, and so their security does not rely on any computational assumption. A secret sharing scheme is *perfect* if every set of players is qualified or forbidden.

Blakley [3] and Shamir [31] presented the first secret sharing schemes. These schemes are perfect, and are called *threshold* secret sharing schemes because the qualified subsets are those whose size is greater than a certain threshold. In a perfect secret sharing scheme, the length in bits of each share must be greater than or equal to the length of the secret. The schemes attaining this bound, as the ones of Blakley and Shamir [3,31], are called *ideal*.

In order to circumvent this restriction on the length of the shares, Blakley and Meadows [4] relaxed the privacy requirements of the threshold schemes.

O. Farràs — Supported by the Spanish Government through a Juan de la Cierva grant and TIN2014-57364-C2-1-R, by the European Union through H2020-ICT-2014-1-644024, and by the Government of Catalonia through Grant 2014 SGR 537.

© Springer International Publishing Switzerland 2016
A. Beckmann et al. (Eds.): CiE 2016, LNCS 9709, pp. 89–98, 2016.
DOI: 10.1007/978-3-319-40189-8_9

They presented the first non-perfect schemes, which were called *ramp* secret sharing schemes. The structure of a ramp scheme is described by two thresholds t and r with $t < r$: Every subset with at most t players is forbidden, every subset with at least r players is qualified, and the other subsets have partial information about the secret value. The length of every share is $1/(r - t)$ times the length of the secret, which is also optimal [28]. The threshold and the ramp schemes in [3,4,31] are *linear*, that is, they can be described in terms of linear maps over a finite field. Because of their efficiency and their homomorphic properties, linear secret sharing schemes play a fundamental role in several areas of cryptography such as secure multiparty computation and distributed cryptography (see [1]).

Blakley and Meadows showed that if we allow some subsets to obtain partial information about the secret value, we can then construct schemes with shorter shares. Their construction motivated the search for efficient non-perfect secret sharing schemes, the exploration of their limitations, and their use in cryptographic applications. The connections between information theory, matroid theory and secret sharing found by Fujishige [20,21], Brickell and Davenport [5], and Csirmaz [13], are powerful tools in the study of perfect secret sharing schemes. The extension of these connections to the non-perfect case was initiated twenty years ago [25,27,30], and has been improved in recent works [15–18,32,33]. Many other works such as [7,8,10,11,14,22,23] presented efficient non-perfect secret sharing schemes with interesting cryptographic applications.

This paper is a survey of the recent advances in non-perfect secret sharing schemes. Section 2 is dedicated to the definition of secret sharing and the main efficiency measures. The main constructions of non-perfect schemes are presented in Sect. 3, putting special emphasis on the linear ones. We present the connection between secret sharing schemes and polymatroids and the resulting lower bounds in Sect. 4. Finally, in Sect. 5, we present the results on ideal secret sharing schemes and their connection with matroids.

2   Secret Sharing Schemes

In this survey we use an information-theoretic definition of secret sharing that was introduced in [15,16]. This definition is very general and covers both perfect and non-perfect secret sharing schemes. The secret and the shares of the scheme are defined as discrete random variables, and the information known about the secret is measured using the Shannon entropy function. Before presenting the definition, we need to introduce some notation.

Given a discrete random vector $S = (S_i)_{i \in Q}$ and a set $X \subseteq Q$, the Shannon entropy of the random variable $S_X = (S_i)_{i \in X}$ is denoted by $H(S_X)$. In addition, for such random variables, one can consider the *mutual information* $I(S_X : S_Y) = H(S_X) + H(S_Y) - H(S_{X \cup Y})$. The reader is referred to [12] for a textbook on information theory, and to [1,29] for a survey on secret sharing.

Definition 1. *Let Q be a finite set of* players, *let $p_0 \in Q$ be a distinguished* player, *which is called* dealer, *and take $P = Q \setminus \{p_0\}$. A secret sharing scheme* *on the set P is a discrete random vector $\Sigma = (S_i)_{i \in Q}$ such that $H(S_{p_0}) > 0$ and $H(S_{p_0}|S_P) = 0$.*

In this definition, the random variable S_{p_0} corresponds to the *secret value*, while the random variables $(S_i)_{i \in P}$ correspond to the *shares* of the secret value of the players in P. We use the access function of the scheme [15,16] to describe the amount of information that the shares provide about the secret value.

Definition 2. *An* access function *on a set P is a monotone increasing function $\Phi : \mathcal{P}(P) \to [0,1]$ with $\Phi(\emptyset) = 0$ and $\Phi(P) = 1$.*

Definition 3. *The* access function Φ *of a secret sharing scheme $\Sigma = (S_i)_{i \in Q}$ is the map $\Phi : \mathcal{P}(P) \to [0,1]$ defined by*

$$\Phi(X) = \frac{I(S_{p_0} : S_X)}{H(S_{p_0})}.$$

If $\Phi(X) = 1$, then $I(S_{p_0} : S_X) = H(S_{p_0})$, which implies that the secret value is determined by the shares of the players in X. If $\Phi(X) = 0$, the random variables S_{p_0} and S_X are independent and the shares of the players in X therefore do not provide any information about the secret. For this reason, a set $X \subseteq P$ is called *forbidden* for Φ if $\Phi(X) = 0$, and it is called *qualified* for Φ if $\Phi(X) = 1$. If every subset is forbidden or qualified, we say that the scheme is *perfect*. Accordingly, an access function is said to be *perfect* if it only takes the values 0 and 1. The *gap* of an access function Φ is defined as the minimum distance between a qualified and a forbidden subset, that is, $\mathrm{gap}(\Phi) = \min\{|Y \smallsetminus X| : \Phi(X) = 0, \Phi(Y) = 1\}$. We say that an access function is *rational* if it takes values only in $\mathbb{Q}$.

As a measure of the efficiency of a scheme, we consider the ratio of the maximum length in bits of the shares to the length of the secret value. We use the Shannon entropy as an approximation of the shortest binary codification.

Definition 4. *The* information ratio $\sigma(\Sigma)$ *of a secret sharing scheme $\Sigma = (S_i)_{i \in Q}$ is defined by*

$$\sigma(\Sigma) = \frac{\max_{i \in P} H(S_i)}{H(S_{p_0})}.$$

There exists a secret sharing scheme for every access function [15]. However, the general constructions provide schemes that are not efficient for most of the access functions. Namely, for most of the access functions, the information ratios of the resulting schemes are exponential in the number of players. This is also the situation in perfect secret sharing schemes [2]. The need of more efficient schemes motivated the search for lower bounds on the information ratio. The *optimal information ratio* $\sigma(\Phi)$ of an access function Φ is the infimum of the information ratios of the secret sharing schemes realizing Φ. We say that a secret sharing scheme for Φ is *optimal* if its information ratio attains $\sigma(\Phi)$.

3 Constructions

In this section we present the main non-perfect secret sharing constructions. We present the family of linear secret sharing schemes, the constructions for uniform access functions, and a general method to construct a secret sharing scheme for any access function.

3.1 Linear Secret Sharing Schemes

Definition 5. *Let $\mathbb{K}$ be a finite field and let ℓ be a positive integer. In a $(\mathbb{K}, \ell)$-linear secret sharing scheme, the random variables $(S_i)_{i \in Q}$ are given by surjective $\mathbb{K}$-linear maps $S_i : E \to E_i$, where the uniform probability distribution is taken on E and $\dim E_{p_0} = \ell$.*

We present the linear secret sharing scheme of Blakley and Meadows [4] in Example 1. In the original paper, the scheme was presented using polynomial notation.

Example 1. Let $P = \{1, \ldots, n\}$, and let t, r be integers with $0 \leq t < r \leq n$. Define $g = r - t$. Let $\mathbb{K}$ be a finite field with $|\mathbb{K}| \geq n + g$, and let $y_1, \ldots, y_g, x_1, \ldots, x_n$ be different elements in $\mathbb{K}$. Define $\mathbf{x}_i = (1, x_i, \ldots, x_i^{r-1}) \in \mathbb{K}^r$ for $i \in P$ and $\mathbf{y}_i = (1, y_i, \ldots, y_i^{r-1}) \in \mathbb{K}^r$ for $i = 1, \ldots, g$. Consider the secret sharing scheme $\Sigma = (S_i)_{i \in Q}$ determined by the linear mappings

- $S_{p_0} : \mathbb{K}^r \to \mathbb{K}^g : \mathbf{u} \mapsto (\mathbf{u} \cdot \mathbf{y}_1, \ldots, \mathbf{u} \cdot \mathbf{y}_g)$, and
- $S_i : \mathbb{K}^r \to \mathbb{K} : \mathbf{u} \mapsto \mathbf{u} \cdot \mathbf{x}_i$ for $i \in P$,

and take the uniform probability distribution on $\mathbb{K}^r$.

In a $\mathbb{K}$-linear secret sharing scheme $(S_i)_{i \in Q}$, the random variable S_X is uniform on its support for every $X \subseteq Q$. Because of that, $H(S_X) = \log |\mathbb{K}| \cdot \operatorname{rank} S_X$, and hence $I(S_{p_0} : S_X) = \log |\mathbb{K}| \cdot (\operatorname{rank} S_{p_0} + \operatorname{rank} S_X - \operatorname{rank} S_{X \cup \{p_0\}})$. Therefore, in order to determine the access function of a linear scheme, we just need to compute the ranks of the linear mappings. Taking into account the properties of the Vandermonde matrix, the computation of the access function Φ of the scheme Σ in Example 1 is straightforward. It can be described as follows.

- $\Phi(X) = 0$ if $|X| \leq t$,
- $\Phi(X) = (|X| - t)/g$ if $t < |X| < r$, and
- $\Phi(X) = 1$ if $|X| \geq r$.

This access function is called the (t, r)-*ramp* access function, and the schemes with access functions of this kind are called *ramp* secret sharing schemes. The scheme is $(\mathbb{K}, g)$-linear, it has information ratio $1/g$, and it is optimal [28]. This scheme is an extension of the *threshold* secret sharing scheme of Shamir [31], which corresponds to the case $g = 1$. A scheme has t-*privacy* if the subsets of size at most t are forbidden, and it has r-*reconstruction* if the subsets of size greater than or equal to r are qualified. The (t, r)-*ramp* secret sharing schemes have t-privacy, r-reconstruction, and the subsets of size greater than t and smaller than r obtain a certain amount of information about the secret, which increases at a constant rate $1/g$.

Linear secret sharing schemes have homomorphic properties that are useful for many cryptographic applications. If we share two secrets s_1 and s_2 among a set of players using a linear scheme Σ, each player can obtain a share of $s_1 + s_2$ by just computing the sum of the two shares he holds. Namely, each of these sums is a share of $s_1 + s_2$ by Σ. Moreover, the sharing and reconstruction operations of linear secret sharing schemes are efficient [16]. The reconstruction of the secret value by a set of players can be obtained by solving the linear equations defined by their shares.

We can describe the shares and the secret of the scheme in Example 1 as evaluations of a polynomial of degree at most $r - 1$ at different points. Hence, the secret and the shares are elements of a codeword of a Reed-Solomon code. This connection can be generalized, and we can obtain linear secret sharing schemes from any linear code. Roughly speaking, given a linear code, we can define linear secret sharing schemes in which the shares and the secret are subsets of coordinates of the codewords (see [7, 16] for more details). For many cryptographic applications, the requirements on the secret sharing schemes are t-privacy and r-reconstruction for certain values of t and r, and there are no restrictions for the subsets of size between t and r. In this case, the construction of schemes from linear codes is especially relevant, because r and t can be estimated from the minimum distance of the code and from the minimum distance of its dual, respectively [7].

The scheme in Example 1 has the restriction that $|\mathbb{K}|$ must be greater than or equal to $n + g$. It may be a drawback if we want to share a short secret among a large set of players. Several works [7, 8, 22, 23] study this problem and present non-perfect linear schemes that are defined over smaller fields that can substitute the ramp schemes. They are constructed from algebraic geometry codes and random codes, and have multiplicative properties, such as the one in Example 1. Multiplicative non-perfect schemes are building blocks of several secure multiparty computation protocols [7, 11, 19]. Linear codes have also been used to construct linear schemes with more efficient sharing and secret reconstruction [10, 14].

3.2   Uniform Access Functions

Ramp access functions belong to the family of uniform access functions. We say that an access function is *uniform* if its values only depend on the size of the subset. Yoshida et al. [33] found optimal linear secret sharing schemes for every rational uniform access function. Farràs et al. [15] provided an alternative proof using different techniques, as well as constructions for the non-rational case. The constructions in both works are based on the following observation. Let Σ_1 be a $(\mathbb{K}, \ell_1)$-linear scheme and let Σ_2 be a $(\mathbb{K}, \ell_2)$-linear scheme for some ℓ_1 and ℓ_2. Let Φ_1 and Φ_2 be the respective access functions. If we take a secret $s \in \mathbb{K}^{\ell_1 + \ell_2}$ and we share the first ℓ_1 coordinates with Σ_1 and the remaining ℓ_2

coordinates with Σ_2, we obtain a $(\mathbb{K}, \ell_1 + \ell_2)$-linear secret sharing scheme whose access function is a convex combination of Φ_1 and Φ_2, and its information ratio is also a convex combination of $\sigma(\Sigma_1)$ and $\sigma(\Sigma_2)$. By combining ramp secret sharing schemes in this way, it is possible to construct optimal schemes for every uniform access function [15,33]. A drawback of these optimal schemes is that the secret can be very large for certain access functions. Namely, the schemes are $(\mathbb{K}, \ell)$-linear for a reasonably large finite field $\mathbb{K}$ and attain the optimal information ratio, but ℓ can be very large. Indeed, as in the perfect case, the secret sharing schemes with optimal information ratio are not necessarily the most efficient ones. Constructions of non-optimal linear secret sharing schemes with shorter shares for every rational uniform access functions were presented in [15,32].

This problem leads to the search for bounds on the length the shares instead of bounds on the information ratio. For ramp access functions, constructing linear secret sharing schemes over finite fields of size smaller than $n + r - t - 1$ is an open problem that is related to the maximum distance separable codes conjecture. Cascudo, Cramer and Xing [6] studied the family of secret sharing schemes with t-privacy and r-reconstruction. They found general lower bounds on the length of the shares that only depend on t, r and n; and specific lower bounds for $(\mathbb{K}, \ell)$-linear secret sharing schemes with information ratio $1/\ell$.

3.3   General Constructions

The access function of a linear secret sharing scheme is a rational function. Conversely, every rational access function admits a linear secret sharing scheme.

Theorem 1 [15]. *Every access function admits a secret sharing scheme. Moreover, every rational access function admits a $(\mathbb{K}, \ell)$-linear secret sharing schemes for every finite field $\mathbb{K}$ and large enough ℓ.*

Next, we present a sketch the proof of this theorem, which is constructive. For a rational access function Φ, the secret of the scheme is set to be in $\mathbb{K}^\ell$, where ℓ is determined by Φ. The smaller the minimum of the non-zero values of $|\Phi(X \cup \{y\}) - \Phi(X)|$ for $X \subseteq P$ and $y \notin P \smallsetminus X$ is, the larger ℓ is. For every subset of players $X \subseteq P$, we define a secret sharing scheme Σ_X that shares among X a part of the secret that is proportional to $\Phi(X)$. Then, for every X in a certain family of subsets $\mathcal{A} \subseteq \mathcal{P}(P)$, the secret is shared independently by means of Σ_X. Access functions with non-rational values require constructions that are not linear. The general construction in [15] follows the same idea, but the schemes Σ_X are substituted by schemes in which the distribution of the secret, which is chosen in a special way, may be not uniform.

Ishai et al. [24] studied the representation of rational access functions for a fixed secret size. By using Markov chain techniques, they presented non-linear constructions for every uniform rational access function, and another construction for every rational access function. For every perfect access function Φ, Benaloh and Leichter [2] presented a way to transform any monotone Boolean formula for Φ to a secret sharing scheme for Φ. This construction has not been extended to the non-perfect case yet.

4 Lower Bounds on the Information Ratio

Despite the limitations described above, the information ratio is the most common parameter for the study of the efficiency of secret sharing schemes. Many results on the information ratio are based on information inequalities, including Shannon inequalities, rank inequalities, and Non-Shannon inequalities. Csirmaz [13] found a connection between perfect secret sharing schemes and polymatroids that translates the search for the optimal information ratio into combinatorial and linear programming problems. The description of the structure of non-perfect secret sharing schemes by means of their access function allowed to extend these techniques to the non-perfect case [15].

Definition 6. *A polymatroid is a pair $\mathcal{S} = (Q, f)$ formed by a finite set Q, the ground set, and a rank function $f \colon \mathcal{P}(Q) \to \mathbb{R}$ satisfying the following properties.*

- *$f(\emptyset) = 0$.*
- *f is monotone increasing: if $X \subseteq Y \subseteq Q$, then $f(X) \le f(Y)$.*
- *f is submodular: $f(X \cup Y) + f(X \cap Y) \le f(X) + f(Y)$ for every $X, Y \subseteq Q$.*

If f is integer valued and $f(X) < |X|$ for every $X \subseteq Q$, then $\mathcal{S}$ is called a matroid.

Fujishige [20,21] showed that for every random vector $(S_i)_{i \in Q}$, the map $h \colon \mathcal{P}(Q) \to \mathbb{R}$ defined by $h(X) = H(S_X)$ is the rank function of a polymatroid with ground set Q. Since secret sharing schemes are random vectors, every secret sharing scheme defines a polymatroid. Conversely, it is not known which polymatroids are defined by secret sharing schemes. From certain polymatroids we can define an access function as follows.

Definition 7. *Let Φ be an access function on P and let $\mathcal{S} = (Q, f)$ be a polymatroid. Then $\mathcal{S}$ is a Φ-polymatroid if*

$$\Phi(X) = \frac{f(p_0) + f(X) - f(\{p_0\} \cup X)}{f(p_0)}$$

for every $X \subseteq P$.

The polymatroid defined by a secret sharing scheme Σ with access function Φ is a Φ-polymatroid. For perfect access functions Φ, Martí-Farré and Padró [26] found a lower bound on $\sigma(\Phi)$ that is called $\kappa(\Phi)$ and that is based on the connection between secret sharing schemes and polymatroids. The parameter $\kappa(\Phi)$ was later generalized to the non-perfect case in [15] as follows.

Definition 8. *For every access function Φ on P, we define*

$$\kappa(\Phi) = \inf\{\sigma_0(\mathcal{S}) : \mathcal{S} \text{ is a } \Phi\text{-polymatroid}\},$$

where $\sigma_0(\mathcal{S}) = \max_{x \in P} f(x)/f(p_0)$.

Note that for every secret sharing scheme Σ, the polymatroid $\mathcal{S}$ associated to Σ satisfies $\sigma_0(\mathcal{S}) = \sigma(\Sigma)$. Hence, $\kappa(\Phi) \leq \sigma(\Phi)$ for every access function Φ. In the next lemma, we present a basic lower bound on κ [15,27,30], which is attained by the scheme in Example 1.

Lemma 1. *For every access function Φ, $\kappa(\Phi) \geq 1/gap(\Phi)$.*

The search for general lower bounds and the search for ideal access functions (see Sect. 5) induced the study of the family of the access functions with constant increment. In the next theorem we present the best known lower bound on the information ratio of secret sharing schemes [16], which is an extension of a result of Csirmaz [13, Theorem 3.2].

Definition 9. *An access function Φ on P has* constant increment μ *if $\Phi(X \cup \{y\}) - \Phi(X) \in \{0, \mu\}$ for every $X \subset P$ and $y \in P$. In this situation, $\mu = 1/k$ for some positive integer k.*

Theorem 2. *For every positive integer k and for infinitely many positive integers n, there exists an access function Φ_n with constant increment $1/k$ on a set of size $n + k - 1$ satisfying $\kappa(\Phi_n) \geq n/(2k \log n)$.*

For every uniform access function Φ, $\kappa(\Phi) = \sigma(\Phi)$ [9,15]. However, in general, κ is not a tight bound on σ. Indeed, $\kappa(\Phi) \leq n$ for every access function Φ [16]. Up to now, all the lower bounds on the information ratio of non-perfect secret sharing schemes are a consequence of the Shannon information inequalities.

5   Ideal Secret Sharing Schemes

A perfect secret sharing scheme Σ is *ideal* if $\sigma(\Sigma) = 1$. In this case, we say that its access function is also *ideal*. Brickell and Davenport [5] proved that every ideal perfect secret sharing scheme determines a matroid. Because of this result, matroids became a crucial tool for the study and for the construction of efficient secret sharing schemes. The connection between non-perfect secret sharing schemes and matroids, and the notion of ideality in the non-perfect case have been studied in [16–18,25,30]. The following definition of ideal secret sharing scheme was introduced in [16], and it is equivalent to the definitions in [25,30].

Definition 10. *A secret sharing scheme $\Sigma = (S_i)_{i \in Q}$ is* ideal *if its access function has constant increment $1/k$ for some $k > 0$ and $H(S_i) = H(S_{p_0})/k$ for every $i \in P$. In this case, its access function is called* ideal *as well.*

If Φ is an ideal access function, then Φ has constant increment $1/\text{gap}(\Phi)$. Next, we extend the notion of matroid port.

Definition 11. *Let P, P_0 be a pair of nonempty, disjoint sets and $\mathcal{M} = (P \cup P_0, r)$ a matroid such that $r(P_0) = |P_0| = $ and $r(P) = r(P \cup P_0)$. The* port *of the matroid $\mathcal{M}$ at the set P_0 is the access function Φ on P defined by*

$$\Phi(X) = \frac{r(P_0) + r(X) - r(P_0 \cup X)}{r(P_0)}.$$

Now we can state the extension of the Brickell and Davenport Theorem [5] to the non-perfect case [18].

Theorem 3. *The access function of every ideal secret sharing scheme is a generalized matroid port.*

If ϕ is a port of a matroid $\mathcal{M} = (P \cup P_0, r)$ at P_0, and $\mathcal{M}$ is $\mathbb{K}$-linearly representable for a finite field $\mathbb{K}$, then ϕ admits an ideal linear secret sharing scheme (see [16]). However, as in the perfect case, not all matroid ports admit ideal secret sharing schemes. The characterization of the matroid ports that are ideal access functions is an open problem. If an access function Φ has constant increment k and it is not a matroid port at a set of size k, then $\kappa(\Phi) \geq 3/(2k)$ [17].

References

1. Beimel, A.: Secret-sharing schemes: a survey. In: Chee, Y.M., Guo, Z., Ling, S., Shao, F., Tang, Y., Wang, H., Xing, C. (eds.) IWCC 2011. LNCS, vol. 6639, pp. 11–46. Springer, Heidelberg (2011)
2. Benaloh, J.C., Leichter, J.: Generalized secret sharing and monotone functions. In: Goldwasser, S. (ed.) CRYPTO 1988. LNCS, vol. 403, pp. 27–35. Springer, Heidelberg (1990)
3. Blakley, G.R.: Safeguarding cryptographic keys. In: AFIPS Conference Proceedings, vol. 48, pp. 313–317 (1979)
4. Blakley, G.R., Meadows, C.: Security of ramp schemes. In: Blakely, G.R., Chaum, D. (eds.) CRYPTO 1984. LNCS, vol. 196, pp. 242–268. Springer, Heidelberg (1985)
5. Brickell, E.F., Davenport, D.M.: On the classification of ideal secret sharing schemes. J. Cryptology **4**, 123–134 (1991)
6. Cascudo, I., Cramer, R., Xing, C.: Bounds on the threshold gap in secret sharing and its applications. IEEE Trans. Inf. Theory **59**, 5600–5612 (2013)
7. Chen, H., Cramer, R., Goldwasser, S., de Haan, R., Vaikuntanathan, V.: Secure computation from random error correcting codes. In: Naor, M. (ed.) EUROCRYPT 2007. LNCS, vol. 4515, pp. 291–310. Springer, Heidelberg (2007)
8. Chen, H., Cramer, R., de Haan, R., Pueyo, I.C.: Strongly multiplicative ramp schemes from high degree rational points on curves. In: Smart, N.P. (ed.) EURO-CRYPT 2008. LNCS, vol. 4965, pp. 451–470. Springer, Heidelberg (2008)
9. Chen, Q., Yeung, R.W.: Two-partition-symmetrical entropy function regions. In: ITW, pp. 1–5 (2013)
10. Cramer, R., Damgård, I.B., Döttling, N., Fehr, S., Spini, G.: Linear secret sharing schemes from error correcting codes and universal hash functions. In: Oswald, E., Fischlin, M. (eds.) EUROCRYPT 2015. LNCS, vol. 9057, pp. 313–336. Springer, Heidelberg (2015)
11. Cramer, R., Damgård, I.B., de Haan, R.: Atomic secure multi-party multiplication with low communication. In: Naor, M. (ed.) EUROCRYPT 2007. LNCS, vol. 4515, pp. 329–346. Springer, Heidelberg (2007)
12. Cover, T.M., Thomas, J.A.: Elements of Information Theory. Wiley, New York (1991)
13. Csirmaz, L.: The size of a share must be large. J. Cryptology **10**, 223–231 (1997)
14. Druk, E., Ishai, Y.: Linear-time encodable codes meeting the Gilbert-Varshamov bound and their cryptographic applications. In: ITCS, pp. 169–182 (2014)

15. Farràs, O., Hansen, T., Kaced, T., Padró, C.: Optimal non-perfect uniform secret sharing schemes. In: Garay, J.A., Gennaro, R. (eds.) CRYPTO 2014, Part II. LNCS, vol. 8617, pp. 217–234. Springer, Heidelberg (2014)
16. Farràs, O., Hansen, T., Kaced, T., Padró, C.: On the information ratio of non-perfect uniform secret sharing schemes. Cryptology ePrint Archive 2014/124 (2014)
17. Farràs, O., Martín, S., Padró, C.: A note on ideal non-perfect secret sharing schemes. Cryptology ePrint Archive 2016/348 (2016)
18. Farràs, O., Padró, C.: Extending Brickell-Davenport theorem to non-perfect secret sharing schemes. Des. Codes Cryptogr. **74**(2), 495–510 (2015)
19. Franklin, M., Yung, M.: Communication complexity of secure computation. In: STOC, pp. 699–710 (1992)
20. Fujishige, S.: Polymatroidal dependence structure of a set of random variables. Inf. Control **39**, 55–72 (1978)
21. Fujishige, S.: Entropy functions and polymatroids–combinatorial structures in information theory. Electron. Comm. Japan **61**, 14–18 (1978)
22. Geil, O., Martin, S., Martínez-Peñas, U., Matsumoto, R., Ruano, D.: On asymptotically good ramp secret sharing schemes (2015). arXiv.org/abs/1502.05507
23. Geil, O., Martin, S., Matsumoto, R., Ruano, D., Luo, Y.: Relative generalized hamming weights of one-point algebraic geometric codes. IEEE Trans. Inform. Theory **60**, 5938–5949 (2014)
24. Ishai, Y., Kushilevitz, E., Strulovich, O.: Lossy chains and fractional secret sharing. In: STACS 2013. LIPICS, vol. 20, pp. 160–171 (2013)
25. Kurosawa, K., Okada, K., Sakano, K., Ogata, W., Tsujii, S.: Nonperfect secret sharing schemes and matroids. In: Helleseth, T. (ed.) EUROCRYPT 1993. LNCS, vol. 765, pp. 126–141. Springer, Heidelberg (1994)
26. Martí-Farré, J., Padró, C.: On secret sharing schemes, matroids polymatroids. J. Math. Cryptol. **4**, 95–120 (2010)
27. Okada, K., Kurosawa, K.: Lower bound on the size of shares of nonperfect secret sharing schemes. In: Safavi-Naini, R., Pieprzyk, J.P. (eds.) ASIACRYPT 1994. LNCS, vol. 917, pp. 33–41. Springer, Heidelberg (1995)
28. Ogata, W., Kurosawa, K., Tsujii, S.: Nonperfect secret sharing schemes. In: Zheng, Y., Seberry, J. (eds.) AUSCRYPT 1992. LNCS, vol. 718, pp. 56–66. Springer, Heidelberg (1993)
29. Padró, C.: Lecture Notes in Secret Sharing. Cryptology ePrint Archive 2012/674
30. Paillier, A.: On ideal non-perfect secret sharing schemes. In: Christianson, B., Crispo, B., Lomas, M., Roe, M. (eds.) Security Protocols 1997. LNCS, vol. 1361, pp. 207–216. Springer, Heidelberg (1998)
31. Shamir, A.: How to share a secret. Commun. ACM **22**, 612–613 (1979)
32. Yoshida, M., Fujiwara, T.: Secure construction for nonlinear function threshold ramp secret sharing. In: IEEE International Symposium on Information Theory, ISIT 2007, pp. 1041–1045 (2007)
33. Yoshida, M., Fujiwara, T., Fossorier, M.: Optimum general threshold secret sharing. In: Smith, A. (ed.) ICITS 2012. LNCS, vol. 7412, pp. 187–204. Springer, Heidelberg (2012)

A Computational Approach
to the Borwein-Ditor Theorem

Aleksander Galicki and André Nies[⊠]

Department of Computer Science, University of Auckland, Auckland, New Zealand
andre@cs.auckland.ac.nz

Abstract. Borwein and Ditor (Canadian Math. Bulletin 21 (4), 497–498, 1978) proved the following. Let $\mathcal{A} \subset \mathbb{R}$ be a measurable set of positive measure and let $\langle r_m \rangle_{m \in \omega}$ be a null sequence of real numbers. For almost all $z \in \mathcal{A}$, there is m such that $z + r_m \in \mathcal{A}$.

In this note we mainly consider the case that $\mathcal{A}$ is Π_1^0 and the null sequence $\langle r_m \rangle_{m \in \omega}$ is computable. We show that in this case every Oberwolfach random real $z \in \mathcal{A}$ satisfies the conclusion of the theorem. We extend the result to finitely many null sequences. The conclusion is now that for almost every $z \in \mathcal{A}$, the same m works for each null sequence.

We indicate how this result could separate Oberwolfach randomness from density randomness.

1   Introduction

Our paper is based on the following result, which extends a previous weaker result by Kestelman [11].

Theorem 1 (D. Borwein and S. Z. Ditor [4], Theorem 1(i)). *Suppose $\mathcal{A} \subset \mathbb{R}$ is a measurable set of positive measure and $\langle r_m \rangle_{m \in \omega}$ is a sequence of real numbers converging to 0. For almost all $z \in \mathcal{A}$, there is an m such that $z + r_m \in \mathcal{A}$.*

Since one can consider the tails of a given null sequence of reals, for almost every $z \in \mathcal{A}$ there are in fact infinitely many m such that $z + r_m \in \mathcal{A}$. (This is the form in which they actually stated the result). We thank Ostaszewski for pointing out the Borwein-Ditor theorem to Nies during his visit at the London School of Ecomonics in June 2015. Ostaszewski's 2007 book with Bingham provides some background related to topology [3].

Nies' colloquium at LSE was about the study of effective versions of "almost everywhere" theorems via algorithmic randomness. The goal for that direction of study is to pin down the level of algorithmic randomness needed for a point x so that the conclusion of a particular effective version of the theorem holds. For instance, Pathak et al. [15] study effective versions of the Lebesgue differentiation theorem, Brattka et al. [5] look at the a.e. differentiability of nondecreasing functions, Galicki and Turetsky [10] study the a.e. differentiability of Lipschitz functions on $\mathbb{R}^n$, and Miyabe et al. [13] consider the Lebesgue density theorem, recalled in Theorem 4 below.

© Springer International Publishing Switzerland 2016
A. Beckmann et al. (Eds.): CiE 2016, LNCS 9709, pp. 99–104, 2016.
DOI: 10.1007/978-3-319-40189-8_10

Unless stated otherwise, we will consider effectively closed (i.e. Π_1^0) sets $\mathcal{A} \subseteq$ $\mathbb{R}$. Without also imposing an effectiveness condition on the null sequences, the points z for which the Borwein-Ditor property holds for all Π_1^0 sets are precisely the 1-generics. Recall that a real is 1-generic if it is not on the boundary of any Σ_1^0 set.

Proposition 2. $z \in \mathbb{R}$ is 1-generic $\Longleftrightarrow$ for every Π_1^0 set $\mathcal{A} \subset \mathbb{R}$ containing z and for every null sequence of real numbers $\langle r_m \rangle_{m \in \omega}, z \in \mathcal{A} + r_m$ for some m.

Proof. ($\Rightarrow$) Suppose $z \notin \mathcal{A} + r_m$ for all m. Then z belongs to the boundary of the complement of $\mathcal{A}$, $\mathcal{B} = \mathbb{R} \setminus \mathcal{A}$, a Σ_1^0 class.
($\Leftarrow$) Suppose $z \in \mathbb{R}$ is not 1-generic. Then it belongs to the boundary of some Σ_1^0 set $\mathcal{B} \subset \mathbb{R}$. Let $\langle z_m \rangle_{m \in \omega}$ be a sequence of points in $\mathcal{B}$ converging to z. Define $r_m = z - z_m$ for all m. Then $\mathcal{A} = \mathbb{R} \setminus \mathcal{B}$ is a Π_1^0 class, $z \in \mathcal{A}$, $\langle r_m \rangle_{m \in \omega}$ is a sequence converging to 0, and $z \notin \mathcal{A} + r_m$ for all m.

2 Comparison of Lebesgue Density and the Borwein-Ditor Property

The definitions below follow [2]. Let λ denote Lebesgue measure on $\mathbb{R}$.

Definition 3. *We define the lower Lebesgue density of a set $\mathcal{C} \subseteq \mathbb{R}$ at a point z to be the quantity*

$$\underline{\varrho}(\mathcal{C}|z) := \liminf_{\gamma, \delta \to 0^+} \frac{\lambda([z - \gamma, z + \delta] \cap \mathcal{C})}{\gamma + \delta}.$$

Note that $0 \leq \underline{\varrho}(\mathcal{C}|z) \leq 1$.

Theorem 4 (Lebesgue [12]). *Let $\mathcal{C} \subseteq \mathbb{R}$ be a measurable set. Then $\underline{\varrho}(\mathcal{C}|z) = 1$ for almost every $z \in \mathcal{C}$.*

The Borwein-Ditor theorem is analogous to the Lebesgue density theorem. Both results say that for almost every point in a measurable class there are, in a specific sense, many arbitrarily close other points in the class.

An open set $\mathcal{C}$ clearly has lower Lebesgue density 1 at each of its members. Thus, the simplest non-trivial case is when $\mathcal{C}$ is closed. We say that a real $z \in [0, 1]$ is a density-one point if $\underline{\varrho}(\mathcal{C}|z) = 1$ for every effectively closed class $\mathcal{C}$ containing z. Similar to the implication ($\Rightarrow$) of Proposition 2, every 1-generic is a density-one point. So being a density-one point is by itself not a randomness notion, and neither is the Borwein-Ditor property for effectively closed sets. In both cases, to remedy this one has to add as an additional condition that the real is Martin-Löf random.

Definition 5. *Let $z \in \mathbb{R}$ be ML-random. We say that z is density random if z is also a density-one point.*

Definition 6. *Let z be ML-random. We say that z is* Borwein-Ditor (BD) *random if for each Π_1^0 set $\mathcal{A} \subseteq \mathbb{R}$ with $z \in \mathcal{A}$, and each computable null sequence of reals $\langle r_m \rangle_{m \in \omega}$, there is an m such that $z + r_m \in \mathcal{A}$.*

Neither of the two randomness notions is equivalent to ML-randomness, because the least element of a non-empty effectively closed set of ML-randoms is neither density random nor BD-random. Density randomness is in fact known to be stronger than difference randomness (i.e. ML-randomness together with Turing incompleteness) by Bienvenu et al. [2] together with Day and Miller [6]. Much less is known at present about the placement of BD-randomness within the established notions.

3   Oberwolfach Randomness Implies BD Randomness

> Ah! the ancient pond
> as a frog takes the plunge
> sound of water
> (Matsuo Basho)

To simplify notation, we identify the unit interval with Cantor space $^\omega 2$ in what follows, ignoring dyadic rationals. For a string σ, as usual by $[\sigma]$ we denote the corresponding basic dyadic interval; for example [101] denotes the interval $[5/8, 3/4]$.

Bienvenu et al. [1] introduced Oberwolfach (OW) randomness, and also gave the following equivalent definition. A *left-c.e. bounded test* is a descending sequence $\langle \mathcal{V}_n \rangle$ of uniformly Σ_1^0 classes in Cantor space such that for some nondecreasing computable sequence of rationals $\langle \beta_s \rangle$ with $\beta = \sup_s \beta_s < \infty$, we have $\lambda(\mathcal{V}_n) \leq \beta - \beta_n$ for all n. Z is *OW-random* iff Z passes each such test in the sense that $Z \notin \bigcap_n \mathcal{V}_n$.

OW-randomness implies density randomness [1]. The converse implication remains unknown. The question is intriguing. All the equivalent characterizations of OW-randomness are within the, by now, almost classical framework of computability and randomness [7,14]. For instance, if Z is ML-random, Z is OW-random iff it does not compute every K-trivial set [1]. On the other hand, the seemingly very close notion of density randomness is defined analytically, and mainly has analytical characterizations such as via differentiability of interval-c.e. functions in [13, Theorem 4.2].

Using Theorem 1 it is easy to check that weak 2-randomness implies BD-randomness. We show that the much weaker notion of OW-randomness already implies BD-randomness.

While density and BD randomness are analogous, it seems unlikely that density implies BD. This provides evidence that OW-randomness is strictly stronger than density randomness.

Theorem 7. *Let Z be Oberwolfach random. Then Z is BD random.*

Proof. Suppose we are given a Π_1^0 class $\mathcal{P} \subseteq {}^\omega 2$ with $Z \in \mathcal{P}$, and a computable null sequence of reals $\langle r_m \rangle_{m \in \omega}$. We may assume that $r_m \le 2^{-m}$. Let $\langle \sigma_m \rangle_{m \in \omega}$ be a computable prefix-free sequence of strings such that $\mathcal{S} = {}^\omega 2 \setminus \mathcal{P} = [\{\sigma_m \colon m \in \mathbb{N}\}]^{\prec}$, and let $\mathcal{S}_m = [\sigma_0, \ldots, \sigma_{m-1}]^{\prec}$, the class of all bit sequences extending one of the σ_i. Let $q(m) = 1 + \max(m, \max_{i < m} |\sigma_m|)$. Now define a left-c.e. bounded test by

$$G_m = \bigcap_{i \le q(m)} (\mathcal{S} + r_m) \setminus \mathcal{S}_m.$$

Clearly this is a descending sequence of uniformly Σ_1^0 sets. Let

$$\beta = \lambda \mathcal{S} \text{ and } \beta_m = \lambda \mathcal{S}_m - m2^{-m}$$

so that $\beta = \sup_m \beta_m$.

Claim. $\lambda G_m \le \beta - \beta_m$.

We actually show this bound for $\mathcal{S} + r_{q(m)} \setminus \mathcal{S}_m$ instead of G_m. Since $A \setminus C \subseteq (A \setminus B) \cup (B \setminus C)$ for sets A, B, C, and by the translation invariance of λ,

$$\lambda(\mathcal{S} + r_{q(m)} \setminus \mathcal{S}_m) \le \lambda(\mathcal{S} \setminus \mathcal{S}_m) + \lambda(\mathcal{S}_m + r_{q(m)} \setminus \mathcal{S}_m).$$

Recall that $r_k \le 2^k$. Hence by definition of $q(m)$, for each $i < m$ we have

$$\lambda([\sigma_m] + r_{q(m)} \setminus [\sigma_m]) \le 2^{-q(m)}.$$

Therefore $\lambda(\mathcal{S}_m + r_{q(m)} \setminus \mathcal{S}_m) \le m2^{-q(m)} \le m2^{-m}$ as required for the claim.

If $Z + r_n \notin \mathcal{P}$ for each n then $Z \in \bigcap_m G_m$, so Z is not OW-random.

We note that this proof works in much greater generality for an abelian group $(S, +)$ that is also a computable probability space (S, μ) with a translation invariant measure, such that $\lim_{r \to 0} \mu((A + r) \triangle A) = 0$ effectively for every basic open set A. For instance, the general theorem also applies to Cantor space with the usual ultrametric and the group structure of the 2-adic integers $(\mathbb{Z}_2, +)$.

Finally, similar to [4] we extend the foregoing theorem to the case of finitely many null sequences, and show that for an OW-random Z, one position works for all of them. This is in the spirit of multiple recurrence in ergodic theory, initiated by Furstenberg and others in the 1970s [9].

Theorem 8. *Let Z be Oberwolfach random. For each Π_1^0 class $\mathcal{P} \subseteq {}^\omega 2$ with $Z \in \mathcal{P}$, and k many computable null sequences of reals $\langle r_{m,v} \rangle_{m \in \omega}$, $0 \le v < k$, there is m such that $Z + r_{m,v} \in \mathcal{P}$ for each $v < k$.*

Proof. We may assume that $r_{m,v} \le 2^{-m}$ for each v. Let $\langle \sigma_m \rangle_{m \in \omega}$, $\mathcal{S}_m$, $q(m)$ and β_m be defined as above. Let

$$G_m = \bigcap_{i \le q(m)} \bigcup_{v < k} (\mathcal{S} + r_{i,v}) \setminus \mathcal{S}_m.$$

Claim. $\lambda G_m \leq k(\beta - \beta_m)$.

We actually show this bound for $\bigcup_{v<k}(\mathcal{S} + r_{q(m),v}) \setminus \mathcal{S}_m$ instead of G_m. By the translation invariance of λ,

$$\lambda \bigcup_{v<k}(\mathcal{S} + r_{q(m),v}) \setminus \mathcal{S}_m \leq k\lambda(\mathcal{S} \setminus \mathcal{S}_m) + \lambda(\bigcup_{v<k} \mathcal{S}_m + r_{q(m),v} \setminus \mathcal{S}_m).$$

By the definition of $q(m)$, we have

$$\lambda(\bigcup_{v<k} \mathcal{S}_m + r_{q(m),v} \setminus \mathcal{S}_m) \leq km2^{-q(m)} \leq km2^{-m},$$

which establishes the claim.

If for each n there is $v < k$ such that $Z + r_{n,v} \notin \mathcal{P}$, then $Z \in \bigcap_m G_m$, so Z is not OW-random.

4 Open Questions

Due to the novelty of the concept of BD-randomness, a number of natural questions remain; they are not necessarily hard. Our first two questions have been tried by a number of researchers; the third has not been considered in any detail so far.

Question 9. Does density randomness imply BD randomness?

Question 10. Does BD-randomness imply difference randomness?

Question 11. Does lowness for BD-randomness coincide with lowness for ML-randomness?

Miyabe et al. [13, Theorem 2.6] show that lowness for density randomness coincides with lowness for ML-randomness. The containment "$\subseteq$" is immediate from the result of Downey et al. [8] that Low(W2R, MLR) = Low(MLR) (where W2R is the class of weakly 2-randoms). This proof works the same for BD-randomness. Thus, low for BD-random implies low for ML-random. However, the proof of the converse containment in the case of density randomness cannot be adapted in any obvious way, because we now have to consider a null sequence computable in the oracle.

Acknowledgement. Research supported by the Marsden fund of New Zealand and the Lion foundation of New Zealand.

References

1. Bienvenu, L., Greenberg, N., Kučera, A., Nies, A., Turetsky, D.: Coherent randomness tests and computing the K-trivial sets. J. Eur. Math. Soc. **18**, 773–812 (2016)
2. Bienvenu, L., Hölzl, R., Miller, J., Nies, A.: Denjoy, Demuth, and Density. J. Math. Log. **14**, 1–35 (2014). 1450004
3. Bingham, N., Ostaszewski, A.: Homotopy and the Kestelman-Borwein-Ditor Theorem. London School of Economics and Political Science, London (2007)
4. Borwein, D., Ditor, S.: Translates of sequences in sets of positive measure. Can. Math. Bull. **21**(4), 497–498 (1978)
5. Brattka, V., Miller, J., Nies, A.: Randomness and differentiability. Trans. AMS **368**, 581–605 (2016). arXiv version at http://arxiv.org/abs/1104.4465
6. Day, A.R., Miller, J.S.: Density, forcing and the covering problem. Math. Res. Lett. **22**(3), 719–727 (2015)
7. Downey, R., Hirschfeldt, D.: Algorithmic Randomness and Complexity, pp. 1–855. Springer, Berlin (2010)
8. Downey, R., Nies, A., Weber, R., Yu, L.: Lowness and Π_2^0 nullsets. J. Symb. Log. **71**(3), 1044–1052 (2006)
9. Furstenberg, H.: Recurrence in Ergodic Theory and Combinatorial Number Theory. Princeton University Press, Princeton (2014)
10. Galicki, A., Turetsky, D.: Randomness and differentiability in higher dimensions. Notre Dame J. of Formal logic (2017, to appear). arXiv preprint arXiv:1410.8578
11. Kestelman, H.: The convergent sequences belonging to a set. J. London Math. Soc. **1**(2), 130–136 (1947)
12. Lebesgue, H.: Sur les intégrales singulières. Ann. Fac. Sci. Toulouse Sci. Math. Sci. Phys. (3) **1**, 25–117 (1909)
13. Miyabe, K., Nies, A., Zhang, J.: Using almost-everywhere theorems from analysis to study randomness. Bull. Symb. Log. (2016, to appear)
14. Nies, A.: Computability and Randomness. Oxford Logic Guides, vol. 51. Oxford University Press, Oxford (2009). Paperback version 2011
15. Pathak, N., Rojas, C., Simpson, S.G.: Schnorr randomness and the Lebesgue differentiation theorem. Proc. Am. Math. Soc. **142**(1), 335–349 (2014)

Semantic Security and Key-Privacy with Random Split of St-Gen Codes

Danilo Gligoroski[1(✉)] and Simona Samardjiska[2]

[1] Department of Telematics,
NTNU, The Norwegian University of Science and Technology,
Trondheim, Norway
danilog@item.ntnu.no
[2] Faculty of Computer Science and Engineering, UKIM, Skopje, Macedonia
simona.samardjiska@finki.ukim.mk

Abstract. Recently we have defined Staircase-Generator codes (St-Gen codes) and their variant with a random split of the generator matrix of the codes. One unique property of these codes is that they work with arbitrary error sets. In this paper we analyze the semantic security against chosen plaintext attack (IND-CPA) and key-privacy i.e. indistinguishability of public keys under chosen plaintext attack (IK-CPA) of the encryption scheme with random split of St-Gen codes. In a similar manner as it was done by Nojima et al. and later by Yamakawa et al. we show that padding the plaintext with a random bit-string provides IND-CPA and IK-CPA in the standard model. The difference with McEliece scheme is that with our scheme the length of the padded random string is significantly shorter.

Keywords: Public key cryptography · Code based cryptosystems · Semantic security · Key-privacy

1 Introduction

The idea about semantic security against chosen-plaintext attack (i.e., *indistinguishability against chosen-plaintext attack (IND-CPA)*) for a public-key cryptosystem (PKC) was initially presented by Goldwasser and Micali in [4]. By replacing the deterministic encryption with probabilistic one, they showed the existence of public key schemes where the ciphertext does not leak any useful information about the plaintext (except its length). Later, in the work of Belare et al. [2] the semantic security against chosen-plaintext attack was systematized in a broader security perspective in relation with other security notions in public-key encryption schemes. Then, in 2001 we got a definition for yet another security notion: *key-privacy* or *anonymity* in public-key schemes. It was introduced by Bellare et al. in [1]. In a nutshell key-privacy asks that an adversary receiving a ciphertext is not able to determine which specific public-key, out of a set of known public keys was used to produce that ciphertext. Under the assumption that the Decision Diffie-Hellman problem is hard, they have showed that

© Springer International Publishing Switzerland 2016
A. Beckmann et al. (Eds.): CiE 2016, LNCS 9709, pp. 105–114, 2016.
DOI: 10.1007/978-3-319-40189-8_11

El Gamal scheme provides anonymity i.e. key-privacy under chosen-plaintext attack (CPA) and that the Cramer-Shoup scheme provides stronger security, i.e., it provides anonymity under chosen-ciphertext attack (CCA). They have also showed that neither the classical RSA scheme nor the RSA-OAEP does not provide key-privacy. All these schemes do not belong to the so-called family of "post-quantum" crypto schemes since they are vulnerable to attacks with quantum computers.

The McEliece public key scheme [6] was published in 1978 and is based on the theory of error-correcting codes and the NP-hardness of the problem of decoding random linear codes. It is considered as a post-quantum scheme. However, the original scheme does not provide neither CPA nor CCA security, it provides even less a key-privacy. A conversion of McEliece scheme that offers CCA security was proposed by Kobara and Imai in [5] in the random oracles model. The weaker security of CPA, but in the so-called standard model where there is no reference to the random oracles, for a modified McEliece scheme was proposed by Nojima et al. in [8] and later based on that work Yamakawa et al. in [12] showed that Nojima's modification provides also a key-privacy.

Recently we proposed an encryption and signature variant of the McEliece scheme based on Staircase-Generator matrix, a list decoding algorithm, and generalized error sets in [3]. Soon after its initial eprint publication, a distinguisher that distinguishes its public key from random matrices was proposed [11], and recently a very similar distinguishing strategy and an Information Set Decoding (ISD) attack was presented as a full and practical key recovery attack by Moody and Perlner [7]. While the public key schemes based on Staircase-Generator matrices and a list decoding strategy have succumbed to the distinguishing attacks, there are some useful applications for the technique of matrix-embedding in steganography where the Staircase-Generator matrices are not public, but private [9]. In such cases, the matrix-embedding technique with Staircase-Generator matrices almost achieves the information theoretical bound with codelengths that are the smallest known in the literature.

In order to thwart distinguishing and ISD attacks of [7,11] to the encryption scheme defined in [3] we proposed to split and replace the public generator matrix into s randomly generated matrices [10]. With the splitting we made distinguishing attacks improbable to mount, i.e., the probability of the attacker obtaining conditions under which a distinguisher or an ISD attack can be mounted, is close to zero.

2   Definition of Staircase-Generator Codes and Random Split of Their Generator Matrix

Throughout the paper, we will denote by $\mathcal{C} \subseteq \mathbb{F}_2^n$ a binary (n, k) code of length n and dimension k. We will denote the generator matrix of the code by G, and $wt(\mathbf{x})$ will denote the Hamming weight of the word $\mathbf{x}$. We recall some of the basic definitions and properties for St-Gen codes from [3,9] and the types of errors used.

Definition 1. *Let $k_i, n_i \in \mathbb{N}$, and let $k = k_1 + k_2 + \cdots + k_w$ and $n = k + n_1 + n_2 + \cdots + n_w$. Further, let B_i be a random binary matrix of dimension $\sum_{j=1}^{i} k_j \times n_i$. A linear binary (n, k) code $\mathcal{C}$ with the following generator matrix in standard form:*

$$
G = \begin{pmatrix} & & & & & n_1 & & n_2 & & & \\ & & & & k_1 \Big\{ & B_1 & & & & & \\ & & I_k & & & & B_2 & & & & \\ & & & & k_2 \Big\{ & & & \cdots & & B_w & \\ & & & & & 0 & & \cdots & & & \end{pmatrix}
\tag{1}
$$

is called Staircase-Generator code (St-Gen code).

Definition 2. *Let ℓ be a positive integer and let $p_d \in \mathbb{F}_2[x_1, x_2, \ldots, x_\ell]$ be a multivariate polynomial of degree $\geqslant 2$. We say that E_ℓ is an error set if it is the solution set of p_d, i.e., $E_\ell = \{\mathbf{e} \in \mathbb{F}_2^\ell \mid p_d(\mathbf{e}) = 0\}$. We will refer to p_d as the defining polynomial.*

We define the density of the error set E_ℓ to be $D(E_\ell) = |E_\ell|^{1/\ell}$. We will refer to the integer $\ell > 0$ as the granulation of E_ℓ.

In [3] it was proven that if two error sets $E_{\ell_1} \subseteq \mathbb{F}_2^{\ell_1}$, $E_{\ell_2} \subseteq \mathbb{F}_2^{\ell_2}$, have the same density ρ, then $D(E_{\ell_1} \times E_{\ell_2}) = \rho$.

Example 1. 1. Let $E_2 = \{\mathbf{x} \in \mathbb{F}_2^2 \mid wt(\mathbf{x}) < 2\} = \{(0,0), (0,1), (1,0)\}$. Then the error set can be described using the defining polynomial $p_d = x_1 x_2$, and for the density of the error set we have $D(E_2) = |E_2|^{1/2} = 3^{1/2}$.

2. Let $E_4 = \{\mathbf{x} \in \mathbb{F}_2^4 \mid 2 \leq wt(\mathbf{x}) \leq 3\}$. Then, the defining polynomial for E_4 is $p_d = 1 + x_1 x_2 + x_1 x_3 + x_1 x_4 + x_2 x_3 + x_2 x_4 + x_3 x_4$ and the density is $D(E_4) = D(E_4^m) = (\sum_{i=2}^{3} \binom{4}{i})^{1/4} = 10^{1/4}$ for any positive integer m.

The decoding of St-Gen codes relies on the technique of *list decoding*. In list decoding, the decoder is allowed to output a list of possible messages one of which is correct. List decoding can handle a greater number of errors than that allowed by unique decoding. In order for the decoding to be efficient, the size of the resulting list has to be polynomial in the code length. The following Proposition from [3] determines the parameters of a St-Gen code that provide an efficient decoding.

Proposition 1 [3]. *Let $\mathcal{C}$ be any binary (n, k) code and $E \subset \mathbb{F}_2^n$ be an error set of density ρ. Let $\mathbf{w}$ be any word of length n, $W_E = \{\mathbf{w} + \mathbf{e} \mid \mathbf{e} \in E\}$ and let $\mathcal{C}_{W_E}$ denote the set of codewords in W_E. Suppose there exists a codeword $\mathbf{c} \in W_E$. Then the expected number of codewords in $W_E \setminus \{\mathbf{c}\}$ is approximately $\rho^n 2^{k-n}$ for large enough n and k.*

Let E_ℓ be an error set with density ρ where ℓ divides n and $m = n/\ell$. We recall Algorithm 1 from [3], that is an efficient algorithm for decoding a code $\mathcal{C}$, that corrects errors from the set E_ℓ^m.

Algorithm 1. Decoding
___
Input: Vector $\mathbf{y} \in \mathbb{F}_2^n$, and generator matrix G of the form (1).

Output: A list $L_w \subset \mathbb{F}_2^k$ of valid decodings of $\mathbf{y}$.

Procedure:
Let $K_i = k_1 + \cdots + k_i$. Represent $\mathbf{x} \in \mathbb{F}_2^k$ as $\mathbf{x} = \mathbf{x}_1 \parallel \mathbf{x}_2 \parallel \cdots \parallel \mathbf{x}_w$ where each $\mathbf{x}_i$ has length k_i. Similarly, represent $\mathbf{y} \in \mathbb{F}_2^n$, as $\mathbf{y} = \mathbf{y}_0 \parallel \mathbf{y}_1 \parallel \mathbf{y}_2 \parallel \cdots \parallel \mathbf{y}_w$, where each $\mathbf{y}_i$ has length n_i and $|\mathbf{y}_0| = k$. We further identify $\mathbf{y}_0$ with $\mathbf{y}_0 = \mathbf{y}_0[1] \parallel \mathbf{y}_0[2] \parallel \cdots \parallel \mathbf{y}_0[w]$, where each $\mathbf{y}_0[i]$ is of length k_i.

During decoding, we will maintain lists $L_1, L_2, \ldots, L_w$ of possible decoding candidates of length K_i.

Step 0: Set a temporary list $T_0 = L_0$ to contain all possible decodings of the first k_1 coordinates of $\mathbf{y}$:
$$T_0 \leftarrow \{\mathbf{x}' = \mathbf{y}_0[1] + \mathbf{e} \mid \mathbf{e} \in E^{k_1/\ell}\}.$$

Step $1 \le i \le w$: Perform list-decoding to recover a list of valid decodings:
For each candidate $\mathbf{x}' \in T_{i-1} \subset \mathbb{F}_2^{K_i}$, add to L_i all the candidates for which $\mathbf{x}'B_i + \mathbf{y}_i \in E^{n_i/\ell}$:
$$L_i \leftarrow \{\mathbf{x}' \in T_{i-1} \mid \mathbf{x}'B_i + \mathbf{y}_i \in E^{n_i/\ell}\}. \tag{2}$$

If $i < w$ then create the temporary list T_i of candidates of length K_{i+1} from L_i:
$$T_i \leftarrow \{\mathbf{x}' \parallel (\mathbf{y}_0[i+1] + \mathbf{e}) \mid \mathbf{x}' \in L_i, \ \mathbf{e} \in E^{k_{i+1}/\ell}\}. \tag{3}$$

Return: L_w.

2.1   Random Split of the Staircase-Generator Matrix

A key parameter of the public-key encryption scheme where we split the staircase-generator matrix is the number of splits s, the number of summands the generator matrix of the code is split in. This parameter further determines the nature of the error used during encryption. We have the following:

Definition 3. *Let $E_\ell \subset \mathbb{F}_2^\ell$ be an error set of granulation ℓ and let s denote the number of splits. The s-tuple $ErrorSplit = (e_1, \ldots, e_s)$, where $e_i \in \mathbb{F}_2^\ell, i \in \{1, \ldots, s\}$ is called A Valid Error Split for E_ℓ if the sum of its elements permuted with any permutation $\sigma_i \in S_\ell$ is an element of E_ℓ i.e. it holds that $e = \sum_{i=1}^{s} \sigma_i(e_i) \in E_\ell$. The set of all valid error splits is denoted as $ValidErrorSplits$ and its size with V, i.e., $V = |ValidErrorSplits|$.*

Example 2. Let $\ell = 4$, $E_\ell = \{(0,0,0,0), (0,0,1,1), (0,1,0,1), (0,1,1,0), (1,0,1,1), (1,1,0,0), (1,1,0,1), (1,1,1,0), (1,1,1,1)\}$ and $s = 4$. The 4-tuple $ErrorSplit = ((1,0,0,0), (1,1,1,1), (1,1,1,1), (1,1,0,1))$ is *a valid error split for E_ℓ because the sum of all its elements permuted by any of all possible $4! = 24$ permutations always gives an element in E_ℓ.*

A formal description of the scheme is given through the next four algorithms for key generation, error set generation, encryption and decryption. Note that Algorithm 3 is run only once at the time of the initialization of the system with parameters ℓ, E_ℓ, s. Even more, in practice, this set can be pre-calculated and publicly available.

Algorithm 2. Key Generation

Parameters: Let $\ell | n$, $m = n/\ell$ and $E \subset \mathbb{F}_2^\ell$ be an error set of granulation ℓ and density ρ. Let s be the number of splits.

Key generation:
The following matrices make up the private key:
- A generator matrix G of a binary (n, k) code of the form (1).
- An invertible matrix $S \in \mathbb{F}_2^{k \times k}$.
- An array of permutation matrices $P_1, P_2, \ldots, P_s$ created as follows:
  1. Select a permutation π on $\{1, 2, \ldots, m\}$, and let $P \in \mathbb{F}_2^{n \times n}$ be the permutation matrix induced by π, so that for any $\mathbf{y} = \mathbf{y}_1 \parallel \mathbf{y}_2 \parallel \ldots \parallel \mathbf{y}_m \in (\mathbb{F}_2^\ell)^m$:

$$\mathbf{y}P = \mathbf{y}_{\pi(1)} \parallel \mathbf{y}_{\pi(2)} \parallel \cdots \parallel \mathbf{y}_{\pi(m)}, \tag{4}$$

i.e., P only permutes the m substrings of $\mathbf{y}$ of length ℓ.
  2. For $i := 1$ to s:

Select randomly m permutations $\sigma_j^i \in \mathcal{S}_\ell$, $j \in \{1, \ldots, m\}$.
Let P_i be defined by
$$\mathbf{y}P_i = \sigma_1^i(\mathbf{y}_{\pi(1)}) \parallel \sigma_2^i(\mathbf{y}_{\pi(2)}) \parallel \cdots \parallel \sigma_m^i(\mathbf{y}_{\pi(m)}),$$
where $\sigma_j^i(\mathbf{x}) = \sigma_j^i(x_1, x_2, \ldots, x_\ell)$.

The public key is formed as follows:

Generate uniformly at random $s - 1$ matrices $G_1, \ldots, G_{s-1}$ of size $k \times n$ over $\mathbb{F}_2$.
Set $G_s = G + G_1 + \cdots + G_{s-1}$.
For all $i \in \{1, 2, \ldots, s\}$, set $G_{\text{pub}}^i = SG_iP_i$.

Public key: $G_{\text{pub}}^1, \ldots, G_{\text{pub}}^s$.
Private key: S, G and $P_1, P_2, \ldots, P_s$.

Algorithm 3. Valid Error Splits (ℓ, E_ℓ, s)

Input: Granulation ℓ, error set E_ℓ, number of splits s.
Output: A set $ValidErrorSplits$ of all possible valid error splits.
1: Set $ValidErrorSplits \leftarrow \emptyset$
2: **for** all $(e_1, \ldots, e_s) \in (\mathbb{F}_2^\ell)^s$ **do**
3:      **if** $\sum_{i=1}^s \sigma_i(e_i) \in E_\ell, \forall (\sigma_1, \ldots, \sigma_s) \in (\mathcal{S}_\ell)^s$ **then**
4:          Add $(e_1, \ldots, e_s)$ to $ValidErrorSplits$.
5:      **end if**
6: **end for**
7: Return $ValidErrorSplits$.

Algorithm 4. Encryption $(\mathbf{m}, G_{\text{pub}}^1, \ldots, G_{\text{pub}}^s, ValidErrorSplits)$

Input: Message to be encrypted $\mathbf{m}$ the public key $G_{\text{pub}}^1, \ldots, G_{\text{pub}}^s$ and a set $ValidErrorSplits$ of all possible valid error splits.
Output: A ciphertext $\mathbf{c} = (\mathbf{c}_1, \ldots, \mathbf{c}_s)$.
1: Set $\mathbf{c}_i = \mathbf{m}G_{\text{pub}}^i + \mathbf{e}_i$, $i = 1, \ldots, s$, where $\mathbf{e}_i = (e_{1,i}, \ldots, e_{\frac{n}{\ell},i})$ and where $(e_{j,1}, \ldots, e_{j,s})$, $j = 1, \ldots, \frac{n}{\ell}$ are randomly drawn from $ValidErrorSplits$.
2: Return $\mathbf{c} = (\mathbf{c}_1, \ldots, \mathbf{c}_s)$

Algorithm 5. Decryption $(\mathbf{c}, S, G, P_1, P_2, \ldots, P_s)$

Input: Ciphertext $\mathbf{c}$, matrix S, the generator matrix G and the permutation matrices $P_1, P_2, \ldots, P_s$.
Output: A decrypted message $\mathbf{m}$.
1: Set $\mathbf{c}_i' = \mathbf{c}_i P_i^{-1}$
2: Set $\mathbf{c}' = \sum_{i=1}^s \mathbf{c}_i'$
3: Set $\mathbf{m}'$ as the output of Algorithm 1 (List decoding of $\mathbf{c}'$ with generator matrix G).
4: Set $\mathbf{m} = \mathbf{m}'S^{-1}$
5: Return $\mathbf{m}$

The initial proposal for encryption scheme that uses St-Gen codes [3], is vulnerable to an Information Set Decoding (ISD) attack [7,11].

In a very recent analysis by Moody and Perlner [7] a modification of Stern's algorithm was provided, dedicated to cryptanalysis of the scheme in [3]. We refer the reader to [7] for details, and here we mention that complexity of the attack is in general given by $ISD_{St} = Pr_{St}^{-1} \cdot Cost_{St}$ where Pr_{St}^{-1} is the probability of success, and $Cost_{St}$ the cost of finding the low weight codeword.

In [10] we gave a detailed security analysis how and why the random split of the generator matrix prevents from ISD attacks. We also give several practical parameter sets. Based on all that we give the following conjecture:

Conjecture 1. The public key $G_{\text{pub}}^1, \ldots, G_{\text{pub}}^s$ produced by Algorithm 1 is indistinguishable from a set of s random $[n, k]$ codes and inverting the encryption with $G_{\text{pub}}^1, \ldots, G_{\text{pub}}^s$ without the knowledge of the private key S, G and $P_1, P_2, \ldots, P_s$ is infeasible in polynomial time.

3   Achieving IND-CPA and IK-CPA by Padding the Plaintext with a Random Bit-String

In this section we apply the ideas described by Nojima et al. in [8] and by Yamakawa et al. in [12] for the McEliece scheme, to show that our encryption scheme with random split of the generator matrix can achieve semantic security against chosen plaintext attack (IND-CPA) and key-privacy, i.e., indistinguishability of public keys under chosen plaintext attack (IK-CPA).

Firs we give a definition of indistinguishability of encrypted data against the chosen plaintext attack (CPA) as it is given in [2].

Definition 4 (IND-CPA [2]). *Let a PKE scheme be the following tuple of polynomial-time algorithms: PKE = (Gen, Enc, Dec).*

1. *On input of security parameter κ, key generation algorithm $Gen(1^\kappa)$ outputs the set of private-key and public-key, (pk, sk).*
2. *Given (pk, sk), a polynomial-time adversary $\mathcal{A}$ chooses two equal-length plaintexts m_0, m_1, $(m_0 \neq m_1)$, and sends them to the encryption oracle.*
3. *Encryption oracle (algorithm) randomly flips coin $b \in \{0,1\}$, to encrypt $Enc(pk, m_b) = c$.*
4. *Given target ciphertext c, adversary $\mathcal{A}$ outputs $b' \in \{0,1\}$, where the advantage of success probability over random guess is defined as follows:*

$$\mathbf{Adv}_{\mathcal{A}}^{ind-cpa}(\kappa) = Pr[b' = 0|b = 0] + Pr[b' = 1|b = 1]. \tag{5}$$

If $\mathbf{Adv}_{\mathcal{A}}^{ind-cpa}(\kappa)$ is negligible, then, we say underlying PKE is IND-CPA secure. Here "negligible" means that for any constant const, there exists $k_0 \in \mathbb{N}$, s.t. for any $\kappa > k_0$, $\mathbf{Adv}$ is less than $\left(\frac{1}{\kappa}\right)^{const}$.

As we said, in order to achieve IND-CPA in a standard model, Nojima et al. [8] proposed a random prepadding for messages in McEliece scheme, i.e., instead of encrypting messages m to encrypt messages $[r|m]$ where r is a random prepadding. However, in order to achieve more than 2^{80} security or more than 2^{128} security, r should have a significant length. More concretely for the McEliece code $(2048, 1289, 69)$ to achieve security of 2^{85}, out of 1289 bits the random prepadding r has to have 1161 bits and only 128 bits are left for m. For the other code $(4096, 2560, 128)$, to achieve security of 2^{131}, out of 2560 bits the random prepadding r has to have 2048 bits and only 512 bits are left for m.

In what follows we investigate the IND-CPA security of the approach of encrypting messages in a form $\mathbf{m} = [r|m]$ for our scheme with a random split of St-Gen matrix (we abbreviate the name of that system as RRS-St-Gen - Randomized Random Split of St-Gen - in the mathematical formulas that refer to that system). By having this form for $\mathbf{m}$ the encryption gets the following form: The ciphertext is the s-tuple $\mathbf{c} = (\mathbf{c}_1, \ldots, \mathbf{c}_s)$ where

$$\mathbf{c}_i = \mathbf{m}G_{\text{pub}}^i + \mathbf{e}_i = (rG_{\text{pub}_1}^i + \mathbf{e}_i) + mG_{\text{pub}_2}^i, \tag{6}$$

where $G_{\text{pub}}^i = \begin{bmatrix} G_{\text{pub}_1}^i \\ G_{\text{pub}_2}^i \end{bmatrix}$, $\mathbf{e}_i = (e_{1,i}, \ldots, e_{\frac{n}{l},i})$ and where $(e_{j,1}, \ldots, e_{j,s})$, $j = 1, \ldots, \frac{n}{l}$ are randomly drawn from $ValidErrorSplits$.

Theorem 1. *Let RRS-St-Gen is a randomized public key encryption scheme as defined in Algorithms 1, 2, 3, 4 and 5 with the following parameters: n, k, l, s, E_l, $ValidErrorSplits$, $V = |ValidErrorSplits|$, that encrypts messages $\mathbf{m} = [r|m]$ where r is a random prepadding. If the s-tuple of matrices $(G_{pub}^1, \ldots, G_{pub}^s)$ is indistinguishable from random, and the inversion of the encryption with $G_{pub}^1, \ldots, G_{pub}^s$ without the knowledge of the private key S, G and $P_1, P_2, \ldots, P_s$ is infeasible in polynomial time, then the advantage of an adversary given by relation (5) is:*

$$\mathbf{Adv}_{A-RRS-St-Gen}^{ind-cpa}(\kappa) = 2\left(\left(1 - \frac{1}{2^{|r|}}\right)\left(\frac{V}{2^{s\cdot l}}\right)^{\frac{n}{l}}\left(1 - \left(\frac{V}{2^{s\cdot l}}\right)^{\frac{n}{l}}\right) + \right.$$
$$\left. + \frac{1}{2^{|r|}}\left(1 - \left(\frac{V}{2^{s\cdot l}}\right)^{\frac{n}{l}}\right)\right) \tag{7}$$

Proof. First note that the assumption about the infeasibility of the inversion of the encryption is a crucial one. If the adversary is capable to invert the encryption, it will simply obtain the whole random prepadding and will guess the value of b' with probability 1. The necessity of the assumption for the indistinguishability from random matrices is due to the attacks that reveal the secret key of the scheme and then is also connected with the inversion of the encryption. We discuss additionally these two assumptions at the end of this proof.

We will compute the probability $Pr[b' = 0|b = 0]$ and due to the symmetry, the probability for $Pr[b' = 1|b = 1]$ has the same value. First let us note that every $\mathbf{c}_i$ have $\frac{n}{l}$ substrings of length l bits. For a concrete value of $\mathbf{m} = [r|m]$ the encryption procedure randomly picks $\frac{n}{l}$ elements from set of s-tuples $ValidErrorSplits$. Note also that the set $ValidErrorSplits$ is a subset of all possible l-bit s-tuples which number is $2^{s \cdot l}$. In case when m_0 was encrypted there are two possible and disjunctive events that can lead the adversary to make the right guess $b' = 0$. Those two events are the following:

– $Event_1 : Event_{1,1} \cap Event_{1,2} \cap Event_{1,3}$
  1. $Event_{1,1}$: The adversary made a wrong guess about the prepadded random value r. $Pr(Event_{1,1}) = 1 - \frac{1}{2^{|r|}}$;
  2. $Event_{1,2}$: The adversary computes $\mathbf{c}_i^{(m_0)} = rG_{\mathrm{pub}_1}^i + m_0 G_{\mathrm{pub}_2}^i$ and for all $\frac{n}{l}$ chunks in all $\mathbf{c}_i$, there exist a valid error split in $ValidErrorSplits$ as a connection between $\mathbf{c}_i^{(m_0)}$ and $\mathbf{c}_i$. $Pr(Event_{1,2}) = \left(\frac{V}{2^{s \cdot l}}\right)^{\frac{n}{l}}$;
  3. $Event_{1,3}$: The adversary computes $\mathbf{c}_i^{(m_1)} = rG_{\mathrm{pub}_1}^i + m_1 G_{\mathrm{pub}_2}^i$, and there is at least one chunk for which there is no valid error split in $ValidErrorSplits$. $Pr(Event_{1,3}) = \left(1 - \left(\frac{V}{2^{s \cdot l}}\right)^{\frac{n}{l}}\right)$.

– $Event_2 : Event_{2,1} \cap Event_{2,2}$
  1. $Event_{2,1}$: The adversary made a correct guess about the prepadded random value r. $Pr(Event_{2,1}) = \frac{1}{2^{|r|}}$;
  2. $Event_{2,2}$: The adversary computes $\mathbf{c}_i^{(m_1)} = rG_{\mathrm{pub}_1}^i + m_1 G_{\mathrm{pub}_2}^i$, and there is at least one chunk for which there is no valid error split in $ValidErrorSplits$. $Pr(Event_{2,2}) = \left(1 - \left(\frac{V}{2^{s \cdot l}}\right)^{\frac{n}{l}}\right)$.

Composing probabilities for all events we get:

$$Pr[b' = 0|b = 0] = Pr(Event_1) + Pr(Event_2)$$
$$= Pr(Event_{1,1})Pr(Event_{1,1})Pr(Event_{1,1}) +$$
$$+ Pr(Event_{2,1})Pr(Event_{2,2})$$
$$= \left(1 - \frac{1}{2^{|r|}}\right)\left(\frac{V}{2^{s \cdot l}}\right)^{\frac{n}{l}}\left(1 - \left(\frac{V}{2^{s \cdot l}}\right)^{\frac{n}{l}}\right) + \frac{1}{2^{|r|}}\left(1 - \left(\frac{V}{2^{s \cdot l}}\right)^{\frac{n}{l}}\right)$$

Since the case $Pr[b' = 1|b = 1]$ is symmetrical, the total $\mathbf{Adv}_{A-RRS-St-Gen}^{ind-cpa}(\kappa)$ is:

$$\mathbf{Adv}_{A-RRS-St-Gen}^{ind-cpa}(\kappa) = 2\left(\left(1 - \frac{1}{2^{|r|}}\right)\left(\frac{V}{2^{s \cdot l}}\right)^{\frac{n}{l}}\left(1 - \left(\frac{V}{2^{s \cdot l}}\right)^{\frac{n}{l}}\right) + \right.$$
$$\left. + \frac{1}{2^{|r|}}\left(1 - \left(\frac{V}{2^{s \cdot l}}\right)^{\frac{n}{l}}\right)\right)$$

A direct conclusion from the security analysis in [10] is that if the length of the message that is encrypted with the scheme is bigger than 256 bits, than the claims in the Conjecture 1 about the indistinguishability of the public key from random matrices and the infeasibility of the inversion of the encryption are plausible. So, for any concrete instantiation of the scheme, the security levels that are achieved with the adversary advantage defined in the relation (3) have to give lengths of the random prepadding to be more than 256 bits. ▫

By the final part of the proof of Theorem 1 we obtain that our scheme achieves the IND-CPA security level of 2^{80} for $|r| = 264$ and the security level of 2^{128} for $|r| = 424$. So these values being higher than 256 are in accordance with the security analysis in [10] and the Conjecture 1. Moreover, they are still significantly lower than the lengths of the prepadded random part in the modified McEliece scheme.

For the key-privacy issue we use the same approach as Yamakawa et al. have in [12].

Definition 5 (IK-CPA [1]). *Let a PKE scheme be the following tuple of polynomial-time algorithms: PKE = (Gen, Enc, Dec). The security of key-privacy is defined as follows.*

1. *On input of security parameter κ, key generation algorithm $Gen(1^\kappa)$ outputs two independent sets of key pairs (pk_0, sk_0), (pk_1, sk_1).*
2. *Given (pk_0), (pk_1), a polynomial-time adversary $\mathcal{A}$ chooses a plaintext m and sends them to the encryption oracle.*
3. *Encryption oracle randomly flips coin $b \in \{0, 1\}$, to output $Enc_{pk_b}(m) = c$.*
4. *Given target ciphertext c, adversary $\mathcal{A}$ outputs $b' \in \{0, 1\}$, where the advantage of success probability over random guess is defined as follows:*

$$\mathbf{Adv}_{\mathcal{A}}^{ik-cpa}(\kappa) = Pr[b' = 0 | b = 0] + Pr[b' = 1 | b = 1]. \tag{8}$$

If $\mathbf{Adv}_{\mathcal{A}}^{ik-cpa}(\kappa)$ is negligible, then, we say underlying PKE is IK-CPA secure.

While the modeling of IK-CPA is not the same as IND-CPA, the value about $\mathbf{Adv}_{\mathcal{A}}^{ik-cpa}(\kappa)$ is the same as for IND-CPA case. Thus we have the following theorem that we give without a proof:

Theorem 2. *Let RRS-St-Gen is a randomized public key encryption scheme as defined in Algorithms 1, 2, 3, 4 and 5 with the following parameters: n, k, l, s, E_l, ValidErrorSplits and $V = |ValidErrorSplits|$, encrypting messages $\mathbf{m} = [r|m]$ where r is a random prepadding. If the s-tuple of matrices $(G_{pub}^1, \ldots, G_{pub}^s)$ is indistinguishable from random, and the inversion of the encryption with $G_{pub}^1, \ldots, G_{pub}^s$ without the knowledge of the private key S, G and $P_1, P_2, \ldots, P_s$ is infeasible in polynomial time, then the advantage of an adversary given by relation (8) is:*

$$\mathbf{Adv}_{\mathcal{A}-RRS-St-Gen}^{ik-cpa}(\kappa) = \mathbf{Adv}_{\mathcal{A}-RRS-St-Gen}^{ind-cpa}(\kappa) \tag{9}$$

4   Conclusions

We have presented a public key encryption scheme based on St-Gen codes and its variant where we split and replace the public generator matrix into s randomly generated matrices. The split strategy is used to thwarts the ISD attacks on the encryption scheme. Then, we showed that randomized version of the encryption scheme offers semantic security against chosen plaintext attack (IND-CPA) and offers key-privacy, i.e., offers indistinguishability of public keys under chosen plaintext attack (IK-CPA) in the standard model. The difference with McEliece scheme is that with our scheme the length of the prepadded random string is significantly shorter. It remains as a next goal to investigate the modification of the scheme for achieving the stronger securities of chosen-ciphertext attacks (CCA and CCA2) both with and without the random oracle model.

Acknowledgement. We would like to thank the anonymous reviewers for their valuable remarks that significantly improved the quality of this paper.

References

1. Bellare, M., Boldyreva, A., Desai, A., Pointcheval, D.: Key-privacy in public-key encryption. In: Boyd, C. (ed.) ASIACRYPT 2001. LNCS, vol. 2248, p. 566. Springer, Heidelberg (2001)
2. Bellare, M., Desai, A., Pointcheval, D., Rogaway, P.: Relations among notions of security for public-key encryption schemes. In: Krawczyk, H. (ed.) CRYPTO 1998. LNCS, vol. 1462, p. 26. Springer, Heidelberg (1998)
3. Gligoroski, D., Samardjiska, S., Jacobsen, H., Bezzateev, S.: McEliece in the world of Escher. Cryptology ePrint Archive: Report 2014/360 (2014). http://eprint.iacr.org/
4. Goldwasser, S., Micali, S.: Probabilistic encryption. J. Comput. Syst. Sci. **28**(2), 270–299 (1984)
5. Kobara, K., Imai, H.: Semantically secure McEliece public-key cryptosystems - conversions for McEliece PKC. In: Kim, K. (ed.) PKC 2001. LNCS, vol. 1992, pp. 19–35. Springer, Heidelberg (2001)
6. McEliece, R.J.: A Public-Key System Based on Algebraic Coding Theory, pp. 114–116. Jet Propulsion Lab (1978). DSN Progress Report 44
7. Moody, D., Perlner, R.: Vulnerabilities of "McEliece in the World of Escher". Cryptology ePrint Archive: Report 2015/966 (2015). http://eprint.iacr.org/
8. Nojima, R., Imai, H., Kobara, K., Morozov, K.: Semantic security for the McEliece cryptosystem without random oracles. Des. Codes Crypt. **49**(1–3), 289–305 (2008)
9. Samardjiska, S., Gligoroski, D.: Approaching maximum embedding efficiency on small covers using staircase-generator codes. In: 2015 IEEE International Symposium on Information Theory (ISIT), pp. 2752–2756, June 2015
10. Samardjiska, S., Gligoroski, D.: An Encryption Scheme Based on Random Split of St-Gen Codes. Cryptology ePrint Archive: Report 2016/202 (2016). https://eprint.iacr.org/2016/202
11. Sendrier, N., Tillich, J.-P.: Private communication, October 2014
12. Yamakawa, S., Cui, Y., Kobara, K., Hagiwara, M., Imai, H.: On the Key-Privacy Issue of McEliece Public-Key Encryption. In: Boztaş, S., Lu, H.-F.F. (eds.) AAECC 2007. LNCS, vol. 4851, pp. 168–177. Springer, Heidelberg (2007)

The Typical Constructible Object

Mathieu Hoyrup[✉]

LORIA, Inria Nancy Grand Est, Villers-lès-Nancy, France
mathieu.hoyrup@inria.fr

1 Introduction

Baire Category is an important concept in mathematical analysis. It gives a notion of large set, hence a way of identifying the properties of typical objects. One of the most important applications of Baire Category is to provide a way of proving the existence of objects with specified properties without having to give an explicit construction, showing at the same time that these properties are prevalent. For instance it has been extensively used in mathematical analysis to better understand and separate classes of real functions such as analytic and smooth functions (see [9] for a wide range of applications of the Baire Category Theorem in analysis).

This note is about Baire Category in constructive or computable analysis. This subject has been studied in many different directions, for instance in reverse mathematics [3], constructive mathematics [2] or computable analysis [1]. In these fields, one is often interested in studying the properties of "constructible" objects, to separate classes of constructible objects, and to identify the effectiveness of mathematical proofs.

Here we are interested in a particular question: how to apply Baire Category inside classes of constructible objects? Such classes are very small in the sense of Baire Category, as they are countable, so strictly speaking Baire Category cannot be applied inside them. However one can adapt Baire Category and save part of it in these small worlds, and this note illustrates this. We will focus on 4 particular classes of constructible objects, depicted in Fig. 1:

- The class C of computable subsets of $\mathbb{N}$.
- The class CE of c.e. subsets of $\mathbb{N}$.
- The class LCE of left-c.e. real numbers in $[0,1]$.
- The class Δ_2^0 of subsets of $\mathbb{N}$ that are computable relative to the halting problem.

The goal is, for each class of constructible objects, to find an analog of Baire Category to define a notion of large set, of typical or generic object. Why is it interesting? It helps understanding which parts of classical mathematics are still available in a constructive setting. From a more practical perspective, it gives a way to prove the existence of constructible objects with specified properties, avoiding as in the classical setting explicit constructions by using a simpler argument.

© Springer International Publishing Switzerland 2016
A. Beckmann et al. (Eds.): CiE 2016, LNCS 9709, pp. 115–123, 2016.
DOI: 10.1007/978-3-319-40189-8_12

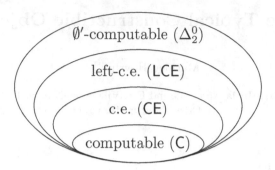

Fig. 1. Four classes of constructible subsets of $\mathbb{N}$

For each class $\mathscr{C}$ of objects, we want to investigate the following question:

What does the typical object of $\mathscr{C}$ look like?

The way to do this is to define a notion of small subclass of $\mathscr{C}$. Such a notion should satisfy the following conditions:

Axiom 1. Every singleton $\{A\}$ with $A \in \mathscr{C}$ is small in $\mathscr{C}$,
Axiom 2. The class $\mathscr{C}$ is not small in itself,
Axiom 3. Effective countable unions of small sets are small, for some notion of effectiveness,
Axiom 4. In the hierarchy depicted in Fig. 1, a subclass is small in the superclass. In other words, a typical object of the superclass does not belong to the subclass.

A measure-theoretical approach to this problem has been developed by many authors using resource-bounded measure and dimension theory, in particular resource-bounded martingales (see [11,12] for instance). These methods were mainly applied to complexity classes, but not to classes of enumerable objects such as c.e. sets or left-c.e. reals: whether such a development can be done is an interesting problem.

Here we adopt the topological approach of Baire Category. For each class it is done by defining the analog of a nowhere dense subclass by means of effectivization. The **meager** subclasses are then generated by the nowhere dense classes: they are the subsets of effective countable unions of nowhere dense classes. More precisely in an effective version,

1. Nowhere dense sets are effective, i.e., describable in some way by programs,
2. Countable unions $\bigcup_i A_i$ are effective, i.e., there is a single program that given i as input, describes A_i in the way specified in 1.

Hence to obtain a notion of meager subclass one simply has to define what is an *effective nowhere dense set*. Axiom 3 will be automatically satisfied.

1.1 Notations and Background

The set of finite binary strings is denoted by 2^*. The Cantor space $2^{\mathbb{N}}$ is the space of infinite binary sequences, also identified to subsets of $\mathbb{N}$ or reals numbers in $[0, 1]$. If A is an infinite binary sequence then $A \upharpoonright n$ is the prefix of A of length n. For each finite binary string $u \in 2^*$, the cylinder $[u] \subseteq 2^{\mathbb{N}}$ is the class of infinite binary extensions of u. The Cantor space is endowed with the topology generated by the cylinders: the open classes $\mathcal{U} \subseteq 2^{\mathbb{N}}$ are the unions of cylinders. A subclass of $2^{\mathbb{N}}$ is nowhere dense if it is disjoint from a dense open class. A subclass is meager if it is a union of nowhere dense classes.

2 Computable Objects

Let C be the class of computable subsets of $\mathbb{N}$.

An ordinary nowhere dense set is a set that is disjoint from a dense open set. A natural effective version is then to require the dense open set to be effective, i.e., expressible as a union $\bigcup_{u \in A}[u]$ where $A \subseteq 2^*$ is a computably enumerable set.

We then say that a class is **meager in** C if it is contained in an effective union of complements of dense open sets. A class is **co-meager in** C if it contains an effective intersection of dense open sets.

This notion of nowhere dense set makes the Baire Category theorem computable:

Theorem 2.1 (Baire Category Theorem in C**).** *Every class that is co-meager in* C *is dense in* C*, i.e., contains computable elements in every cylinder.*

In other words, the class C is not meager in itself and Axiom 2 is satisfied.

Observe that Axiom 1 is also satisfied, i.e., a singleton is meager in C: if $A \subseteq \mathbb{N}$ is computable then $2^{\mathbb{N}} \setminus \{A\}$ is a dense effective open class. Every subclass that can be effectively listed is also meager in C, for instance the class P of polynomial-time decidable problems.

Example. Consider a map T from $[0, 1]$ to $[0, 1]$ that is computable. Think of T as a dynamical system: if $x \in [0, 1]$ is the state of the system at time t then $T(x)$ is the state at time $t + 1$. An initial state x_0 induces a trajectory defined by $x_{t+1} = T(x_t)$. What do typical trajectories look like?

Now imagine that one simulates T on a computer, computing the trajectory starting from some state x_0. Of course the computer can only manipulate computable real numbers, so only the computable part of the dynamical system can be observed on the computer, namely the map $T : [0, 1] \cap \mathsf{C} \to [0, 1] \cap \mathsf{C}$ (here C is the set of computable real numbers). What do typical trajectories of the restricted system look like? Are they representative of the original system over $[0, 1]$?

If the system has a dense trajectory then one can show that all the typical trajectories are dense (i.e., the set of initial states inducing a dense trajectory is co-meager), and the computable Baire Category theorem directly implies that the typical trajectories of the restricted system are also dense, in particular there exists a computable dense trajectory.

3   Computable Relative to the Halting Problem

Let Δ_2^0 be the class of subsets of $\mathbb{N}$ that are computable relative to the halting problem.

One can relativize the Computable Baire Category theorem to any oracle, in particular to the halting problem, in a straightforward way. It gives for free a notion of nowhere dense class in Δ_2^0: it is a class that is disjoint from a dense open class that is effective relative to the halting problem. Effective unions can be equivalently taken relative to the halting problem or not (the two notions are equivalent by the relativized s-m-n theorem).

Again, Axioms 1 and 2 are satisfied, i.e., the class Δ_2^0 is not meager in itself and every singleton is meager in Δ_2^0. Moreover Axiom 4 is also satisfied as the subclass LCE is meager in Δ_2^0: the halting problem can effectively list LCE, which is then an effective union of singletons, hence is meager in Δ_2^0.

We now present a particular class that is co-meager in Δ_2^0 that has received a lot of attention.

3.1   1-Generic Sets

In Sect. 2 we mentioned that an ordinary nowhere dense class is a class that is disjoint from a dense open set, which naturally induces the first effective version. A nowhere dense class can equivalently be defined as a subset of the boundary of an arbitrary (i.e., not necessarily dense) open set, which gives another possible effective version: a subset of the boundary of an effective open set.

It gives a strictly weaker notion of nowhere dense class: in general the boundary of an effective open class is not disjoint from a dense effective open class. However, it is always disjoint from a dense open class that is effective *relative to the halting problem*. As a result, it is meager in Δ_2^0. As the effective open classes can be effectively enumerated, taking the union of their boundaries gives a class that is meager in Δ_2^0. Its complement is known as the class of **1-generic** subsets of $\mathbb{N}$, and is co-meager in Δ_2^0. In particular it is non-empty and even dense. It was introduced by Jockush [8] in order to simplify constructions in recursion theory, in the same way as Baire Category simplifies proofs of existence results in mathematical analysis.

While the class of 1-generic sets is just one particular class that is co-meager Δ_2^0, it happens that it captures many interesting co-meager classes in Δ_2^0, in the sense that it contains them. In other words, many typical properties are already satisfied by the 1-generic sets. Indeed being 1-generic is a kind of universal property as it is about *every* effective open set.

For instance, we mentioned that the subclass LCE $\subsetneq \Delta_2^0$ is meager in Δ_2^0, i.e., typical Δ_2^0 sets are not left-c.e. Actually, no 1-generic set is left-c.e. Indeed, if $x \in [0,1]$ is left-c.e. then the interval $[0,x)$ is an effective open set that is dense along x, i.e., contains x in its boundary.

In the same way as Baire Cateory provides a simple way of proving existence results without giving explicit constructions, 1-genericity can be easily used to prove existence of Δ_2^0-sets with prescribed properties, and is an alternative to explicit constructions.

A famous example is Kleene-Post's theorem, on the way to the solution to Post's problem:

Theorem 3.1 (Kleene and Post [10]**).** *There exist two Turing-incomparable Δ_2^0-sets, i.e. two Δ_2^0-sets that are not computable relative to each other.*

Actually for a typical Δ_2^0-set A, its two halves $A_0 = \{n \in \mathbb{N} : 2n \in A\}$ and $A_1 = \{n \in \mathbb{N} : 2n + 1 \in A\}$ are Turing-incomparable Δ_2^0-sets. Moreover,

Theorem 3.2 (Jockush [8]**).** *Every 1-generic set has Turing-incomparable halves.*

Proof. This can be proved very easily: given a Turing machine M, let $\mathcal{U} = \{A : \exists n, M^{A_0}(n) = 0$ but $n \in A_1\}$. $\mathcal{U}$ is an effective open class and if A_1 is Turing reducible to A_0 via M then A belongs to the boundary of U: adding an arbitrary large element to A_1 makes A fall in $\mathcal{U}$ (possible when A_1 is co-infinite). Hence if A is 1-generic, A does not belong to the boundary of $\mathcal{U}$ and as A_1 is easily co-infinite, M does not compute A_1 relative to A_0.

Hence instead of constructing a Δ_2^0-set with the specific property, one simply has to check that the property is co-meager in Δ_2^0, or even that it is captured by 1-genericity. The relativized computable Baire Category theorem gives the existence result for free.

4   Enumerable Objects

As we saw, the case of computable sets is a straightforward effectivization of the ordinary, non-effective setting, from which the case of Δ_2^0-sets is a straightforward relativization.

We now turn our attention to intermediate classes of objects. What is a small class inside the class CE of c.e. subsets of $\mathbb{N}$? What is a typical c.e. set? The same questions can be asked for left-c.e. sets, Π_1^0-classes, etc.

4.1   C.e. Sets

First observe that as $\mathsf{C} \subseteq \mathsf{CE}$, we can declare every class that is meager in C to be meager in CE. Axiom 1 is satisfied, i.e., every singleton is meager in CE: (i) if $A \subseteq \mathbb{N}$ is finite then it is computable so as before $\{A\}$ is meager as its complement is a dense effective open set, (ii) if A is infinite then the class $\mathcal{U} = \{B \subseteq \mathbb{N} : A \not\subseteq B\}$ is a dense effective open set that does not contain A. However

Axiom 4 is not satisfied as C is not meager in CE, so we have to add new meager classes in CE.

The idea is to weaken the definition of 1-genericity. Let us recall that a set $A \subseteq \mathbb{N}$ is 1-generic if it does not belong to the boundary of any effective open class $\mathcal{U} \subseteq 2^{\mathbb{N}}$; A belongs to the boundary of $\mathcal{U}$, or $\mathcal{U}$ is dense along A, means that $\mathcal{U}$ contains sets arbitrarily close to A, i.e., in every cylinder $[A \upharpoonright n]$.

Definition 4.1. *We say that A belongs to the* **down-closure** *of $\mathcal{U}$, or that $\mathcal{U}$ is* **dense above A**, *if $\mathcal{U}$ contains sets $B \supseteq A$ arbitrarily close to A, i.e., in every cylinder $[A \upharpoonright n]$. We say that A belongs to the* **down-boundary** *of $\mathcal{U}$ if A belongs to the down-closure of $\mathcal{U}$ but not to $\mathcal{U}$.*

Observe that the down-boundary of $\mathcal{U}$ is contained in the boundary of $\mathcal{U}$, so if $\mathcal{U}$ is an effective open set then its down-boundary is meager in Δ_2^0. We declare that its down-boundary is already meager in CE. It gives a notion of genericity in the class CE.

Definition 4.2. *A set A is* **generic from above** *if it belongs to every effective open class that is dense above A. In other words, A is generic from above if it does not belong to the down-boundary of any effective open class.*

This notion is equivalent to the notion of p-genericity introduced by Ingrassia [7]. We then prove a Baire Category theorem in CE.

Theorem 4.3 (Baire Category in CE [5,7]). *Let $\mathcal{U}_n \subseteq 2^{\mathbb{N}}$ be dense uniformly effective open classes. The class of c.e. sets in $\bigcap_n \mathcal{U}_n$ that are generic from above is dense in $2^{\mathbb{N}}$.*

As a result, the class CE is not meager in itself and Axiom 2 is satisfied. Moreover Axiom 4 is satisfied, i.e., the subclass C is meager in CE: (i) the class of co-finite sets is already meager in C as it can be effectively listed, (ii) if $A \subseteq \mathbb{N}$ is co-infinite and computable then it is not generic from above, as the effective open set $2^{\mathbb{N}} \setminus \{A\}$ is dense above A.

Genericity from above is a weakening of 1-genericity, that is sometimes sufficient for our purpose. For instance the simple argument showing that the two halves of a 1-generic set are Turing-incomparable (Theorem 3.2) immediately applies to co-infinite sets that are generic from above. As a result, Theorem 4.3 implies the solution to Post problem invented by Friedberg and Muchnik.

Theorem 4.4 (Friedberg-Muchnik   [4,13]). *There exist two Turing-incomparable c.e. sets.*

Additionally, we can say that having Turing-incomparable halves is a typical property of c.e. sets, and every c.e. set that is generic from above has this property.

Obverse that it does not give an alternative proof of Friedberg-Muchnik's theorem, as Theorem 4.3 is showed using the priority method with finite injury, the method invented by Friedberg and Muchnik to prove their result. However, many constructions using a simple form of the priority method are captured by the Baire Category theorem in CE. For instance, every co-infinite set A that is generic from above is not autoreducible: there is no Turing functional that for each $n \in \mathbb{N}$, decides $n \in A$ given $A \setminus \{n\}$ as oracle. Not all finite injury arguments are captured by Theorem 4.3: for instance Ingrassia [7] proved the existence of a p-generic (i.e., co-infinite and generic from above) c.e. set that is Turing equivalent to the halting set, hence is not a solution to Post's problem.

4.2   Left-C.e. Reals

We can adapt the previous definitions in a straightforward way, replacing the inclusion ordering over $2^{\mathbb{N}}$ by the lexicographic ordering. Identifying subsets of $\mathbb{N}$ with real numbers in $[0, 1]$, we get the natural ordering of real numbers.

Definition 4.5. *We say that $x \in [0, 1]$ belongs to the* **left-closure** *of $\mathcal{U} \subseteq [0, 1]$, or that $\mathcal{U}$ is* **dense on the right of** *x, if $\mathcal{U}$ contains reals $y \geq x$ arbitrarily close to x, i.e., in every interval $[x, x+\epsilon)$. We say that x belongs to the* **left-boundary** *of $\mathcal{U}$ if x belongs to the left-closure of $\mathcal{U}$ but not to $\mathcal{U}$.*

For instance, if $0 < a < 1$ then the left-boundary of $(a, 1]$ is $\{a\}$ while the left-boundary of $[0, a)$ is empty. In particular, while every left-c.e. real $a > 0$ belongs to the boundary of the effective open set $[0, a)$, preventing it from being 1-generic, it does not belong to its left-boundary.

We then declare the left-boundary of an effective open class to be meager in LCE, and we get a notion of genericity.

Definition 4.6. *A real x is* **generic from the right** *if it belongs to every effective open class that is dense on the right of x. In other words, x is generic from the right if x does not belong to the left-boundary of any effective open class.*

Again we have a Baire Category theorem in LCE.

Theorem 4.7 (Baire Category in LCE). *The class of left-c.e. reals that are generic from the right is non-empty and dense.*

Observe that we do not need to intersect with an effective family of dense open sets as in Theorem 4.3, as every dense set is also dense on the right of every $x < 1$. It means that every real that is generic from the right is weakly-1-generic, i.e., belongs to every dense effective open set. The construction of a weakly-1-generic left-c.e. real presented in [14] actually builds a real that is generic on the right, hence proves Theorem 4.7.

Again the theorem implies that Axiom 2 is satisfied, i.e., that LCE is not meager in itself. Axiom 4 is satisfied as CE is meager in LCE: we know from the previous section that every c.e. set it outside some dense effective open set.

However this time Axiom 1 is *not* satisfied! By definition a left-c.e. real that is generic on the right does not belong to the left-boundary of any effective open class. We need to add more nowhere dense classes. However we were unable to identify a natural way of doing it.

4.3   Other Classes of Objects

In [5] we introduce a way of defining a notion of genericity for more general classes of enumerable objects and prove a Baire Category theorem for such classes. For instance, one can define what is a generic Π_1^0 subset of $2^{\mathbb{N}}$. We apply this method to obtain a result in computable analysis related to the non-computability of the ergodic decomposition theorem. In [6] we give other applications of this method showing that many complicated constructions in recursion theory can be more easily obtained by choosing the suitable topology on the space of objects, and using the corresponding Baire Category theorem on that space.

5   Conclusion

For many classes of constructible objects it is possible to define a notion of meager subclass and a corresponding notion of genericity. A typical element of the class is then generic and automatically satisfies many interesting properties. The relevance of these notions can be measured in two ways that oppose to each other:

- A Baire Category theorem should hold, i.e., the class should not be meager in itself. Said differently, the notion of genericity should not be too strong.
- The notion of genericity should be strong enough to capture many useful interesting properties.

When these conditions are satisfied, existence results become easy to derive.

In this note, the Baire Category theorem for classes of enumerable objects is proved using the simplest form of priority method with finite injury. A future direction would be to define weaker notions of meager subclass, or stronger notions of genericity, capturing more advanced methods from recursion theory such as the priority method with infinite injury.

References

1. Brattka, V., Hendtlass, M., Kreuzer, A.P.: On the uniform computational content of the Baire category theorem (2015). CoRR arXiv:1510.01913
2. Bridges, D., Ishihara, H., Vîță, L.: A new constructive version of Baire's theorem. Hokkaido Math. J. **35**(1), 107–118 (2006)
3. Brown, D.K., Simpson, S.G.: The Baire category theorem in weak subsystems of second-order arithmetic. J. Symb. Logic **58**(2), 557–578 (1993)
4. Richard, M.: Friedberg: two recursively enumerable sets of incomparable degrees of unsolvability (solution of Post's problem, 1944). Proc. Natl. Acad. Sci. U.S.A. **43**(2), 236–238 (1957)
5. Hoyrup, M.: Irreversible computable functions. In: Mayr, E.W., Portier, N. (eds.) 31st International Symposium on Theoretical Aspects of Computer Science (STACS 2014), STACS 2014, 5–8 March 2014, Lyon, France, LIPIcs, vol. 25, pp. 362–373. Schloss Dagstuhl - Leibniz-Zentrum fuer Informatik (2014)

6. Hoyrup, M.: Genericity of weakly computable objects. Theory of Computing Systems (2016, to appear)
7. Ingrassia, M.A.: P-genericity for recursively enumerable sets. Ph.D. thesis, University of Illinois at Urbana-Champaign (1981)
8. Carl, G.: Jockush: simple proofs of some theorems on high degrees. Can. J. Math. **29**, 1072–1080 (1977)
9. Jones, S.H.: Applications of the Baire category theorem. Real Anal. Exch. **23**(2), 363–394 (1999)
10. Kleene, S.C., Emil, L.: Post: the upper semi-lattice of degrees of recursive unsolvability. Ann. Math. **59**(3), 379–407 (1954)
11. Jack, H.: Lutz: category and measure in complexity classes. SIAM J. Comput. **19**(6), 1100–1131 (1990)
12. Lutz, J.H.: Effective fractal dimensions. Mathe. Logic Q. **51**(1), 62–72 (2005)
13. Muchnik, A.A.: On the unsolvability of the problem of reducibility in the theory of algorithms. Dokl. Akad. Nauk SSSR **108**, 194–197 (1956)
14. Nies, A.: Computability and Randomness. Oxford Logic Guides. Oxford University Press, Oxford (2009)

Computability in Symbolic Dynamics

Emmanuel Jeandel[1,2,3]($\boxtimes$)

[1] Université de Lorraine, LORIA, UMR 7503, 54506 Vandoeuvre-lès-Nancy, France
emmanuel.jeandel@loria.fr
[2] CNRS, LORIA, UMR 7503, 54506 Vandoeuvre-lès-Nancy, France
[3] Inria, 54600 Villers-lès-Nancy, France

Abstract. We give an overview of the interplay between computability and symbolic dynamics.

A multidimensional shift of finite type (SFT) is a set of colorings of $\mathbb{Z}^d$ given by local rules. SFTs are one of the most fundamental objects in symbolic dynamics [LM95], and are well understood when $d = 1$ where they can be studied using finite automata theory. The situation becomes drastically different in dimension 2, where they are sometimes called tilings of the (discrete) plane, as almost any natural question about them becomes undecidable [Ber64, Rob71].

The uncomputability of many properties has for a long time being seen as a hurdle in the study of multidimensional symbolic dynamics. Douglas Lind [Lin04] has in particular described multidimensional SFTs as "The Swamp of Undecidability. It's a place you don't want to go".

In recent years, the position has changed, as many results have proven that it is actually possible to understand quite well many properties of multidimensional dynamical systems, as long as one accepts that the answer might involve computability theory.

We present here a few examples of this phenomenon. The focus of the first few sections is on one-dimensional symbolic dynamical systems given by computable constraints, and we show how these constraints translate into computability obstructions on their dynamics. In the last part, we explain how these results may be translated to multidimensional dynamical systems given by finite means, using strong embedding theorems.

1 Definitions

We start with a few relevant definitions.

Let A be a finite alphabet. We denote by A^* the set of finite words over the alphabet A, and by $A^{\mathbb{Z}}$ the set of biinfinite words over the alphabet A. The empty word will be denoted by ϵ. Given a finite word $w = w_0 w_1 \ldots w_{n-1}$ (whose length n is denoted $|w|$) and a biinfinite word $u = \ldots u_{-1} u_0 u_1 \ldots$, we say that w appears in u (or that u contains w) if there exists some position k s.t. $u_{i+k} = w_i$ for all $0 \leq i < n$. We will also use the notion "w appears in u" for u a finite word, with a similar definition.

© Springer International Publishing Switzerland 2016
A. Beckmann et al. (Eds.): CiE 2016, LNCS 9709, pp. 124–131, 2016.
DOI: 10.1007/978-3-319-40189-8_13

Given a set $\mathcal{F} \subseteq A^*$ of words, the *subshift* defined by $\mathcal{F}$ is the set of all biinfinite words where no word of $\mathcal{F}$ appears. We usually denote by $X_{\mathcal{F}}$ the subshift defined by $\mathcal{F}$.

Example 1. Let $A = \{0, 1\}$. Let $\mathcal{F} = \{00\}$. Then $X_{\mathcal{F}}$ is the set of biinfinite words that do not contain two consecutive symbols 0. For $\mathcal{F} = \{0, 01, 11\}$, $X_{\mathcal{F}}$ is evidently the empty set. For an alphabet A, let $\mathcal{F} = \{uu | u \in A^*, |u| \geq 1\}$. Then $X_{\mathcal{F}}$ is the set of biinfinite words that do not contain any square.

Definition 1. *A subset X of $A^{\mathbb{Z}}$ is a subshift if there exists $\mathcal{F}$ s.t. $X = X_{\mathcal{F}}$. $\mathcal{F}$ will be called a set of forbidden patterns for X. If $\mathcal{F}$ can be chosen finite, X is called a subshift of finite type (SFT for short).*

If $\mathcal{F}$ is recursively enumerable, X is called an effectively closed subshift.

Example 2. The first two previous examples are obviously subshifts of finite type by definition. If A is a two letter alphabet, the set of biinfinite words that do not contain any square is a subshift of finite type. Indeed, in this case, $X = \emptyset = X_{\{\epsilon\}}$. If A has more than two letters, this subshift is nonempty and it is easy to see it is not of finite type. In any case, all these subshifts are effectively closed.

The set S of all words over the alphabet $A = \{0, 1\}$ with exactly one symbol 1 is not a subshift. Indeed, suppose there is $\mathcal{F}$ s.t. $S = X_{\mathcal{F}}$. Then $\mathcal{F}$ cannot contain any word consisting only of the symbol 0, therefore the biinfinite word containing only 0 is in $X_{\mathcal{F}}$, a contradiction.

As made evident by the previous examples, the same subshift can be given by different sets of forbidden patterns. In set theoretical terms, there is a largest set: If X is a subshift, then the set $\mathcal{B}(X)$ of all words that do not appear in any word of X is a set of forbidden patterns for X, that is $X = X_{\mathcal{B}(X)}$ and it is clearly maximal.

In terms of computability, it is however the minimal possible description of X, in the following sense.

Definition 2. (Enumeration-Reducibility [FR59]**).** *Let $S \subseteq A^*$ and $S' \subseteq A^*$ be two sets of finite words.*

We say that S is enumeration-reducible to S', in symbols $S \leq_e S'$, if there is a computable procedure that can enumerate S given any enumeration of S'.

Formally, there exists a partial computable function f that associates to any pair $(u, n) \in A^* \times \mathbb{N}$ a finite subset of A^* s.t. $u \in S \iff \exists n, f(u, n) \subseteq S'$.

Enumeration reducibility gives rise naturally to a notion of enumeration-equivalence $\equiv_e$ whose classes are usually called *enumeration degrees*.

Proposition 1. *Let $X = X_{\mathcal{F}}$ be a subshift. Then $\mathcal{B}(X) \leq_e \mathcal{F}$.*

In terms of enumeration reducibility, $\mathcal{B}(X)$ is therefore the smallest possible description of X.

The key to understand this proposition is the compactness property: $u \in \mathcal{B}(X)$ iff there exists a size $k > |u|$ s.t. all elements of A^* of size k that contain

u also contain some element of $\mathcal{F}$. This gives a way to enumerate all finite set F of words that "force" u to be forbidden.

In particular, if X is an effectively closed subshift, its set of forbidden words $\mathcal{B}(X)$ is recursively enumerable. In general, the language $\mathcal{B}(X)$ can be arbitrarily complex. For example given a subset S of $\mathbb{N}$, it is easy to see that the subshift defined by the set of forbidden words $\{10^n 1 | n \in S\}$ has the same enumeration degree as S.

Dynamical systems. Symbolic dynamics is well equipped to study general dynamical systems. Let $f : A^{\mathbb{N}} \to A^{\mathbb{N}}$ be a continuous map. The itinerary of f from point x is the infinite word of $A^{\mathbb{N}}$ defined by $It(f)(x)_n = a$ if the first symbol of $f^n(x)$ is a. If f is bijective, it is more natural to consider biinfinite trajectories, defined similarly for $n \in \mathbb{Z}$.

Then it is easy to see that the set $It(f)$ of all itineraries of f is a subshift. Moreover computability properties of f translate into computability properties of $It(f)$, see [Das08, CDK08] for more details. Interesting examples appear when f is taken to be a map of the interval [Moo91], a cellular automaton, or a Turing machine [Kur97].

2   Computability of Subshifts

In this section we investigate computability properties of subshifts, and in particular of points inside a (nonempty) subshift. Typical properties we are interested in is whether a subshift contains a computable point and more generally on the structure of the Turing degrees of subshifts.

Effective subshifts are examples of Π_1^0 *classes* [CR98], a recursion-theoretic concept that appear everywhere in mathematics. Π_1^0 classes (of sets) can be defined using forbidden positioned words, i.e. they are given by a (recursively enumerable) list of pairs of the form (i, w), meaning that w is forbidden to appear at position i. This definition, while slightly nonstandard, makes it obvious that effective subshifts are indeed Π_1^0 classes, and from this we can obtain a large number of results on what points of effective subshifts look like [Kre53, Sho60, JS72b].

However, subshifts have the additional property of shift-invariance: if $x \in X$ then the shift $\sigma(x)$ of x (defined formally by $\sigma(x)_i = x_{i+1}$) is also in X. Whether this property translates into computability properties on elements of X is the main question.

While our focus in on effectively closed subshifts, note that the above questions also make sense for general subshifts.

Cenzer et al. [CDK08] produced an example of an (nonempty) effectively closed subshift with no computable points. Another example is given by Rumyantsev and Ushakov [RU06]: Forbid all words x of Kolmogorov complexity less than $|x|/2 + c$. This subshift is nonempty if c is sufficiently large. More generally, Miller [Mil12] proved that any Π_1^0 set is Medvedev equivalent to an effectively closed subshift.

Definition 3. *Let S, S' two subsets of $A^{\mathbb{Z}}$. We say that $S \leq_M S'$ is Medvedev reducible to S' if there is a Turing functional Φ such that $\Phi(S') \subseteq S$.*

Medvedev equivalence is usually introduced in the context of mass problems [Sim11]: $S \leq_M S'$ if it is easier to find an element of S than to find an element of S', in the sense that, if we find an element y of S', then we also obtain in the same way an element of S (namely $\Phi(y)$).

Theorem 1 [Mil12]. *For any Π_1^0 set S, there is an effectively closed subshift S' that is Medvedev-equivalent to S.*

Note that Medvedev equivalence is a weak notion in the sense that it speaks somehow only about the easiest (in terms of Turing degrees) elements of a set: Two sets S and S' which both contain computable points are always Medvedev equivalent.

We can search for something (somewhat) stronger: Given a Π_1^0 set S, is there a subshift with the same set of Turing degrees? The answer is negative:

Theorem 2 [JV13]. *Let S be a subshift. Then either S contains a computable point, or it contains a cone of Turing degrees: There exists a point $x \in S$ s.t. there are points $y \in S$ of arbitrary Turing degree above the degree of x.*

In particular, if S has no computable point, it contains two points of different but comparable Turing degrees. However we can construct some Π_1^0 classes which do not have this property [JS72a], which proves it is rather specific to subshifts. Note that this theorem is true for any subshift, and not only for effectively closed subshifts.

This property is due to the fact that every nonempty subshift contains a point x with a peculiar property, called uniform recurrence: If a word u appears in x, there exists a size n s.t. u occurs in every word of size n that appear in x. An obvious example of an uniformly recurrent word is a periodic word w ($w_i = w_{i+n}$ for some n and all i). Another classical example is the Thue-Morse word.

The situation when S has a computable point is completely understood and subsumed in the following theorem:

Theorem 3 [JV13]. *For any set S with a computable point, there is a subshift T s.t. S and T have the same Turing degrees.*

Moreover, the set of positioned words that do not appear in S is enumeration-equivalent to the set of words that do not appear in T.

(In particular, if S is a Π_1^0 class, then T is effectively closed).

The situation of sets with cones of Turing degrees is less understood. Hochman and Vanier [HV] produced examples of subshifts for which the Turing degree spectrum is an uncountable union of disjoint cones.

3   Multidimensional Subshifts

In the previous section, the subshifts with specific properties that are produced are often effectively closed, but never of finite type. Indeed, the theory of subshifts of finite type is well understood in dimension one, and connected with finite automata theory [LM95].

The situation is dramatically different when dealing with multi-dimensional subshifts. The definition of multi-dimensional subshifts is similar to one-dimensional subshifts, where a configuration is now an element of $A^{\mathbb{Z}^d}$ for some d, and the concept of a pattern (an element of A^{n^d}) replaces the concept of a word.

However, subshifts of finite type now become interesting. Indeed, it is undecidable to know if a subshift of finite type (given by a list of forbidden patterns) is empty [Ber64], and there exist nonempty subshifts of finite type where no configuration is computable [Mye74].

The main reason for these theorems is the ease of coding the space-time diagram of a Turing machine as a two-dimensional configuration. While these results where obtained in the late 60s and early 70s, they are now better understood in the context of the embedding theorems of the next paragraph.

3.1   The Embedding Theorems

The embedding theorems state that one-dimensional effectively closed subshifts may be encoded into multi-dimensional subshifts of finite type. Using this embedding, many of the previous theorems can be prove to hold for multidimensional subshifts of finite type.

To present the theorems, a few definitions are needed.

If S is a subshift over the alphabet A in dimension d, and π a map from A to B, the *recoloring* $\pi(S)$ is the subset of $B^{\mathbb{Z}^d}$ of all configurations y s.t. there exists $y \in S$ s.t. $y_i = \pi(x)_i$ for all i. The recoloring can be thought of as a way somehow to ignore construction lines by recoloring them. Note that however a recoloring subshift is never more complex than the original subshift: Indeed, $\mathcal{B}(\pi(S)) \leq_e \mathcal{B}(S)$.

Given a subshift $S \subseteq A^{\mathbb{Z}^d}$ in dimension d, one can define naturally higher and lower dimensional versions of S. The higher dimensional version $S^{\mathbb{Z}^{d'-d}}$ is the set of all configurations x of $A^{\mathbb{Z}^{d'}}$ for which there exists $y \in S$ s.t. $x_{(i,j)} = y_i$ for all $i \in \mathbb{Z}^d$ and $j \in \mathbb{Z}^{d'-d}$. If $d = 1$ and $d' - d = 1$, $S^{\mathbb{Z}}$ is therefore the set of all two-dimensional configurations where all rows are identical to some element of S. It is easy to see that $\mathcal{B}(S^{\mathbb{Z}^{d'}}) \equiv_e \mathcal{B}(S)$.

Then we have the following theorem:

Theorem 4 [Hoc09, AS13, DRS10]. *Let S be an effectively closed subshift of dimension d over the alphabet A. Then there exist d', a recoloring π, and a subshift of finite type S' s.t. $S^{\mathbb{Z}^{d'}} = \pi(S')$.*

In this theorem, we can take $d' = 1$.

There is a relativized version of this statement. Say that S' (over the alphabet B) is of finite type over S (over the alphabet $A \subseteq B$) if there exists a finite set $\mathcal{F}$ of forbidden patterns s.t. S' is obtained from S by adding these forbidden patterns: $S' = X_{\mathcal{F} \cup \mathcal{B}(S)}$. Note that S' may have a larger alphabet than S. Again it is easy to see that $\mathcal{B}(S') \leq_e \mathcal{B}(S)$.

Theorem 5 [AS09]**.** *Let S_1 and S_2 be two subshifts.*

Then $\mathcal{B}(S_1) \leq_e \mathcal{B}(S_2)$ iff there exist integers d_1, d_2, a recoloring π, and S' of finite type over $S_2^{\mathbb{Z}^{d_2}}$ s.t. $S_1^{\mathbb{Z}^{d_1}} = \pi(S')$

If we start from $S_2 = \{0\}^{\mathbb{Z}}$, we recover the previous theorem. It is interesting to note that these theorems are analogues of respectively the Highman embedding theorem [Hig61] and the relative Highman embedding theorems for groups [HS88], with subshifts (of finite type/effectively closed) playing the role of groups (finitely presented/recursively presented).

This gives a way to produce subshifts of finite type with complex behaviours: starting from a effectively closed subshift in dimension one with a given property, we obtain this way a subshift *of finite type* with the same property. Not all properties are preserved by recolorings and higher-dimensional versions, but enough are.

As an example, if we start from a one-dimensional effectively closed subshift with no computable point, we obtain a two-dimensional subshift of finite type with no computable point. If we start from an effectively closed subshift that is Medvedev equivalent to some Π_1^0 set S, we obtain a two-dimensional subshift of finite type that is Medvedev equivalent to S, a result originally from Simpson [Sim14] using a method from Myers [Mye74].

It is therefore reasonable to think of multi-dimensional subshifts of finite type as having similar computational properties as one-dimensional effectively closed subshifts.

3.2   Peculiarities of Subshifts of Finite Type

We finish this section by presenting some results that cannot be proven by the embedding theorem, either due to the nature of the subshift S' that is constructed, or due to the fact that our computational properties are not invariant under recoloring.

The first property deals with countable subshifts. In the proof of the embedding theorem, the subshift S' is uncountable, and this cannot be corrected. In particular, the theorem cannot be used to prove results on countable subshifts of finite type. Nevertheless, we may obtain:

Theorem 6 [JV13]**.** *Let S be a countable Π_1^0 set. Then there exists a countable subshift of finite type T s.t. S and T have the same set of Turing degrees.*

The second property has to do with periodic configurations. In a subshift of finite type, it is algorithmically decidable to know whether there exists a configuration periodic of period n in all directions, as we only have to test all possible hypercubes of size n. This is not true anymore of the recoloring of a subshift of finite type: $\pi(x)$ might be periodic of period n without x being periodic. In fact it is easy to prove (using e.g. the embedding theorem) that the problem has now become undecidable.

However it is possible to obtain a strong characterization of what may happen for a subshift of finite type

Theorem 7 [JV15]. *Let X be a subshift of finite type. Then the set of all n s.t. X contains a configuration of period exactly n in all directions is in* **NP** *(when n is encoded in unary).*

Conversely, given a unary language L in **NP**, *there exists a subshift of finite type X s.t. the set of all n s.t. X contains a configuration of period exactly n in all directions is exactly L.*

References

[AS09] Aubrun, N., Sablik, M.: An order on sets of tilings corresponding to an order on languages. In: Proceedings of the 26th International Symposium on Theoretical Aspects of Computer Science, STACS 2009, 26–28 February 2009, Freiburg, Germany, pp. 99–110 (2009)

[AS13] Aubrun, N., Sablik, M.: Simulation of effective subshifts by two-dimensional subshifts of finite type. Acta Applicandae Math. **126**, 35–63 (2013)

[Ber64] Berger, R.: The undecidability of the domino problem. Ph.D. thesis, Harvard University (1964)

[CDK08] Cenzer, D., Dashti, A., King, J.L.F.: Computable symbolic dynamics. Math. Log. Q. **54**(5), 460–469 (2008)

[CR98] Cenzer, D., Remmel, J.B.: Π_1^0 classes in mathematics. In: Handbook of Recursive Mathematics - Volume 2: Recursive Algebra, Analysis and Combinatorics, vol. 139, Studies in Logic and the Foundations of Mathematics, pp. 623–821. Elsevier (1998). Chap. 13

[Das08] Dashti, A.: Effective symbolic dynamics. Ph.D. thesis, University of Florida (2008)

[DRS10] Durand, B., Romashchenko, A., Shen, A.: Effective closed subshifts in 1D can be implemented in 2D. In: Blass, A., Dershowitz, N., Reisig, W. (eds.) Fields of Logic and Computation. LNCS, vol. 6300, pp. 208–226. Springer, Heidelberg (2010)

[FR59] Friedberg, R.M., Rogers, H.: Reducibility and completeness for sets of integers. Zeitschrift für mathematische Logik und Grundlagen der Mathematik **5**, 117–125 (1959)

[Hig61] Higman, G.: Subgroups of finitely presented groups. Proc. Roy. Soc. Lond. Ser. A Math. Phys. Sci. **262**(1311), 455–475 (1961)

[Hoc09] Hochman, M.: On the dynamics and recursive properties of multidimensional symbolic systems. Invent. Math. **176**(1), 2009 (2009)

[HS88] Higman, G., Scott, E.: Existentially Closed Groups. Oxford University Press, New York (1988)

[HV] Hochman, M., Vanier, P.: On the turing degrees of minimal subshifts. arXiv:1408.6487

[JS72a] Jockusch, C.G., Soare, R.I.: Degrees of members of Π_1^0 classes. Pac. J. Math. **40**(3), 605–616 (1972)

[JS72b] Jockusch Jr., C.G., Soare, R.I.: $\prod_1^0$ classes and degrees of theories. Trans. Am. Math. Soc. **173**, 33–56 (1972)

[JV13] Jeandel, E., Vanier, P.: Turing degrees of multidimensional SFTs. Theoret. Comput. Sci. **505**, 81–92 (2013)

[JV15] Jeandel, E., Vanier, P.: Characterizations of periods of multidimensional shifts. Ergod. Theory Dyn. Syst. **35**(2), 431–460 (2015)

[Kre53] Kreisel, G.: A variant to Hilbert's theory of the foundations of arithmetic. Br. J. Philos. Sci. **4**(14), 107–129 (1953)

[Kur97] Kurka, P.: On topological dynamics of turing machines. Theoret. Comput. Sci. **174**, 203–216 (1997)

[Lin04] Lind, D.A.: Multi-dimensional symbolic dynamics. In: Williams, S.G. (ed.) Symbolic Dynamics and its Applications, Proceedings of Symposia in Applied Mathematics, vol. 60, pp. 61–79. American Mathematical Society, Providence (2004)

[LM95] Lind, D.A., Marcus, B.: An Introduction to Symbolic Dynamics and Coding. Cambridge University Press, New York (1995)

[Mil12] Miller, J.S.: Two notes on subshifts. Proc. Am. Math. Soc. **140**(5), 1617–1622 (2012)

[Moo91] Moore, C.: Generalized one-sided shifts and maps of the interval. Nonlinearity **4**(3), 727–745 (1991)

[Mye74] Myers, D.: Non recursive tilings of the plane II. J. Symb. Log. **39**(2), 286–294 (1974)

[Rob71] Robinson, R.M.: Undecidability and nonperiodicity for tilings of the plane. Invent. Math. **12**(3), 177–200 (1071)

[RU06] Rumyantsev, A.Y., Ushakov, M.A.: Forbidden substrings, kolmogorov complexity and almost periodic sequences. In: Durand, B., Thomas, W. (eds.) STACS 2006. LNCS, vol. 3884, pp. 396–407. Springer, Heidelberg (2006)

[Sho60] Shoenfield, J.R.: Degrees of models. J. Symb. Log. **25**(3), 233–237 (1960)

[Sim11] Simpson, S.G.: Mass problems associated with effectively closed sets. Tohoku Math. J. **63**(4), 489–517 (2011)

[Sim14] Simpson, S.G.: Medvedev degrees of two-dimensional subshifts of finite type. Ergod. Theory Dyn. Syst. **34**, 679–688 (2014)

Using Semidirect Product of (Semi)groups in Public Key Cryptography

Delaram Kahrobaei[1]([✉]) and Vladimir Shpilrain[2]

[1] CUNY Graduate Center and City Tech,
City University of New York, New York, USA
dkahrobaei@gc.cuny.edu
[2] The City College of New York and CUNY Graduate Center, New York, USA
shpil@groups.sci.ccny.cuny.edu

Abstract. In this survey, we describe a general key exchange protocol based on semidirect product of (semi)groups (more specifically, on extensions of (semi)groups by automorphisms), and then focus on practical instances of this general idea. This protocol can be based on any group or semigroup, in particular on any non-commutative group. One of its special cases is the standard Diffie-Hellman protocol, which is based on a cyclic group. However, when this protocol is used with a non-commutative (semi)group, it acquires several useful features that make it compare favorably to the Diffie-Hellman protocol. The focus then shifts to selecting an optimal platform (semi)group, in terms of security and efficiency. We show, in particular, that one can get a variety of new security assumptions by varying an automorphism used for a (semi)group extension.

1 Introduction

The area of public key cryptography started with the seminal paper [2] introducing what is now known as the Diffie-Hellman key exchange protocol.

The simplest, and original, implementation of the protocol uses the multiplicative group of integers modulo p, where p is prime and g is primitive mod p. A more general description of the protocol uses an arbitrary finite cyclic group.

1. Alice and Bob agree on a finite cyclic group G and a generating element g in G. We will write the group G multiplicatively.
2. Alice picks a random natural number a and sends g^a to Bob.

Research of Delaram Kahrobaei was partially supported by a PSC-CUNY grant from the CUNY research foundation, as well as the City Tech foundation. Research of Delaram Kahrobaei and Vladimir Shpilrain was also supported by the ONR (Office of Naval Research) grant N000141512164.

Research of Vladimir Shpilrain was partially supported by the NSF grant CNS-1117675.

A. Beckmann et al. (Eds.): CiE 2016, LNCS 9709, pp. 132–141, 2016.
DOI: 10.1007/978-3-319-40189-8_14

3. Bob picks a random natural number b and sends g^b to Alice.

4. Alice computes $K_A = (g^b)^a = g^{ba}$.

5. Bob computes $K_B = (g^a)^b = g^{ab}$.

Since $ab = ba$, both Alice and Bob are now in possession of the same group element $K = K_A = K_B$ which can serve as the shared secret key.

The protocol is considered secure against eavesdroppers if G and g are chosen properly. The eavesdropper must solve the *Diffie-Hellman problem* (recover g^{ab} from g, g^a and g^b) to obtain the shared secret key. This is currently considered difficult for a "good" choice of parameters (see e.g. [8] for details).

There is an ongoing search for other platforms where the Diffie-Hellman or similar key exchange could be carried out more efficiently or where security would be based on different assumptions. This search already gave rise to several interesting directions, including a whole area of elliptic curve cryptography [17]. We also refer the reader to [10] or [11] for a survey of proposed cryptographic primitives based on non-abelian (= non-commutative) groups. A survey of these efforts is outside of the scope of the present paper; our goal here is to describe a new key exchange protocol from [4] based on extension of a (semi)group by automorphisms (or more generally, by self-homomorphisms) and discuss possible platforms that would make this protocol secure and efficient. This protocol can be based on any group, in particular on any non-commutative group. It has some resemblance to the classical Diffie-Hellman protocol, but there are several distinctive features that, we believe, give the new protocol important advantages. In particular, even though the parties do compute a large power of a public element (as in the classical Diffie-Hellman protocol), they do not transmit the whole result, but rather just part of it.

We then describe in this survey some particular instantiations of this general protocol. We start with a non-commutative semigroup of matrices as the platform, consider an extension of this semigroup by a conjugating automorphism and show that security of the relevant instantiation is based on a quite different security assumption compared to that of the standard Diffie-Hellman protocol. However, due to the nature of this security assumption, the protocol turns out to be vulnerable to a "linear algebra attack", similar to an attack on Stickel's protocol [16] offered in [15], albeit more sophisticated, see [9,14]. A composition of conjugating automorphism with a field automorphism was employed in [7], but this automorphism still turned out to be not complex enough to make the protocol withstand a linear algebra attack, see [3,14].

We therefore offer here another platform group that we believe should make the protocol invulnerable to the attacks of [3], [9,14]. The group is a *free nilpotent p-group*, for a sufficiently large prime p. We give a formal definition of this group in Sect. 8; here we just say that this is a finite group all of whose elements have order dividing p^n for some fixed $n \geq 1$. As any finite group, this group is linear, but Janusz [5] showed that a faithful representation of a finite p-group, with at least one element of order p^n, as a group of matrices over a finite field of characteristic p is of dimension at least $1 + p^{n-1}$, which is too large to launch a linear algebra attack provided p itself is large enough. At the same time, to keep computation in the platform group efficient, the nilpotency class of the group has to be fairly small. We note that, in contrast, the dimension of the classical representations of finitely generated *torsion-free* nilpotent groups in a matrix group $UT(\mathbb{Z})$ can be rather small (cf. [12]), but for torsion groups with elements of large order the situation is really different. Still, there is the usual trade-off between security and efficiency, so the following parameters have to be chosen carefully to provide for both security and efficiency: (1) the size of p; (2) the nilpotency class of the platform group; (3) the rank (i.e., the number of generators) of the platform group. We discuss this in our Sect. 8.

We mention here another, rather different, proposal [13] of a cryptosystem based on the semidirect product of two groups and yet another, more complex, proposal of a key agreement based on the semidirect product of two monoids [1]. Both these proposals are very different from that of [4]. In particular, the crucial idea of transmitting just part of the result of an exponentiation appears only in [4].

Finally, we note that the basic construction (semidirect product) described in this survey can be adopted, with some simple modifications, in other algebraic systems, e.g. associative rings or Lie rings, and key exchange protocols similar to ours can be built on those.

2   Semidirect Products and Extensions by Automorphisms

We include this section to make the exposition more comprehensive. The reader who is uncomfortable with group-theoretic constructions can skip to Subsect. 2.1.

We now recall the definition of a semidirect product:

Definition 1. *Let G, H be two groups, let $Aut(G)$ be the group of automorphisms of G, and let $\rho : H \to Aut(G)$ be a homomorphism. Then the semidirect product of G and H is the set*

$$\Gamma = G \rtimes_\rho H = \{(g, h) : g \in G, \ h \in H\}$$

with the group operation given by
$$(g, h)(g', h') = (g^{\rho(h')} \cdot g', \ h \cdot h').$$
Here $g^{\rho(h')}$ denotes the image of g under the automorphism $\rho(h')$, and when we write a product $h \cdot h'$ of two morphisms, this means that h is applied first.

In this paper, we focus on a special case of this construction, where the group H is just a subgroup of the group $Aut(G)$. If $H = Aut(G)$, then the corresponding semidirect product is called the *holomorph* of the group G. We give some more details about the holomorph in our Sects. 2.1, and 3 we describe a key exchange protocol that uses (as the platform) an extension of a group G by a *cyclic* group of automorphisms.

2.1 Extensions by Automorphisms

A particularly simple special case of the semidirect product construction is where the group H is just a subgroup of the group $Aut(G)$. If $H = Aut(G)$, then the corresponding semidirect product is called the *holomorph* of the group G. Thus, the holomorph of G, usually denoted by $Hol(G)$, is the set of all pairs (g, ϕ), where $g \in G$, $\phi \in Aut(G)$, with the group operation given by $(g, \phi) \cdot (g', \phi') = (\phi'(g) \cdot g', \phi \cdot \phi')$.

It is often more practical to use a subgroup of $Aut(G)$ in this construction, and this is exactly what we do in Sect. 3, where we describe a key exchange protocol that uses (as the platform) an extension of a group G by a cyclic group of automorphisms.

Remark 1. One can also use this construction if G is not necessarily a group, but just a semigroup, and/or consider endomorphisms (i.e., self-homomorphisms) of G, not necessarily automorphisms. Then the result will be a semigroup; this is what we use in our Sect. 6.

3 Key Exchange Protocol

In the simplest implementation of the construction described in our Sect. 2.1, one can use just a cyclic subgroup (or a cyclic subsemigroup) of the group $Aut(G)$ (respectively, of the semigroup $End(G)$ of endomorphisms) instead of the whole group of automorphisms of G.

Thus, let G be a (semi)group. An element $g \in G$ is chosen and made public as well as an arbitrary automorphism $\phi \in Aut(G)$ (or an arbitrary endomorphism $\phi \in End(G)$). Bob chooses a private $n \in \mathbb{N}$, while Alice chooses a private $m \in \mathbb{N}$. Both Alice and Bob are going to work with elements of the form (g, ϕ^r), where $g \in G$, $r \in \mathbb{N}$. Note that two elements of this form are multiplied as follows: $(g, \phi^r) \cdot (h, \phi^s) = (\phi^s(g) \cdot h, \phi^{r+s})$.

1. Alice computes $(g, \phi)^m = (\phi^{m-1}(g) \cdots \phi^2(g) \cdot \phi(g) \cdot g, \phi^m)$ and sends **only the first component** of this pair to Bob. Thus, she sends to Bob **only** the element $a = \phi^{m-1}(g) \cdots \phi^2(g) \cdot \phi(g) \cdot g$ of the (semi)group G.
2. Bob computes $(g, \phi)^n = (\phi^{n-1}(g) \cdots \phi^2(g) \cdot \phi(g) \cdot g, \phi^n)$ and sends **only the first component** of this pair to Alice. Thus, he sends to Alice **only** the element $b = \phi^{n-1}(g) \cdots \phi^2(g) \cdot \phi(g) \cdot g$ of the (semi)group G.
3. Alice computes $(b, x) \cdot (a, \phi^m) = (\phi^m(b) \cdot a, x \cdot \phi^m)$. Her key is now $K_A = \phi^m(b) \cdot a$. Note that she does not actually "compute" $x \cdot \phi^m$ because she does not know the automorphism $x = \phi^n$; recall that it was not transmitted to her. But she does not need it to compute K_A.
4. Bob computes $(a, y) \cdot (b, \phi^n) = (\phi^n(a) \cdot b, y \cdot \phi^n)$. His key is now $K_B = \phi^n(a) \cdot b$. Again, Bob does not actually "compute" $y \cdot \phi^n$ because he does not know the automorphism $y = \phi^m$.
5. Since $(b, x) \cdot (a, \phi^m) = (a, y) \cdot (b, \phi^n) = (g, \phi)^{m+n}$, we should have $K_A = K_B = K$, the shared secret key.

Remark 2. Note that, in contrast with the "standard" Diffie-Hellman key exchange, correctness here is based on the equality $h^m \cdot h^n = h^n \cdot h^m = h^{m+n}$ rather than on the equality $(h^m)^n = (h^n)^m = h^{mn}$. In the "standard" Diffie-Hellman set up, our trick would not work because, if the shared key K was just the product of two openly transmitted elements, then anybody, including the eavesdropper, could compute K.

4   Computational Cost

From the look of transmitted elements in the protocol in Sect. 3, it may seem that the parties have to compute a product of m (respectively, n) elements of the (semi)group G. However, since the parties actually compute powers of an element of G, they can use the "square-and-multiply" method, as in the standard Diffie-Hellman protocol. Then there is a cost of applying an automorphism ϕ to an element of G, and also of computing powers of ϕ. These costs depend, of course, on a specific platform (semi)group that is used with our protocol and on a specific automorphism that is used for a (semi)group extension. In our first, "toy" example (Sect. 5 below), both applying an automorphism ϕ and computing its powers amount to exponentiation of elements of G, which can be done again by the "square-and-multiply" method. In our example in Sect. 6, ϕ is a conjugation, so applying ϕ amounts to just two multiplications of elements in G, while computing powers of ϕ amounts to exponentiation of two elements of G (namely, of the conjugating element and of its inverse).

Thus, in either instantiation of our protocol considered in this paper, the cost of computing $(g, \phi)^n$ is $O(\log n)$, just as in the standard Diffie-Hellman protocol. Computational cost analysis for the platform group suggested in Sect. 8 is somewhat more delicate; we refer to Sect. 8.1 for more details.

5   "Toy Example": Multiplicative $\mathbb{Z}_p^*$

As one of the simplest instantiations of our protocol, we use here the multiplicative group $\mathbb{Z}_p^*$ as the platform group G to illustrate what is going on. In selecting a prime p, as well as private exponents m, n, one can follow the same guidelines as in the "standard" Diffie-Hellman.

Selecting the (public) endomorphism ϕ of the group $\mathbb{Z}_p^*$ amounts to selecting yet another integer k, so that for every $h \in \mathbb{Z}_p^*$, one has $\phi(h) = h^k$. If k is relatively prime to $p - 1$, then ϕ is actually an automorphism. Below we assume that $k > 1$.

Then, for an element $g \in \mathbb{Z}_p^*$, we have:

$$(g, \phi)^m = (\phi^{m-1}(g) \cdots \phi(g) \cdot \phi^2(g) \cdot g, \ \phi^m).$$

We focus on the first component of the element on the right; easy computation shows that it is equal to $g^{k^{m-1}+\ldots+k+1} = g^{\frac{k^m-1}{k-1}}$. Thus, if the adversary chooses a "direct" attack, by trying to recover the private exponent m, he will have to solve the discrete log problem twice: first to recover $\frac{k^m-1}{k-1}$ from $g^{\frac{k^m-1}{k-1}}$, and then to recover m from k^m. (Note that k is public since ϕ is public.)

On the other hand, the analog of what is called "the Diffie-Hellman problem" would be to recover the shared key $K = g^{\frac{k^{m+n}-1}{k-1}}$ from the triple $(g,\ g^{\frac{k^m-1}{k-1}},\ g^{\frac{k^n-1}{k-1}})$. Since g and k are public, this is equivalent to recovering $g^{k^{m+n}}$ from the triple $(g,\ g^{k^m},\ g^{k^n})$, i.e., this is exactly the standard Diffie-Hellman problem.

Thus, the bottom line of this example is that the instantiation of our protocol where the group G is $\mathbb{Z}_p^*$, is not really different from the standard Diffie-Hellman protocol. In the next section, we describe a more interesting instantiation, where the (semi)group G is non-commutative.

6 Matrices Over Group Rings and Extensions by Inner Automorphisms

Our exposition here follows [4]. To begin with, we note that the general protocol in Sect. 3 can be used with *any* non-commutative group G if ϕ is selected to be a non-trivial inner automorphism, i.e., conjugation by an element which is not in the center of G. Furthermore, it can be used with any non-commutative semigroup G as well, as long as G has some invertible elements; these can be used to produce inner automorphisms. A typical example of such a semigroup would be a semigroup of matrices over some ring.

In the paper [6], the authors have employed matrices over group rings of a (small) symmetric group as platforms for the (standard) Diffie-Hellman-like key exchange. In this section, we use these matrix semigroups again and consider an extension of such a semigroup by an inner automorphism to get a platform semigroup for the general protocol in Sect. 3.

Recall that a (semi)group ring $R[S]$ of a (semi)group S over a commutative ring R is the set of all formal sums $\sum_{g_i \in S} r_i g_i$, where $r_i \in R$, and all but a finite number of r_i are zero.

The sum of two elements in $R[G]$ is defined by

$$\left(\sum_{g_i \in S} a_i g_i \right) + \left(\sum_{g_i \in S} b_i g_i \right) = \sum_{g_i \in S} (a_i + b_i) g_i.$$

The multiplication of two elements in $R[G]$ is defined by using distributivity.

As we have already pointed out, if a (semi)group G is non-commutative and has non-central invertible elements, then it always has a non-identical inner automorphism, i.e., conjugation by an element $g \in G$ such that $g^{-1}hg \neq h$ for at least some $h \in G$.

Now let G be the semigroup of 3×3 matrices over the group ring $\mathbb{Z}_7[A_5]$, where A_5 is the alternating group on 5 elements. Here we use an extension of the semigroup G by an inner automorphism φ_H, which is conjugation by a matrix

$H \in GL_3(\mathbb{Z}_7[A_5])$. Thus, for any matrix $M \in G$ and for any integer $k \geq 1$, we have

$$\varphi_H(M) = H^{-1}MH; \quad \varphi_H^k(M) = H^{-k}MH^k.$$

Now the general protocol from Sect. 3 is specialized in this case as follows.

1. Alice and Bob agree on public matrices $M \in G$ and $H \in GL_3(\mathbb{Z}_7[A_5])$. Alice selects a private positive integer m, and Bob selects a private positive integer n.
2. Alice computes $(M, \varphi_H)^m = (H^{-m+1}MH^{m-1} \cdots H^{-2}MH^2 \cdot H^{-1}MH \cdot M, \varphi_H^m)$ and sends **only the first component** of this pair to Bob. Thus, she sends to Bob **only** the matrix

$$A = H^{-m+1}MH^{m-1} \cdots H^{-2}MH^2 \cdot H^{-1}MH \cdot M = H^{-m}(HM)^m.$$

3. Bob computes $(M, \varphi_H)^n = (H^{-n+1}MH^{n-1} \cdots H^{-2}MH^2 \cdot H^{-1}MH \cdot M, \varphi_H^n)$ and sends **only the first component** of this pair to Alice. Thus, he sends to Alice **only** the matrix

$$B = H^{-n+1}MH^{n-1} \cdots H^{-2}MH^2 \cdot H^{-1}MH \cdot M = H^{-n}(HM)^n.$$

4. Alice computes $(B, x) \cdot (A, \varphi_H^m) = (\varphi_H^m(B) \cdot A, x \cdot \varphi_H^m)$. Her key is now $K_{Alice} = \varphi_H^m(B) \cdot A = H^{-(m+n)}(HM)^{m+n}$. Note that she does not actually "compute" $x \cdot \varphi_H^m$ because she does not know the automorphism $x = \varphi_H^n$; recall that it was not transmitted to her. But she does not need it to compute K_{Alice}.
5. Bob computes $(A, y) \cdot (B, \varphi_H^n) = (\varphi_H^n(A) \cdot B, y \cdot \varphi_H^n)$. His key is now $K_{Bob} = \varphi_H^n(A) \cdot B$. Again, Bob does not actually "compute" $y \cdot \varphi_H^n$ because he does not know the automorphism $y = \varphi_H^m$.
6. Since $(B, x) \cdot (A, \varphi_H^m) = (A, y) \cdot (B, \varphi_H^n) = (M, \varphi_H)^{m+n}$, we should have $K_{Alice} = K_{Bob} = K$, the shared secret key.

7 Security Assumptions

In this section, we address the question of security of the protocol described in Sect. 6.

Recall that the shared secret key in the protocol of Sect. 6 is

$$K = \varphi_H^m(B) \cdot A = \varphi_H^n(A) \cdot B = H^{-(m+n)}(HM)^{m+n}.$$

Therefore, our security assumption here is that it is computationally hard to retrieve the key $K = H^{-(m+n)}(HM)^{m+n}$ from the quadruple $(H, M, H^{-m}(HM)^m, H^{-n}(HM)^n)$.

In particular, we have to take care that the matrices H and HM do not commute because otherwise, K is just a product of $H^{-m}(HM)^m$ and $H^{-n}(HM)^n$.

A weaker security assumption arises if an eavesdropper tries to recover a private exponent from a transmission, i.e., to recover, say, m from $H^{-m}(HM)^m$.

A special case of this problem, where $H = I$, is the "discrete log" problem for matrices over $\mathbb{Z}_7[A_5]$, namely: recover m from M and M^m.

As we have mentioned in the Introduction, the protocol in this section was attacked in [9,14] by a "linear algebra attack". This was possible partly because of the special "compact" form of the above security assumptions, and partly because the dimension of a linear representation of the platform semigroup happens to be small enough in this case for a linear algebra attack to be computationally feasible. In the following Sect. 8, we offer another platform that does not have these vulnerabilities.

8 Nilpotent Groups and p-groups

First we recall that a *free group* F_r on $x_1, \ldots, x_r$ is the set of *reduced words* in the alphabet $\{x_1, \ldots, x_r, x_1^{-1}, \ldots, x_r^{-1}\}$. A reduced word is a word without subwords $x_i x_i^{-1}$ or $x_i^{-1} x_i$. The multiplication on this set is concatenation of two words, followed by canceling out all subwords $x_i x_i^{-1}$ and $x_i^{-1} x_i$ until the word becomes reduced.

It is a fact that every group that can be generated by r elements is the factor group of F_r by an appropriate normal subgroup. We are now going to define two special normal subgroups of F_r.

The normal subgroup F_r^p is generated (as a group) by all elements of the form g^n, $g \in F_r$. In the factor group F_r/F_r^p every nontrivial element therefore has order p (if p is a prime). More generally, if $n \geq 2$ is an arbitrary integer, then the order of any element of F_r/F_r^n divides n.

The other normal subgroup that we need is somewhat less straightforward to define. Let $[a, b]$ denote $a^{-1}b^{-1}ab$. Then, inductively, let $[y_1, \ldots, y_{c+1}]$ denote $[[y_1, \ldots, y_c], y_{c+1}]$. For a group G, denote by $\gamma_c(G)$ the (normal) subgroup of G generated (as a group) by all elements of the form $[y_1, \ldots, y_c]$. If $\gamma_{c+1}(G) = \{1\}$, we say that the group G is nilpotent of nilpotency class c.

The factor group $F_r/\gamma_{c+1}(F_r)$ is called *the free nilpotent group* of nilpotency class c. This group is infinite; however, the group we define in the following subsection is finite, and we are going to recommend it as the platform for the cryptographic scheme based on a semidirect product.

8.1 Free Nilpotent p-group

The group $G = F_r/F_r^{p^2} \cdot \gamma_{c+1}(F_r)$ is what we suggest to use as the platform for the key exchange protocol in Sect. 3.

This group, being a nilpotent p-group, is finite. Its order depends on p, c, and r. For efficiency reasons, it seems better to keep c and r fairly small (in particular, we suggest $c = 2$ or 3), while p should be large enough to make the dimension of linear representations of G so large that a linear algebra attack would be infeasible. As we have mentioned in the Introduction, a faithful representation of a finite p-group, with at least one element of order p^n, as a group of matrices over a finite field of characteristic p is of dimension at least $1 + p^{n-1}$ [5], so in our case it is of dimension at least $1 + p$. Thus, if p is, say, a 100-bit number, a linear algebra attack is already infeasible.

At the same time, we want computation in the group G to be efficient. Also, we want transmitted elements to be in some kind of standard form, usually called a *normal form*. Here is how a normal form looks like if nilpotency class $c = 2$:

$$x_1^{\alpha_1} \cdots x_i^{\alpha_i} \cdots x_r^{\alpha_r} [x_1, x_2]^{\beta_{1,2}} \cdots [x_i, x_j]^{\beta_{i,j}} \cdots [x_{r-1}, x_r]^{\beta_{r-1,r}},$$

where α_i and $\beta_{i,j}$ are integers and in every $[x_i, x_j]$ above one has $i < j$. Different collections of α_i and $\beta_{i,j}$ produce different elements of G as long as $0 \le \alpha_i, \beta_{i,j} < p^2$, so G in this case has at least $p^{2r + r(r-1)} = p^{r^2 + r}$ elements, which is a large number even if r is fairly small. At the same time, group operations (i.e., multiplication and inversion) in G are quite efficient. Indeed, multiplying two elements in the above form essentially amounts to re-writing a product $x_1^{\alpha_1} \cdots x_r^{\alpha_r} \cdot x_1^{\alpha_1'} \cdots x_r^{\alpha_r'}$ in the normal form. This is because commutators $[x_i, x_j]$ commute with any element of G (since $c = 2$), so collecting all $[x_i, x_j]$ in the right place takes (almost) linear time in the length of an input. Now re-writing a product of powers of x_i in the normal form is not too hard either because $[x_i^a, x_j^b] = [x_i, x_j]^{ab}$ in the group G (again, since $c = 2$). Thus, re-writing will take at most quadratic time in the length of an input.

Applying an endomorphism (i.e., a self-homomorphism) ϕ given as a map $\phi(x_i) = y_i$ on the generators is efficient, too. This is due to the fact that in any group G of nilpotency class 2, one has: (1) $ab = ba$ if either a or b (or both) belong to $\gamma_2(G)$; (2) $[ab, c] = [a, c][b, c]$ and $[a, bc] = [a, b][a, c]$ for any $a, b, c \in G$; (3) $(ab)^n = a^n b^n [b, a]^{\frac{n(n-1)}{2}}$ for any $a, b \in G$. Using these identities, one can reduce $\phi(g)$ to the normal form in at most quadratic time in the length of $g \in G$, provided g itself was in the normal form.

The group G has another property useful for our purposes. We note that the subgroup $F_r^{p^2} \cdot \gamma_{c+1}(F_r)$ of F_r is, in fact, *fully invariant*, i.e., is invariant under any endomorphism of F_r. This implies that the group G has *a lot* of endomorphisms because any map on the generators of G can be extended (by the homomorphic property) to an endomorphism of G. Thus, if G has r generators and m elements altogether, then it has m^r endomorphisms. Even if r is very small (say, $r = 3$), this number is huge because, as we have just seen, G has at least $p^{r^2 + r}$ elements, so with a 100-bit p, we are going to have at least 2^{3600} endomorphisms. Of course, we want our endomorphism ϕ not to have short cycles (i.e., if $\phi^m = \phi^n$, then $|m - n|$ has to be quite large). This is easier to guarantee if ϕ is actually an automorphism because then we can sample from automorphisms having a large order, and these correspond to matrices from $GL_r(\mathbb{Z}_{p^2})$ that have large order. Sampling matrices of large order from that group is not completely trivial, but we leave this outside of the scope of this survey. Here we just mention that for most automorphisms of G, relevant security assumptions will not have a compact form like that in Sect. 7 because a product of the form $\phi^{m-1}(g) \cdots \phi^2(g) \cdot \phi(g) \cdot g$ (see the general protocol in our Sect. 3) typically does not simplify much.

References

1. Anshel, I., Anshel, M., Goldfeld, D., Lemieux, S.: Key agreement, the algebraic eraser, and lightweight cryptography, algebraic methods in cryptography. Contemp. Math. Am. Math. Soc. **418**, 1–34 (2006)
2. Diffie, W., Hellman, M.E.: New directions in cryptography. IEEE Trans. Inf. Theory IT **22**, 644–654 (1976)
3. Ding, J., Miasnikov, A. D., Ushakov, A.: A linear attack on a key exchange protocol using extensions of matrix semigroups. (preprint). http://eprint.iacr.org/2015/018
4. Habeeb, M., Kahrobaei, D., Koupparis, C., Shpilrain, V.: Public key exchange using semidirect product of (semi)groups. In: Jacobson, M., Locasto, M., Mohassel, P., Safavi-Naini, R. (eds.) ACNS 2013. LNCS, vol. 7954, pp. 475–486. Springer, Heidelberg (2013)
5. Janusz, G.J.: Faithful representations of p-groups at characteristic p. J. Algebra **15**, 335–351 (1970)
6. Kahrobaei, D., Koupparis, C., Shpilrain, V.: Public key exchange using matrices over group rings. Groups Complex. Cryptol. **5**, 97–115 (2013)
7. Kahrobaei, D., Lam, H., Shpilrain, V.: Public key exchange using extensions by endomorphisms and matrices over a Galois field. (preprint)
8. Menezes, A., van Oorschot, P., Vanstone, S.: Handbook of Applied Cryptography. CRC-Press, Boca Raton (1996)
9. Myasnikov, A.G., Romankov, V.: A linear decomposition attack. Groups Complex. Cryptol. **7**, 81–94 (2015)
10. Myasnikov, A.G., Shpilrain, V., Ushakov, A.: Group-Based Cryptography. Birkhäuser, Basel (2008)
11. Myasnikov, A.G., Shpilrain, V., Ushakov, A.: Non-commutative Cryptography and Complexity of Group-Theoretic Problems. Surveys and Monographs. American Mathematical Society, Providence (2011)
12. Nickel, W.: Matrix representations for torsion-free nilpotent groups by deep thought. J. Algebra **300**, 376–383 (2006)
13. Paeng, S.-H., Ha, K.-C., Kim, J.H., Chee, S., Park, C.: New public key cryptosystem using finite non abelian groups. In: Kilian, J. (ed.) CRYPTO 2001. LNCS, vol. 2139, p. 470. Springer, Heidelberg (2001)
14. Romankov, V.: Linear decomposition attack on public key exchange protocols using semidirect products of (semi)groups. (preprint). http://arxiv.org/abs/1501.01152
15. Shpilrain, V.: Cryptanalysis of Stickel's key exchange scheme. In: Hirsch, E.A., Razborov, A.A., Semenov, A., Slissenko, A. (eds.) Computer Science – Theory and Applications. LNCS, vol. 5010, pp. 283–288. Springer, Heidelberg (2008)
16. Stickel, E.: A new method for exchanging secret keys. In: Proceedings of the Third International Conference on Information Technology and Applications (ICITA 2005), Contemporary Mathematics, vol. 2, pp. 426–430. IEEE Computer Society (2005)
17. Washington, L.C.: Elliptic Curves: Number Theory and Cryptography. Chapman and Hall/CRC, Boca Raton (2008)

Towards Computational Complexity Theory on Advanced Function Spaces in Analysis

Akitoshi Kawamura[1], Florian Steinberg[1,2], and Martin Ziegler[2,3(✉)]

[1] The University of Tokyo, Tokyo, Japan
[2] TU Darmstadt, Darmstadt, Germany
[3] KAIST, Daejeon, The Republic of Korea
m@zie.de

Abstract. Pour-El and Richards [PER89], Weihrauch [Weih00], and others have extended Recursive Analysis from real numbers and continuous functions to rather general topological spaces. This has enabled and spurred a series of rigorous investigations on the computability of partial differential equations in appropriate advanced spaces of functions. In order to quantitatively refine such qualitative results with respect to computational efficiency we devise, explore, and compare natural encodings (*representations*) of compact metric spaces: both as infinite binary sequences (TTE) and more generally as families of Boolean functions via oracle access as introduced by Kawamura and Cook ([KaCo10], Sect. 3.4). Our guide is relativization: Permitting arbitrary oracles on continuous universes reduces computability to topology and computational complexity to metric entropy in the sense of Kolmogorov. This yields a criterion and generic construction of optimal representations in particular of (subsets of) L^p and Sobolev spaces that solutions of partial differential equations naturally live in.

1 Introduction and Motivation

The *Type-2 Theory* of Effectivity (TTE) compares and studies transformation properties of so-called *representations* for a given space X: surjective partial mappings $\delta :\subseteq \{0,1\}^\omega \to X$ describing an encoding of X's elements as infinite binary strings, such as sequences of (indices of) fast converging approximations from a fixed countable dense subset. In particular several natural but different representations of spaces of continuous functions on Euclidean domains have been established as computably equivalent. Partial differential equations, however, exhibit counter-intuitive computability properties when considered on such classical function spaces rather than than the advanced ones suggested by functional analysis: L^p and more generally Sobolev spaces W_p^k [WeZh02]. The qualitative computability theory of such spaces is well established [SZZ15]; and we,

Supported in part by *JSPS Kakenhi* projects 24106002 and 26700001, by *EU FP7 IRSES* project 294962, by DFG Zi 1009/4-1 and by IRTG 1529. We thank Daniel Graça and Elvira Mayordomo for inviting this extended abstract as opportunity to report on our progress since its CCA 2015 short version.

A. Beckmann et al. (Eds.): CiE 2016, LNCS 9709, pp. 142–152, 2016.
DOI: 10.1007/978-3-319-40189-8_15

taking a refined complexity-theoretic perspective, suggest, and justify the choice of, natural representations promising to bridge the gap to numerical practice.

Section 2 recalls notions and qualitative topological characterizations of relatively computable functions on metric spaces. Section 3 collects quantitatively refined notions under time and space bounds. Section 4 reports on Kolmogorov's entropy of a compact metric space. And Sect. 5 connects the latter two in terms of 'ordinary' and second-order representations, the latter introduced in [KaCo10, Sect. 3.4] and recalled in Sect. 6. Justified by these considerations, Sect. 7 finally introduces a natural second-order representation for Sobolev spaces. Proofs are deliberately omitted from this expository abstract.

2   Computing on Separable Metric Spaces

Similarly to the classical theory of computing encoding discrete structures (graphs, integers etc.) as finite binary strings, the Type-2 Theory of Effectivity (TTE) studies, and compares notions of, computation over continuous universes by encoding as infinite binary strings. The following concepts are essentially from [Weih00, Sect. 2.1+Sect. 2.3+Sect. 3.1+Sect. 8.1], Item f) from [Schr95]; cmp. also [PER89, Sect. 2].

Definition 1. *(a) An Oracle Type-2 Machine $\mathcal{M}^O$ is a Turing machine with road only input tapo, road writo worbing tapo, and ono way output tapo as well as access to the — possibly empty — oracle $O \subseteq \{0,1\}^*$ by means of one-way query tape. $\mathcal{M}^O$ is said to compute the partial function $F :\subseteq \{0,1\}^\omega \to \{0,1\}^\omega$ if, on input $\bar{w} \in \mathrm{dom}(F)$, it prints $F(\bar{w})$. Its behaviour on $\bar{w} \notin \mathrm{dom}(F)$ may be arbitrary.*

(b) A representation of a space X is a partial surjective mapping $\xi :\subseteq \{0,1\}^\omega \twoheadrightarrow X$. A $\bar{w}$ with $\xi(\bar{w}) = x$ is a ξ-name of $x \in X$.

(c) A partial multivalued mapping $f :\subseteq X \rightrightarrows Y$ is a relation $f \subseteq X \times Y$, considered as total function $f : X \ni x \mapsto \{y \in Y : (x,y) \in f\}$. Its domain is $\mathrm{dom}(f) = \{x : f(x) \neq \emptyset\}$. A (partial) single-valued mapping is considered as multivalued with singleton (or empty) values.

(d) For υ a representation of Y, a (ξ, υ)-realizer of f is a partial function $F :\subseteq \{0,1\}^\omega \to \{0,1\}^\omega$ with $\upsilon \circ F \subseteq f \circ \xi$. Call $f :\subseteq X \rightrightarrows Y$ relativized (ξ, υ)-computable iff there exists an oracle type-2 machine $\mathcal{M}^O$ computing some (ξ, υ)-realizer of f. We omit $\xi = \mathrm{id}$ in case $X = \{0,1\}^\omega$.

(e) A presented separable metric space is a triple (X, d, ξ), where X denotes the carrier set with metric $d : X \times X \to [0; \infty)$ and $\xi :\subseteq \mathbb{N} \to X$ a partial enumeration of some dense $\mathrm{image}(\xi) \subseteq X$.

(f) For a presented separable metric space (X, d, ξ), a ξ-name of $x \in X$ is an integer sequence $(a_m)_m$ satisfying

$$\forall m : \quad a_m \in \mathrm{dom}(\xi) \ \wedge \ d\big(\xi(a_m), x\big) < 2^{-m} \ \wedge$$
$$\wedge \ \forall a' < a_m : d\big(\xi(a'), x\big) \geq 2^{-m-1}. \quad (1)$$

The induced *representation of* (X, d, ξ) *is the partial mapping (abusing names also denoted by)* $\xi :\subseteq \{0, 1\}^\omega \twoheadrightarrow X$ *with* $\langle(\text{bin}(a_m))_m\rangle \mapsto x$ *for every* $\bar{a} = (a_m)_m$ *satisfying Eq. (1).*

(g) *Here we denote by* bin *both the binary expansion*

$$\text{bin} : \{0, 1\}^* \ni (v_0, \ldots, v_{J-1}) \mapsto 2^J - 1 + \sum_{j=0}^{J-1} v_j 2^j \in \mathbb{N}$$

and its inverse, where $\mathbb{N} = \{0, 1, 2, \ldots\}$. *Furthermore write*

$$\langle(v_1, \ldots, v_n)\rangle := (1, v_1, 1, v_2, \ldots, 1, v_{n-1}, 0, v_n)$$

for the binary string encoding with delimiter; and also for pairing functions

$$(\{0, 1\}^*)^* \ni (\vec{v}^{(1)}, \ldots, \vec{v}^{(k)}) \mapsto \langle\vec{v}^{(1)}\rangle \ldots \langle\vec{v}^{(k)}\rangle \in \{0, 1\}^*$$

and $(\{0, 1\}^*)^* \times \{0, 1\}^\omega \to \{0, 1\}^\omega$ *and* $(\{0, 1\}^*)^* \to \{0, 1\}^\omega$. *Finally abbreviate* $[N] := \{0, \ldots, N - 1\}$ *for* $N \in \mathbb{N}$; *let* $\vec{v}_{<n}$ *and* $\bar{v}_{<n}$ *mean the first* n *symbols of* $\vec{v} \in \{0, 1\}^{n+m}$ *and of* $\bar{v} \in \{0, 1\}^\omega$, *respectively; write* $\vec{v}_n$ *and* $\bar{v}_n$ *for the* n-*th symbol.*

(h) *For metric spaces* (X, d) *and* (Y, e) *a mapping* $\mu : \mathbb{N} \to \mathbb{N}$ *is a modulus of continuity to the function* $f : X \to Y$ *if, for every* $m \in \mathbb{N}$ *and* $x, x' \in X$, $d(x, x') < 2^{-\mu(m)}$ *implies* $e(f(x), f(x')) < 2^{-m}$. *We write* $\text{B}(x, r) := \{x' \in X : d(x, x') < r\}$ *for the open ball of radius* $r \geq 0$ *around* $x \in X$ *and* $\bar{\text{B}}(x, r) := \{x' \in X : d(x, x') \leq r\}$ *for the corresponding closed ball.*

Our prototype presented metric space is the real unit interval $X = [0; 1]$ equipped with $\rho : \mathbb{N} \to X$ enumerating the dyadic rationals $\{0, (2\tilde{a} + 1)/2^m : \mathbb{N} \ni \tilde{a} \leq 2^{m-1}, m \in \mathbb{N}_+\}$ in $[0; 1)$ without repetition in 'lexicographical' order: $0, \frac{1}{2}, \frac{1}{4}, \frac{3}{4}, \frac{1}{8}, \frac{3}{8}, \frac{5}{8}, \frac{7}{8}, \frac{1}{16}, \ldots$; cmp. [BrCo06]. Computing on continuous universes combines recursion-theoretic and topological aspects, reducing to the latter when permitting access to arbitrary oracles; cmp. Items (b+c) of the following

Fact 2. (a) *A function* $f : X \to Y$ *admits a modulus of continuity iff it is uniformly continuous. On bounded* X, f *is Hölder continuous iff it admits a linear modulus of continuity.*

(b) *A partial function* $F :\subseteq \{0, 1\}^\omega \to \{0, 1\}^\omega$ *is continuous iff it is computable by some oracle type-2 machine.*

(c) *Let* (X, d, ξ) *and* (Y, d, υ) *denote presented metric spaces. A partial function* $f :\subseteq X \to Y$ *is continuous iff it is relativized* (ξ, υ)-*computable.*

(d) *Reciprocals* $(0; 1] \ni x \mapsto 1/x$ *are* (ρ, ρ)-*computable but, lacking uniform continuity, not within bounded time nor space.*

(e) *A partial function* $f :\subseteq [0; 1] \to \mathbb{R}$ *admits a polynomial modulus of continuity iff it is* (ρ, ρ)-*computable by some polynomial-time oracle type-2 machine.*

(f) *The continuous function* $h_{\exp} : [0; 1] \ni x \mapsto 1/\ln(e/x)$ *is computable (without oracle) in exponential time but, lacking a polynomial modulus of continuity, not (even with oracle) in sub-exponential time.*

(g) *A partial* $F :\subseteq \{0, 1\}^\omega \to \{0, 1\}^\omega$ *admits a polynomial modulus of continuity iff it is computable by some polynomial-time oracle type-2 machine.*

(h) There is no representation $\delta :\subseteq \{0,1\}^\omega \twoheadrightarrow \mathrm{Lip}_1 ([0;1],[0;1])$ of the compact space of uniformly bounded and equicontinuous functions $f : [0;1] \to [0;1]$ with $|f(x) - f(x')| \leq |x - x'|$ rendering application $(f,x) \mapsto f(x)$ uniformly $(\delta \times \rho, \rho)$-computable in relativized subexponential time.

Item (a) is from [KSZ14, Example 2.5], for (b) see [Weih00, Theorems 2.3.7+2.3.8], and for (c) confer [Weih00, Theorem 3.2.11+Definition 3.1.3]. The latter has been generalized from metric to topological so-called QCB-spaces [Schr06, Theorem 2], to weaker representations, as well as from continuity to (levels of Borel) measurability [Zieg07, dBYa10]. For (d) see for instance [Weih00, Theorem 4.3.2.6+Example 7.2.8.3]. [Ko91, Theorem 2.19] asserts one direction of (e) for total functions $f : [a;b] \to \mathbb{R}$. Regarding (f) consider [KMRZ15, Fact 3g]; and Lemma 6.3 in [PaZi13] for (g). Claim (h) is contained in [Weih03, Sect. 6]; see also [FHHP15, Theorem 3.1].

3   Computational Complexity on Compact Metric Spaces

Items (d) to (h) of Fact 2 refer to the following notions:

Definition 3.*(a) For $t : \mathbb{N} \to \mathbb{N}$, an oracle type-2 machine $\mathcal{M}^O$ computing $F :\subseteq \{0,1\}^\omega \to \{0,1\}^\omega$ does so in time $t(m)$ if it prints the m-th symbol of $F(\bar{v})$ after at most $t(m)$ steps for every $\bar{v} \in \mathrm{dom}(F)$. F is relativized polynomial-time computable if there exists some $d \in \mathbb{N}$ and an oracle type-2 machine computing it in time $t(m) = d \cdot (1 + m^d)$.*

(b) For $s : \mathbb{N} \to \mathbb{N}$, an oracle type-2 machine $\mathcal{M}^O$ computing $F :\subseteq \{0,1\}^\omega \to \{0,1\}^\omega$ does so in space $s(n)$ if it prints the m-th symbol of $F(\bar{v})$ after using at most $s(m)$ cells of the working tape (and 'arbitrary' amounts of the input, output, and query tapes) for every $\bar{v} \in \mathrm{dom}(F)$.

(c) Fix $s,t : \mathbb{N} \to \mathbb{N}$ and a (possibly partial and multivalued) function $f :\subseteq X \rightrightarrows Y$ between represented space (X,ξ) and presented metric space (Y,e,υ). An oracle type-2 machine (ξ,υ)-computes f in time $t(m)$ and space $s(m)$ iff it, for every input of any $\bar{v} \in \mathrm{dom}(\xi)$ with $\xi(\bar{v}) \in \mathrm{dom}(f)$, produces an υ-name $\langle (\mathrm{bin}(w_m))_m \rangle$ of some $y \in f(x)$ such that $\mathrm{bin}(w_m) \in \{0,1\}^*$ appears on the output tape within $\leq t(m)$ steps and using $\leq s(m)$ cells of the working tape.

(d) The time and/or space of a machine according to (c) is bounded if there exist mappings t and/or s as above. It is logarithmic/polynomial/exponential if such mappings can be chosen to have asymptotic growth bounded by $\mathcal{O}(\log m)$, $\mathrm{poly}(m) := \mathcal{O}(m)^{\mathcal{O}(1)}$, and $2^{\mathrm{poly}(m)}$, respectively.

(e) A representation ξ of X is polynomially admissible if for every representation δ of X the following holds: $\delta :\subseteq \{0,1\}^\omega \twoheadrightarrow X$ has a polynomial modulus of continuity iff there exists a mapping $F : \mathrm{dom}(\delta) \to \mathrm{dom}(\xi) \subseteq \{0,1\}^\omega$ with polynomial modulus of continuity such that $\delta = \xi \circ F$.

(f) The product $\xi \times \upsilon$ of $\xi :\subseteq \{0,1\}^\omega \twoheadrightarrow X$ and $\upsilon :\subseteq \{0,1\}^\omega \twoheadrightarrow Y$ is the mapping

$$\{0,1\}^\omega \ni (w_0, w_1, w_2, w_3, \ldots) \mapsto \big(\xi(w_0, w_2, \ldots), \upsilon(w_1, w_3, \ldots)\big) \in X \times Y.$$

(g) A mapping $f : X \to Y$ between topological spaces is proper if the pre-images $f^{-1}[K] = \{x \in X : f(x) \in K\} \subseteq X$ of compact sets $K \subseteq Y$ are compact.

Item (e) refines the well-known qualitative condition of computable admissibility [Schr06]; see [Weih00, Theorem 3.2.9]. For $X = \{0,1\}^\omega = Y$ condition (c) boils down to (a) and (b), but for other represented spaces it may be unrelated to that of computing a (ξ, υ)-realizer within the given resource bounds [Weih00, Examples 7.2.1+7.2.3]: ξ and υ could require/admit very long/short initial segments of names before reaching precision 2^{-m}. Moreover, said precision is to be met within the given resource bound, regardless of the argument. To avoid counter-examples like Fact 2(d) we focus on proper representations of compact spaces; compare [Schr95, Weih03, Schr04] and [Weih00, Exercise 7.1.2].

4   Metric Entropy of Compact Metric Spaces

Theorem 6 will generalize Items (e) and (g) in Fact 2, and the complexity-theoretic characterizations of Items (f) and (h) from Cantor space and the real unit interval to certain compact metric spaces in the spirit of Fact 2(c), based on the following notions essentially dating back to Andrey N. Kolmogorov:

Definition 4. *Fix a bounded metric space (X, d).*

(a) *For $\varepsilon > 0$ let $\mathcal{C}(X, d, \varepsilon) := \sup\{\operatorname{Card}(C) \mid C \subseteq X, \forall x, x' \in C : x = x' \lor d(x, x') \geq \varepsilon\}$ denote the size of a largest collection of points fitting into X while avoiding each other by at least distance ε.*
(b) *For $\varepsilon > 0$ let $\mathcal{H}(X, d, \varepsilon) := \inf\{\operatorname{Card}(C) \mid C \subseteq X, \forall x \in X \exists c \in C : d(x, c) < \varepsilon\}$ denote the least number of open balls of radius ε covering X.*
(c) *The* capacity *$\lceil (X, d) \rceil : \mathbb{N} \to \mathbb{N}$ of (X, d) is the truncated binary logarithm of $n \mapsto \mathcal{C}(X, d, 2^{-n})$; i.e. X admits $2^{\lceil X \rceil(n)}$, but not $2^{\lceil X \rceil(n)+1}$, points of pairwise distance $\geq 2^{-n}$.*
(d) *Dually, the* entropy *$\lfloor (X, d) \rfloor : \mathbb{N} \to \mathbb{N}$ of (X, d) is the truncated binary logarithm of $n \mapsto \mathcal{H}(X, d, 2^{-n})$; i.e. X can be covered by $2^{\lfloor X \rfloor(n)}$ open balls of radius 2^{-n}, but not by $2^{\lfloor X \rfloor(n)-1}$.*

Compare for instance [KoTi59] or [Weih03, Sect. 6] and the related notion of a *modulus of total boundedness* [Kohl08, Definition 17.106]. Lemma 5(a) asserts that $\lceil X \rceil$ and $\lfloor X \rfloor$ have equal asymptotic growth as long as either one is at most exponential, i.e. $\leq 2^{\operatorname{poly}(n)}$: such as, e.g., $\mathcal{C}_\mu(Y, [0; 1])$ for both μ and $\lceil Y \rceil$ polynomials according to Item d) of the following

Lemma 5. *(a) Suppose $C \subseteq X$ is maximal w.r.t. $\subseteq$ satisfying $\forall x, x' \in C : x = x' \lor d(x, x') \geq \varepsilon$. Then $\bigcup_{c \in C} \operatorname{B}(c, \varepsilon) = X$.*
For (X, d) totally bounded, $\mathcal{C}(X, d, \cdot)$ and $\mathcal{H}(X, d, \cdot)$ are non-increasing total functions $(0; \infty) \to \mathbb{N}$ satisfying $\mathcal{H}(X, d, \varepsilon) \leq \mathcal{C}(X, d, \varepsilon) \leq \mathcal{H}(X, d, \varepsilon/2)$. In particular it holds $\lfloor (X, d) \rfloor(n) \leq \lceil (X, d) \rceil(n) \leq \lfloor (X, d) \rfloor(n+1)$.

(b) The finite set $X := \{1, 2, \ldots, 2^k\}$ of integers has constant capacity and entropy $\lceil X \rceil(n) \equiv k \equiv \lfloor X \rfloor(n)$. The Euclidean cube/torus $[0; 2^k)^d$, equipped with the maximum norm, has capacity $\lceil [0; 2^k)^d \rceil(n) = (n + k) \cdot d = \lfloor [0; 2^k)^d \rfloor(n)$ and thus polynomial entropy.

(c) The compact space from Fact 2(h) equipped with the supremum norm has asymptotically exponential capacity and entropy: $\lceil \mathrm{Lip}_1([0; 1], [0; 1]) \rceil(n) = \Theta(2^n) = \lfloor \mathrm{Lip}_1([0; 1], [0; 1]) \rfloor(n)$. The same holds for $\mathrm{Lip}_1([0; 1], [0; 1]) \subseteq L^p$ equipped with the norm $f \mapsto \|f\|_p := \sqrt[p]{\int_0^1 |f(t)|^p\, dt}$ for any fixed $p \geq 1$.

(d) Suppose totally bounded (X, d) has diameter $\mathrm{diam}(X) := \sup\{d(x, x') : x, x' \in X\} \leq 1$ and super-logarithmic yet at most exponential entropy $\lfloor X \rfloor : \mathbb{N} \to \mathbb{N}$. Moreover fix some strictly increasing $\mu : \mathbb{N} \to \mathbb{N}$. W.r.t. sup-norm the space $C_\mu(X, [0; 1]) := \{f : X \to [0; 1] \text{ has modulus of continuity } \mu\}$ has $\log \lfloor C_\mu(X, [0; 1]) \rfloor(n) = \Theta\big(\lfloor X \rfloor(\mu(n \pm \Theta(1)))\big)$. A set $Y \subseteq C(X, [0; 1])$ is relatively compact iff it belongs to $C_\mu(X, [0; 1])$ for some μ.

(e) Cantor space $2^\omega = \{0, 1\}^\omega$, equipped with the metric $\beta(\bar{v}, \bar{w}) := 2^{-\min\{n : v_n \neq w_n\}}$ has linear capacity $\lceil (2^\omega, \beta) \rceil(n) = n + 1$; equipped with the topologically equivalent metric $\beta'(\bar{v}, \bar{w}) := 1/(1 + \min\{n : v_n \neq v_m\})$ on the other hand it has exponential capacity $\lceil (2^\omega, \beta') \rceil(n) = 2^n - 1$.

(f) However whenever d and d' are strongly equivalent metrics on X in the sense that $d' \cdot 2^{-c} \leq d \leq d' \cdot 2^c$ holds for some $c \in \mathbb{N}$ (such as in case X lives in some finite-dimensional normed real vector space), their induced capacities and entropies differ by at most a constant shift, i.e., it holds $\forall n \geq c$:

$$\lceil (X, d') \rceil(n - c) \leq \lceil (X, d) \rceil(n) \leq \lceil (X, d') \rceil(n + c),$$
$$\lfloor (X, d') \rfloor(n - c) \leq \lfloor (X, d) \rfloor(n) \leq \lfloor (X, d') \rfloor(n + c)$$

(g) Let (X, d) and (Y, e) be compact metric spaces and $f : X \to Y$ have modulus of continuity μ. Then the image $f[X] \subseteq Y$ has entropy $\lfloor F[X] \rfloor \leq \lfloor X \rfloor \circ \mu$.

Item (d) quantitatively refines the classical Arzelá-Ascoli Theorem; cmp. [Weih03, Theorem 6.7.3].

5 Relativized Complexity and Entropy

The entropy/capacity of compact metric spaces essentially determines the relativized computational complexity of functions on them:

Theorem 6. For a compact metric space (X, d) the following are equivalent:

(i) X has polynomially bounded entropy: $\lfloor X \rfloor(m) \leq p(m)$ for some $p \in \mathbb{N}[m]$.

(ii) X has a proper representation $\delta :\subseteq \{0, 1\}^\omega \twoheadrightarrow X$ with polynomial modulus of continuity.

(iii) X admits a representation δ rendering the following parameterized partial/fuzzy/ soft equality test relativized $\delta \times \delta$-computable in time polynomial in m:

$$X \times X \times \mathbb{N} \ni (x, y, m) \mapsto 1^\omega \text{ for } x = y, \quad \mapsto 0^\omega \text{ for } d(x, y) \geq 2^{-m}. \quad (2)$$

(iv) There exists a representation δ of X rendering Equation (2) relativized $\delta \times \delta$-computable in space logarithmic in m.

(v) There exists a representation δ of X rendering the metric $d : X \times X \to [0; \infty)$ relativized $(\delta \times \delta, \rho)$-computable in polynomial time

(vi) or in logarithmic space.

(vii) There exists a representation δ of X and polynomial $q \in \mathbb{N}[m]$ such that every 1-Lipschitz function $f : X \to [0; 1]$ is relativized (δ, ρ)-computable in time $q(m)$

(viii) or space $\mathcal{O}\big(\log q(m)\big) = \mathcal{O}(\log m)$.

Perhaps surprisingly, the same holds with δ replaced, in Items (iii) to (viii), by a *second-order* representation Δ of X in the following sense:

6 Second-Order Complexity Theory

According to Fact 2(h) compact space $\mathrm{Lip}_1([0; 1], [0; 1])$ does not admit a complexity-wise reasonable representation, i.e., encoding as infinite binary strings $\{0, 1\}^\omega \cong 2^{\{1\}^*}$: essentially due to their restriction to sequential access which requires 'skipping' over the f's (approximate) values at many arguments $f(x')$ before reaching the desired $f(x)$; whereas function arguments in practice provide oracle-like random access to their values. This has been formalized by encoding real function arguments as oracles [KaCo10].

Remark 7 *Classical oracles are decision problems $O \subseteq \{0, 1\}^*$, that is, they return a single bit. Function oracles on the other hand return finite strings, that is, they correspond to elements of Baire space $\mathbb{N}^\mathbb{N}$ encoded in binary as mappings $\varphi : \{0, 1\}^* \to \{0, 1\}^*$. Now if the answer $\vec{w} = \varphi(\vec{v})$ to a query $\vec{v}$ is 'long', an oracle machine $\mathcal{M}^\varphi$ arguably should be allotted more time than the same when run with an oracle ψ giving 'short' answers. This leads to second-order polynomial resource bounds; see [KaCo96]. Function oracles of polynomial length, on the other hand, can be encoded into decision oracles queried bitwise (and in particular satisfying effective polynomial boundedness [KaPa15]).*

For the purpose of this work we focus on the latter:

Definition 8.*(a) A second-order representation of a space X is a partial surjective mapping $\Xi :\subseteq 2^{\{0,1\}^*} \twoheadrightarrow X$, where 2^Y denotes the set of all subsets $O \subseteq Y$, each identified with its characteristic function $1_O : Y \to \{0, 1\}$.*

(b) An oracle Type-2 machine with variable/generic oracle is called contingent and denoted $\mathcal{M}^?$. It computes a partial function $F :\subseteq 2^{\{0,1\}^*} \times \{0, 1\}^\omega \to \{0, 1\}^\omega$ if, for every $(O, \vec{v}) \in \mathrm{dom}(F)$, $\mathcal{M}^O$ on input $\vec{v}$ prints $F(O, \vec{v})$. It does so in logarithmic/polynomials/exponential time/space if the n-th symbol of $F(O, \vec{v})$ appears within such resource bounds of time/work tape cells, independently of $(O, \vec{v}) \in \mathrm{dom}(F)$ while permitting unbounded use of the input, output, and query tapes. More precisely $\mathcal{M}^?$ may peruse a fixed-depth stack of write-only query tapes where an oracle call refers to, and purges, the top one.

(c) *Fix a second-order represented space (X, Ξ), (first-order) represented space (Y, υ), and presented metric space (Z, e, ζ) with induced representation ζ. A contingent oracle machine $\mathcal{M}^?$ (Ξ, υ, ζ)-computes $f :\subseteq X \times Y \rightrightarrows Z$ in time $t(m)$ and space $s(m)$ iff, for every $O \in \mathrm{dom}(\Xi)$ and $\bar{v} \in \mathrm{dom}(\upsilon)$ with $(\Xi(O), \upsilon(\bar{v})) \in \mathrm{dom}(f)$, $\mathcal{M}^O$ on input $\bar{v}$ produces a ζ-name $\langle (\mathrm{bin}(w_m))_m \rangle$ of some $z \in f(\Xi(O), \upsilon(\bar{v}))$ such that $\mathrm{bin}(w_m) \in \{0, 1\}^*$ appears on the output tape within at most $t(m)$ steps and using at most $s(m)$ cells of the working tape (and again 'arbitrary' amounts of the input, output, and query tapes).*

(d) *For second-order representations $\Xi :\subseteq 2^{\{0,1\}^*} \twoheadrightarrow X$ and $\Upsilon :\subseteq 2^{\{0,1\}^*} \twoheadrightarrow Y$, their (binary) product $\Xi \times \Upsilon$ is the mapping*

$$2^{\{0,1\}^*} \ni O \mapsto \left(\Xi(\{\vec{v} : 0\vec{v} \in O\}), \Upsilon(\{\vec{w} : 1\vec{v} \in O\}) \right) \in X \times Y.$$

(e) *Following up on Definition 1f), the second-order representation induced by a presented metric space (X, d, ξ) is the mapping*

$$\Xi :\subseteq 2^{\{0,1\}^*} \ni \left\{ \langle \mathrm{bin}(2^m), \mathrm{bin}(2^j) \rangle : \mathrm{bin}(a_m)_j = 1 \right\} \mapsto x \in X$$

for every ξ-name $\bar{a} = (a_m)_m \in \mathbb{N}^\omega$ of x.

(f) *Fix presented metric spaces (X, d, ξ) and (Y, e, υ) with induced (first-order) representations ξ and υ. Justified by Fact 2(c), equip (any fixed compact subset $\mathcal{Z}$ of) the space $C(X, Y)$ of continuous total functions $f : X \to Y$ with the following second-order representation υ^ξ: Let*

$$O = \big\{ \langle \mathrm{bin}(a), \mathrm{bin}(2^m), \mathrm{bin}(2^j) \rangle \; : \; a \in \mathrm{dom}(\xi),$$
$$\big\langle \mathrm{bin}\left(\varphi(a, m) \right) \big\rangle_j = 1 \big\} \subseteq \{0, 1\}^*$$

be an υ^ξ-name of $f \in C(X, Y)$ for every mapping $\varphi : \mathrm{dom}(\xi) \times \mathbb{N} \subseteq \mathbb{N} \times \mathbb{N} \to \mathrm{dom}(\upsilon) \subseteq \mathbb{N}$ where $(\varphi(a, m))_m$ is an υ-name of $f(\xi(a))$.

So Item (c) is about functions with ordinary/first-order represented co-domain, (g) with second-order ones. And υ^ξ according to Item (f) encodes (approximations in terms of the dense sequence in Y given by υ to) the values of f on the dense sequence in X given by ξ; cmp. [KaPa15, top of p. 8]. The fixed-depth stacks and subtle semantics of query tapes have been well justified in the discrete setting [Wils88, Buss88, ACN07] as well as in computational analysis [KaOt14]. Definition 8(f) does not (yet) incorporate quantitative information about continuity of f. In the case $\mathcal{Z} = \mathrm{Lip}_1([0; 1], [0; 1])$ the representation here called ρ^ρ is (equivalent to one) well-known [KaCo10, KORZ12, FHHP15, FeZi15]; and renders application $(f, x) \mapsto f(x)$ computable in polynomial-time.

7 Representing L^p and Sobolev Spaces

Recall that, for compact $X \subseteq \mathbb{R}^d$, $L^p(X) = W^{0,p}(X)$ consists of all measurable (but not necessarily continuous) functions $f : X \to \mathbb{R}$ such that

$\|f\|_p < \infty$; and, more generally, $W^{k,p}(X)$ of all f whose weak partial derivatives $\partial^{\vec{j}} f := \partial x_1^{j_1} \cdots \partial x_d^{j_d} f$ up to order $|\vec{j}| := j_1 + \cdots + j_d \leq k$ belong to $L^p(X)$, equipped with the norm $\|f\|_{k,p} := \max_{|\vec{j}| \leq k} \|\partial^{\vec{j}} f\|_p$. By Lemma 5(c) and the extension of Theorem 6, no (first or) second-order representation of $\mathrm{Lip}_1([0;1], [0;1]) \subseteq L_p[0;1]$ can simultaneously render both $(f, g) \mapsto |f - g|$ and $f \mapsto \|f\|_p$ polynomial-time computable.

Definition 9.(a) *Inspired by Definition 1(h) call* $\mu : \mathbb{N} \to \mathbb{N}$ *an* L^p-*modulus of*
$f \in L^p[0;1]$ *if* $\sqrt[p]{\int_0^1 |f(t+h) - f(t)|^p \, dt} < 2^{-m}$ *whenever* $|h| < 2^{-\mu(m)}$, *with the convention* $f(t) \equiv 0$ *for* $t \notin [0;1]$.
(b) *Abbreviate* $W_\mu^{k,p}[0;1] := \{ f \in W^{k,p}[0;1] : \partial^k f \text{ has } L^p\text{-modulus } \mu \}$.
(c) *Let* Ξ *denote the second-order representation of* $L^1[0;1] \supseteq W^{k,p}[0;1]$ *s.t. a* Ξ-*name of* f *is a* ρ^ω-*name of the continuous* $[0;1] \ni s \mapsto \int_0^s f(t) \, dt \in \mathbb{R}$.

By Fréchet-Kolmogorov, $Y \subseteq L^p([0;1])$ is relatively compact iff there exists some μ with $Y \subseteq W_\mu^{0,p}[0;1]$: Our convention of extending f with zero asserts $W_\mu^{0,p}$ to be bounded by $2^{\mu(0)}$. Although harder than Lemma 5d), we can prove

Theorem 10. *Fix polynomial-time computable* $p \geq 1$ *and strictly increasing* μ.

(a) $\log \lfloor W_\mu^{0,p}([0;1]) \rfloor (n) = \mu(n \pm \Theta(1))$.
(b) *For any fixed polynomial* μ, *the embedding* $W_\mu^{1,p}[0;1] \hookrightarrow C_{n \mapsto \mu(n+1)}[0;1]$ *is well-defined and* (Ξ, ρ^ω)-*computable in polynomial time*.
(c) *For any fixed* $k \in \mathbb{N}$ *and polynomial* μ, *differentiation* $\partial : W_\mu^{k+1,p}[0;1] \to W_\mu^{k,p}[0;1]$ *is well-defined and* (Ξ, Ξ)-*computable in polynomial time*.
(d) *For any fixed* $k \in \mathbb{N}$ *and polynomial* μ, *the embedding* $W_\mu^{k+1,p}[0;1] \hookrightarrow W^{k,p}[0;1]$ *is* (Ξ, Ξ)-*computable in polynomial time*.

References

[ACN07] Aehlig, K., Cook, S., Nguyen, P.: Relativizing small complexity classes and their theories. In: Duparc, J., Henzinger, T.A. (eds.) CSL 2007. LNCS, vol. 4646, pp. 374–388. Springer, Heidelberg (2007)

[BrCo06] Braverman, M., Cook, S.A.: Computing over the reals: foundations for scientific computing. Not. AMS **53**(3), 318–329 (2006)

[Buss88] Buss, J.F.: Relativized alternation and space-bounded computation. J. Comput. Syst. Sci. **36**, 351–378 (1988)

[dBYa10] de Brecht, M., Yamamoto, A.: Topological properties of concept spaces. Inf. Comput. **208**(4), 327–340 (2010)

[CSV13] Chaudhuri, S., Sankaranarayanan, S., Vardi, M.Y.: Regular real analysis. In: Proceedings of 28th Annual IEEE Symposium on Logic in Computer Science (LiCS2013), pp. 509–518 (2013)

[FeZi15] Férée, H., Ziegler, M.: On the Computational Complexity of Positive Linear Functionals on C[0;1]. In: Kotsireas, I.S., Rump, S.M., Yap, C.K. (eds.) MACIS 2015. LNCS, vol. 9582, pp. 489–504. Springer, Heidelberg (2016). doi:10.1007/978-3-319-32859-1_42

[FHHP15] Férée, H., Hainry, E., Hoyrup, M., Péchoux, R.: Characterizing polynomial time complexity of stream programs using interpretations. Theor. Comput. Sci. **595**, 41–54 (2015)

[GrKi14] Gregoriades, V., Kihara, T.: Recursion and effectivity in the decomposability conjecture. (2014, submitted). arXiv:1410.1052

[Hert04] Hertling, P.: A BanachMazur computable but not Markov computable function on the computable real numbers. Ann. Pure Appl. Log. **132**, 227–246 (2004)

[KaCo96] Kapron, B.M., Cook, S.A.: A new characterization of type-2 feasibility. SIAM J. Comput. **25**(1), 117–132 (1996)

[KaCo10] Kawamura, A., Cook, S.A.: Complexity theory for operators in analysis. In: Proceedings of 42nd Annual ACM Symposium on Theory of Computing (STOC 2010); Full Version in ACM Transactions in Computation Theory, vol. 4 no. 2, article 5 (2012)

[KaOt14] Kawamura, A., Ota, H.: Small complexity classes for computable analysis. In: Csuhaj-Varjú, E., Dietzfelbinger, M., Ésik, Z. (eds.) MFCS 2014, Part II. LNCS, vol. 8635, pp. 432–444. Springer, Heidelberg (2014)

[KaPa15] Kawamura, A., Pauly, A.: Function spaces for second-order polynomial time. In: Beckmann, A., Csuhaj-Varjú, E., Meer, K. (eds.) CiE 2014. LNCS, vol. 8493, pp. 245–254. Springer, Heidelberg (2014)

[KMRZ15] Kawamura, A., Müller, N., Rösnick, C., Ziegler, M.: Computational benefit of smoothness: parameterized bit-complexity of numerical operators on analytic functions and Gevrey's hierarchy. J. Complex. **31**(5), 689–714 (2015)

[Ko91] Ko, K.-I.: Computational Complexity of Real Functions. Birkhäuser, Boston (1991)

[Kohl08] Kohlenbach, U.: Applied Proof Theory. Springer, Heidelberg (2008)

[KORZ12] Kawamura, A., Ota, H., Rösnick, C., Ziegler, M.: Computational complexity of smooth differential equations. In: Rovan, B., Sassone, V., Widmayer, P. (eds.) MFCS 2012. LNCS, vol. 7464, pp. 578–589. Springer, Heidelberg (2012)

[KoTi59] Kolmogorov, A.N., Tikhomirov, V.M.: $\mathcal{E}$-Entropy and $\mathcal{E}$-Capacity of Sets in Functional Spaces. Uspekhi Mat. Nauk **14**(2), 3–86 (1959). also pp. 86–170 in Selected Works of A.N. Kolmogorov vol. III (Shiryayev, A.N. Ed.), Nauka (1993) and Springer (1987)

[KSZ14] Kawamura, A., Steinberg, F., Ziegler, M.: Complexity of Laplace's and Poisson's Equation. Bulletin of Symbolic Logic **20**(2), 231 (2014). Full version in Mathem. Structures in Computer Science (2016)

[PaZi13] Pauly, A., Ziegler, M.: Relative computability and uniform continuity of relations. J. Log. Anal. **5**, 1–39 (2013)

[PER89] Pour-El, M.B., Richards, I.: Computability in Analysis and Physics. Springer, Heidelberg (1989)

[Schr95] Schröder, M.: Topological spaces allowing type-2 complexity theory. In: Workshop on Computability and Complexity in Analysis, Informatik-Berichte 190, FernUniversität Hagen (1995)

[Schr04] Schröder, M.: Spaces allowing type-2 complexity theory revisited. Math. Logic Q. **50**, 443–459 (2004)

[Schr06] Schröder, M.: Admissible representations in computable analysis. In: Beckmann, A., Berger, U., Löwe, B., Tucker, J.V. (eds.) CiE 2006. LNCS, vol. 3988, pp. 471–480. Springer, Heidelberg (2006)

[SZZ15] Sun, S.M., Zhong, N., Ziegler, M.: On the Computability of the Navier-Stokes Equation. In: Beckmann, A., Mitrana, V., Soskova, M. (eds.) Evolving Computability. LNCS, vol. 9136, pp. 334–342. Springer, Heidelberg (2015)

[Weih00] Weihrauch, K.: Computable Analysis. Springer, Heidelberg (2000)

[Weih03] Weihrauch, K.: Computational complexity on computable metric spaces. Math. Log. Q. **49**(1), 3–21 (2003)

[WeZh02] Weihrauch, K., Zhong, N.: Is wave propagation computable or can wave computers beat the turing machine? Proc. London Math. Soc. **85**(2), 312–332 (2002)

[Wils88] Wilson, C.B.: A measure of relativized space which is faithful with respect to depth. J. Comput. Syst. Sci. **36**, 303–312 (1988)

[Zieg07] Ziegler, M.: Real hypercomputation and continuity. Theory Comput. Syst. **41**, 177–206 (2007)

Ergodicity of Noisy Cellular Automata: The Coupling Method and Beyond

Irène Marcovici[✉]

Institut Élie Cartan de Lorraine, Université de Lorraine, Nancy, France
irene.marcovici@univ-lorraine.fr

Abstract. When perturbating a cellular automaton by a random noise (positive probability of error, for each cell independently), the system is generally expected to be ergodic, meaning that during its evolution, it eventually forgets about its initial condition. For a high noise, this can be shown by coupling. However, for a small noise, ergodicity is often very difficult to prove. We present extensions of the coupling method to small noises when the cellular automaton has some specific properties (hardcore exclusion, nilpotency, permutivity).

Consider a set of cells indexed by $\mathbb{Z}$, each cell containing a letter from a finite symbol set S. A *cellular automaton* (CA) is a dynamical system which acts locally and synchronously on the configuration space $S^{\mathbb{Z}}$. When the updates are random, we obtain a *probabilistic cellular automaton* (PCA): at each time step, the new content of each cell is randomly chosen, independently of the others, according to a distribution given by the states in a finite neighbourhood of the cell. Examples of PCA are given by *noisy* CA: the updates are governed by a deterministic rule, which is perturbated by errors with a positive probability.

A PCA is said to be *ergodic* if it forgets its initial condition, meaning that it has a unique and attractive invariant measure. A variety of tools have been developed to study the ergodicity of PCA. But most of them only allow to handle PCA for which the transition probability to any state given any neighbourhood states is large enough. In particular, ergodicity is often very difficult to prove for noisy CA, even in cases where it appears clear from heuristics or simulations.

In Sect. 1, we recall the coupling method and the notion of envelope PCA, which gives a general framework to prove ergodicity for the high noise regime. Details and more references can be found in a joint publication with A. Bušić and J. Mairesse [1]. The next sections are devoted to three examples of families of CA for which some specific tools allow to prove the ergodicity for small noise. In Sect. 2, we present new results on the noisy version of the *hardcore* CA. These results stem from a joint work with J.B. Martin, motivated by the study of a percolation game, together with A.E. Holroyd [3]. Then, in Sects. 3 and 4, we consider perturbations of *nilpotent* and *permutive* CA. These sections are based on a work initiated with S. Taati and carried on together with M. Sablik. It is interesting to note that nilpotent CA and permutive CA present opposite behaviours within the rich zoology of CA. While nilpotent CA reach in

© Springer International Publishing Switzerland 2016
A. Beckmann et al. (Eds.): CiE 2016, LNCS 9709, pp. 153–163, 2016.
DOI: 10.1007/978-3-319-40189-8_16

bounded time a configuration where a single symbol remains, *bipermutive* CA are expected to converge to the uniform distribution on $S^{\mathbb{Z}}$ from a large class of initial distributions (*randomization* phenomenon).

1   Definitions

Let S be a finite set of *symbols*. We equip the set $S^{\mathbb{Z}}$ of *configurations* with the product topology. For a finite subset $K \subset \mathbb{Z}$ and an element $y \in S^K$, the set $[y_K] = \{z \in S^{\mathbb{Z}} ; \forall k \in K, z_k = y_k\}$ is called a *cylinder* of *base* K, and we denote by $\mathcal{C}(K)$ the set of all cylinders of base K. The set of probability distributions on $S^{\mathbb{Z}}$ for the Borelian σ-algebra is denoted by $\mathcal{M}(S^{\mathbb{Z}})$. For a distribution $\mu \in \mathcal{M}(S^{\mathbb{Z}})$, we denote by $\mu[y_K]$ the probability of the cylinder $[y_K]$.

Let $m \geq 1$ be an integer, and let $n_1, \ldots, n_m \in \mathbb{Z}$. A *local rule* of *neighbourhood* $\mathcal{N} = \{n_1, \ldots, n_m\}$ is a function $f : S^m \to S$. The *cellular automaton* (CA) of local rule f is the function $F : S^{\mathbb{Z}} \to S^{\mathbb{Z}}$ defined by:

$$\forall x \in S^{\mathbb{Z}}, \ \forall k \in \mathbb{Z}, \ F(x)_k = f(x_{k+n_1}, \ldots, x_{k+n_m}).$$

We denote by $\sigma : S^{\mathbb{Z}} \to S^{\mathbb{Z}}$ the *shift* map, defined by $\sigma(x)_k = x_{k-1}$. By Curtis-Hedlund-Lyndon theorem, CA are exactly continuous functions that commute with σ.

For *probabilistic cellular automata* (PCA), the local rule is a function $\varphi : S^m \to \mathcal{M}(S)$, where $\mathcal{M}(S)$ denotes the set of probability distributions on S. From a configuration $x \in S^{\mathbb{Z}}$, cell k is updated by a symbol chosen according to the distribution $\varphi(x_{k+n_1}, \ldots, x_{k+n_m})$, independently for different cells. A PCA can be viewed as a Markov chain on $S^{\mathbb{Z}}$. The evolution of a PCA is described by a family of random variables $(X^t)_{t \geq 0}$, where X^t represents the configuration at time t when iterating the dynamics from the (deterministic or random) initial condition X^0. Formally, the PCA of local rule φ can also be seen as the function

$$\Phi : \mathcal{M}(S^{\mathbb{Z}}) \to \mathcal{M}(S^{\mathbb{Z}})$$
$$\mu \mapsto \Phi\mu$$

defined on cylinder sets by:

$$\Phi\mu[y_K] = \sum_{[x_{K+\mathcal{N}}] \in \mathcal{C}(K+\mathcal{N})} \mu[x_{K+\mathcal{N}}] \prod_{k \in K} \varphi(x_{k+n_1}, \ldots, x_{k+n_m})(y_k),$$

for any probability distribution $\mu \in \mathcal{M}(S^{\mathbb{Z}})$. If the initial configuration X^0 has distribution μ_0, then at time t, configuration X^t has distribution $\Phi^t \mu_0$.

Definition 1. *Let Φ be a PCA. The distribution $\pi \in \mathcal{M}(S^{\mathbb{Z}})$ is invariant for Φ if $\Phi\pi = \Phi$. The PCA Φ is ergodic if it has a unique invariant distribution π and if for any initial distribution $\mu_0 \in \mathcal{M}(S^{\mathbb{Z}})$, $(\Phi^t \mu_0)_{t \geq 0}$ converges (weakly) to π.*

In this article, we will focus on specific families of PCA obtained when adding random and independent errors in the updates of a deterministic CA.

Let F be a deterministic CA, and let Φ be a PCA with same symbol set and same neighbourhood as F. We say that Φ is an ε-*perturbation* of F if its local function is such that for all $x_1, \ldots, x_m \in S, \varphi(x_1, \ldots, x_m)(f(x_1, \ldots, x_m)) \geq 1-\varepsilon$, meaning that there is a deviation from F with probability at most ε.

Let us consider a map $R : S \to \mathcal{M}(S)$, and let F be a CA of local rule f. The *noisy version* of the CA F with noise R is the PCA of local rule $\varphi = R \circ f$. Starting from a configuration $x \in S^{\mathbb{Z}}$, the CA F is first applied, and then, for each cell k independently, the symbol at cell k is replaced by a symbol distributed according to the distribution $R(F(x)_k)$. We denote simply by $R_{i,j}$ the probability $R(i)(j)$ for a symbol i to be changed into symbol j. The noise is said to be *positive* if for all $i, j \in S$, $R_{i,j} > 0$. The matrix R is a stochastic matrix. The noise *preserves the uniform distribution* on S if for all $j \in S$, $\sum_{i \in S} R_{i,j} = 1$, meaning that the matrix R is doubly stochastic.

We will also pay special attention to *elementary PCA*, of symbol set $S = \{0, 1\}$ and neighbourhood $\mathcal{N} = \{0, 1\}$. They are characterized by four parameters: the probabilities $\theta_{ij} = \varphi(i, j)(1)$ to update a cell by the symbol 1 if its neighbourhood is in state i, j, for $i, j \in S^2$. If we further assume that $\theta_{01} = \theta_{10}$, the general tools that have been developed to prove the ergodicity allow to handle more than 90 % of the volume of the cube defined by the parameter space [2]. But for example, the noisy hardcore CA we study in Sect. 3 belongs to an open domain of the cube where none of those criteria is valid.

2 The Coupling Method

2.1 The Envelope PCA

Intuitively, a PCA is ergodic if it "forgets" its initial condition. In some cases, it is possible to prove the ergodicity in a constructive way, by making evolve simultaneously the trajectories from different initial conditions, using a common source of randomness, and showing that the evolutions of all these trajectories are asymptotically the same.

The envelope PCA allows to systematize this idea of *coupling*. Instead of running the PCA from different initial configurations, we define a new PCA on an extended alphabet, containing a symbol ? representing sites whose values are not known (*i.e.* which may differ between the different copies) and we run it from a single initial configuration containing only the symbol ?. Each time we are able to make the different copies match on a cell, the symbol ? is replaced by the state $q \in S$ on which the different copies agree. An evolution of the envelope PCA thus encodes a coupling of different copies of the original PCA, with a symbol ? denoting sites where the copies disagree. If the density of symbols ? converges to 0 when time goes to infinity, it means that the PCA is ergodic.

Let us assume that Φ is a PCA defined on a binary symbol set $S = \{0, 1\}$, and let $\tilde{S} = \{0, 1, ?\}$. We define a partial order on $\tilde{S}$ by $0 \prec ? \succ 1$. The *envelope*

PCA $\tilde{\Phi}$ of Φ is the PCA of neighbourhood $\mathcal{N}$ and local function $\tilde{\varphi} : \tilde{S}^m \to \mathcal{M}(\tilde{S})$ defined for $q \in S$ by

$$\tilde{\varphi}(y_1, \ldots, y_m)(q) = \min\{\varphi(x_1, \ldots, x_m)(q)\,;\; x_1 \preceq y_1, \ldots, x_m \preceq y_m\},$$

where in the expression above, $x_1, \ldots, x_m$ are taken in S. The probability of a transition to the symbol ? is then given by:

$$\tilde{\varphi}(y_1, \ldots, y_m)(?) = 1 - \tilde{\varphi}(y_1, \ldots, y_m)(0) - \tilde{\varphi}(y_1, \ldots, y_m)(1).$$

From a configuration $y \in \tilde{S}^{\mathbb{Z}}$, cell k is thus updated by the symbol $q \in S$ with the minimum of the probabilities of transition to the symbol q for Φ, taken over all the values of the neighbourhood of cell k that are compatible with the unknown cells of y. With the remaining probability, the cell is updated by a ?.

Proposition 1. *If the density $\tilde{\Phi}^t \delta_{?^{\mathbb{Z}}}[?]$ of symbols ? at time t starting from the initial configuration $?^{\mathbb{Z}}$ converges to 0 as $t \to \infty$, then the PCA Φ is ergodic.*

The fact that symbols ? die out is equivalent to the ergodicity of the envelope PCA $\tilde{\Phi}$, but the ergodicity of the original PCA Φ does not imply the ergodicity of $\tilde{\Phi}$. Note also that the definition of the envelope PCA can be extended to sets of symbols having more than two elements [1].

2.2   Ergodicity Criterion

In this section, we still consider a binary PCA Φ and its envelope PCA $\tilde{\Phi}$. In the evolution of the envelope PCA, at each time step, a cell is updated by the symbol ? only if it has at least one neighbour in state ?, and in that case, it becomes a ? with probability at most:

$$
\begin{aligned}
p_? &= \tilde{\varphi}(?, \ldots, ?)(?) \\
&= 1 - \min_{x_1, \ldots, x_m \in S} \varphi(x_1, \ldots, x_m)(0) - \min_{x_1, \ldots, x_m \in S} \varphi(x_1, \ldots, x_m)(1) \\
&= \max_{x_1, \ldots, x_m \in S} \varphi(x_1, \ldots, x_m)(0) - \min_{x_1, \ldots, x_m \in S} \varphi(x_1, \ldots, x_m)(0) \\
&= \max_{x_1, \ldots, x_m \in S} \varphi(x_1, \ldots, x_m)(1) - \min_{x_1, \ldots, x_m \in S} \varphi(x_1, \ldots, x_m)(1).
\end{aligned}
$$

This quantity measures how much the probability transitions depend on the value of the neighbourhood.

Let us consider the oriented graph G describing the dependences between sites in the space-time diagram of the PCA. The set of vertices of G is $\mathbb{Z} \times \mathbb{N}$, and there is an edge from (k, t) to $(\ell, t+1)$ if $k \in \ell + \mathcal{N}$. For a given parameter $p \in (0, 1)$, the *directed site percolation* on G consists in declaring each site to be open with probability p, independently for different sites. One can show that there is a critical value $p_c(\mathcal{N}) \in (0, 1)$, such that if $p < p_c(\mathcal{N})$, then there is almost surely no infinite open (oriented) component (note that $p_c(\mathcal{N}) \geq 1/\operatorname{Card} \mathcal{N}$).

By dominating the process of symbols ? in the space-time diagram of the envelope PCA by a directed site percolation of parameter $p_?$, one proves that if $p_? < p_c(\mathcal{N})$, then the symbols ? die out. Next proposition follows.

Proposition 2. *Let $p_c(\mathcal{N})$ be the critical value of the two-dimensional directed site percolation of neighbourhood $\mathcal{N}$. If $p_? < p_c(\mathcal{N})$, then $\tilde{\Phi}^t \delta_{?^{\mathbb{Z}}}[?] \xrightarrow[t \to +\infty]{} 0$, so that the PCA Φ is ergodic.*

For a noisy CA with noise R on a binary symbol set, one can check that unless the CA is constant, $p_? = |R_{0,1} - R_{1,1}|$, which shows ergodicity for $R_{0,1}$ and $R_{1,1}$ close enough to each other.

For elementary PCA, we have $p_c(\mathcal{N}) \approx 0.7$. If $p_? = \max_{i,j \in S^2} \theta_{ij} - \min_{i,j \in S^2} \theta_{ij}$ is smaller than this critical value, then the PCA is ergodic.

2.3   Coupling from the Past

If the PCA $\tilde{\Phi}$ is ergodic, that is, if $\tilde{\Phi}^t \delta_{?^{\mathbb{Z}}}[?] \xrightarrow[t \to +\infty]{} 0$, then the unique invariant distribution of π can be sampled *exactly* by *coupling from the past* (Propp-Wilson method). We define the update function $h : \tilde{S}^m \times (0,1) \to \tilde{S}$ of $\tilde{\Phi}$ by:

$$h(y_1, \ldots, y_m, u) = \begin{cases} \mathbf{0} & \text{if } 0 \leq u \leq \tilde{\varphi}(y_1, \ldots, y_k)(\mathbf{0}), \\ \mathbf{1} & \text{if } 1 - \tilde{\varphi}(y_1, \ldots, y_k)(\mathbf{1}) \leq u \leq 1, \\ ? & \text{otherwise.} \end{cases}$$

This function has the property that if $(U_k)_{k \in \mathbb{Z}}$ is a family of independent random variables, uniformly distributed on $(0,1)$, then for any $y \in \tilde{S}^{\mathbb{Z}}$, the configuration $(Z_k)_{k \in \mathbb{Z}}$ defined by $Z_k = h(y_{k+n_1}, \ldots, y_{k+n_m}, U_k)$ is distributed according to $\tilde{\Phi}\delta_y$.

Let K be a finite subset of $\mathbb{Z}$. We draw a sequence $(u_{k,-t})_{k \in \mathbb{Z}, t \geq 0}$ of independent random samples, uniformly distributed on $(0,1)$. We iterate the envelope PCA $\tilde{\Phi}$ from time $-T$ to time 0, starting with the configuration $?^{\mathbb{Z}}$ at time $-T$, and always using the sample $u_{k,-t}$ to update cell k at time $-t$, with the help of the update function h. If for some time T, the resulting configuration obtained at time 0 with this procedure is such that there are no symbols $?$ on the cells of K, then the symbols observed on K are distributed according to the marginal of the distribution π on K.

After having iterated the envelope PCA from some time $-T$, the effect of starting from time $-(T+1)$ can only be to change some $?$ in the space time diagram into symbols $\mathbf{0}$ and $\mathbf{1}$ (once a symbol $\mathbf{0}$ or $\mathbf{1}$ appears at a cell $(k, -t)$ of the space time diagram, it is fixed). The fact that $\tilde{\Phi}^t \delta_{?^{\mathbb{Z}}}[?] \to 0$ ensures that there exists almost surely a time T such that at time $t = 0$, there are no more symbols $?$ on the cells of K. Note that when implementing this sampling procedure, it is enough to consider only the cells that are in the *dependence cone* of K.

3   The Noisy Hardcore CA

Let us consider the elementary PCA Φ defined by the parameters $\theta_{00} = 1 - p$, $\theta_{01} = \theta_{10} = \theta_{11} = q$, with $0 < p + q \leq 1$. This PCA is the noisy version of the deterministic CA of local rule $f(i, j) = (1 - i)(1 - j)$, with a noise defined by $R_{1,0} = p$ and $R_{0,1} = q$. The rule f can be seen as an exclusion rule: a cell becomes a **1** if it has no neighbour in state **1**, and a **0** otherwise. For Φ, after applying that deterministic dynamics, each **1** is changed into a **0** with probability p, and each **0** is changed into a **1** with probability q, independently.

Until recently, the question of ergodicity of Φ was a repertoried open problem. This PCA is closely related to the enumeration of directed lattice animals, which are classical objects in combinatorics. It also appears in the study of a percolation game [3]. With the notations of Sect. 2.2, we have $p_? = 1 - p - q$. Thus, the criterion of Proposition 2 provides the ergodicity of Φ for $p + q > 0.3$ or so, but is of no help for smaller values of $p + q$. In that case, the comparison with oriented percolation is too rough to prove that symbols **?** die out. Nevertheless, one can prove the following (see [3] for complete proof in the case $p = 0$ or $q = 0$).

Proposition 3. *The noisy hardcore CA, that is, the elementary PCA Φ defined by $\theta_{00} = 1 - p$, $\theta_{01} = \theta_{10} = \theta_{11} = q$, is ergodic for any parameters $p + q < 1$.*

Proof (sketch). The local function of $\tilde{\Phi}$ is given by the following probability transitions (time is going up, and symbol $*$ represents any element of $\tilde{S}$).

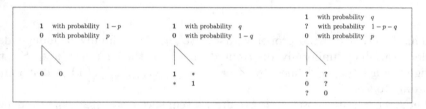

The envelope PCA $\tilde{\Phi}$ is itself the noisy version of the deterministic CA $\tilde{F}$ (which is the deterministic CA obtained when taking $p = q = 0$ in the table above), with a noise $\tilde{R}$ that changes any symbol into a **0** with probability p, and into a **1** with probability q.

For a given configuration in $\tilde{S}^{\mathbb{Z}}$, let us weight the occurrences of the symbols **?** as follows:

- if a **?** is followed by the pattern **01**, then it receives weight 3;
- if a **?** is followed by a **0** and then by something other than a **1**, it receives weight 2;
- otherwise, a **?** receives weight 1.

One can prove that the weight can only decrease under the action of the deterministic CA $\tilde{F}$. Precisely, if μ is a shift-invariant and reflection-invariant distribution on $\tilde{S}^{\mathbb{Z}}$, then $\tilde{F}\mu[?01] + \tilde{F}\mu[?0] + \tilde{F}\mu[?] \leq \mu[?01] + \mu[?0] + \mu[?]$.

We now add the random noise $\tilde{R}$, and consider the PCA $\tilde{\Phi}$. Let μ be an invariant distribution of $\tilde{\Phi}$. By some computations, one can prove that we necessary have $\mu[?] = 0$, since otherwise, we would get $\tilde{\Phi}\mu[?01] + \tilde{\Phi}\mu[?0] + \tilde{\Phi}\mu[?] < \mu[?01] + \mu[?0] + \mu[?]$, which would be in contradiction with the fact that μ is an invariant distribution of $\tilde{\Phi}$.

Thus, there is no shift-invariant and reflection-symmetric stationary distribution in which the symbol ? appears with positive probability. Let us iterate $\tilde{\Phi}$ from the configuration $?^{\mathbb{Z}}$. By a coupling argument, the density $\tilde{\Phi}^t \delta_{?^{\mathbb{Z}}}[?]$ is decreasing with t. If it was not converging to 0, then we could extract a convergent subsequence of the Cesàro sums of $\tilde{\Phi}^t \delta_{?^{\mathbb{Z}}}$ and obtain an invariant distribution μ of $\tilde{\Phi}$ satisfying $\mu[?] > 0$, which is not possible. Thus, $\tilde{\Phi}^t \delta_{?^{\mathbb{Z}}}[?] \xrightarrow[t \to +\infty]{} 0$, and Φ is ergodic.

4 Perturbating a Nilpotent CA

Let F be a *nilpotent* CA. It means that there exists an integer N such that F^N is a constant function, equal to $\alpha^{\mathbb{Z}}$ for some symbol $\alpha \in S$.

If F is not the constant function equal to $\alpha^{\mathbb{Z}}$, then for an ε-perturbation of F with small ε, the value $p_?$ is close to 1. Thus, Proposition 2 (and its analogous for larger symbol sets) cannot be used to prove the ergodicity of an ε-perturbation of F. In that case, the envelope PCA as defined in Sect. 2.1 is not an adapted tool. Nevertheless, once again, the coupling method can be used to prove the ergodicity.

Proposition 4. *Let F be a nilpotent CA. There exists $\varepsilon_c > 0$ such for $\varepsilon < \varepsilon_c$, any ε-perturbation of F is ergodic.*

Proof. Let $\varepsilon > 0$, and let Φ_ε be an ε-perturbation of F. We prove that if ε is small enough, we can couple all the trajectories of Φ_ε.

Let K be a finite subset of points of $\mathbb{Z}$. We consider a configuration $(x_{k,0})_{k \in \mathbb{Z}}$ obtained at time $t = 0$, after iterating Φ_ε from a given time in the past, using a sequence $(u_{k,-t})_{k \in \mathbb{Z}, t \geq 0}$ of independent samples, uniformly distributed in $(0,1)$, and an update function having the following property: if $u_{k,-t} > \varepsilon$, then cell k is updated according to the local rule of the deterministic CA F, while if $u_{k,-t} \leq \varepsilon$, the value may differ. We prove that almost surely, there exists a time $T > 0$ such that the evolutions from all the possible starting configurations at time $-T$ provide the same sequence $(x_{k,0})_{k \in K}$ at time 0.

We define recursively the sets $\mathcal{N}_i$ by $\mathcal{N}_0 = \{0\}$, and $\mathcal{N}_{i+1} = \mathcal{N}_i + \mathcal{N} = \{a + b ;\ a \in \mathcal{N}_i, b \in \mathcal{N}\}$ for $i \geq 0$, so that $\mathcal{N}_t$ is the neighbourhood of F^t.

For $k \in \mathbb{Z}$, and times $t, i \geq 0$, we define $V_i(k, -t) = \{(k + \ell, -t - i), \ell \in \mathcal{N}_i\}$. It is the set of cells at time $-t - i$ in the space-time diagram from which the state of cell k at time $-t$ may depend.

We also introduce $W(k, -t) = \bigcup_{0 \leq i \leq N-1} V_i(k, -t)$, where we recall that N is such that F^N is constant, equal to $\alpha^{\mathbb{Z}}$.

We call a cell $(k, -t)$ an *error* if $u_{k,-t} \leq \varepsilon$. Since F^N is a constant function, if there is no error in $W(k, -t_0)$, then the value $x_{k,-t_0}$ of cell k at time $-t_0$ does not depend on the value of the configuration at time $-t_0 - N$ (we have $x_{k,-t_0} = \alpha$ in all cases).

For $k \in \mathbb{Z}$, $t \geq 0$, let us define the set:

$$E(k, -t) = \begin{cases} \emptyset & \text{if there is no error in } W(k, -t), \\ V_N(k, -t) & \text{if there is at least one error in } W(k, -t). \end{cases}$$

We define recursively the sets $(A_i)_{i \geq 0}$ by $A_0 = K \times \{0\}$, and:

$$A_{i+1} = E(A_i) = \bigcup_{(k,-t) \in A_i} E(k,-t) \text{ for } i \geq 0.$$

Note that if $(k, -t) \in A_i$, then $t = iN$. We have the following property: if A_i is empty, then if we iterate Φ_ε from time $-iN$ to time 0, using the samples $(u_{k,-t})_{k \in \mathbb{Z}, 0 \leq t < iN}$, the values $(x_{k,0})_{k \in K}$ obtained on K at time 0 do not depend on the choice of the configuration $(x_{k,-iN})_{k \in \mathbb{Z}} \in S^{\mathbb{Z}}$ from which we start at time $-iN$.

In the figure below, errors are represented by red dots. Blue domains represent cells that are known to be in state α (there are no errors affecting them in the last N time steps). Black domains represent cells for which we need further information in the past to determine their state.

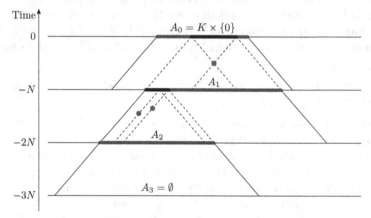

Let us prove that if ε is small enough, then almost surely, there exists an integer after which all the sets A_i are empty.

We set $m_i = \mathrm{Card}\, \mathcal{N}_i$. Let $(\ell, -t)$ be an error, with $t = iN+j$, $0 \leq j \leq N-1$. We have $(\ell, -t) \in W(k, -iN)$ if and only if $k \in \ell - \mathcal{N}_j$. Thus, the number of points $(k, -iN)$ such that $(\ell, -t)$ is an error of $W(k, -iN)$ is bounded by $m_j \leq m_{N-1}$. If there is an error in $W(k, -iN)$, then $\mathrm{Card}\, E(k, -iN) = m_N$. It follows that a given error has a contribution of at most $L = m_{N-1} m_N$ points to A_{i+1}.

Let $M = m_0 + m_1 + \ldots + m_{N-1}$. We have $\mathrm{Card}\, W(k, -t) = M$ for any $k \in \mathbb{Z}, t \geq 0$. The number of points of $\bigcup_{k \in A_i} W(k, -iN)$ is thus smaller than $(\mathrm{Card}\, A_i) \times M$, and each point is an error with probability ε, independently. Consequently, $\mathrm{Card}\, A_{i+1}$ is bounded by the sum of $(\mathrm{Card}\, A_i) \times M$ independent random variables, whose value is L with probability ε, and 0 with probability $1 - \varepsilon$. If $\varepsilon < 1/LM$, a comparison with a branching process proves that there is extinction: almost surely, the sets A_i are eventually empty. Consequently, Φ_ε is ergodic (note that the bound given for ε is rough and can certainly be improved).

5   Noisy Permutive Cellular Automata

Let F be a CA of neighbourhood $\mathcal{N} = \{\ell, \ell + 1, \ldots, r\}$ and local function $f :$ $S^m \to S$, with $m = r - \ell + 1 \geq 2$. We say that F is *left-permutive* (resp. *right-permutive*) if, for all $w = w_{\ell+1} \cdots w_r \in S^{m-1}$, the mapping:

$$\tau_w : S \to S$$
$$a \mapsto f(aw) \quad (\text{resp. } f(wa))$$

is bijective. A CA is *permutive* if it is either left or right-permutive. It is *bipermutive* if it is both left and right-permutive. For example, if $S = \mathbb{Z}_n$ and $a, b, c \in \mathbb{Z}_n$, the *affine* CA defined by $f(x, y) = ax + by + c$ is left-permutive (resp. right-permutive) if a (resp. b) is invertible in $\mathbb{Z}_n$.

Let F be a permutive CA. Using the bijections τ_w one can prove that F is surjective. For deterministic CA, surjectivity is equivalent to preserving the uniform distribution λ on $S^{\mathbb{Z}}$ (that is, the product of the uniform distribution on S). Next proposition shows that when adding a noise preserving λ, the PCA indeed converges to λ. The proof below is adapted from a work of Vasilyev [2,4].

Proposition 5. *Let F be a permutive CA, and let R be a positive noise preserving the uniform distribution. The noisy version Φ of F with noise R is ergodic, and its unique invariant distribution is the uniform distribution λ.*

Proof. We will prove that for any $N \in \mathbb{N}$, and any initial distribution μ on $S^{\mathbb{Z}}$, the marginal distribution of $\Phi^t \mu$ on $K = \{-N, \ldots, N\}$ converges exponentially to the uniform Bernoulli distribution on S^K, that we denote by λ_K. Precisely, we will prove that for any $N \in \mathbb{N}$, there exists $\theta < 1$ such that for any distribution μ on $S^{\mathbb{Z}}$, we have: $\forall t \geq 0$, $\|(\Phi^t \mu)_{|K} - \lambda_K\|_1 \leq 2\theta^t$, where for $u : S^K \to \mathbb{R}$, $\|u\|_1 = \sum_{x \in S^K} |u(x)|$.

Let us first assume that F is left-permutive and that $\mathcal{N} = \{0, 1, \ldots, r\}$, and let $w \in S^r$. By permutivity of F, we have a bijection

$$\sigma_w : S^K \longrightarrow S^K$$
$$x \longmapsto f(xw),$$

where we still denote by f the map from $S^{K \cup \{N+1, \ldots, N+r\}}$ to S^K induced by the local function of the CA F.

So, when fixing the word w as a boundary condition on the right of K, the noisy CA Φ maps a word $x \in S^K$ to a random word Z_K distributed according to a product distribution with marginal distribution $R(y_k)$ at site $k \in K$, where $y = \sigma_w(x)$. From a given $x \in S^K$, we denote by $P_w(x, z)$ the probability for Z_K to be equal to some $z \in S^K$, so that: $P_w(x, z) = \prod_{k \in K} R_{y_k, z_k}$.

Recall that λ_K is the uniform distribution on S^K. The map σ_w being bijective, it preserves λ_K. By assumption, the noise R also preserves the uniform distribution, so that we obtain $P_w \lambda_K = \lambda_K$.

For any $w \in S^r$, the transition matrix P_w is positive. Therefore, there exists $\theta_w < 1$ such that for any probability distributions ν, ν' on S^K, we have

$$||P_w \nu - P_w \nu'||_1 \le \theta_w ||\nu - \nu'||_1,$$

the above inequality being true in particular for $\theta_w = 1 - \varepsilon_w$, where $\varepsilon_w = \min\{P_w(i,j) \, ; \, i,j \in S\}$.

Let us set $\theta = \max\{\theta_w \, ; \, w \in S^r\}$. It follows that for any sequence $(w_t)_{t \ge 0}$ of words of S^r, we have: $||P_{w_{t-1}} \ldots P_{w_1} P_{w_0} \nu - P_{w_{t-1}} \ldots P_{w_1} P_{w_0} \nu'||_1 \le \theta^t ||\nu - \nu'||_1$. In particular, for $\nu' = \lambda_K$, we obtain that for any distribution ν on S^K and any sequence $(w_t)_{t \ge 0}$ of words of S^r, $||P_{w_{t-1}} \ldots P_{w_1} P_{w_0} \nu - \lambda_K||_1 \le \theta^t ||\nu - \lambda_K||_1 \le 2\theta^t$.

Time			
$t = 3$		$V_3 \sim P_{w_2} P_{w_1} P_{w_0} \nu$	w_3
$t = 2$		$V_2 \sim P_{w_1} P_{w_0} \nu$	w_2
$t = 1$		$V_1 \sim P_{w_0} \nu$	w_1
$t = 0$		$V_0 \sim \nu$	w_0
	$-N$		N $N+r$

Let now μ be a distribution on $S^{\mathbb{Z}}$. When iterating Φ, it induces a random sequence of words $(W_t)_{t \ge 0}$ on $\{N+1, \ldots, N+r\}$. Using the above inequality, we get:

$$\forall t \ge 0, \; ||(\Phi^t \mu)_{|K} - \lambda_K||_1 \le \max_{w_0, \ldots, w_{t-1} \in S^r} ||P_{w_{t-1}} \ldots P_{w_1} P_{w_0} \mu_{|K} - \lambda_K||_1 \le 2\theta^t.$$

If the neighbourhood of F is not of the form $\mathcal{N} = \{0, 1, \ldots, r\}$, then there exists $s \in \mathbb{Z}$ such that $F \circ \sigma^s$ is a left-permutive CA having a neighbourhood of that form. The noisy version of $F \circ \sigma^s$ is $\Phi \circ \sigma^s$, and the previous inequality provides: $||((\Phi \circ \sigma^s)^t \mu')_{|K} - \lambda_K||_1 \le 2\theta^t$, for any distribution μ'. In particular, for $\mu' = \sigma^{-st} \mu$, since Φ and σ commute, we obtain: $||(\Phi^t \mu)_{|K} - \lambda_K||_1 \le 2\theta^t$, which ends the proof. The right-permutive case is analogous.

In a collaboration still in progress with S. Taati and M. Sablik, we investigate the ergodicity of more general noisy *surjective* CA.

Concerning elementary PCA, it is still a challenging open question whether they are ergodic as soon as $\theta_{ij} \in (0, 1)$ for all $i, j \in S$.

Acknowledgments. The author thanks warmly S. Taati for the joint work on Sects. 4 and 5 and for his careful reading, and M. Sablik for fruitful discussions. This article is also based on a collaboration with J.B. Martin (Sect. 3) and on a previous work with A. Bušić and J. Mairesse (Sect. 2).

References

1. Bušić, A., Mairesse, J., Marcovici, I.: Probabilistic cellular automata, invariant measures, and perfect sampling. Adv. Appl. Probab. **45**(4), 960–980 (2013)
2. Dobrushin, R.L., Kryukov, V.I., Toom, A.L.: Stochastic Cellular Systems: Ergodicity, Memory, Morphogenesis. Nonlinear science. Manchester University Press, Manchester (1990)
3. Holroyd, A.E., Marcovici, I., Martin, J.B.: Percolation games, probabilistic cellular automata, and the hard-core model (2015). http://arxiv.org/abs/1503.05614
4. Vasilyev, N.B.: Bernoulli and Markov stationary measures in discrete local interactions. In: Developments in Statistics, vol. 1, pp. 99–112. Academic Press, New York (1978)

Types in Programming Languages, Between Modelling, Abstraction, and Correctness

Extended Abstract

Simone Martini[✉]

Dipartimento di Informatica–Scienza e Ingegneria,
Università di Bologna and INRIA, Bologna, Italy
simone.martini@unibo.it

Keywords: Types · Programming languages · History of computing ·
Abstraction mechanisms

1 Introduction

The notion of *type* to designate a class of values, and the operations on those values, is a central feature of any modern programming language. In fact, we keep calling them *programming* languages, but the part of a modern language devoted to the actual specification of the control flow (that is, programming *stricto sensu*) is only a fraction of the language itself, and two different languages are not much apart under that perspective. What "makes a language" are much more its modelling capabilities to describe complex relations between portions of code and between data. In a word, the central part of a language is made by the abstraction mechanisms it provides to model its application domain(s), all issues the language theorist may well group together in the type chapter of a language definition.

The conquest of the summit by the notion of type is the result of a rather slow process in the history of programming languages. In a previous paper [21] we have sketched some of the earliest history, observing that the concept of type we understand nowadays is not the same it was perceived in the sixties, and that it was largely absent (as such) in the programming languages of the fifties. While the technical term "type" arrives on the scene at the end of the fifties (for sure in the report on Algol 58 [26])[1], the use of types as a modelling tool for the "objects of the real world" is the contribution of the sixties (in particular under the influence of McCarthy [22] and Hoare [15]), which will materialize

---

[1] The very first use of the term "type" in programming is probably Curry's [7], to distinguish between memory words containing instructions (*"orders"*) and those containing data (*"quantities"*). These reports by Curry, as reconstructed by [10], contain a surprising, non-trivial mathematical theory of programs, up to a theorem analogous to the "well-typed expressions do not go wrong" of [23]! Despite G.W. Patterson's review on JSL 22(01), 1957, 102–103, we do not know of any influence of this theory on subsequent developments of programming languages.

© Springer International Publishing Switzerland 2016
A. Beckmann et al. (Eds.): CiE 2016, LNCS 9709, pp. 164–169, 2016.
DOI: 10.1007/978-3-319-40189-8_17

in languages like Algol W [30] or Pascal. Moreover, we observed in [21] that the notion of "type" of programming languages, which we now conflate with the concept of the same name of mathematical logic, is instead relatively independent from the logical tradition, until the Curry-Howard isomorphism [18] will make an explicit bridge between them. The connection between these two concepts remains anonymous for a long time—some of the people knew very well the other field, and it is certain that, from mid sixties, the mathematical logic work started influencing programming languages (we think, among other, to Landin, Scott, Strachey, Hoare, McCarthy, Morris etc.). But there is no explicit, mutual recognition—concepts and formal systems are systematically re-discovered in the two fields. The first explicit connection we know of, in a non technical, but explicit, way is [16].

The present paper will elaborate on this story, focusing on that fundamental period covering the seventies and the early eighties. It is there that the types become the cornerstone of the programming language design, passing first from the abstract data type (ADT) movement and blossoming then into the object-oriented paradigm. This will also be the occasion to reflect on how it could have been possible that a concept like ADTs, with its clear mathematical semantics, neat syntax, and straightforward implementation, could have given way to objects, a lot dirtier from any perspective the language theorist may take.

2   Modeling and Correctness

A central issue of the story we have told in [21] is the provision of a language mechanism to introduce new data types, in an extensible way (that is, different from a palette of types fixed at language design time). This materializes in the two related proposals of *records and typed references* [15], and Simula's *classes* [9]. A further, important realization is that a type is given not only by its class of values, but it is defined *together* the operations acting on those values (see, e.g., [8][2]). Simula was an extension of Algol 60 designed for discrete event simulation. One of the main concepts in Simula is the *class*[3]—the specification of both data and operations, which may have several dynamic instances (*objects* in the modern terminology, or *processes* in Simula I), whose life is not required to be dynamically nested. Under this view, Simula classes are a good candidate as a language mechanism for the definition of such types, since they permit the simultaneous definition of data and operations. However, this has to be seen together with the need to enforce some level of correctness, at a syntactic (and if possible, static) way—a central feature of what Priestley [27] calls the "Algol

---

[2] "A type is a class of values. Associated with each type there are a number of operations which apply to such values".

[3] "Class" is, however, the terminology of Simula 67; in Simula I they are called *activities*.

research programme[4]", where the design of programming languages should assist (or even guide) the programmer in avoiding bugs or, worse, unintended behaviors in a program. While Simula classes may be used to define types as "data plus operations", they do not provide the necessary *abstraction* to enforce correctness, because the language does not distinguish between a type and its implementation (between, say, a stack, and the list used to implement that stack: a stack may be incautiously manipulated by any operation on lists). The need for this abstraction is the core of much literature on data and programming languages in the early seventies.

3   Abstract Data Types

The search for economic and terse linguistic constructs comes together with the need for the definition of a precise semantics for those constructs. Parnas' seminal [25] introduces the term *information hiding*, meaning that a stable interface towards the rest of the program should protect the design choices which are bound to change (which may thus evolve without affecting the other parts of the program). In Parnas' view this is a general design methodology, which applies to types, modules, packages, etc. It is a crucial forward step of the programming language community that this hiding should be enforced by linguistic abstraction mechanisms, and not merely guaranteed by a design methodology. Looking at the published literature (e.g., Morris [24], Hoare [17], Goguen [12], Liskov and Zilles [20], Reynolds [29]—who also explicitly introduces the expression "representation independence"[5]—, Guttag [14], etc.), we see that around 1972–1973 the time is ripe for a substantial achievement. If Hoare [17] uses an axiomatic settings, starting from Goguen [12] the (informal) abstraction requirement is described semantically by *freeness*—a data type is, in its mathematical semantics, a free algebra over a set of constructors (that is, non-interpreted function names). Freeness means that one cannot uses implementation dependent information on a value, because a values is simply an inductive construction over the constructors—hence abstraction. Moreover, this provides a powerful proof-technique on programs—structural induction [3,4]. Finally, equations on terms—and hence all the good properties of equational logic—provide the axiomatic semantics needed to distinguish between types (algebras) over isomorphic sets

---

[4] Under this term Priestly refers to the "coherent and comprehensive research programme within which the Algol 60 report had the status of a paradigmatic achievement, in the sense defined by the historian of science Thomas Kuhn. This research programme established the first theoretical framework for studying not only the design of programming languages, but also the process of software development." Therefore, are grouped under this broad term the developments of structured programming, of software engineering à la Dijkstra, of the formal description of programming language semantics, etc.

[5] A language provides representation independence if two correct implementations of a single specification of an ADT are observationally indistinguishable by the clients of these types.

of constructors, but with different behavior (stacks and queues, say). The notion of abstract data type will make into programming languages with CLU [19], which will have a significant impact on subsequent languages.

At the end of the seventies, thus, it seems that types in programming languages are in a successful, and positive position—type abstraction mechanisms made into good (albeit essentially academic) programming languages, and their linguistic constructs come with a clear and mathematically sound formal semantics. However, in a sort of *coup de théâtre*, at the peak of their success ADTs will have to give way to another concept—objects, a much dirtier mechanism, able to enforce a lot less correctness than ADTs. Objects, and not ADTs will be the real players of the programming languages of the following decade(s).

4   Objects

To understand the fall of ADTs we must contrast correctness with flexibility. The utopia is to have them both at the same time, but realism tells us that they represent, most of the time, conflicting aims. ADTs provide abstraction at the expenses of (reuse and, then, of) compatibility. We cannot give here all the details of an example of the problem[6]. Suppose only to have an ADT T with an operation f, which is then extended into a new type $T1$, sharing the same values of T but with an extended set of operations and for which the operation f is also redefined. It is now natural (and convenient) to assume that the language enforces that $T1$ is compatible with T (a value of $T1$ may appear in any context requiring T). The problem now is that in an expression like $f(t)$ the choice of which code for f will be executed (the one for T or the one for $T1$) depends on the static context, that is, on the static type of t. Thus, if t (of static type T) references a value $t1$ of type $T1$ (which may correctly happen, in view of the compatibility of $T1$ with T), it is the erroneous f to be applied to a value of $t1$—abstraction breaks, which means that compatibility of $T1$ and T must be abandoned, which results in a drastic programming burden.

From this perspective the solution is easy—allow for a *dynamic* choice for the selection of the code for f in $f(t)$, depending not from the static typing of t, but from the actual type of the value referenced by t. In the programming language jargon, do not use functions and overloading, and use instead methods and dynamic lookup. Object oriented programming may be seen as the result of this observation. It is a paradigm where: (i) there is mechanism which, under certain conditions, supports the inheritance of the implementation of certain operations from other, analogous constructs; (ii) there is a notion of compatibility defined in terms of the operations admissible for a certain construct; (iii) operations on values are dynamically selected on the basis of the "actual type" of the arguments to which they are applied. These features *together* allow for the flexibility of the paradigm when used in actual programming of large scale systems. But, at the same time, these features *together* cannot be given semantics in the clean framework of algebraic types, at least in their simple formulation. Providing a

---

[6] For a pedagogical discussion, see [11], Chap. 10.

sound semantics and a formal treatment of objects will be a challenge for almost fifteen years (see, e.g., [1,2,28] for the references therein). Finally, we will have to explain why types are so central for objects, in view of the *lack* of types in languages like Smalltalk [13]. The distinction made in [6] (and the discussion in [5]) will guide our reflection.

5 Conclusions

The history of computer science is innervated by the continuous tension between formal beauty and technological effectiveness. Types in programming languages are an evident example of this dialectics. They are introduced for a better verification of the correctness of programs, and yet—contrary to mathematical logic—they must be experienced by the working programmer as an enabling feature[7], allowing for simpler writing of programs.

In its formal approach, computer science never used ideological glasses (types per se; constructive mathematics per se; linear logic per se; etc.), but exploited what it found useful for the design of more elegant, economical, usable artifacts. This eclecticism (or even anarchism, in the sense of epistemological theory) is one of the distinctive traits of the discipline, and one of the reasons of its success.

References

1. Abadi, M., Cardelli, L.: A semantics of object types. In: Ninth Annual IEEE Symposium on Logic in Computer Science, pp. 332–341 (1994)
2. Abadi, M., Cardelli, L.: A Theory of Objects. Springer, New York (1996)
3. Burstall, R.: Proving properties of programs by structural induction. Comput. J. **12**(1), 41–48 (1969)
4. Burstall, R., Landin, P.J.: Programs and their proofs: an algebraic approach. Mach. Intell. **4**, 17–43 (1969)
5. Cook, W.R.: Object-oriented programming versus abstract data types. In: de Bakker, J.W., de Roever, W.P., Rozenberg, G. (eds.) Foundations of Object-Oriented Languages. LNCS, vol. 489, pp. 151–178. Springer, Heidelberg (1990)
6. Cook, W.R., Hill, W., Canning, P.S.: Inheritance is not subtyping. In: Proceedings of the 17th ACM SIGPLAN-SIGACT Symposium on Principles of Programming Languages, POPL 1990, pp. 125–135. ACM, New York, NY, USA (1990)
7. Curry, H.B.: On the composition of programs for automatic computing. Technical report Memorandum 10337, Naval Ordnance Laboratory (1949)
8. Dahl, O.-J., Hoare, C.A.R.: Hierarchical program structures. In: Structured Programming, Chap. 3, pp. 175–220. Academic Press (1972)
9. Dahl, O.-J., Nygaard, K.: Simula: an ALGOL-based simulation language. Commun. ACM **9**(9), 671–678 (1966)
10. De Mol, L., Carlé, M., Bullyinck, M.: Haskell before Haskell: an alternative lesson in practical logics of the ENIAC. J. Logic Comput. **25**(4), 1011–1046 (2015)
11. Gabbrielli, M., Martini, S.: Programming Languages: Principles and Paradigms. Undergraduate Topics in Computer Science. Springer, Heidelberg (2010)

---

[7] An expression of Vladimir Voevodsky.

12. Goguen, J.: Some comments on data abstraction. Notes for a course at ETH Zurich (1973)
13. Goldberg, A., Kay, A.: Smalltalk-72 instruction manual. Technical report SSL 76-6. Learning Research Group, Xerox Palo Alto Research Center (1976)
14. Guttag, J.: The specification and application to programming of abstract data types. Ph.D. thesis, University of Toronto (1975)
15. Hoare, C.A.R.: Record handling. ALGOL Bull. **21**, 39–69 (1965)
16. Hoare, C.A.R.: Notes on data structuring. In: Structured Programming, Chap. 2, pp. 83–174. Academic Press (1972)
17. Hoare, C.A.R.: Proof of correctness of data representation. Acta Informatica **1**, 271–281 (1972)
18. Howard, W.A.: The formulae-as-types notion of construction. In: Seldin, J.P., Hindley, J.R. (eds.) To H.B. Curry: Essays on Combinatory Logic Lambda Calculus and Formalism, pp. 479–490. Academic Press, Cambridge (1980)
19. Liskov, B., Snyder, A., Atkinson, R., Schaffert, C.: Abstraction mechanisms in CLU. Commun. ACM **20**(8), 564–576 (1977)
20. Liskov, B., Zilles, S.: Programming with abstract data types. In: Proceedings of the ACM SIGPLAN Symposium on Very High Level Languages, pp. 50–59. ACM (1972)
21. Martini, S.: Several types of types in programming languages. Paper Presented at HAPOC 2015, Pisa (2015)
22. McCarthy, J.: A basis for a mathematical theory of computation, preliminary report. In: Papers Presented at the May 9–11, 1961, Western Joint IRE-AIEE-ACM Computer Conference, IRE-AIEE-ACM 1961 (Western), pp. 225–238, New York, NY, USA. ACM (1961)
23. Milner, R.: A theory of type polymorphism in programming. J. Comput. Syst. Sci. **17**(3), 348–375 (1978)
24. Morris, J.H.: Types are not sets. In: Proceedings of the 1st Annual ACM SIGACT-SIGPLAN Symposium on Principles of Programming Languages, POPL 1973, pp. 120–124. ACM, New York, NY, USA (1973)
25. Parnas, D.L.: On the criteria to be used in decomposing systems into modules. CACM **15**(2), 1053–1058 (1972)
26. Perlis, A.J., Samelson, K.: Preliminary report: international algebraic language. Commun. ACM **1**(12), 8–22 (1958)
27. Priestley, M.: A Science of Operations. Machines, Logic and the Invention of Programming. Springer, Heidelberg (2011)
28. Reddy, U.S.: Objects of closures: abstract semantics of object oriented languages. In: ACM Conference on Lisp and functional programming. ACM (1988)
29. Reynolds, J.C.: Towards a theory of type structure. In: Robinet, B. (ed.) Programming Symposium. LNCS, vol. 19, pp. 408–423. Springer, London (1974)
30. Sites, R.L.: Algol W reference manual. Technical report STAN-CS-71-230, Computer Science Department, Stanford University (1972)

AFCAL and the Emergence of Computer Science in France: 1957–1967

Pierre-Éric Mounier-Kuhn[1] and Maël Pégny[2(✉)]

[1] CNRS, Université Paris-Sorbonne, Maison des Sciences de l'Homme, Paris, France
`mounier@msh-paris.fr`
[2] CNRS, Université de Paris 1 Panthéon-Sorbonne, IHPST, Paris, France
`maelpegny@gmail.com`

Abstract. Founded in 1957, the *Association Française de Calcul* (AFCAL) was the first French society dedicated mainly to numerical computation. Its rapid growth and amalgamation with sister societies in related fields (Operations Research, Automatic Control) in the 1960s resulted in changes of its name and purpose, including the invention and adoption of the term *informatique* in 1962–1964, then in the adoption of *cybernétique* in 1967. Our paper aims at explicating the motives of its creation, its evolving definition and the functions it fulfilled. We seek to understand how this association, altogether a learned and a professional society, contributed to the emergence and recognition of Computing as an academic discipline in France. The main sources are the scattered surviving records of AFCAL, conserved in the archives of the Observatoire de Paris, of the *Institut de Mathématiques appliquées de Grenoble* (IMAG) and of the CNRS' *Institut Blaise Pascal* in Paris, as well as AFCAL's first congress and journal, *Chiffres*.

1 Introduction

The *Association Française de Calcul* (AFCAL) was founded in 1957 by French academics and industry engineers who wanted to join forces, to form a specialist community promoting applied mathematics and computing technologies in the context of the post-war modernization. With this technoscientific and political agenda, the new society soon attracted hundreds of members and support from various organizations.

It would be anachronistic to retrospectively tag AFCAL as a champion of "computer science". It focused initially on numerical analysis, a sub-discipline of applied mathematics, and on the use of calculating machines. The idea that electronic computers and their programs could become a new, distinct scientific academic field did not emerge until the 1960s. One of the main questions addressed in this paper is precisely how AFCAL participated in this emergence: what convergence of agenda and which internal tensions came to shape digital computation as a new academic discipline[1].

---

[1] This question is addressed in a broader scope in Mounier-Kuhn (2010a).

A. Beckmann et al. (Eds.): CiE 2016, LNCS 9709, pp. 170–181, 2016.
DOI: 10.1007/978-3-319-40189-8_18

The archives of the association disappeared accidentally in a flood in the 1980s, making our research a bit difficult. The main sources are scattered surviving records of AFCAL, conserved in a variety of institutions: the archives of the *Observatoire de Paris* keep a box of papers from AFCAL's early years[2], those of the *Institut de Mathématiques appliquées de Grenoble* and of the CNRS *Institut Blaise Pascal* in Paris contain correspondences and reports exchanged between founding members. There are certainly more records waiting to be discovered in other organizations or with private persons. Published sources include AFCAL journal *Chiffres* (from 1958) and its conferences (from 1960). Finally, the first meeting on the history of computing in France included several papers addressing the evolution of French societies in this field[3].

In this sketch of our paper[4], we will highlight three episodes of the association's beginnings: the circumstances and motives of its creation; its participation to the international congress ICIPS and the international association IFIP; the functions of the association and their evolutions. In this last part, we will examine the role of the association in the emergence of *informatique* as a new scientific discipline.

2   The Emergence of a Computing Community

While the French academic spheres in the post-war period were dominated by pure mathematics, a growing demand for applied mathematics was expressed by physicists, and by various branches of engineering. From the late 1940s on, several young university professors responded to this demand by creating courses in numerical analysis and computing laboratories (Mounier-Kuhn 2012). In the early 1950s, electronic calculators were acquired from manufacturers and installed in these laboratories or, on a larger scale, in big organizations such as *Électricité de France* (E.D.F.) and aeronautics companies. In 1955, the first stored-program computers entered service in France, considerably reinforcing these facilities and offering new perspectives to mathematical modeling as well as to data processing.

Within two years, a series of learned societies were founded in related fields, a veritable blooming of professional communities active in Operations Research (SOFRO), Automatic Control (AFRA), Measure and Regulation (AFIC) and Computing (AFCAL). Simultaneously, scientists with political connections established a governmental council, CSRSPT (*Conseil Supérieur de la Recherche Scientifique et du Progrès Technique*), in order to support the modernization of research and higher education. This body identified automatic control, applied mathematics and computing as a priority sector, under the label *Cybernétique*.

---

[2] *Archives de l'Observatoire de Paris*, Ms 1061 II-2-D.

[3] See the papers by Anne Brygoo, Jean Carteron, Colette Hoffsaës and Félix Paoletti in the proceedings of *Colloque sur l'histoire de l'informatique en France*, 1988.

[4] The present text is only a provisional account of a work in progress, and should not be quoted without the authors' permission.

2.1 Technoscientific Networking: Applied Mathematics as a Modernization Tool

One of the main actors of this movement was Jean Kuntzmann, an algebraist turned militant of applied mathematics, who had created a small laboratory of numerical analysis at the university of Grenoble to meet the needs of the local *Institut Polytechnique*[5]. His correspondance reveals his role as the initiator of the creation of AFCAL and sheds light on his motivations.

Kuntzmann's action to foster useful mathematics was supported by the local academic authorities and by the Ministry of Education which favored initiatives toward vocational studies. Most of the financial resources of his laboratory came from contracts with civilian or military clients who appreciated his expertise and, in turn, brought stimulating problems. The synergy between funding, research, teaching and practical computing generated a cumulative growth process. In the mid-1950s, Kuntzmann was recognized in France and abroad as a leading specialist in his field. He was consulted by junior professors who wanted to set up their own computing laboratories (in Lille, Nancy, Barcelona and other universities). His laboratory attracted visiting teachers and doctoral students.

Having systematically informed the major French users about the services his laboratory could provide, and received acknowledgements of their interest, Kuntzmann had built an interactive network of colleagues and partners. In April 1956, he began contacting other French specialists to suggest an association.

His correspondents, men of his generation or younger engineers and professors, shared his project and responded positively. This first poll allowed Kuntzmann to solicit and obtain the patronage of academic notabilities at a higher hierarchical level, the dean of the Paris faculty of science and the director of *Institut Henri Poincaré*[6]. He did not explain his purpose in detail, as these men obviously knew the usefulness of a learned society: Kuntzmann's correspondence contains no negative answer, not a single letter from a colleague objecting to his project.

Already a member of several learned societies, Kuntzmann had a precise idea of the one he wanted to create. He belonged to the *Société française de mathématiques*, which was still too general-purpose and exclusively academic to be of real use in his project. He also participated in the International Association for Analog Calculation, which by contrast covered only part of the domain. The closest model was the American Association for Computing Machinery (ACM) with its specialized journal, to which Kuntzmann subscribed.

In Spring 1956, Kuntzmann also became a member of the German *Gesellschaft für Angewandte Mathematik und Mechanik* (GAMM). Kuntzmann could not attend the inauguration of the PERM computer in May 1956, but, always keen on completing his documentation on the state of the art, the Grenoblois ordered the proceedings of the important Darmstadt conference held

---

[5] Our choice of J. Kuntzmann as a key actor is based on his role as a science entrepreneur and the availability of his correspondence conserved at the *Archives départementales de l'Isère* (Grenoble), henceforth ADI.

[6] J. Kuntzmann's letters to J. Pérès and R. Darmois, May 3 1956 (ADI IMAG 4).

the previous year[7]. The GAMM existed since the 1920s and, with its institutional weight, could be a model of what Kuntzmann and his partners wanted to create in France, where no learned society existed in applied mathematics until 1955.

It is revealing that Kuntzmann aimed to create a new society rather than join the cybernetician community, which had been popular since Norbert Wiener's manifesto in 1948, and which possessed its own organization. While Kuntzmann sent one of his doctoral students at the International Congress of Cybernetics (Namur, Belgium, 1956) and ordered its proceedings, he explained in a harsh letter why he refused to join the *Association Internationale de Cybernétique*: "I have always thought that the grandiose term of Cybernetics covers very diverse matters. Some are mathematical, others are technical. Unfortunately, on top of that comes some philosophy, and even certainly babble (*fumisterie*). Either Cybernetics shall fall into oblivion or, some day, one will have to separate what is robust from what is not. I reserve my adhesion (and my participation in meetings) for the moment when this decantation is achieved[8]."

We know that this rejection of Cybernetics, or of what it had become in the mid-1950s, was shared by other French computing pioneers. They believed Cybernetics could not provide the identity of their emerging profession.

Kuntzmann knew that an informal group existed already in Paris, gathering periodically a handful of specialists and distinguished amateurs in a seminar on numerical calculation. Somehow its lack of institutional strategy and of scientific ambition reflected the low status of reckoning in the past decades. Kuntzmann courteously suggested to integrate the group in his future association to broaden its scale and scope. The merger was completed swiftly at a meeting in Paris, during a conference on scientific management. It avoided a possible conflict and brought a mailing list of 200 corresponding members to the nascent AFCAL[9]. The *Groupe de Calcul Numérique* held its seminars at the Paris *Institut d'Astrophysique*, and the AFCAL naturally followed suit, obtaining an office and a mailbox in this convenient venue. More importantly, the director of the Institut d'Astrophysique, André Danjon, an energetic and open-minded astronomer, agreed to become AFCAL's first president, bringing his own legitimacy and political connections.

Danjon had sent his young astronomers to learn programming on the IBM650 computer at IBM France, and was aware of the potential of the new machines. Beyond the technical importance of calculation in science, we can suggest the hypothesis of a solidarity between emerging disciplines. Just like computing in the years to come, astrophysics had to assert its legitimacy against the conservatism of traditional astronomy, which was entirely mathematical and had

---

[7] J. Kuntzmann's letters to J. Heinhold, TH München, June 5 1956 (ADI IMAG 4). On the Darmstadt conference, see Petzold (2004).

[8] J. Kuntzmann's letter to AIC, October 2 1956 (ADI IMAG 4).

[9] J. Kuntzmann's letter to É. de Lacroix de Lavalette, May 4 1956 (ADI IMAG 4); and É. de Lacroix de Lavalette's historical report on his *Groupe de calcul numérique*, 1995. Private archives of the first author.

reasons to view astrophysics as an uncertain science, lacking rigor, perhaps not a science at all: "Basically [for them], astrophysics could not be considered a science" (*À l'extrême, l'astrophysique ne pouvait être considérée comme une science*[10]).

Under Danjon's tutelage, AFCAL had two vice-presidents: Prof. Jean-André Ville, a mathematician who had just been put in charge of a computing center at the Paris faculty of science; and Jean Carteron, a Polytechnique engineer who headed the computing center at the R&D Department within E.D.F., and was equally active in founding the association (EDF sponsored AFCAL generously) (Mounier-Kuhn 2010b). Kuntzmann reserved for himself the responsibility of the future journal, a position of scientific gate-keeper in the new field.

Kuntzmann's work to promote modern theories and practices in applied mathematics was in synergy with the efforts of the computer industry to market their machines and their applications, and with the need of mathematical modeling among users. The lists of AFCAL members (200 in 1957, some 500 six years later) and of the institutions where they worked offer a fairly complete picture of French applied mathematics[11]: higher education and research, laboratories of major public or private companies, computer makers, consultants in mathematical and programming expertise. These various specialists felt the need to share their experiences and to promote their methods, which was the primary function of the association. It took nearly a year to create the association, as each founding member had many other urgent tasks. The constitutive assembly (May 31 1957), beyond adopting the charter, heard lectures on the present computing needs and resources among users, on recent machines' capabilities, and on the search for better programming methods that AFCAL should foster, a program covering most of the purposes of the association.

In June 1957, Kuntzmann set to create its journal, *Chiffres*, that is to find a suitable publisher and to hunt for good papers. The journal would also include sections devoted to scientific news, the announcement of courses and seminars, reviews of books and of conferences... In October, Carteron urged Kuntzmann to "give life as soon as possible to the Association" by organizing meetings, and both men gave it all the publicity they could in France and abroad. By November, Kuntzmann had formed an editorial board and gathered sufficient material for a first issue of *Chiffres*.

From its very beginning on, the association had an international dimension. Firstly, its membership soon included mathematicians working in Switzerland (Charles Blanc, Pierre Banderet, etc.), in Italy (particularly at the Olivetti-Bull joint venture) and in Belgium (Vitold Belevitch, Jacques Burniat). About 10 % of AFCAL members in the early 1960s worked outside of France. Its audience still remained limited to French-speaking countries or readers, as its publications provided initially no English abstracts.

---

[10] Evry Schatzmann's interview with J.-F. Picard, February 24 1987.
[11] *Annuaire 1963 de l'AFCALTI. Archives privées de J. Carteron.*

3   Creating an Internationale of Computing: IFIP and Its Scientific Agenda

Soon after its creation, the association contributed to establish an internationale of computing specialists. In December 1957, AFCAL was contacted by an American engineer and entrepreneur, Isaac L. Auerbach. As a member of the US Association for Computing Machinery (ACM), he planned to come to Paris in order to organize an international meeting of Information Processing societies from different countries. Isaac Auerbach had just left the Burroughs Corporation to form his own consultant company. Establishing international contacts to help Auerbach Associates conduct market surveys was certainly an incentive. Yet the main rationale for this internationalization was, as in other learned societies, to exchange information and gain political weight collectively.

The French reacted positively and agreed to meet with Auerbach and to cooperate for the preparation of the conference[12]. Just as in the creation process of AFCAL, the movement to join forces and cooperate appears to have been quite smooth: it was somewhat obvious to all actors, and it hardly needed more explanation than a few words in a letter. In June 1958, in Paris again, an ad-hoc committee gathered computer experts from most developed countries to organize the conference.

The International Conference of Information Processing Societies (ICIPS) met in June 1959 at the UNESCO building in Paris, gathering some 1800 participants from 38 countries during a week. For many French scientists, it was the first opportunity to hear foreign scholars (particularly Americans) present their work, and to discover the state of the art in programming techniques and in non-numerical applications of computers, an important milestone in their evolution towards research in computer science. Beside processor architecture or computing and coding methods, sessions were devoted to machine translation, pattern recognition, machine learning or future perspectives in computing. This diffusion of information was precisely the purpose pursued by Isaac Auerbach and Pierre Auger, the physicist who headed the French delegation at Unesco[13].

The ICIPS conference gave birth to the International Federation for Information Processing (IFIP) (see Auerbach 1986; Carteron 1996). From then on, IFIP set up a similar worldwide conference every three years in a major city (Munich in 1962, New York in 1965, etc.). The mere existence of this global learned society and of its conferences, comparable with those in astronomy or geophysics, contributed to establish a scientific status for information processing. Such a status was explicitly targeted since its creation, as highlighted by the general

---

[12] Correspondence between I. Auerbach, A. Danjon, J. Carteron, J. Kuntzmann and J. Ville, December 7–12 1957 (ADI IMAG 5) and *Archives de l'Observatoire de Paris*, Ms 1061 II-2-D.

[13] The 600 pages proceedings were published by the UNESCO both in Paris (1960) and London (1960). Pierre Auger happened to coin the term *Traitement de l'information*, translated from "Information processing". The journal of the congress can be found at http://unesdoc.unesco.org/images/0015/001537/153718fb.pdf.

secretary of the 1959 conference, Pierre Auger: "It is absolutely indispensable to have an international organization to respond to the needs of the science of Information Processing[14]".

Moreover, in subsequent years the IFIP created specific committees on specific problems: Software theory and practice, Algorithmic languages and calculi, Formal description of programming concepts, Programming languages, Programming methodology. IFIP thus not only constituted an international computing community, it also provided it with a scientific roadmap of research directions.

Being essentially French-speaking was not a serious handicap for the association, at a time when many mathematicians could at least read French as well as German. The association was thus part, from its beginnings, of an international family of sister societies with which it actively interacted. J. Carteron became IFIP's treasurer and general secretary. AFCAL members participated in the various IFIP committees and working groups, discussing research and education problems and drawing inspiration from foreign experiences (Mounier-Kuhn 2014).

4   Tensions Over a New Discipline

4.1   Applied Mathematics vs Data Processing

Operations research was excluded from AFCAL's scope, as it already had its own society. So was all work exclusively related to hardware, which belonged to the *Société Française des Electriciens*, the French member of IEEE, and could be published in a new journal, *Automatisme*. As for data processing, it was a set of office techniques remote from any scientific formalization. The association constituted the study of computation as a specific, hierarchized object, where calculating machines were scientific instruments applying branches of mathematics.

Since the very beginning of AFCAL, however, a tension had appeared between two conceptions of its scope: one centered on applied mathematics, the other pushing to include data processing. And of course this would grow along with the progressive diversification of the society in the following decade. In 1957 already, Carteron had written a letter to president Danjon on the definition of AFCAL's domain. Carteron recommended not to separate numerical computing and information processing: "For example, problems of sorting on magnetic tapes, or of machine translation of language, are part of our preoccupations, as well as the technical aspects of automatic data processing problems[15]". The fact that he mentioned these particular problems reveals that he was remarkably aware of recent research directions. There was also a sociological rationale in his proposal: with the advent of electronic computers, data processing would soon become too complex for traditional punch-card technicians; it would require

---

[14] *Journal du Congrès*, No. 6, p. 2 (italics ours).

[15] J. Carteron's letter to A. Danjon, December 6 1957 (*Archives de l'Observatoire de Paris*, Ms 1061 II-2-D).

engineers trained at a higher level, who could be eligible among the advanced elite of AFCAL members.

At the other end of the spectrum, Kuntzmann, like other academic numerical analysts, wanted to involve the association in militantism for a reform of mathematical education, in order to give more room to applied mathematics. In several letters and reports, he advocated for a restructuring of mathematical categories, which would acknowledge the importance of "algorithmic mathematics". Even if he was soon to include a machine translation team in his institute, and courses on data-processing techniques in its curriculum, his priority always remained applied mathematics. Even decades later he kept considering *informatique* a technical application of mathematics.

In 1960, this debate became a central topic at AFCAL's annual assembly. The members who wanted to broaden the association's scope to data processing were led by Jean Carteron, Henri Boucher, a computer pioneer in the French Ministry of Defence, and Philippe Dreyfus, then a sort of sales evangelist for Bull computers who had already an international reputation. They succeeded in adding to the association's name the initials "TI" (*traitement de l'information*). Thus renamed, AFCALTI asserted the unity of a field broadened beyond applied mathematics, precisely at the time when it organized a joint conference with SOFRO (Operations Research) and the French chapter of The Institute of Management Sciences (TIMS). This change was still consolidated as Carteron was elected president as successor to Danjon. The journal's title became *Chiffres-Revue française de traitement de l'information*.

4.2   Toward a New Scientific Discipline

At the 1962 annual assembly, a neologism, *informatique*, was defined by Philippe Dreyfus and Robert Lattès as the "technique of the logical and automatic processing of information, the support of human knowledge and communication" (see Ph. Dreyfus 1962). Note that various attempts had already been made to name the computer field, including *Informatik* in Germany as early as in 1957. Ph. Dreyfus later explained the invention of *informatique* by the simple need he felt to name his professional occupation, encompassing all types of applications of electronic computers[16]. But there was another reason: Dreyfus had just left Bull to create a software and service bureau firm, in partnership with Robert Lattès, a brilliant mathematician who worked simultaneously for the French atomic energy authority (*Commissariat à l'Énergie Atomique*) and for a mathematical consultancy firm, SEMA. *Informatique* was the brand name of their company, just as, at the same time across the Atlantic, Walter Bauer founded Informatics Inc.

According to the president of AFCALTI, Carteron, *informatique* covered "all disciplines covered by calculation and information processing, from the logical design of computers to numerical analysis through machine translation." (see Carteron 1963) Defined as a "technique" in 1962, *informatique* thus evolved into a set of "disciplines".

---

[16] P. Mounier-Kuhn's interview with Ph. Dreyfus, September 21 2007.

A further step was taken in the proceedings of the third congress of the association (CNRS 1963). In the brief introduction, the editors asserted that programming was no longer a mere means to communicate with the machine, but that it turned into "a science of languages extending from algebra to decision theory", allowing to define "the modes of expression as well as the construction of compilers [...]. Finally, considerations about logic allowed to monitor the evolution of ideas concerning theoretical *informatique*". *Science*, *informatique théorique*: these bold assertions were thus launched, with no explanatory development, likely to stimulate the discussion and to accelerate an evolution.

It is also a good case for a geography of science. The fact that this happened in Toulouse is not irrelevant. None in Grenoble would have written such provoking phrases under Kuntzmann's direction. The Toulouse computer scientists, by contrast, were mainly physicists, engineers or applied mathematicians coming from marginal positions. Contrary to Kuntzmann, they did not worry about their mathematical legitimacy, and felt more free to embrace a new academic identity. The next year, Dreyfus and Lattès hammered their message again. Aiming at a broader readership than AFCALTI members, they chose Le Monde to publish an article-manifesto: "A new discipline: *Informatique*" (Lattès and Dreyfus 1964). The context was favourable: several French universities had already created specialized degrees, IBM had just launched its "third generation" computers, operating systems were now a central issue for computer scientists, and the takeover of the French company Bull by General Electric had triggered reactions in industry and government spheres.

The headlines described *informatique* as a broad set of sciences and techniques, reflecting the variety of AFCALTI's interests. Then Dreyfus and Lattès recalled their definition of 1962, but assertively added a much more scientific dimension: "More than a simple technique, computing is a discipline, a science-interface (*une science-carrefour*) covering a large and disparate sector as much technical as scientific; it is also a mental attitude in approaching problems." They insisted that computing had become a fully-fledged research sector, with specific matters of its own "such as software and certain technologies". In conclusion the article stressed the growing importance of computing for society and economy, which required its integration in all parts of higher education and executive training, "to raise generations of men adapted to the mode of thought now required by the machine."

By that time, *informatique* was being gradually adopted in French common vocabulary, coexisting with other terms, and triggering discussions (within and outside the association) about its content, its meaning, its position between science and technology. The controversy opposed mainly those who thought in terms of applied mathematics and those who asserted the central place now occupied by computers in different activities. Branch or application of mathematics? Or technique, or even science of computers? or of information? The initial definition given in 1962 was tentative, and anyone was entitled to propose variants of it. This debate would eventually continue over several decades, drawing a dividing line within academic spheres, and shaping its institutional organisation.

4.3 The Evolving Functions of the Association and the Institutionalization of Computing

Finally, we examine how the definition of the functions of the association evolved over time, and draw lessons on the emergence of Computing as a new professional field in France.

The association had multiple functions from the start. Its founders knew that their field and its economic impact were growing fast. One of their initial goals was to increase awareness of the potential applications of computation among institutional actors; some members such as J. Kuntzmann also wanted to involve the association in their campaign for a reform of mathematical education, which would give more room to applied mathematics and numerical computation.

But the association was initially dominated by academics and engineers, and its primary goal was research. A key issue in the definition of its function was thus the definition of its scientific object. This definition underwent important, and somewhat tortuous evolutions during its first ten years. In the beginning, it was dedicated to computation, its methods and the relevant mathematical developments. Analog computation was not excluded, but numerical computation was central.

But as we have seen (Sect. 4.1), as computing applications kept growing, a more applied point of view gained traction within the association, and it became more and more inclusive. This expansion was justified by the common interests shared by the different specialists, and by the extremely rapid growth of computer applications. This rate of growth helps to explain the frantic pace of institutional evolution, as the association underwent three amalgamations in less than a decade.

If the term "amalgamation" is adequate at an institutional level, it is nevertheless misleading at the level of research practices. The series of institutional amalgamations did not erase the differences between the various intellectual traditions involved. Considering its scientific object, the evolution of the association might be better described as a movement of "expansion-subdivision". In its March-April 1967 issue, the *Chiffres* journal was divided in two collections, one for theory and methodology, the other for practice. In 1968 it was further divided into three different collections: mathematics, computer science, operations research. In the 70s, the subdivision process continued, as the association itself was subdivided into subsections and work groups (Brygoo 1988). The association was thus a place where different subdisciplines would interact, not a melting pot where those subdisciplines would lose their original identities.

The association general assembly, and *Chiffres* were also places where *informatique* was proposed and discussed as a disciplinary term. Less successful neologisms, such as the 'economic cybernetics' (*cybernétique économique*), reveal the fluidity of the new professional domain, and the efforts made to cope with it. As Gingras (1991) put it, scientific associations are a privileged place for specialists of a given domain to articulate a discourse on their activity and intellectual identity, both for themselves and for outsiders. Terminological innovation is thus a way to define and defend the identity of a new group of specialists. The association was thus one of the institutions where the professional identity of *informaticien* emerged.

To conclude, let us use this story to test the idea that the creation of scientific associations can be a criterium for the emergence of new disciplines. When it was renamed AFIRO, the association was a scientific institution at least including computer science as part of its object. According to this criterium, it played a part in the institutionalization of computer science. But a couple of remarks should be made in order to understand more precisely this institutionalization process. We must distinguish between disciplines *qua* intellectual tradition and disciplines *qua* institution. As Gayon (2015) remarked, an intellectual tradition like evolutionary theory could exist for decades without being institutionalized, and natural history is still an academic institution without being an autonomous intellectual tradition anymore. Furthermore, for an association to be the institutionalization of intellectual tradition X, the intellectual tradition X has to be identified as such by actors. In a typical scenario, an intellectual tradition would be formed before it would be equipped with institutions such as scientific societies.

But at the beginning of the AFCAL, it was not obvious that computation, even if its fast development demanded an institutional response, would represent a distinct intellectual tradition, other than just a branch of applied mathematics. Second, even when the intellectual tradition of computer science started to emerge, it did not lead the association to focus exclusively on computer science - quite the contrary. The word *informatique* was included in its title precisely when the scope of the association was broadened to include *recherche opérationnelle* (operations research), then automatics and control. The association thus defined and covered a much broader field of expertise and applications than the academic discipline of computer science in France would in the near future, since Operations Research and Automatics have autonomous academic institutionalizations outside of computer science departments.

Far from being just an institutionalization of an already existing academic tradition, the association was a forum where, through collaborative work, tensions over agenda and terminological debate, the conscience of the existence of a new discipline, computer science, was going to emerge. Using the existence of the association as a static criterium for the institutionalization of an academic tradition masks its dynamic role in the very constitution of this intellectual tradition.

5   Conclusion

The creation of the AFCAL, and much of its evolution, was application driven. The huge expansion of computer applications in the first years of its existence is the driving force behind the growth of its membership, and the fluidity of its scope. It can be also seen as a decisive factor in the definition of the various association functions. Spreading the knowledge about potential applications of the new technologies among institutional actors, reforming the mathematical education, gathering specialists coming from different domains around a new research object: all these actions can be understood as a fast institutional response to a rapidly evolving field of high socio-economical impact.

References

Auberbach, I.L.: The start of IFIP-personal recollections. Ann. Hist. Comput. **8**(2), 180–192 (1986)

Brygoo, A.: L'AFCET et l'informatique à travers les éditoriaux des bulletins. In: Colloque sur l'histoire de l'informatique en France, 3–5 Mai 1988, vol. 1 (1988)

Carteron, J.: Du Calcul à l'Informatique et à la recherche opérationnelle. Chiffres-RFTI, no. 4, p. XII (1963)

Carteron, J.: Memories of bygone times In: Zemanek, H. (ed.) 36 Years of IFIP. IFIP (1996). http://www.ifip.or.at/36years/a21cart2.html

Chatelin, P. (ed.): Colloque sur l'histoire de l'informatique en France. IMAG, Grenoble (1988)

CNRS: Troisième Congrès de l'AFCALTI, Toulouse, Dunod, Paris, 14–17 Mai 1963 (1965)

Dreyfus, P.: L'informatique. Chiffres-RFTI, no. 1, pp. XII–XIV (1963). (Article first published in Gestion, June 1962)

Dreyfus, P., Lattès, R.: L'homme et les calculatrices électroniques. Une discipline neuve: l'informatique, Le Monde, 9 Juillet 1964, p. 9 (1964)

Gayon, J.: Biologie et philosophie de la biologie: paradigmes. In: Hoquet, T., Merlin, F. (eds.) Précis de philosophie de la biologie, Paris Vuibert (2014)

Gingras, Y.: L'institutionnalisation de la recherche en milieu universitaire et ses effets. Sociologie et sociétés **23**(1), 41–54 (1991)

Mounier-Kuhn, P.: L'émergence d'une science. L'informatique en France, de la 2e Guerre mondiale au Plan Calcul. Paris, Presses de l'Université Paris-Sorbonne (2010a)

Mounier-Kuhn, P.: Jean Carteron. IEEE Ann. Hist. Comput. **32**, 82–89 (2010b)

Mounier-Kuhn, P.: Computer science in French universities: early entrants and latecomers. Inf. Cult. J. Hist. **47**(4), 414–456 (2012)

Mounier-Kuhn, P.: Algol in France: from universal project to embedded culture. IEEE Ann. Hist. Comput. **36**(4), 6–25 (2014)

Petzold, H.: Eine Informatiktagung vor der Gründung der Informatik: Die Darmstädter Konferenz von 1955. In: Seising, R., Folkerts, M., Hashagen, U. (eds.) Form, Zahl, Ordnung. Studien zur Wissenschafts- und Technikgeschichte, pp. 759–782. Franz Steiner Verlag, Wiesbaden (2004)

UNESCO: Traitement numérique de l'information: comptes-rendus des travaux de la conférence internationale sur le traitement numérique de l'information, Unesco, Paris, Dunod, Paris, 15–20 Juin 1959 (1960)

UNESCO: Information Processing: Proceedings of the International Conference on Information Processing. UNESCO, Butterworths, London (1960)

Computable Reductions and Reverse Mathematics

Reed Solomon$^{(\boxtimes)}$

University of Connecticut, Storrs, CT 06268, USA
david.solomon@uconn.edu

Abstract. Recent work in reverse mathematics on combinatorial principles below Ramsey's theorem for pairs has made use of a variety of computable reductions to give a finer analysis of the relationships between
these principles. We use three concrete examples to illustrate this work,
survey the known results and give new negative results concerning RT^1_k,
SRT^2_ℓ and COH. Motivated by these examples, we introduce several variations of ADS and describe the relationships between these principles
under Weihrauch and strong Weihrauch reductions.

1 Introduction

The general project of reverse mathematics is to formalize mathematical theorems in second order arithmetic and to determine which set-theoretic axioms
are required to prove them. Typically, we take RCA_0, which includes the non-
induction axioms of PA as well as the Δ^0_1-comprehension and Σ^0_1-induction
schemes, to be the base subsystem of second order arithmetic over which we
prove our equivalences. RCA_0 has the advantage that it allows a reasonable
amount of coding and that theorems provable in RCA_0 roughly correspond to
theorems for which existential objects can be found computably in the parameters. More specifically, an ω-model satisfies RCA_0 if and only if the second
order part of the model is closed under Turing reducibility and the Turing join.
The ω-models of other subsystems of second order arithmetic can be described
in similarly natural computability-theoretic terms. These connections allow us
to prove results in reverse mathematics using techniques and intuitions from
computability theory.

While we will primarily be concerned with ω-models, it is worth noting the
importance of non-ω-models. They play a crucial role in, for example, measuring
levels of induction or conservation. More importantly for our purposes, they
are used to separate Ramsey's theorem for pairs (RT^2_2) from its stable version
(SRT^2_2). In the next section, we will define RT^2_2 and SRT^2_2 precisely, but for now
we use them to make a motivational point. From a computability standpoint, the
natural first step to separate these principles would be to prove that for every Δ^0_2
set D, either D or $\overline{D}$ contains an infinite low set. Unfortunately, this statement

R. Solomon—I would like to thank the anonymous referee for many helpful sugges-
tions to improve the presentation of the material in this article.

A. Beckmann et al. (Eds.): CiE 2016, LNCS 9709, pp. 182–191, 2016.
DOI: 10.1007/978-3-319-40189-8_19

is false (see [6]), so this method fails at its initial step. However, Chong et al. [4] constructed a non-ω-model of RCA_0 in which this property holds and they used this model to prove that SRT_2^2 does not imply RT_2^2 over RCA_0.

One of the underlying motivations for the work presented here is the question of whether SRT_2^2 and RT_2^2 can be separated by an ω-model. Because the fundamental property of the Chong, Slaman and Yang model is false in any ω-model of RCA_0, it appears that new ideas are needed. There are a range of tools from proof theory and computability theory to compare the strengths of theorems and many of these tools give a finer analysis than the analysis given by provable equivalence over RCA_0. In this article, we will use several concrete examples to survey recent results in this direction using reductions which are more commonly used in computable analysis.

We begin by restricting our attention to ω-models and to combinatorial principles which have the Π_2^1 form

$$\forall X\,(\Phi(X) \to \exists Y \Psi(X, Y))$$

where $\Phi(X)$ and $\Psi(X, Y)$ are arithmetic. We refer to a statement P of this form as a *problem* and we refer to sets X which satisfy $\Phi(X)$ as *instances* of P. Given an instance X of P, we call a witness Y such that $\Psi(X, Y)$ a *solution* to X.

Given two problems P and Q of this form, consider how we might show that P holds in every ω-model of $\mathsf{RCA}_0 + Q$. Fix an ω-model of $\mathsf{RCA}_0 + Q$ and recall that the second order part of this model is a Turing ideal, so it is closed under Turing reducibility and the Turing join. We fix an instance X of P in this ideal and try to construct an instance $\widehat{X}$ of Q which is also in the ideal. The simplest way to ensure that $\widehat{X}$ is in the ideal is to make it computable from X. Because $\widehat{X}$ is in the ideal, it must have a solution $\widehat{Y}$ in the ideal and, because the ideal is closed under the join operation, $\widehat{Y} \oplus X$ is also in the ideal. If we are lucky, we can use $\widehat{Y} \oplus X$ to compute a solution Y to X which, because the ideal is closed under Turing reducibility, will also be in the ideal.

Many proofs of implications $Q \to P$ in RCA_0 proceed in this manner. However, it is worth noting that many variations are possible. For example, the proof could be more complicated in the sense that it requires us to solve several (possibly nested) instances of Q before computing a solution to X. On the other hand, the proof could be simpler in the sense that the solution $\widehat{Y}$ to $\widehat{X}$ might already be a solution to X. More generally, $\widehat{Y}$ might compute a solution to X without needing to reference the original instance X of P.

The following definition gives a framework in which we can begin to address these finer questions about exactly how the proof of P from Q (over ω-models of RCA_0) proceeds. In the next section, we will illustrate these reductions with a number of concrete examples. Hirschfeldt and Jockusch [9] give many additional examples as well as more general reduction procedures to extend this type of analysis.

Definition 1. *Let P and Q be problems.*

1. *P is ω-reducible to Q, denoted $P \leq_\omega Q$, if every ω-model of $\mathrm{RCA}_0 + Q$ is a model of P. (This reducibility has also been called computable entailment in a number of earlier articles).*
2. *P is computably reducible to Q, denoted $P \leq_c Q$, if every instance X of P computes an instance $\widehat{X}$ of Q such that if $\widehat{Y}$ is any solution to $\widehat{X}$, then there is a solution Y to X computable from $X \oplus \widehat{Y}$.*
3. *P is strongly computably reducible to Q, denoted $P \leq_{sc} Q$, if every instance X of P computes an instance $\widehat{X}$ of Q such that if $\widehat{Y}$ is any solution to $\widehat{X}$, then there is a solution Y to X computable from $\widehat{Y}$.*
4. *P is Weihrauch reducible to Q, denoted $P \leq_W Q$, if there are Turing functionals Φ and Δ such that if X is an instance of P, then Φ^X is an instance of Q and if $\widehat{Y}$ is any solution to Φ^X, then $\Delta^{X \oplus \widehat{Y}}$ is a solution to X.*
5. *P is strongly Weihrauch reducible to Q, denoted $P \leq_{sW} Q$, if there are Turing functionals Φ and Δ such that if X is an instance of P, then Φ^X is an instance of Q and if $\widehat{Y}$ is any solution to Φ^X, then $\Delta^{\widehat{Y}}$ is a solution to X.*

It is straightforward to see that the following implications hold between these reductions. Hirschfeldt and Jockusch [9] prove that none of the given arrows reverse.

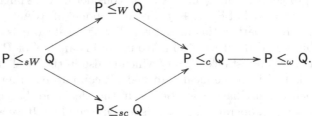

2   Three Examples

In this section, we give three examples of relationships between principles below Ramsey's theorem for pairs to illustrate these reducibilities. To fix notation, let $[\omega]^2$ denote the set of pairs $\langle x, y \rangle$ with $x < y$. We frequently use k to denote the set $\{0, \ldots, k-1\}$.

Definition 2. *A k-coloring of $[\omega]^2$ is a function $c : [\omega]^2 \to k$ and we write $c(x, y)$ in place of $c(\langle x, y \rangle)$. We say that c is stable if for every x, there is a color $i < k$ such that for every sufficiently large y, $c(x, y) = i$. That is, for every x, $\lim_y c(x, y)$ exists. For a stable coloring c, we say that x has limit color i if $\lim_y c(x, y) = i$.*

Given a coloring $c : [\omega]^2 \to k$, we say that $H \subseteq \omega$ is homogeneous if there is a color i such that $c(x, y) = i$ for all $x, y \in H$ with $x < y$. In this case, we say that H is homogeneous for color i. Similarly, if c is a stable coloring, we say that H is limit homogeneous if there is a color i such that $\lim_y c(x, y) = i$ for all

$x \in H$. When we are dealing with more than one coloring, we will refer to sets which are homogenous for the coloring c as c-homogeneous sets.

Definition 3. *We define the following versions of Ramsey's theorem.*

- *Ramsey's theorem for pairs (RT^2_2). Every coloring $c : [\omega]^2 \to 2$ has an infinite homogeneous set.*
- *Stable Ramsey's theorem for pairs (SRT^2_2). Every stable coloring $c : [\omega]^2 \to 2$ has an infinite homogeneous set.*
- *Limit Ramsey's theorem for pairs (D^2_2). Every stable coloring $c : [\omega]^2 \to 2$ has an infinite limit homogeneous set.*
- *Ramsey's theorem for singletons (RT^1_k). Every coloring $c : \omega \to k$ has an infinite homogeneous (i.e. monochromatic) set. (This principle is often called the pigeonhole principle.)*

Each of these statements can be written in the Π^1_2 form given above. In each case, the instances of the problem are the colorings of the appropriate type and the solutions to a given problem c are the infinite homogeneous sets of the appropriate type. We introduce one final principle which will be used in our third example.

Definition 4. *A set Y is cohesive for a sequence $\langle X_n \mid n \in \omega \rangle$ of subsets of ω if for every n, either $Y \cap X_n$ or $Y \cap \overline{X_n}$ is finite.*

- *Cohesive principle (COH). For every sequence $\langle X_n \mid n \in \omega \rangle$ of subsets of ω, there is an infinite cohesive set Y.*

Example 1. We consider the principles D^2_2 and SRT^2_2. D^2_2 arises naturally in computability theory by considering a set $D \leq_T 0'$ with a fixed Δ^0_2 approximation $f(x, s)$. By restricting the domain of $f(x, s)$ to values $x < s$, we can view f as a 2-coloring of $[\omega]^2$. The infinite limit homogeneous sets of this coloring are exactly the infinite subsets of D and of $\overline{D}$.

To determine the relationship between SRT^2_2 and D^2_2 under these reductions, notice that for a stable coloring $c : [\omega]^2 \to 2$, every homogeneous set is limit homogeneous (but not conversely). It follows that $\mathsf{D}^2_2 \leq_{sW} \mathsf{SRT}^2_2$ by letting the functionals Φ and Δ be the identity. That is, given an instance c of D^2_2, we map c to itself but view it as an instance of SRT^2_2. From any solution H to c as an instance of SRT^2_2 (i.e. an infinite homogeneous set H), we map H to itself but view it as a solution to the original instance c of D^2_2 (i.e. an infinite limit homogeneous set).

The non-trivial direction is to use D^2_2 to solve instances of SRT^2_2. A proof that $\mathsf{RCA}_0 + \mathsf{D}^2_2$ implies SRT^2_2 over ω-models goes as follows. Fix a stable coloring $c : [\omega]^2 \to 2$ as an instance of SRT^2_2. We regard c as an instance of D^2_2 and obtain an infinite limit homogeneous set L of color i. To produce a homogeneous set of color i, we thin L to $H = \{h_0, h_1, \ldots\}$. Let h_0 be the least element of L. Having defined h_n, let h_{n+1} be the least element of L such that $h_n < h_{n+1}$ and $c(h_j, h_{n+1}) = i$ for all $j \leq n$. (This proof works more generally in $\mathsf{RCA}_0 + \mathsf{D}^2_2$, but there is an induction issue. It uses $B\Sigma_2^0$ and hence relies on the fact that D_2^2 implies $B\Sigma_2^0$ as shown by Chong et al. [3].)

This proof shows that $\mathsf{SRT}_2^2 \leq_c D_2^2$. From an instance c of SRT_2^2, we compute an instance of D_2^2 (via the identity function) and then use an arbitrary D_2^2 solution L together with the original SRT_2^2 coloring c to compute an SRT_2^2 solution H to c. However, notice that the computation of H from L is non-uniform because it depends on knowing the limit color for L and that the thinning process uses the original SRT_2^2 instance c. Thus this proof leaves open the questions of whether $\mathsf{SRT}_2^2 \leq_W D_2^2$ (can the proof be made uniform?) and whether $\mathsf{SRT}_2^2 \leq_{sc} D_2^2$ (is the reference to the original coloring c necessary to compute H from L?). Both of these reductions fail.

Theorem 5 (Dzhafarov [7]). $\mathsf{SRT}_2^2 \not\leq_W D_2^2$ and $\mathsf{SRT}_2^2 \not\leq_{sc} D_2^2$.

The proof of Theorem 5 uses a generalization of Seetapun's method of cone avoidance for RT_2^2 and in fact shows something stronger. Let $D_{<\infty}^2$ be the problem for which the instances are given by pairs $\langle c, k \rangle$ where $c : [\omega]^2 \to k$ is a stable coloring and the solutions to $\langle c, k \rangle$ are the infinite limit homogeneous sets. The stronger result from Dzhafarov [7] is that $\mathsf{SRT}_2^2 \not\leq_W D_{<\infty}^2$ and $\mathsf{SRT}_2^2 \not\leq_{sc} D_{<\infty}^2$.

Example 2. We consider the principles RT_k^1 for varying numbers of colors. These principles differ in a significant way from those in the first example. Given $c : \omega \to k$, there is an infinite homogeneous set H computable from c by non-uniformly fixing a color i for which there are infinitely many x such that $c(x) = i$. Therefore, when comparing RT_k^1 and RT_ℓ^1, we immediately have $\mathsf{RT}_k^1 \leq_c \mathsf{RT}_\ell^1$ because $\leq_c$ allows non-uniformity and access to the original RT_k^1 coloring. Furthermore, if $k \leq \ell$, then every k-coloring is an ℓ-coloring and we immediately have $\mathsf{RT}_k^1 \leq_{sW} \mathsf{RT}_\ell^1$. Therefore, we only consider potential reductions from RT_k^1 to RT_ℓ^1 when $\ell < k$ and the reduction is stronger than $\leq_c$.

It is instructive to consider a combinatorial proof of RT_3^1 from RT_2^1. Fix a coloring $c : \omega \to 3$ and use it to define a coloring $d : \omega \to 2$ by $d(x) = 0$ if $c(x) = 0$ or $c(x) = 1$ and $d(x) = 1$ if $c(x) = 2$. Let H be an infinite d-homogeneous set. If H is d-homogeneous for color 1, then it is a c-homogeneous set for color 2 and we are done. On the other hand, if H is d-homogeneous for color 0, then we need another application of RT_2^1 to get our c-homogeneous set. Let d_0 be the restriction of d to H. Applying RT_2^1 to d_0 gives an infinite d_0-homogeneous set $H_0 \subseteq H$ which is clearly c-homogeneous as well. This proof does not use the original coloring to thin out a homogeneous set, but it does have a split between cases (so is non-uniform) and it potentially uses two applications of RT_2^1 to solve a single instance of RT_3^1. The following theorem collects a number of recent results related to the connection between these principles.

Theorem 6. *The following negative results hold for all $0 < \ell < k$.*

- $\mathsf{RT}_k^1 \not\leq_{sW} \mathsf{RT}_\ell^1$ *(Dorais et al. [5]).*
- $\mathsf{RT}_k^1 \not\leq_W \mathsf{RT}_\ell^1$ *(independently by Hirschfeldt and Jockusch [9], by Patey [11] and by Rakotoniaina [12]).*
- $\mathsf{RT}_k^1 \not\leq_{sc} \mathsf{RT}_\ell^1$ *(Dzhafarov [7]).*

The most interesting result concerning these principles comes from refining the forcing techniques to remove any computable dependence between the instances of RT^1_k and RT^1_ℓ.

Theorem 7 (independently by Hirschfeldt and Jockusch [9] and by Patey [11]). *For any $\ell < k$, there is a coloring $c : \omega \to k$ such that every coloring $d : \omega \to \ell$ (computable from c or not) has an infinite homogeneous set H which does not compute any infinite c-homogeneous set.*

The proof of Theorem 7 given in Patey [11] shows that the coloring $c : \omega \to k$ can even be made low.

Example 3. We consider the relationship between COH and the principles above. The interest in these relationships is partly motivated by the fact that RCA_0 proves the equivalence of RT^2_2 and SRT^2_2+COH. Cholak et al. [2] introduced this equivalence as an important tool for constructing solutions to instances of RT^2_2 in two steps by using COH to pass from a general 2-coloring to a stable 2-coloring and then using SRT^2_2 to solve the stable coloring. Therefore, we would like to know whether COH is reducible (by any of these reductions) to SRT^2_k for some $k \geq 2$ or to $SRT^2_{<\infty}$. Partial answers to these questions are given by the following theorem.

Theorem 8 (Dzhafarov [7]). $COH \not\leq_W SRT^2_{<\infty}$ *and* $COH \not\leq_{sc} SRT^2_2$.

Dzhafarov's proof that $COH \not\leq_{sc} SRT^2_2$ uses a new tree labeling method for constructing solutions to SRT^2_2. The version of this method given in [7] is closely tied to coloring pairs and it was left as an open question whether this result could be extended to larger exponents. We will return to this question below.

A second motivation for studying the relationships between COH and other Ramsey principles is that COH can be recast as a sequential version of Ramsey's Theorem for Singletons in which the solution is allowed to make finitely many errors. We say that Y is *almost homogeneous* for a coloring $c : \omega \to k$ if there is a finite set F such that $Y - F$ is homogeneous for c. Under a suitable coding, COH is sW-equivalent to the statement that for every sequence $\langle c_k \mid k \geq 1 \rangle$ of colorings $c_k : \omega \to k$, there is an infinite set Y such that Y is almost homogeneous for every coloring c_k. (For the details of this coding, see [8].) From this perspective, we would like to understand the precise relationship between RT^1_k and SRT^2_ℓ as a stepping stone to determine whether COH can be reduced (by $\leq_{sc}$, $\leq_c$ or $\leq_\omega$) to SRT^2_ℓ or to $SRT^2_{<\infty}$. As above, $RT^1_k \leq_c SRT^2_\ell$ for any k and ℓ because $\leq_c$ allows access to the RT^1_k coloring, and it is not hard to see that $RT^1_k \leq_{sW} SRT^2_\ell$ whenever $k \leq \ell$. Therefore, our interest is in comparing RT^1_k and SRT^2_ℓ when $\ell < k$. The following theorem answers this question for W-reducibility.

Theorem 9 (Hirschfeldt and Jockusch [9]). *For all $\ell < k$, $RT^1_k \not\leq_W SRT^2_\ell$.*

In recent work, we were able to answer a number of the questions left open by the results above. Our main result is to show that $RT^1_k \not\leq_{sc} SRT^2_\ell$ when $\ell < k$ in a strong form analogous to Theorem 7.

Theorem 10 (Dzhafarov et al. [8]). *For all $\ell < k$, there is a coloring c : $\omega \to k$ such that every stable coloring $d : [\omega]^2 \to \ell$ has an infinite homogeneous set which does not compute an infinite c-homogeneous set.*

The technique for proving this theorem involves an extension and a simplification of the tree labeling method used to build solutions to SRT_2^2 in Dzhafarov's proof of $\mathsf{COH} \not\leq_{sc} \mathsf{SRT}_2^2$ in Theorem 8. It remains an open questions whether the coloring c can be low as in Patey's proof of Theorem 7. There are two corollaries below to Theorem 10. The first corollary follows immediately and the second corollary follows by viewing COH as a sequential form of RT_k^1 allowing finitely many errors.

Corollary 11 (Dzhafarov et al. [8]). *For all $\ell < k$, $\mathsf{RT}_k^1 \not\leq_{sc} \mathsf{SRT}_\ell^2$.*

Corollary 12 (Dzhafarov et al. [8]). $\mathsf{COH} \not\leq_{sc} \mathsf{SRT}_{<\infty}^2$.

3   Variations on ADS

In this section, we consider variations of the ascending/descending sequence principle ADS (defined below) which are motivated by the combinatorial relationships between RT_2^2, SRT_2^2 and D_2^2. Throughout this section, when we say $(L, \leq_L)$ is a linear order, we assume that $L \subseteq \omega$. That is, we assume our algebraic structures are coded in the natural numbers. We use $\leq$ to denote the usual order on ω.

Definition 13. *Let $(L, \leq_L)$ be a linear order and let $S \subseteq L$. We say that S is an*

- *ascending sequence if for all $x, y \in S$, $x \leq y$ implies $x \leq_L y$;*
- *descending sequence if for all $x, y \in S$, $x \leq y$ implies $x \geq_L y$;*
- *ascending chain if for all $x \in S$, there are only finitely many $y \in S$ with $y \leq_L x$;*
- *descending chain if for all $x \in S$, there are only finitely many $y \in S$ with $y \geq_L x$.*

Note that if S is an ascending (descending) chain, then S is isomorphic to ω (or ω^*). However, it is given as a chain in L rather than as an ascending (descending) sequence in the sense that the elements of S do not necessarily appear in ascending (descending) order when enumerated in increasing $\leq$-order.

Definition 14. *We define the following principles.*

- *Ascending/descending sequence principle (ADS): Every infinite linear order has an infinite ascending or descending sequence.*
- *Ascending/descending chain principle (ADC): Every infinite linear order has an infinite ascending or descending chain.*

There principles have the required Π_2^1 form given in the introduction and the combinatorial relationship between ADC and ADS is similar to the combinatorial relationship between D_2^2 and SRT_2^2. For example, ADC and ADS have the same instances and an ADS-solution to $(L, \leq_L)$ is also an ADC-solution to $(L, \leq_L)$. Therefore, ADC $\leq_{sW}$ ADS via identity functionals just as $D_2^2 \leq_{sW} SRT_2^2$.

More importantly for our analogy, every ADC-solution S to $(L, \leq_L)$ can be thinned out to an ADS-solution H to $(L, \leq_L)$. To see why, suppose $S = \{s_0, s_1, \ldots\}$ is an infinite ascending chain. We define an infinite ascending sequence $H = \{h_0, h_1, \ldots\}$ as follows. Let h_0 be the $\leq$-least element of S. Having defined h_n, let h_{n+1} be the $\leq$-least element of S such that $h_n < h_{n+1}$ and $h_n <_L h_{n+1}$. A similar argument works in the case when S is a descending chain. As with SRT_2^2 and D_2^2, the computation of H from S is non-uniform (because it requires knowing whether S is ascending or descending) and uses the ADS-instance $(L, \leq_L)$. Therefore, the thinning process shows ADS $\leq_c$ ADC exactly as with $SRT_2^2 \leq_c D_2^2$. Before considering whether this reduction can be made uniform, we introduce two notions of stability in this context.

Definition 15. *Let $(L, \leq_L)$ be a linear order. An element $x \in L$ is called small if there are only finitely many $y \in L$ such that $y <_L x$. Similarly, $x \in L$ is called large if there are only finitely many $y \in L$ such that $x <_L y$. We say that L is stable if every $x \in L$ is either small or large.*

If L is stable, then L is isomorphic to either $\omega + k$, $k + \omega^*$ or $\omega + \omega^*$. The established definition for the stable version SADS of ADS restricts the instances of the problem to linear orders isomorphic to $\omega + \omega^*$, that is, to stable linear orders with infinitely many small elements and infinitely many large elements. In the context of ω-models of RCA_0 or of $\leq_c$-reductions, neglecting to consider linear orders L of type $\omega + k$ or $k + \omega^*$ does not matter because L can (non-uniformly) compute an infinite ascending sequence (in the case of $\omega + k$) or descending sequence (in the case of $k + \omega^*$). In our analogy with SRT_2^2, the orders $\omega + k$ and $k + \omega^*$ correspond to stable colorings $c : [\omega]^2 \to 2$ in which there is a fixed color i such that almost every x has limit color i. However, in the context of uniform reductions, it is important not to discount these trivial cases. To account for these cases, we consider two stable versions of ADS and of ADC.

Definition 16. *We define the following notions of stability for ADS and ADC.*

- SADS: *Every stable linear order L with infinitely many small and large elements has an infinite ascending or descending sequence.*
- SADC: *Every stable linear order L with infinitely many small and large elements has an infinite ascending or descending chain.*
- General-ADS: *Every stable linear order L has an infinite ascending or descending sequence.*
- General-ADC: *Every stable linear order L has an infinite ascending or descending chain.*

It is straightforward to check that these four principles are equivalent under $\leq_c$ and that the following relationships hold under $\leq_{sW}$.

- SADS $\leq_{sW}$ General-ADS $\leq_{sW}$ ADS.
- SADC $\leq_{sW}$ General-ADC $\leq_{sW}$ ADC.
- SADC $\leq_{sW}$ SADS.
- General-ADC $\leq_{sW}$ General-ADS.

Taken together with these reductions, the following theorem gives us a complete picture of the relationships between the six principles introduced in this section under $\leq_W$ and $\leq_{sW}$.

Theorem 17 (Astor et al. [1]). *We have the following negative results concerning* $\leq_W$.

(1) SADS $\not\leq_W$ ADC.
(2) General-ADC $\not\leq_W$ SADS.
(3) ADC $\not\leq_W$ General-ADS.

The first statement in Theorem 17 is proved by a Seetapun-style forcing construction while the second statement is proved with a simpler forcing notion. The last statement follows from the fact that computable instances of SADC have low solutions and that there are computable instances of ADC which have no low solutions (as shown in [10]).

The relationships between these six principles under $\leq_W$ and $\leq_{sW}$ are summarized by the following diagram. The arrows indicate an $\leq_{sW}$ reduction while the missing arrows indicate the failure of a $\leq_W$ reduction.

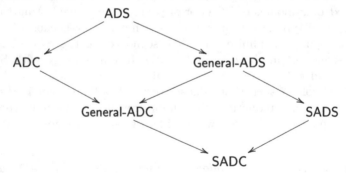

References

1. Astor, E.P., Dzhafarov, D.D., Solomon, R., Suggs, J.: The uniform content of partial and linear orders (in preparation)
2. Cholak, P.A., Jockusch Jr., C.G., Slaman, T.A.: On the strength of Ramsey's theorem for pairs. J. Symb. Log. **66**, 1–55 (2001)
3. Chong, C.T., Lempp, S., Yang, Y.: On the role of collection principles for Σ_2^0 formulas in second-order reverse mathematics. Proc. Am. Math. Soc. **138**, 1093–1100 (2010)

4. Chong, C.T., Slaman, T.A., Yang, Y.: The metamathematics of stable Ramsey's theorem for pairs. J. Am. Math Soc. **27**, 863–892 (2014)
5. Dorais, F.G., Dzhafarov, D.D., Hirst, J.L., Mileti, J.R., Shafer, P.: On uniform relationships between combinatorial problems. Trans. Am. Math. Soc. **368**(2), 1321–1359 (2016)
6. Downey, R., Hirschfeldt, D.R., Lempp, S., Solomon, R.: A Δ_2^0 set with no infinite low set in either it or its complement. J. Symb. Log. **66**, 1371–1381 (2001)
7. Dzhafarov, D.D.: Strong reducibilities between combinatorial principles. J. Symb. Log. (to appear)
8. Dzhafarov, D.D., Patey, L., Solomon, R., Westrick, L.B.: Ramsey's theorem for singletons and strong computable reducibility (submitted)
9. Hirschfeldt, D.R., Jockusch Jr., C.G.: On notions of computability theoretic reduction between Π_2^1 principles. J. Math. Logic (to appear)
10. Hirschfeldt, D.R., Shore, R.A.: Combinatorial principles weaker than Ramsey's theorem for pairs. J. Symb. Log. **72**, 171–206 (2007)
11. Patey, L.: The weakness of being cohesive, thin or free in reverse mathematics. Isr. J. Math. (to appear)
12. Rakotoniaina, T.: The computational strength of Ramsey's theorem. Ph.D. thesis, University of Cape Town (2015)

Contributed Papers

Busy Beavers and Kolmogorov Complexity

Mikhail Andreev[⊠]

Moscow Lomonosov State University, Moscow, Russia
amishaa@mail.ru

Abstract. The idea to find the "maximal number that can be named" can be traced back to Archimedes (see his *Psammit* [1]). From the viewpoint of computation theory the natural question is "which number can be described by at most n bits"? This question led to the definition of the so-called "busy beaver" numbers (introduced by T. Rado). In this note we consider different versions of the busy beaver-like notions defined in terms of Kolmogorov complexity. We show that these versions differ depending on the version of complexity used (plain, prefix, or a priori complexities) and find out how these notions are related, providing matching lower and upper bounds.

1 Introduction

In 1962 Tibor Radó [5] suggested to consider, for each natural n, the maximal integer that can be printed by a terminating computation of a Turing machine that has at most n states. The alphabet of the machine is assumed to be binary (blank and non-blank symbols). The machine starts on the empty tape and stops at some time. After that we count the number of non-blank symbols on the tape. Radó proved that this function grows faster that any computable function (of n). The same is true for other functions defined in a similar way (e.g., the maximal number of steps in a terminating computation of a machine with n states on the empty tape, or the maximal shift of its working head). Still these definitions look too machine-dependent: even small changes in the model (say, allowing two tapes or one-sided tape) could give different (but still fast-growing) functions.

A more invariant approach becomes possible if we use the notions for algorithmic information theory (Kolmogorov complexity theory). We assume here that the reader is familiar with the basic notions of this theory (see, e.g., [7] or [4], or the short introduction in [6]). We consider the maximal number that has complexity at most n, i.e., the maximal number that is an output of some program of length at most n. Here we assume that the programming language is an optimal decompressor in the sense of algorithmic information theory (that leads to a minimal complexity function; see [7] or [4] for the formal definitions). It is easy to show (see, e.g., [7, Sect. 1.2]) that we get the same function (up to $O(1)$-change in the argument) if we consider the maximal running time of the optimal decompressor on programs of length at most n. (The latter definition depends on the choice of interpreter for the optimal programming language and

© Springer International Publishing Switzerland 2016
A. Beckmann et al. (Eds.): CiE 2016, LNCS 9709, pp. 195–204, 2016.
DOI: 10.1007/978-3-319-40189-8_20

the computation model used to define the running time, but for every choice we get the same function up to $O(1)$-change in the argument.)

In other words, we fix optimal (plain) decompressor D and denote the complexity with respect of this decompressor D by $C(\cdot)$ (the *plain Kolmogorov complexity*). Then $B(n) = \max\{N \mid C(N) \leqslant n\}$, so $B(n)$ is the maximum value of D on arguments of length at most n (we consider inputs as binary strings and outputs as natural numbers). Define $BB(n)$ as the maximum computation time for D on the same inputs (for arbitrary fixed machine computing D in arbitrary fixed computation model). As we have mentioned, the following statement holds: $B(n-c) \leqslant BB(n) \leqslant B(n+c)$ for some constant c and for all n (see [7]).

Additive constant in the argument is unavoidable, since the function $C(N)$ is defined only up to an $O(1)$ additive term (when you replace one optimal decompressor by another, an additive $O(1)$ term appears). So we will not distinguish $B(n)$ and $BB(n)$ and will use the notation $B(n)$ in the sequel for this *plain busy beaver function*.

One can repeat the same definitions for prefix-free decompressors and prefix-free Kolmogorov complexity (see [6,7] for the definitions). We define the *prefix busy beaver function* $BP(n) = \max\{N \mid K(N) \leqslant n\}$.

Again one can consider the maximal computation time of an optimal prefix-free decompressor (as defined in [7, Sect. 4.4]) on inputs of size at most n, and again we get two functions that are the same (up to an additive $O(1)$-term in the argument), for the same reasons.[1]

So we may forget about computation time, and consider the functions B and BP defined as explained above. We will compare the growth rate of the functions B and BP and show that these functions are different (B grows faster than BP). We also compare these functions with an intermediate function BP' that will be defined in terms of the a priori probability.

Let us first recall the definition of a priori probability. A priori probability $\mathbf{m}(k)$ of number k can be defined as the k-th term of a maximal (up to multiplicative constant) converging lower semicomputable non-negative series. Levin showed that such a series exists, and proved that $\mathbf{m}(k) = 2^{-K(k)+O(1)}$ (see, e.g., [7, Chap. 4] for the details). Now we consider the modulus of convergence for this series: for every n we find minimal N such that $\sum_{k>N} \mathbf{m}(k) < 2^{-n}$. Denote this N by $BP'(n)$. The difference between BP and BP' can be explained as follows: after $BP(n)$ all terms of the series $\sum \mathbf{m}(k)$ are small enough (less than 2^{-n}), and after $BP'(n)$ the *tail* of this series is small enough. Obviously $BP(n) \leqslant BP'(n)$, or, more accurately, $BP(n) \leqslant BP'(n+c)$ due to $O(1)$ additive

---

[1] One can define prefix complexity in different ways, using prefix-free decompressors (no element of the domain is a prefix of another element of the domain) or prefix-stable decompressors (if $D(x)$ is defined, then $D(y) = D(x)$ for every y that has prefix x). The argument works only for prefix-free decompressors; the problem with the prefix-stable ones is that computation time of a prefix-stable decompressor is not a prefix-stable function. It would be interesting to know whether the result remains true for prefix-stable decompressors.

terms in both definitions (Kolmogorov complexity is defined up to an $O(1)$ additive term, and a priori probability is defined up to an $\Theta(1)$ factor).

This three functions share basic computational properties with classical busy beaver function: they are not computable and grow faster than any computable function. All three functions are computable with oracle for the halting problem (as well as the classical busy beaver function.

In this article we compare growth rates of these functions. Theorem 1 shows that all three functions are relatively close to each other: all three functions are equal up to at most $(1+\varepsilon)\log n$ argument shift. Theorem 2 shows that the bound provided by Theorem 1 is quite tight. For example, one cannot remove ε from the previous statement: a gap greater than $\log n$ appears between BP and BP' for some values of n, as well as between BP' and B for some (other) values of n.

Theorem 1. (i) *There exists a constant c such that $BP(n) \leqslant BP'(n + c)$ and $BP'(n) \leqslant B(n + c)$ for all n.*

(ii) *There exists a constant c such that $B(n) \leqslant BP(n + K(n) + c)$ for all n.*

(iii) *Let (x_n, y_n) be a sequence of pairs of natural numbers such that $x_n \leqslant y_n$, the sequence x_n is lower semicomputable, and the sequence y_n is upper semicomputable. Assume that $\sum_n 2^{x_n - y_n} < +\infty$. Then there exists c such that $B(x_n) \leqslant BP(y_n + c)$ for all n.*

This theorem uses the notion of lower and upper semicomputable sequences. Recall that a sequence y_n of real numbers is *lower semicomputable* if y_n is a (point-wise) limit of some total computable non-decreasing (in k) rational-valued function of two arguments: $y_n = \lim_k y(n, k)$; *upper semicomputability* is defined in a symmetric way using non-increasing functions. If y_n are natural numbers, the function $y(\cdot, \cdot)$ can be chosen in such a way that its values are also natural numbers, and convergence means that for each n the equality $y_n = y(n, k)$ is true for all sufficiently large k. See [7] for the details.[2]

Items (i) and (ii) are rather simple, and (iii) is a more symmetric way to present (ii) (as we will see later). Note that (ii) is a special case of (iii) if we let $(x_n, y_n) = (n, n + K(n))$. Another special case of (iii) is obtained if we let $(x_n, y_n) = (n - K(n), n)$, so $B(n - K(n)) \leqslant BP(n + c)$ for some c and all n.

The statement about $(1 + \varepsilon)\log n$ mentioned above can be obtained as a corollary of (ii) since $K(n) \leqslant (1 + \varepsilon)\log n$ for $\varepsilon > 0$ (note that the series $\sum 2^{-(1+\varepsilon)\log n} = \sum(1/n^{1+\varepsilon})$ converges).

Items (ii) and (iii) are not completely symmetric: why do we add c to the right side, instead of subtracting it from the left side? We can formulate symmetric statements:

(ii') $B(n - K(n) - c) \leqslant BP(n)$ for some c and all n;

(iii') Under the same assumptions as in (iii) we have $B(x_n - c) \leqslant BP(y_n)$ for some c and all n.

These statements are also true; we will return to them after we prove Theorem 1 (they are easy corollaries of it).

---

[2] One may also speak about semicomputability for sequences that have terms $+\infty$ and/or $-\infty$; in this case we allow the values of the function $y(\cdot, \cdot)$ to be infinite.

The next results say that if $\sum_n 2^{x_n - y_n} = +\infty$ (for lower semicomputable x_n and upper semicomputable y_n) then (iii) is not true anymore. Moreover, in this case a large gap may appear both between B and BP' and between BP' and BP (but in different places).

Theorem 2. *Assume that (x_n, y_n) is a sequence of different pairs of natural numbers, $x_n \leqslant y_n$, the sequence x_n is enumerable from below, and the sequence y_n is enumerable from above. Assume also that $\sum 2^{x_n - y_n} = +\infty$. In this case*
  (i) *there exists n such that $B(x_n) > BP'(y_n)$;*
  (ii) *there exists n such that $BP'(x_n) > BP(y_n)$.*

There is no constant c in this theorem (in contrast to the previous one), but one can easily put it on any side (or even both): changing all x_n or all y_n by an additive constant does not change the divergence condition. For example, it is true that *for all c there exists n such that $B(x_n) > BP'(y_n + c)$* or *for all c there exists n such that $BP'(x_n - c) > BP(y_n)$*, and so on.

Using Theorems 1 and 2 one can easily deduce that for every upper semi-computable sequence a_n the following six conditions are equivalent:

- $BP(n) \leqslant BP'(n + a_n + c)$ for some c and for all n;
- $BP'(n) \leqslant B(n + a_n + c)$ for some c and for all n;
- $BP(n) \leqslant B(n + a_n + c)$ for some c and for all n;
- $BP(n - a_n) \leqslant BP'(n + c)$ for some c and for all n;
- $BP'(n - a_n) \leqslant B(n + c)$ for some c and for all n;
- $BP(n - a_n) \leqslant B(n + c)$ for some c and for all n;

Moreover, all these conditions are equivalent to the condition $\sum 2^{-a_n} < +\infty$ (which, in its turn, is equivalent to $a_n \geqslant K(n) - O(1)$, see [7]).

The meaning of Theorems 1 and 2 can be explained as follows. In these results we compare slow-growing functions that are *inverse* to the functions B, BP, and BP'. We show that they are equal up to $(1 + \varepsilon)$ times the logarithm of their values, and that this ε cannot be omitted: without it *both* inequalities between neighbor functions may be violated. As Theorem 1 shows, these big gaps cannot happen at the same places (otherwise the total gap between lowest and highest functions exceeds the upper bound).

Statement (ii) from Theorem 2 has been proven by Gács [3] for the case $(x_n, y_n) = (n - a_n, n)$ (and the general case may be derived as a consequence of this special one, as we will see later), so our main result is item (i) from Theorem 2. Still we provide all the proofs in the next section for uniformity and reader's convenience.

How can we modify our definitions? One can look at the maximal N such that $C(N|n) \leqslant n$ or such that $K(N|n) \leqslant n$. But we do not get new notions in this way: this quantity is still equal to $B(n)$ up to a $O(1)$-change in the argument. Indeed, the conditional complexity $C(x|n)$ is bounded by the unconditional complexity $C(x)$; on the other hand, if $C(N|n) = n$, then the conditional program of length n for N may be considered as a conditional prefix-free program with the same condition n (if n is given as a condition, we know when to stop reading the program of length n). Moreover, this program also can be used as unconditional program for N, since n (its length) is determined by the program. In general, $K(x|\,C(x)) = C(x|\,C(x)) = C(x)$ (up to O(1) additive term), see [7].

To finish our introduction, let us mention that BP' can be equivalently defined as the modulus of convergence for computable non-negative series of rational numbers with Martin-Löf random sums.

Theorem 3. *Let $\sum a_n$ be a computable series of rational non-negative numbers whose sum is Martin-Löf random. Let $N(\varepsilon)$ be the modulus of convergence of this series, i.e., the minimal value of N such that $\sum_{n>N} a_n < \varepsilon$. Then $BP'(n-c) \leqslant N(2^{-n}) \leqslant BP'(n+c)$ for some c and all n.*

The first inequality was proven in [2, Theorem 19], while the second one follows from the definition of the a priori probability (recall that **m** is bigger than any computable converging sequence, up to $O(1)$ factor). In [2] it was also shown that if $N(\varepsilon)$ is the modulus of convergence for some computable converging series $\sum a_n$ with non-negative terms, and $BP(n-c) \leqslant N(2^{-n})$ for some c and all n, then the same property holds for BP' (for a different value of c).

2   Upper Bounds

In this section we prove Theorem 1.

(i) The inequality $BP(n) \leqslant BP'(n+c)$ follows directly from definitions. If we define $\mathbf{m}(n)$ exactly as $2^{-K(n)}$, it is true even without c-term.

Now we prove that $BP'(n) \leqslant B(n+c)$ for some c and for all n. To do this, we construct an algorithm that, given n, enumerates at most 2^n different integers, and the last of them is bigger than $BP'(n)$. The n-bit string that is the bit representation of the item's number in this enumeration, identifies the last number (n is known, being the length of this string), so we get the required inequality. How the enumeration algorithm works? This algorithm approximates all $\mathbf{m}(n)$ from below in parallel; we assume that at every moment only finitely many approximations are not zeros. As soon as the tail of the current approximation for **m**, starting from the last enumerated integer, becomes greater than 2^{-n} (i.e., the current approximation to BP' exceeds the last enumerated integer) we enumerate a new integer that is bigger than all k with non-zero current approximations for $\mathbf{m}(k)$. Obviously this cannot happen more than 2^n times: every time an integer is enumerated, we leave behind total **m**-weight at least 2^{-n}.

(ii) It is well known that $K(x) \leqslant C(x) + K(C(x)) + O(1)$ (for example, see [7, Sect. 4.6]). The following slightly more general statement is also true: if $C(x) \leqslant n$, then $K(x) \leqslant n + K(n) + O(1)$. Let us prove it. Starting with a program for x that has length at most n, we prepend a block of the form $0^k 1$ to it (this block is obviously self-delimited) making the total length exactly $n+2$. Then we prepend a self-delimited code for n (of length $K(n)$), and the result is a self-delimited code for x (decode n first, then read exactly $n+2$ symbols, remove

$0^k 1$ leading block, then use C-decompressor). This generalisation immediately implies that $B(n) \leqslant BP(n + K(n) + c)$ for some c and for all n.

(iii) We will show that this inequality is a consequence of (ii). We start by showing that we can assume x_n and y_n to be computable without loss of generality.

By assumption, the sequences x_n [resp. y_n] are lower [resp. upper] semicomputable. For each n, consider a uniformly computable sequence of pairs (x_n^i, y_n^i) of integers that monotonically converge to (x_n, y_n) as $i \to \infty$. Combine arbitrarily all these sequences into one sequence, leaving only the first appearance of each pair (removing all duplicates). We get a computable sequence $(\tilde{x}_i, \tilde{y}_i)$; every pair (x_n, y_n) appears in this sequence together with finitely many its approximations. Note that $\sum_i 2^{\tilde{x}_i - \tilde{y}_i}$ is at most two times bigger than $\sum_n 2^{x_n - y_n}$: every time a new approximation for x_n or y_n appears, the respective term is the sum is increased by factor 2 or more, so the sum for $\tilde{x}_i, \tilde{y}_i$ is at most twice bigger than the original one, and if the original sum is finite, then the new one is also finite. Note also that the desired inequality for the new sequence implies the same inequality for the original sequence (that is a subsequence of the new one). So we can assume x_n, y_n is computable without loss of generality.

Now assume that a computable sequence (x_i, y_i) is given. Define $f(n) = \min\{y_i - n \mid x_i = n\}$; if n does not appear among x_i, the value $f(n)$ is $+\infty$. The function f is upper semicomputable, and $\sum_n 2^{-f(n)} < +\infty$, since the pairs $(n, n + f(n))$ are guaranteed to appear among (x_i, y_i). So $f(n) \geqslant K(n) - O(1)$. Therefore, $x_i + K(x_i) \leqslant y_i + O(1)$ for the pairs with minimal y_i (for a given x_i) and therefore for all pairs. The function BP increases, so we get $B(x_i) \leqslant BP(x_i + K(x_i) + O(1)) \leqslant BP(y_i + O(1))$ for all pairs. The claim (iii) is proven.

Symmetric results (mentioned above) are also easy to prove:

(ii') $B(n - K(n) - c) \leqslant BP(n)$ for some c and for all n.

(iii') If x_n and y_n satisfy the same assumptions as in (iii), then $B(x_n - c) \leqslant BP(y_n)$ for some c and for all n.

To prove (ii') we use (ii) for a smaller argument: $B(n - K(n) - e) \leqslant$
$\leqslant BP(n - K(n) - e + K(n - K(n) - e) + c)$ holds for some c and all n, e. Now we want to choose the constant e in such a way that the argument in the right hand side is at most n for all n (recall that function BP is monotone):
$n - K(n) - e + K(n - K(n) - e) + c \leqslant n$.
Indeed, $K(n - K(n) - e) \leqslant K(n, K(n)) + K(e) + O(1) \leqslant K(n) + K(e) + O(1)$, and $e - K(e)$ can be made arbitrary large for large enough e (larger than sum of $O(1)$ terms in the inequalities).

To derive (iii') from (ii'), one can use the same technique as used to deduce (iii) from (ii). The only difference is that one should group pairs with the same y_i (instead of x_i, as we did in the proof).

3   Lower Bounds

In this section we prove Theorem 2.

3.1   Proof of the Claim (i)

We have a sequence of different pairs (x_n, y_n) of integers such that $x_n \leqslant y_n$. We assume that x_n is lower semicomputable, y_n is upper semicomputable and $2^{x_n - y_n} = +\infty$. We need to show that there exists n such that $B(x_n) > BP'(y_n)$.

First we will reduce this statement to its special case where $(x_n, y_n) = (n, n + a_n)$, and a_n is some upper semicomputable sequence of natural numbers (the value $+\infty$ is also allowed).

For this reduction we use the same trick as in the previous section. First we replace (x_n, y_n) by its approximations (x_n^i, y_n^i), and then combine all these approximations into one computable sequence by removing the duplicates. The sum of $2^{x_i - y_i}$ may only increase (we add new elements), there are no duplicates (we removed them) and if $B(x_n^i) > BP'(y_n^i)$ then $B(x_n) > BP'(y_n)$ since we use monotone approximations and the busy beaver functions are monotone. So we may assume without loss of generality that the sequence (x_n, y_n) is a computable sequence of different integer pairs.

Let $a_n = \min\{y_i - n \mid x_i = n\}$. The sequence a_n is enumerable from above (since the sequence (x_i, y_i) is computable). Note that $\sum_n 2^{-a_n} \geqslant \frac{1}{2} \sum_i 2^{x_i - y_i}$. Indeed, if we group pairs with $x_i = n$, the sum of this group is bounded by a geometric sequence with common ratio $1/2$, so the sum can be replaced by the maximal element (up to a 2-factor). Therefore, $\sum_n 2^{-a_n} = +\infty$, and all pairs $(n, n + a_n)$ appear among (x_i, y_i), so we get the desired reduction.

Now we use the following lemma: if a_n is an upper semicomputable sequence of integers and $\sum_n 2^{-a_n} = +\infty$, there exists a computable sequence $\tilde{a}_n \geqslant a_n$ such that $\sum_n 2^{-\tilde{a}_n} = +\infty$. Indeed, we can approximate a_n from above until some finite part of the series $\sum 2^{-a_n}$ exceeds 1, then fix the current approximations for this part and call them $\tilde{a}_n$. Then the same argument is used for the tail, etc. This argument show that we may assume without loss of generality that a_n is a computable sequence.

It remains to prove the following statement: if a_n is a computable sequence of integers and $\sum 2^{-a_n} = +\infty$, then there exists n such that $B(n) > BP'(n + a_n)$. In other words, we need to show that there exists some u such that $C(u) \leqslant n$ and $\sum_{i \geqslant u} \mathbf{m}(i) < 2^{-n - a_n}$.

To prove an upper bound for $C(u)$, we need to construct a decompressor that provides a short description for u. However, this gives a bound with some additive constant term, so we need to construct a decompressor D such that for every d there exist n and u such that

$$C_D(u) \leqslant n - d \quad \text{and} \quad \sum_{i \geqslant u} \mathbf{m}(i) < 2^{n - a_n}.$$

where $C_D(u)$ is the minimal length of p such that $D(p) = u$.

To prove this, we use the game technique. Consider a game where Alice plays with Bob. They make alternating moves. Alice enumerates sets $D_0, D_1, \ldots$; at each move she adds finitely many integers to finitely many D_i (so her move is a finite object). The set D_i may contain at most 2^i elements. Bob approximates from below some sequence $\mu(0), \mu(1), \ldots$; initially all $\mu(i)$ are zeros, and at each step Bob may increase finitely many of them by some rational numbers, but the sum $\sum \mu(i)$ should not exceed 1.

Assuming that both players respect the rules, Alice wins if (for limit values of D_i and $\mu(i)$) *for every d there exists n and u such that $u \in D_{n-d}$ and* $\sum_{i \geqslant u} \mu(i) < 2^{-n-a_n}$. One may reformulate this statement eliminating u: for every d there exist n such that

$$\sum_{i \geqslant \max(D_{n-d})} \mu(i) < 2^{-n-a_n}. \tag{*}$$

We will prove that Alice has a computable winning strategy in this game. This implies the desired result. Indeed, we may let Alice use this strategy against the "blind" strategy of Bob that approximates from below the a priori probability function $\mu(i) = \mathbf{m}(i)$. Then the behavior of Alice is computable, the sets D_i are enumerable and we construct a decompressor D that maps k-bit string p into pth element in the enumeration of D_k (in the last sentence binary string p is identified with an integer it represents in the binary notation). This decompressor has the required property.

So why Alice has a computable strategy? She should guarantee the existence of a suitable n for each d. This is done independently for each d; Alice chooses for each d some interval $[l_d, r_d]$ where n with the required properties exist. This intervals are chosen in such a way that there are no collisions (for different d the values of $n - d$ cannot be the same, i.e., the intervals $[l_d - d, r_d - d]$ are disjoint). The intervals should be large enough: the sum of 2^{-a_n} over n in $[l_d, r_d]$ should exceed 2^{d+1} (we will see that this is enough for our purposes). Since we assume that a_n is a computable sequence and $\sum 2^{-a_n} = +\infty$, we can choose $[l_d, r_d]$ in a computable way.

How Alice constructs D_{n-d} for $n \in [l_d, r_d]$? It is done in a straightforward way. Alice chooses some n (say, the minimal value $n = l_d$) and tries to achieve (*) by adding large elements to D_{n-d}. More precisely, if (*) is violated, Alice takes some number k that is greater that all non-zero terms in μ (i.e., $\mu(k') = 0$ for all $k' \geqslant k$) and adds k to D_{n-d}. Then Bob may increase μ-values; as soon as (*) is violated again, Alice repeats this procedure, and so on. At some point (after 2^{n-d} steps) a maximal cardinality of D_{n-d} is reached. But at that time Bob has used at least $2^{n-d}2^{-n-a_n} = 2^{-d-a_n}$ of his reserve (each time a tail of size 2^{-n-a_n} is cut). Then Alice switches to next value of n, and forces Bob to lose or to use 2^{-d-a_n} again for this new value of n. Ultimately Bob will lose the game since the sum of $2^{-d-a_n} = 2^{-d}2^{-a_n}$ over n in $[l_d, r_d]$ exceeds 1. (A technical correction: we required that the limit value of $\sum \mu(i)$ is strictly less that some threshold; it is not enough to know that all the approximations are strictly less than this threshold (only a non-strict inequality is guaranteed). To remedy this problem, we may use an additional factor of 2 — so we require the sum of 2^{-a_n} over $n \in [l_d, r_d]$ to be greater than 2^{d+1}, not 2^d.) Claim (i) is proven.

3.2   Proof of the Claim (ii)

We again consider a sequence of different pairs (x_n, y_n) such that $x_n \leqslant y_n$, the sequence x_n is lower semicomputable, the sequence y_n is upper semicomputable and $\sum 2^{x_n - y_n} = +\infty$. We want to prove (following Gács) that there exists n such that $BP'(x_n) > BP(y_n)$

We can use the same reasoning as in (i) with minor modifications to show that we can assume without loss of generality that $(x_n, y_n) = (n - a_n, n)$ for some computable sequence of non-negative integers a_n with $\sum_n 2^{-a_n} = +\infty$. This time we need to group terms with the same y_i, not x_i. We need to prove then that there exists n such that $BP(n) < BP'(n - a_n)$. In other words, we need to prove that there exist n and u such that $\mathbf{m}(i) < 2^{-n}$ for all $i \geqslant u$, but $\sum_{i \geqslant u} \mathbf{m}(i) > 2^{-n+a_n}$ (all terms in the u-tail are small but their sum is big).

To show that the sum of $\mathbf{m}$-tail is big, we need to construct a lower semicomputable semimeasure for which this sum is big, and then use the maximality of $\mathbf{m}$. Again a constant appears, so we need to prove a stronger statement: *there exists a lower semicomputable semimeasure α such that for every d there are n and u with the following property:*

$$\sum_{i \geqslant u} \alpha(i) > 2^{-n+a_n+d} \quad \text{but} \quad \mathbf{m}(i) < 2^{-n} \text{ for all } i \geqslant u.$$

Again we may use the game approach and imagine that Alice approximates from below some semimeasure α while Bob approximates from below some semimeasure β, and the claim above (with β instead of $\mathbf{m}$) is the winning condition for Alice. We will construct a computable strategy for Alice in this game; applying it against the blind strategy of Bob (who approximates $\mathbf{m}(\cdot)$ from below), we get the required statement.

Let us note first that it is enough to construct (for every d) a winning strategy in the similar game where winning condition is required only for this d. Indeed, we may use $2d$-strategy to win the d-game with $\sum_i \alpha(i) \leqslant 2^{-d}$ (using $2d$-strategy with factor 2^{-d}). Then we can use all the strategies (for d-games for all d) in parallel against Bob and sum up all the increases, since the winning condition is monotone and the strategies can only help each other. In this way Alice keeps the total sum less than $\sum_d 2^{-d} \leqslant 1$ and wins all games.

So how could Alice win the d-game? She should increase her weights gradually by using small weights far away where Bob has only zeros. As soon as her total weight exceeds 2^{-n+a_n+d} for some n, Bob has to react and assign weight at least 2^{-n} for some i. Then Alice continues to increase the weights (on the right of the place used by Bob), and again after 2^{-n+a_n+d} new Alice's weight Bob should react by assigning weight at least 2^{-n} at some other place. If Alice uses this strategy with small weights (see the discussion below) until her total weight reaches 1, and waits each time until Bob violates the winning condition for Alice, we have the following property of Bob's weights:[3]

---

[3] Technically speaking, Bob is obliged to react only if the Alice's tail is strictly greater than 2^{-n+a_n+d}. But this leads only to a constant factor that is not important, so we ignore this problem.

for each n there are at least 2^{n-a_n-d} Bob's weights $\beta(\cdot)$ that exceed 2^{-n}.

Note that Alice's actions are the same for all n; it is Bob who should care about all n and provide a large enough weight at the moments where Alice is in the winning position (for some n).

What is the total weight Bob uses in this process? The property above guarantees that Bob uses at least 2^{-a_n-d} to prevent Alice from winning for given n. The sum of these quantities for all n is infinite according to our assumption (so at some point Bob will be unable to increase the weights). However, the same Bob's move can be useful on different levels (for different values of n), so we need the following technical lemma valid for every series $\sum_i \beta(i)$ with non-negative values:

$$\sum_j 2^{-j} \cdot \#\{i : \beta(i) \geqslant 2^{-j}\} \leqslant 2 \sum_i \beta(i).$$

Indeed, each $\beta(i)$ from the right hand side appears in the left hand side as the sum of 2^{-j} for all j such that $2^{-j} \leqslant \beta(i)$, and this sum does not exceed $2\beta(i)$.

To finish the description of Alice's strategy, we need to say how small should be the weight increases used by Alice. We know that the sum $\sum_n 2^{-a_n-d}$ is infinite, so there is a finite part of this sum that is large (greater than 4, to be exact). Alice then may use weights 2^{-s} where s is some integer greater that all $n + a_n + d$ for n that appear in this finite part.

Acknowledgements. This work was supported by ANR-15-CE40-0016-01 RaCAF grant, and RFBR grant number 16-01-00362.

References

1. Archimedes: The Sand Reckoner. In: Heath, T.L. (ed.) The Works of Archimedes. Dover, New York (1953)
2. Bienvenu, L., Shen, A.: Random semicomputable reals revisited. In: Dinneen, M.J., Khoussainov, B., Nies, A. (eds.) Computation, Physics and Beyond. LNCS, vol. 7160, pp. 31–45. Springer, Heidelberg (2012). http://arxiv.org/pdf/1110.5028v1.pdf
3. Gács, P.: On the relation between descriptional complexity and algorithmic probability. Theoret. Comput. Sci. **22**, 71–93 (1983)
4. Li, M., Vitányi, P.: An Introduction to Kolmogorov Complexity and Its Applications, 3rd edn. Springer, New York (2008). (1st edn. (1993); 2nd edn. (1997), xxiii+790pp. ISBN 978-0-387-49820-1
5. Radó, T.: On non-computable functions. Bell Syst. Tech. J. **41**(3), 877–884 (1962)
6. Shen, A.: Around Kolmogorov complexity: basic notions and results. In: Vovk, V., Papadopoulos, H., Gammerman, A. (eds.) Measures of Complexity: Festschrift for Alexey Chervonenkis, pp. 75–116. Springer, Cham (2015). http://arxiv.org/pdf/1504.04955
7. Vereshchagin, N.K., Uspensky, V.A., Shen, A.: Kolmogorov complexity and algorithmic randomness (In Russian). Draft english translation: http://www.lirmm.fr/~ashen/kolmbook-eng.pdf

The Domino Problem for Self-similar Structures

Sebastián Barbieri[1(✉)] and Mathieu Sablik[2]

[1] LIP, ENS de Lyon – CNRS – INRIA – UCBL – Université de Lyon, Lyon, France
sebastian.barbieri@ens-lyon.fr
[2] Institut de Mathmatiques de Marseille, UMR 7373, Aix-Marseille Université,
Marseille, France
mathieu.sablik@univ-amu.fr

Abstract. We define the domino problem for tilings over self-similar structures of $\mathbb{Z}^d$ given by forbidden patterns. In this setting we exhibit non-trivial families of subsets with decidable and undecidable domino problem.

Introduction

In its original form, the domino problem was introduced by Wang [10] in 1961. It consists of deciding if copies of a finite set of Wang's tiles (square tiles of equal size, not subject to rotation and with colored edges) can tile the plane subject to the condition that two adjacent tiles possess the same color in the edge they share. Wang's student Berger showed undecidability for the domino problem on the plane in 1964 [3] by using a reduction to the halting problem. In 1971, Robinson [8] simplified Berger's proof.

Symbolic dynamics classically studies sets of colorings of $\mathbb{Z}^d$ from a finite set of colors which are closed in the product topology and invariant by translation, such sets are called subshifts. Given a finite set of patterns $\mathcal{F}$ (a pattern is a coloring of a finite part of $\mathbb{Z}^d$), we associate a subshift of finite type $X(\mathcal{F})$ which corresponds to the set of colorings which does not contain any occurrence of patterns in $\mathcal{F}$. The domino problem can therefore be expressed in this setting: given a finite set of forbidden patterns $\mathcal{F}$, is it possible to decide whether the subshift of finite type $X(\mathcal{F})$ is not empty?

It is well known that there exists an algorithm deciding if a subshift of finite type is empty in dimension one [5] and that there is no such algorithm in higher dimensions. The natural question that comes next is: *What is the frontier between decidability and undecidability in the domino problem?*

One way to explore this question is to consider subshifts defined over more general structures, such as finitely generated groups or monoids and ask where the domino problem is decidable. This approach has yielded various result in different structures: Some examples are the hyperbolic plane [6], confirming a conjecture of Robinson [9] and Baumslag-Solitar groups [1]. The conjecture in this direction is that the domino problem is decidable if and only if the group is virtually free. The conjecture is known to hold in the case of virtually nilpotent

© Springer International Publishing Switzerland 2016
A. Beckmann et al. (Eds.): CiE 2016, LNCS 9709, pp. 205–214, 2016.
DOI: 10.1007/978-3-319-40189-8_21

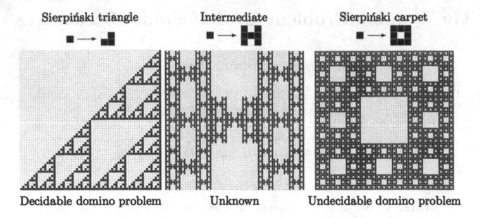

Fig. 1. Some digitalizations of fractal structures and the status of their domino problem

groups [2]. The main idea of the proof of this result is to construct a grid by local rules in order to use the classical result in $\mathbb{Z}^2$.

In this paper we explore another way to delimit the frontier between decidability and undecidability of this problem. In geometry the structures which lie between the line and the plane can have Hausdorff dimension strictly between one and two. In this article we propose a way to define the domino problem in a digitalization of such fractal structures. In Sect. 1 we use self-similar substitutions to define a "fractal" structure where a natural version of the domino problem can be defined. We exhibit a large class of substitutions (including the one which represents the Sierpiński triangle) where the domino problem is decidable (Sect. 2), another class (including the Sierpiński carpet) where the problem is undecidable (Sect. 4) and an intermediate class where the question is still open (see Fig. 1 for an example of each of these classes).

1 Position of the Problem

1.1 Coloring of $\mathbb{Z}^d$ and Local Rules

Given a finite alphabet $\mathcal{A}$, a coloring of $\mathbb{Z}^d$ is called a *configuration*. The set of configurations, denoted $\mathcal{A}^{\mathbb{Z}^d}$, is a compact set according to the usual product topology. A *subshift* is a closed set of configurations which is invariant by the shift action. Given a finite subset $S \subset \mathbb{Z}^d$, a *pattern with support S* is an element p of $\mathcal{A}^S$. A pattern $p \in \mathcal{A}^S$ *appears* in a configuration $x \in \mathcal{A}^{\mathbb{Z}^d}$ if there exists $z \in \mathbb{Z}^d$ such that $x_{z+S} = p$. In this case we write $p \sqsubset x$.

Equivalently, a subshift can be defined with a set of forbidden patterns $\mathcal{F}$ as the set of configurations where no patterns of $\mathcal{F}$ appear. We denote it by $X(\mathcal{F})$. If $\mathcal{F}$ is finite, $X(\mathcal{F})$ is called *subshift of finite type* which can be considered as the set of tilings defined by the local constraints given by $\mathcal{F}$.

1.2 Self-similar Structures

We want to extend the condition of coloring to self-similar structures of $\mathbb{Z}^d$. This means that only some cells can be decorated by elements of $\mathcal{A}$. To formalize that, a structure is coded as a subset of $\{0,1\}^{\mathbb{Z}^d}$ and self-similarity is obtained by a substitution.

Let $\mathcal{A}$ be a finite alphabet. A *substitution* is a function $s : \mathcal{A} \to \mathcal{A}^R$ where $R = [1, l_1] \times \cdots \times [1, l_d]$ is a d-dimensional rectangle. It is naturally extended to act over patterns which have rectangles as support by concatenation. We denote the successive iterations of s over a symbol by s, s^2, s^3 and so on. The subshift generated by a substitution s is the set

$$X_s := \{x \in \mathcal{A}^{\mathbb{Z}^d} | \forall p \sqsubset x, \exists n \in \mathbb{N}, a \in \mathcal{A}, p \sqsubset s^n(a)\}.$$

To obtain self-similar structures, we restrict the notion of substitution to $\{0,1\}$ imposing that the image of 0 consists of a block of 0s. These substitutions are called *self-similar*. Self-similar substitutions represent digitalizations of the iterations of the following procedure: start with the hypercube $[0,1]^d$, subdivide it in a $l_1 \times \cdots \times l_d$ grid and remove the blocks in the positions z of the grid where $s(1)_z = 0$. Then repeat the same procedure with every sub-block.

Example 1. Consider $\mathcal{A} = \{\ ,\blacksquare\}$. The self-similar substitution s such that:

$\square \longrightarrow \boxed{}$ and $\blacksquare \longrightarrow \blacksquare\blacksquare$ is called the Sierpiński triangle substitution and is extended by concatenation as shown in Fig. 2.

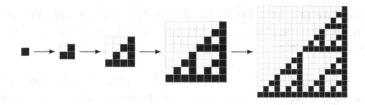

Fig. 2. First four iterations of the Sierpiński triangle substitution.

1.3 Coloring of a Self-similar Structure and Local Rules

Let $\mathcal{A}$ be a finite alphabet where $0 \in \mathcal{A}$ and s be a self-similar substitution. Consider $X_s \subset \{0,1\}^{\mathbb{Z}^d}$ the associated self-similar structure. A configuration $x \in \mathcal{A}^{\mathbb{Z}^d}$ is *compatible* with s if $\pi(x) \in X_s$ where π is a map which sends all elements of $\mathcal{A}\backslash\{0\}$ onto 1 and 0 onto 0. Given a finite set of patterns $\mathcal{F}$ we define the *set of configurations on* X_s *defined by local rules* $\mathcal{F}$ as

$$X_s(\mathcal{F}) = \left\{x \in \mathcal{A}^{\mathbb{Z}^d} : \pi(x) \in X_s \text{ and no pattern of } \mathcal{F} \text{ appears in } x\right\}.$$

1.4   The Domino Problem on Self-similar Structures

The domino problem for $\mathbb{Z}^d$ is defined as the language

$$\text{DP}(\mathbb{Z}^d) = \{\mathcal{F} \text{ finite set of patterns} : X(\mathcal{F}) \neq \emptyset\}.$$

It is the language of all finite sets of patterns over a finite alphabet such that it is possible to construct a configuration without patterns of $\mathcal{F}$.

Classical results which can be found in [5] show that the domino problem for $\mathbb{Z}$ is decidable. In the other hand, we know that for $d > 1$ the domino problem for $G = \mathbb{Z}^d$ is undecidable (see [3,8]). This gap of decidability when the dimension increases motivates us to define the domino problem for structures which lay between those groups. Thus given a self-similar substitution s we introduce the *s-based domino problem* as the language

$$\text{DP}(s) := \{\mathcal{F} \text{ finite set of patterns} : X_s(\mathcal{F}) \neq \{0^{\mathbb{Z}^d}\}\}.$$

That is, $\text{DP}(s)$ is the set of all finite sets of forbidden patterns such that there is at least a configuration containing a non-zero symbol. We assume implicitly that $\mathcal{F}$ does not contain any pattern consisting only of 0s. By a compactness argument, $\text{DP}(s)$ can be equivalently defined as the set of $\mathcal{F}$ such that for arbitrarily big $n \in \mathbb{N}$ the non-zero symbols in $s^n(1)$ can be colored avoiding all patterns in $\mathcal{F}$.

2   Self-similar Structures with Decidable Domino Problem

In this section we present a family of self-similar substitutions such that their domino problem is decidable. In order to present this result in the most general setting, we introduce the channel number of a self-similar substitution.

Let $\mathbb{H} = \{-1, 0, 1\}^d$ and consider the set $\Lambda \subset \{0, 1\}^{\{1,2,3\}^d}$ consisting of all d-dimensional hypercube patterns of side 3 which appear in X_s and that have a 1 in the center. Let $\Lambda_n = s^n(\Lambda)$ be the set of the images of each $q \in \Lambda$ under s^n by concatenation and S_n be the support corresponding to the image of position $(2, \ldots, 2)$ of q under s^n. We define the *n-channel number* $\chi(s, n)$ of s as follows:

$$\chi(s, n) = \max_{p \in \Lambda_n} |\{z \in S_n \mid \exists h \in (z + \mathbb{H}) \cap (supp(p) \backslash S_n), p_z = p_h = 1\}|$$

In other words, it is the maximum number of positions in the support of the pattern $s^n(1)$ such that if we surround it either by blocks of 0 or copies of $s^n(1)$ appearing in X_s there might be two symbols 1, one appearing in $s^n(1)$ and another outside, at distance smaller than 1. We say that s is *channel bounded* if there exists $K \in \mathbb{N}$ such that for all n, $\chi(s, n) \leq K$. The Sierpiński triangle substitution from Fig. 2 is an example of a channel bounded substitution as colorings of $s^n(1)$ can be constructed by pasting three colorings of $s^{n-1}(1)$ and forbidden patterns can only appear locally around the corners.

Theorem 1. *For every channel bounded self-similar substitution s the domino problem DP(s) is decidable.*

Proof. Let $\mathcal{F}$ be a set of forbidden patterns over an alphabet $\mathcal{A}$. It suffices to show that if s is channel bounded, it is possible to calculate $N \in \mathbb{N}$ with the property that the existence of any coloring of $s^N(1)$ with symbols from $\mathcal{A}$ without any subpattern from $\mathcal{F}$ implies $X_s(\mathcal{F}) \neq \{0^{\mathbb{Z}^d}\}$. Indeed, an algorithm could calculate N and try every coloring of $s^N(1)$. If there exists one where no pattern in $\mathcal{F}$ appears it returns that $X_s(\mathcal{F}) \neq \{0^{\mathbb{Z}^d}\}$, otherwise it returns $X_s(\mathcal{F}) = \{0^{\mathbb{Z}^d}\}$.

For simplicity, suppose that $\forall p \in \mathcal{F}, supp(p) \subseteq \mathbb{H}$ and let K be a bound for $\chi(s, n)$ (If the support is $\{-m, \ldots, m\}^d$ we can recalculate a new K). We claim that $N := 2^{|\mathcal{A}|^{(3^d-1)K}}$ suffices. For each $q \in \Lambda$ consider the a coding $J_q = \{j_1, \ldots, j_{k_q}\}$ with $k_q \leq K$ of the positions from the definition of $\chi(s, n)$. That is, J_q codes for all $n \in \mathbb{N}$ the set of positions which matter when considering only q. Any recursive ordering similar to the one given by a Quadtree works. Consider a coloring of $s^n(1)$ without subpatterns in $\mathcal{F}$ and store the symbols of this coloring appearing in J_q as a tuple $(a_{j_1}, a_{j_2}, \ldots a_{j_{k_q}}) \in \mathcal{A}^{J_q}$. Therefore all the information concerning the dependency of this coloring with its possible surroundings can be stored on $|\Lambda|$ tuples. Now, given the set of all colorings of $s^n(1)$ which do not contain any forbidden pattern we can extract the $|\Lambda|$ tuples from each one of them. All this information for the level n is represented as a subset of $\prod_{q \in \Lambda} \mathcal{A}^{J_q}$. By definition this is the only information needed in order to make sure of the existence of a coloring of $s^{n+1}(1)$ with no subpatterns in $\mathcal{F}$. Moreover, the tuples representing those patterns can be obtained from the ones of $s^n(1)$ because the positions from the definition of $\chi(s, n+1)$ necessarily appear in the patterns of $s^n(1)$. This means it is possible to extract pasting rules which can be codified in a function $\mu_s : 2^{\prod_{q \in \Lambda} \mathcal{A}^{J_q}} \to 2^{\prod_{q \in \Lambda} \mathcal{A}^{J_q}}$.

This function codes how to construct the tuples of level $n+1$ from the tuples of level n. Obviously, $\mu_s(\emptyset) = \emptyset$, therefore there are two possibilities: either this function arrives eventually at $\emptyset$ and there are no colorings of $s^m(1)$ for some $m \in \mathbb{N}$ or μ_s cycles and thus it's possible to construct colorings of $s^m(1)$ for arbitrarily big m. By pigeonhole principle this behavior must occur before $|2^{\prod_{q \in \Lambda} \mathcal{A}^{J_q}}| \leq 2^{|\mathcal{A}|^{(3^d-1)K}}$ iterations of μ_s.

3   The Mozes Property for Self-similar Structures

Most of the proofs of the undecidability of the domino problem on $\mathbb{Z}^2$ are based on the construction of a self-similar structure. A Theorem proven by Mozes [7] and later generalized by Goodman-Strauss [4] shows that every $\mathbb{Z}^d$-substitutive subshift is a sofic subshift for $d \geq 2$. This theorem fails for the case $d = 1$. The importance of this result is the fact that multidimensional substitutions can be realized by local rules. In order to present a family of self-similar substitutions with undecidable domino problem we will make use of an analogue of the theorem shown by Mozes.

Definition 1. *A self-similar substitution* s *satisfies the Mozes property if for every substitution* s' *defined over the same rectangle and over an alphabet* $\mathcal{A}$ *containing* 0 *and such that* $\forall a \in \mathcal{A} \backslash \{0\}$, $\pi(s'(a)) = s(1)$ *and* $s'(0) = s(0)$ *there exists an alphabet* $\mathcal{B}$ *containing the symbol* 0, *a finite set of forbidden patterns* $\mathcal{F} \subseteq \mathcal{B}^*_{\mathbb{Z}^d}$ *and a local function* $\Phi : \mathcal{B} \to \mathcal{A}$ *such that* $\Phi(0) = 0$ *and the function* $\phi : \mathcal{B}^{\mathbb{Z}^d} \to \mathcal{A}^{\mathbb{Z}^d}$ *given by* $\phi(x)_z = \Phi(x_z)$ *is surjective from* $X_s(\mathcal{F})$ *to* $X_{s'}$.

In other words, it's the analogue of saying that $X_{s'}$ is a sofic subshift, except that the SFT extension has to be based on X_s. Currently, we have been able to produce a class of substitutions that satisfy the Mozes property but we have not found a characterization of those who do. An example of a substitution without the Mozes property is the one given by ■ ⟶ ▪▪.

An interesting example of a substitution satisfying the Mozes property is the Sierpiński carpet shown in Fig. 3. This substitution it not channel bounded as at least one of the four borders of a coloring of $s^{n-1}(1)$ matter when constructing $s^n(1)$ and thus $\chi(s, n)$ grows exponentially. In fact this substitution belongs to a bigger class which also satisfies the Mozes property. In the next section we introduce this class and use this previous fact to prove that all the substitutions belonging to it have undecidable domino problem.

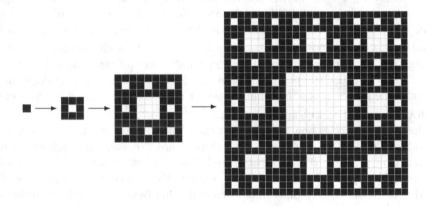

Fig. 3. The first iterations of the Sierpiński carpet substitution.

4   Self-similar Structures Where the Domino Problem Is Undecidable

In this section we present a family of two-dimensional self-similar substitutions with undecidable s-based domino problem. The definition of this class follows.

Definition 2. *A self-similar substitution* s *defined on* $[1, l_1] \times [1, l_2]$ *contains a grid if there are integers* $1 \leq i_1 < i_2 <= l_1$ *and* $1 \leq j_1 < j_2 <= l_2$ *such that* $j \in \{j_1, j_2\}$ *or* $i \in \{i_1, i_2\}$ *implies that* $s(1)_{(i,j)} = 1$.

One example of a self-similar substitution that contains a grid is the Sierpiński carpet. One interesting property of these substitutions is that they satisfy the Mozes property. This follows from a technical construction which uses a layer that looks like a generalized Robinson tiling [8] and stores the information of the simulated substitution and its past hierarchically.

Theorem 2. *All self-similar substitutions which contain a grid satisfy the Mozes property.*

In what remains of this section we show the following theorem:

Theorem 3. *Let s be a self-similar substitution which contains a grid. Then the domino problem $DP(s)$ is undecidable.*

Proof. We claim that an oracle for $DP(s)$ can be used to decide $DP(\mathbb{Z}^2)$. This is enough to conclude, as $DP(\mathbb{Z}^2)$ is undecidable.

Let s be defined on $[1, l_1] \times [1, l_2]$, some values satisfying the grid condition (i_1, i_2) and (j_1, j_2) and consider a substitution s' over the alphabet $\mathcal{A}(s') = \{\blacksquare, \boxminus, \square, 0\}$ given by the following rules: Let $C = \{(i_1, j_1), (i_1, j_2), (i_2, j_1), (i_2, j_2)\}$, $H = \{(i, j)|j \in \{j_1, j_2\}\} \backslash C$ and $V = \{(i, j)|i \in \{i_1, i_2\}\} \backslash C$.

$$s'(\blacksquare)_z = \begin{cases} 0, & \text{if } s(1)_z = 0 \\ \boxminus, & \text{if } z \in H \\ \square, & \text{if } z \in V \\ \blacksquare, & \text{else} \end{cases} \quad s'(\square)_z = \begin{cases} 0, & \text{if } s(1)_z = 0 \\ \square, & \text{if } z \in V \cup C \\ \blacksquare, & \text{else} \end{cases} \quad s'(\boxminus)_z = \begin{cases} 0, & \text{if } s(1)_z = 0 \\ \boxminus, & \text{if } z \in H \cup C \\ \blacksquare, & \text{else} \end{cases}$$

For example, in the case where s is the Sierpiński carpet we get:

For any $y \in X_{s'} \backslash \{0^{\mathbb{Z}^2}\}$ and $n \in \mathbb{N}$ we have $s'^n(\blacksquare) \sqsubseteq y$. Indeed, $\blacksquare$ appears in the image of every symbol $a \in \mathcal{A}(s') \backslash \{0\}$. This implies that for every positive integer n, a $\blacksquare$ must appear at a bounded distance of every non-zero symbol in $s'^n(a)$. This argument extends inductively because if $s'^{n-1}(\blacksquare)$ appears at a bounded distance in every $s'^k(a)$ with $k > n$, it suffices to apply s' to obtain that $s'^n(\blacksquare)$ appears at bounded distance in $s'^{k+1}(a)$.

As s satisfies the Mozes property there exists an alphabet $\mathcal{B}(s')$, a finite set $\mathcal{F}(s') \subset \mathcal{B}(s')^*_{\mathbb{Z}^2}$ and $\Phi : \mathcal{B}(s') \to \mathcal{A}(s')$ such that $\Phi(0) = 0$ and the function $\phi : \mathcal{B}(s')^{\mathbb{Z}^d} \to \mathcal{A}(s')^{\mathbb{Z}^d}$ given by $\phi(x)_z = \Phi(x_z)$ is surjective from $X_s(\mathcal{F})$ to $X_{s'}$.

Consider a finite set of forbidden patterns $\mathcal{F}$ over an alphabet $\mathcal{A}$ defining a $\mathbb{Z}^2$ subshift $X(\mathcal{F})$. Without loss of generality $\mathcal{F}$ contains only patterns with supports $\{(0,0), (1,0)\}$ and $\{(0,0), (0,1)\}$ (one can choose a conjugated version of $X(\mathcal{F})$ satisfying this property by using a higher block code. See [5]).

Finally, consider the alphabet $\mathcal{S} := \mathcal{B}(s') \times (\mathcal{A} \cup \{0\})$ along with the set of forbidden patterns $\mathcal{G}$ given by the union of the following sets:

- Zeros correspond: $\{(0, a) \mid a \in \mathcal{A}\} \cup \{(b, 0) \mid b \in \mathcal{B}(s')\backslash\{0\}\}$.
- First layer forbidden patterns: $\{p \times q \mid p \in \mathcal{F}(s'), q \in \mathcal{A}^{supp(p)}\}$. These forbidden patterns make sure that configurations belonging to the first layer of $X_s(\mathcal{G})$ belong to $X_s(\mathcal{F}(s'))$.
- Horizontal forbidden patterns: let $p \in \mathcal{S}^{\{(0,0),(1,0)\}}$ be denoted by (a, b, c, d) if $p(0,0) = (a, c)$ and $p(1,0) = (b, d)$ and $q \in \mathcal{A}^{\{(0,0),(1,0)\}}$ be denoted by (c, d) if $q(0,0) = c$ and $q(1,0) = d$. The set of horizontal forbidden patterns is $\{(a, b, c, d) \mid (a = \boxminus, b \in \{\boxminus, \boxdot\}$ and $c \neq d)$ or $(a = \boxdot, b = \boxminus$ and $(c, d) \in \mathcal{F})\}$.

- Vertical forbidden patterns: let $p \in \mathcal{S}^{\{(0,0),(0,1)\}}$ be denoted by (a, b, c, d) if $p(0,0) = (a, c)$ and $p(0,1) = (b, d)$ and $q \in \mathcal{A}^{\{(0,0),(0,1)\}}$ be denoted by (c, d) if $q(0,0) = c$ and $q(0,1) = d$. The set of vertical forbidden patterns is given by $\{(a, b, c, d) \mid (a = \boxed{\text{\rotatebox{90}{$\boxminus$}}}, b \in \{\boxed{\text{\rotatebox{90}{$\boxminus$}}}, \boxdot\}$ and $c \neq d)$ or $(a = \boxdot, b = \boxed{\text{\rotatebox{90}{$\boxminus$}}}$ and $(c, d) \in \mathcal{F})\}$.

These rules codify the following idea: $\boxdot$s carry arbitrary symbols from $\mathcal{A}$ in the second layer and the arrows send this information left and up respecting the rules from $\mathcal{F}$, see Fig. 4. By iterating the substitution s it is easy to see that $s^n(1)$ actually contains 2^n vertical and horizontal lines. This means that the intersections of these lines contain symbols of $\mathcal{A}$ which represent a $2^n \times 2^n$ pattern which contains no forbidden pattern from $\mathcal{F}$. Therefore if $X_s(\mathcal{G}) \neq \{0^{\mathbb{Z}^2}\}$ then $X(\mathcal{F}) \neq \emptyset$ by compactness. Conversely if $X(\mathcal{F}) \neq \emptyset$ it is possible to always build the second layer of a point having $s'^n(1)$ in the first layer.

Suppose there is an algorithm for deciding DP(s). Then for any $\mathcal{F}$ defining a $\mathbb{Z}^2$ subshift the alphabet $\mathcal{S}$ and the rules $\mathcal{G}$ can be built in order to decide if $X_s(\mathcal{G}) \neq \{0^{\mathbb{Z}^2}\}$. This is equivalent to deciding if $X(\mathcal{F}) \neq \emptyset$, therefore DP($\mathbb{Z}^2$) can be decided. This yields the desired contradiction.

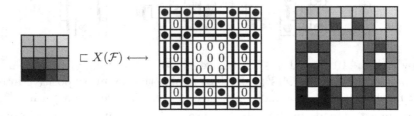

Fig. 4. On the left a pattern from $X(\mathcal{F})$. In the right its coding in $X_s(\mathcal{G})$. The blue squares are arbitrary symbols from $\mathcal{A}$. (Color figure online)

5   Generalizations and Perspectives

Here we present some ideas to generalize previous results in order to advance towards a characterization of the self-similar structures where the domino problem is decidable. In the previous sections the information which allows to simulate grids is transferred through straight lines. We can imagine less rigid possibilities.

5.1   Connectivity

We propose a way to define the directions in which the information can be transferred in a substitution in $\mathbb{Z}^2$. Given a self-similar substitution defined over $[1, l_1] \times [1, l_2]$ we denote by $\mathbb{X}$ the set of coordinates z such that $s(1)_z = 1$. Let $\mathbb{S} = \{(0, -1), (0, 1), (-1, 0), (0, 1)\}$ and $\mathbb{W}$ contains $\{(-1, -1), (1, 1)\}$ if $\{(1, 1), (l_1, l_2)\} \in \mathbb{X}$ and $\{(-1, 1), (1, -1)\}$ if $\{(1, l_2), (l_1, 1)\} \in \mathbb{X}$. We say s admits a rigid (respectively flexible) vertical line at $1 \leq v \leq l_1$ if there is a non-repeating sequence of vertices $(v, 1) = x_1, \ldots x_n = (v, l_2)$ such that the differences $x_j - x_{j-1}$ belong to $\mathbb{S}$ (respectively $\mathbb{W} \cup \mathbb{S}$). We define rigid and flexible horizontal lines for $1 \leq h \leq l_2$ analogously. We also say that two lines are weakly disjoint if they share no consecutive pair of vertices in their path.

According to these notions, we distinguish the following four subclasses:

- s has *bounded connectivity* if s has at most one flexible horizontal and vertical line;
- s has *a isthmus* if $s(1)$ has at least two weakly disjoint flexible lines in one direction and at most one weakly disjoint flexible line in the other direction;
- s has *a weak grid* if $s(1)$ has at least two flexible horizontal lines and two flexible vertical lines which are pairwise weakly disjoint.
- s has *a strong grid* if $s(1)$ has at least two rigid horizontal lines and two rigid vertical lines which are pairwise weakly disjoint.

If s has bounded connectivity the proof of Theorem 1 can be adapted to show decidability. If s has a strong grid it is possible to adapt the proof of Theorem 3 to show the undecidability of the domino problem associated to such substitution, moreover, a generalization of that proof works even in the case of weak grids. Nevertheless we still have no results supporting either direction in the isthmus case. We believe that the Mozes property does not hold in the isthmus case, which would be evidence towards decidability. Figure 5 presents the domino problem of different substitutions according to this classification.

B. Connectivity	Isthmus	Weak grid	Strong grid
DP decidable	Unknown	DP undecidable	DP undecidable

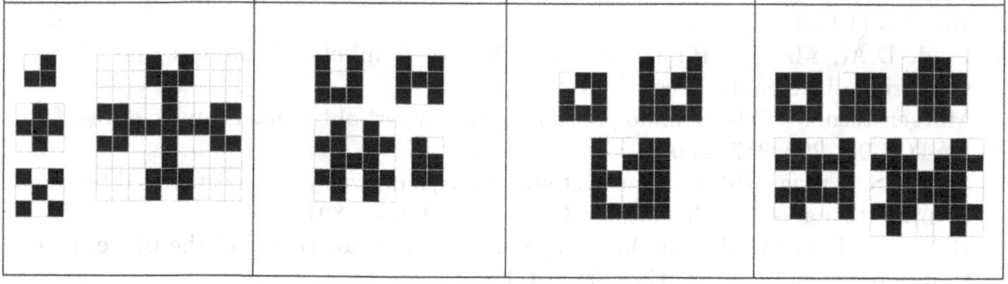

Fig. 5. Some examples of substitution according to this classification

5.2   Concluding Remarks

We introduced the domino problem on self-similar structures in order to understand the frontier between decidability and undecidability when we go from the
line (dimension 1) to the plane (dimension 2). Our results show that there is no
decidability threshold on the Hausdorff dimension. Indeed, there are self-similar
structures with decidable domino problem and Hausdorff dimension arbitrary
near to 2 (obtained by s_n) and self-similar structures with undecidable domino
problem and Hausdorff dimension arbitrary near to 1 (obtained by s'_n).

Thus, the frontier between decidability and undecidability seems more likely
to be based on the presence of a grid where it is possible to implement a computation. To confirm this hypothesis, it remains to study self-similar structures
with an isthmus. In the case of an isthmus the substitution presents an unique
bridge which links different zones. This impedes a Mozes-like [7] or Goodman-
Strauss-like [4] construction. The main problem is that in order to simulate a
substitution there is the need to transfer arbitrarily big amounts of information
by that isthmus. We believe the study of this class of substitutions will certainly
provide new tools to the study of how information can be transfered.

Acknowledgements. We would like to thank the reviewers for their helpful remarks.
This work was partially supported by the ANR project QuasiCool (ANR-12-JS02-011-
01)

References

1. Aubrun, N., Kari, J.: Tiling problems on Baumslag-Solitar groups. Preprint (2013)
2. Ballier, A., Stein, M.: The domino problem on groups of polynomial growth. arXiv
   preprint (2013). arXiv:1311.4222
3. Berger, R.: The undecidability of the domino problem. Am. Math. Soc. **66**, 1–72
   (1966)
4. Goodman-Strauss, C.: Matching rules and substitution tilings. Ann. Math. **147**(1),
   181–223 (1998)
5. Lind, D.A., Marcus, B.: An Introduction to Symbolic Dynamics and Coding.
   Cambridge University Press, New York (1995)
6. Margenstern, M.: The domino problem of the hyperbolic plane is undecidable. Bull.
   EATCS **93**, 220–237 (2007)
7. Mozes, S.: Tilings, substitution systems and dynamical systems generated by them.
   Journal d'Analyse Mathématique **53**(1), 139–186 (1989)
8. Robinson, R.M.: Undecidability and nonperiodicity for tilings of the plane. Inventiones Mathematicae **12**, 177–209 (1971)
9. Robinson, R.M.: Undecidable tiling problems in the hyperbolic plane. Inventiones
   Mathematicae **44**, 159–264 (1978)
10. Wang, H.: Proving theorems by pattern recognition I. Commun. ACM **3**(4), 220–
    234 (1960)

Axiomatizing Analog Algorithms

Olivier Bournez[1]([⊠]), Nachum Dershowitz[2], and Pierre Néron[3]

[1] Laboratoire d'Informatique de l'X (LIX), École Polytechnique, Palaiseau, France
bournez@lix.polytechnique.fr
[2] School of Computer Science, Tel Aviv University, Ramat Aviv, Israel
nachum@cs.tau.ac.il
[3] French Network and Information Security Agency (ANSSI), Paris, France
pierre.neron@ssi.gouv.fr

Abstract. We propose a formalization of generic algorithms that includes analog algorithms. This is achieved by reformulating and extending the framework of abstract state machines to include continuous-time models of computation. We prove that every hybrid algorithm satisfying some reasonable postulates may be expressed precisely by a program in a simple and expressive language.

1 Introduction

In [14], Gurevich showed that any algorithm that satisfies three intuitive "Sequential Postulates" can be step-by-step emulated by an abstract state machine (ASM). These postulates formalize the following intuitions: (I) one is dealing with discrete deterministic state-transition systems; (II) the information in states suffices to determine future transitions and may be captured by logical structures that respect isomorphisms; and (III) transitions are governed by the values of a finite and input-independent set of ground terms. All notions of algorithms for "classical" *discrete-time* models of computation in computer science are covered by this formalization. This includes Turing machines, random-access memory (RAM) machines, and their sundry extensions. The geometric constructions in [18], for example, are loop-free examples of discrete-step continuous-space (real-number) algorithms. The ASM formalization also covers general discrete-time models evolving over continuous space like the Blum-Shub-Smale machine model [1].

However, capturing *continuous-time* models of computation is still a challenge, that is to say, capturing models of computation that operate in continuous (real) time and with real values. Examples of continuous-time models of computations include models of analog machines like the General Purpose Analog

O. Bournez—This author's research was partially supported by a French National Research Agency's grant (ANR-15-CE40-0016-02).

N. Dershowitz—This author's research benefited from a fellowship at the Institut d'Études Avancées de Paris (France), with the financial support of the French National Research Agency's "Investissements d'avenir" program (ANR-11-LABX-0027-01 Labex RFIEA+).

© Springer International Publishing Switzerland 2016
A. Beckmann et al. (Eds.): CiE 2016, LNCS 9709, pp. 215–224, 2016.
DOI: 10.1007/978-3-319-40189-8_22

Computer (GPAC) of Claude Shannon [20], proposed as a mathematical model of the Differential Analyzers, built for the first time in 1931 [7], and used to solve various problems ranging from ballistics to aircraft design – before the era of the digital computer [16]. Others include Pascal's 1642 *Pascaline*, Hermann's 1814 *Planimeter*, as well as Bill Phillips' 1949 water-run *Financephalograph*. Continuous-time computational models also include neural networks and systems built using electronic analog devices. Such systems begin in some initial state and evolve over time; results are read off from the evolving state and/or from a terminal state. More generally, determining which systems can actually be considered to be computational models is an intriguing question and relates to philosophical discussions about what constitutes a programmable machine. Continuous-time computation theory is far less understood than its discrete-time counterpart [4]. Another line of development of continuous-time models was motivated by hybrid systems, particularly by questions related to the hardness of their verification and control. In hybrid systems, the dynamics change in response to changing conditions, so there are discrete transitions as well as continuous ones. Here, models are not seen as necessarily modeling analog machines, but, rather, as abstractions of systems about which one would like to establish properties or derive verification algorithms [4]. Some work on ASM models dealing with continuous-time systems has been accomplished for specific cases [8,9]. Rust [19] specifies forms of continuous-time evolution based on ASMs using infinitesimals. However, we find that a comprehensive framework capturing general analog systems is still wanting.

Our goal is to capture all such analog and hybrid models within one uniform notion of computation and of algorithm. To this end, we formalize a generic notion of continuous-time algorithm. The proposed framework is an extension of [14], as discrete-time algorithms are a simple special case of analog algorithms. (The initial attempt [5] was not fully satisfactory, as no completeness theorem nor general-form result was obtained. Here, we indeed achieve both.) We provide postulates defining continuous-time algorithms, in the spirit of those of [14], and we prove some completeness results. We define a simple notion of an analog ASM program and prove that all models satisfying the postulates have corresponding analog programs (Lemma 16 and Theorem 20). Furthermore, we provide conditions guaranteeing that said program is unique up to equivalence (Theorem 21 and Corollary 22). All of this seamlessly extends the results of [14] to analog and hybrid systems. The proposed framework covers all classes of continuous-time systems that can be modeled by ordinary differential equations or have hybrid dynamics, including the models in [4] and the examples in [5]. It is a first step towards a general understanding of computability theory for continuous-time models, taken in the hope that it will also lead to a formalization of a "Church-Turing thesis" for analog systems in the spirit of what has been achieved for discrete-time models [2,3,10]. Systems with continuous input signals and other means of specifying continuous behavior are left for future work.

Some of our ideas were inspired by the way the semantics of hybrid systems are given in the approach of Platzer [17]. Among attempts at studying the semantics of analog systems within a general framework is [22]. Recent results on comparing analog models include [11]. Soundness and (relative) completeness results for a programming language with infinitesimals have also been obtained in [21]. Applications to verification have been explored [15].

2   General Algorithms

We want to generalize the notion of algorithms introduced by Gurevich in [14] in order to capture not only the sequential case but also continuous behavior. (For lack of place, we assume some familiarity with [14].) However, when evolving continuously, an algorithm can no longer be viewed as a discrete sequence of states, and we need a notion of evolution that can capture both kinds of behavior. This is based on a notion of a *timeline* that corresponds to algorithm execution.

Definition 1 (Time). Time $\mathbb{T}$ *corresponds to a totally ordered monoid: there is an associative binary operation* $+$, *with some neutral element* 0, *and a total relation* $\leq$ *preserved by* $+$: $t \leq t'$ *implies* $t + t'' \leq t' + t''$ *for all* $t'' \in \mathbb{T}$.

An element of $\mathbb{T}$ will be called a *moment*. Examples of time $\mathbb{T}$ are $\mathbb{R}^{\geq 0}$ and $\mathbb{N}$. As expected, $t < t'$ will mean $t \leq t'$ but not $t = t'$.

Definition 2 (Timeline). *A timeline is a subset of* $\mathbb{T}$ *containing* 0. *We let* $\mathbb{I}$ *denote the set of all timelines.*

For a moment $i \in I$ of timeline I, we write $Jump(i)$ if there exists $t \in I$ with $i < t$, and there is no $t' \in I$ with $i < t' < t$. We write $Flow(i)$ otherwise: that means that for all t, $i < t$, there is some in-between $t' \in I$ with $i < t' < t$. A moment i with $Jump(i)$ is meant to indicate a discrete transition. In this case, we write i^+ for the smallest t greater than i. A timeline I is non-Zeno if for any moment $i \in I$, there is a finite number of moments $j \leq i$ with $Jump(j)$. $\mathbb{I}$ is non-Zeno if all its timelines are.

For timelines $\mathbb{I} = \mathbb{R}^{\geq 0}$, for instance, we have $Flow(i)$ for all $i \in \mathbb{I}$. For $\mathbb{I} = \mathbb{N}$, we have $Jump(i)$ for all $i \in \mathbb{I}$, and $i^+ = i + 1$. We intend (for hybrid systems, in particular) to also consider timelines mixing both properties, that is, with $Flow(i)$ for some i and $Jump(i)$ for other i. Formally building such timelines is easy (for example $\bigcup_{n \in \mathbb{N}} [n, n + 0.5]$). All these examples are non-Zeno.

Definition 3 (Truncation). *Given a timeline* $I \in \mathbb{I}$ *and a moment* i *of* I, *the truncated timeline* $I[i]$ *is the timeline defined by* $I[i] = \{t \mid i + t \in I\}$.

With timelines in hand, we can define hybrid dynamical systems.

Definition 4 (Dynamical System). *A dynamical system* $\langle \mathcal{S}, \mathcal{S}_0, \iota, \varphi \rangle$ *consists of the following: (a) a nonempty set (or class)* $\mathcal{S}$ *of states; (b) a nonempty subset (or subclass)* $\mathcal{S}_0 \subseteq \mathcal{S}$, *called* initial *states; (c) a timeline map* $\iota : \mathcal{S} \to \mathbb{I}$, *with* $\mathbb{I}$ *non-Zeno; (d) a trajectory map* $\varphi : (X : \mathcal{S}) \times \iota(X) \to \mathcal{S}$. *We require that, for any state* X *and moments* $i, i + i' \in \iota(X)$, *one has*

$$\varphi(X, 0) = X, \qquad \iota(\varphi(X, i)) = \iota(X)[i], \qquad \varphi(X, i + i') = \varphi(\varphi(X, i), i').$$

Together, the *timeline* and *trajectory* maps associate to each state its future evolution. For a state X, $\iota(X)$ defines the timeline corresponding to the system behavior starting from X, and $\varphi(X)$ defines its concrete evolution by associating to each moment in $\iota(X)$ its corresponding state. The third condition ensures that evolution during $i + i'$ is similar to first evolving during i and then during i'; the preceding condition ensures a similar property for timelines (and ensures consistency of the last condition).

Postulate I. *An algorithm is a dynamical system.*

A vocabulary $\mathcal{V}$ is a finite collection of fixed-arity (possibly nullary) function symbols, some functions of which may be tagged *relational*. A term whose outermost function symbol is relational is termed *Boolean*. We assume that $\mathcal{V}$ contains the scalar (nullary) function *true*. A (first-order) *structure* X of vocabulary $\mathcal{V}$ is a nonempty set S, the *base set (domain)* of X, together with interpretations of the function symbols in $\mathcal{V}$ over S: A j-ary function symbol f is interpreted as a function, denoted $[\![f]\!]_X$, from S^j to S. Elements of S are also called elements of X, or *values*. Similarly, the interpretation of a term $f(t_1, \ldots, t_n)$ in X is recursively defined by $[\![f(t_1, \ldots, t_n)]\!]_X = [\![f]\!]_X([\![t_1]\!]_X, \ldots, [\![t_n]\!]_X)$.

Let X and Y be structures of the same vocabulary $\mathcal{V}$. An *isomorphism* from X onto Y is a one-to-one function ζ from the base set of X onto the base set of Y such that $f(\zeta x_1, \ldots, \zeta x_j) = \zeta x_0$ in Y whenever $f(x_1, \ldots, x_j) = x_0$ in X.

Definition 5 (Abstract Transition System). *An abstract transition system is a dynamical system whose states S are (first-order) structures over some finite vocabulary $\mathcal{V}$, such that the following hold:*

(a) *States are closed under isomorphism, so if $X \in S$ is a state of the system, then any structure Y isomorphic to X is also a state in S, and Y is an initial state if X is.*

(b) *Transformations preserve the base set: that is, for every state $X \in S$, for any $i \in \iota(X)$, $\varphi(X, i)$ has the same base set as X.*

(c) *Transformations respect isomorphisms: if $X \cong_\zeta Y$ is an isomorphism of states $X, Y \in S$, then $\iota(X) = \iota(Y)$ and for all $i \in \iota(X)$, $X_i \cong_\zeta Y_i$, where $X_i = \varphi(X, i)$, and $Y_i = \varphi(Y, i)$.*

Postulate II. *An algorithm is an abstract transition system.*

When $\iota(X)$ is $\mathbb{N}$ (or order-isomorphic to $\mathbb{N}$) for all X, this corresponds precisely to the concepts introduced by [14], considering that $\varphi(X, n) = \tau^{[n]}(X)$.

It is convenient to think of a structure X as a memory of some kind: If f is a j-ary function symbol in vocabulary $\mathcal{V}$, and $\bar{a}$ is a j-tuple of elements of the base set of X, then the pair $(f, \bar{a})$ is called a *location*. We denote by $[\![f(\bar{a})]\!]_X$ its interpretation in X, i.e. $[\![f]\!]_X(\bar{a})$. If $(f, \bar{a})$ is a location of X and b is an element of X then $(f, \bar{a}, b)$ is an *update* of X. When Y and X are structures over the same domain and vocabulary, $Y \setminus X$ denotes the set of updates $\Delta^+ = \{(f, \bar{a}, [\![f(\bar{a})]\!]_Y) \mid [\![f(\bar{a})]\!]_Y \neq [\![f(\bar{a})]\!]_X\}$.

We want instantaneous evolution to be describable by updates:

Definition 6. *An* infinitesimal generator *is* (a) *a function* Δ *that maps states* X *to a set* $\Delta(X)$ *of updates, and* (b) *preserves isomorphisms: if* $X \cong_\zeta Y$ *is an isomorphism of states* $X, Y \in \mathcal{S}$, *then for all updates* $(f, \bar{a}, b) \in \Delta(X)$, *we have an isomorphic update* $(f, \overline{\zeta a}, \zeta b) \in \Delta(Y)$.

We write $Jump(X)$ and say that X is a *jump* when $Jump(0)$ in timeline $\iota(X)$; otherwise, we write $Flow(X)$ and say that it is a *flow*. For states X with $Jump(X)$, the following is natural:

Definition 7. *The* update generator *is the infinitesimal generator defined on jump states* X *as* $\Delta(X) = \Delta^+(X)$, *where* $\Delta^+(X)$ *stands for* $\varphi(X, 0^+) \backslash X$.

To deal with flow states, we will also define some corresponding infinitesimal generator Δ_ψ. Before doing so, let's see how to go from semantics to generators.

An *initial evolution* over S is a function whose domain of definition is a timeline and whose range is S. An initial evolution is said to be *initially constant* if it has a constant prefix: that is to say, there is some $0 < t$ such that the function is constant over $[0 .. t]$.

Definition 8 (Semantics). *A* semantics ψ *over a class* $\mathcal{C}$ *of sets* S *is a partial function mapping initial evolutions over some* $S \in \mathcal{C}$ *to an element of* S.

Remark 9. When $\mathbb{T} = \mathbb{R}^{\geq 0}$, an example of semantics over the class of sets S containing $\mathbb{R}$ is the derivative ψ_{der}, mapping a function f to its derivative at 0 when that exists. When $\mathbb{T} = \mathbb{N}$, an example of semantics over the class of all sets would be the function $\psi_\mathbb{N}$ mapping f to $f(1)$. More generally, when $0 \in \mathbb{T}$ is such that $Jump(0)$, an example of semantics over the class of all sets is the function $\psi_\mathbb{N}$ mapping f to $f(0^+)$.

Consider a semantics ψ over a class of sets S. Let X be a state whose domain is in the class and a location $(f, \bar{a})$ of X. Denote by $Evolution(X, (f, \bar{a}))$ the corresponding initial evolution: that is to say, the function given formally by $Evolution(X, (f, \bar{a})) : t \mapsto [\![f(\bar{a})]\!]_{\varphi(X,t)}$ for $0 \leq t \leq I_1, t \in \iota(X)$, for some $I_1 \in \iota(X)$, with $I_1 = 0^+$ for a jump. We use $\psi[X, f, \bar{a}]$ to denote the image of this evolution under ψ (when it exists).

Definition 10 (Infinitesimal Generator Associated with ψ). *The infinitesimal generator associated with* ψ *maps each state* X, *such that* $\psi[X, f, \bar{a}]$ *is defined for all locations, to the set:* $\Delta_\psi(X) = \{(f, \bar{a}, \psi[X, f, \bar{a}]) \mid (f, \bar{a})$ *is a location of* X, $Evolution(X, (f, \bar{a}))$ *is not initially constant*$\}$.

The update generator Δ^+ (see Definition 7) is the infinitesimal generator associated with the semantics $\psi_\mathbb{N}$ (of Remark 9) over flow states.

From now on, we assume that some semantics ψ is fixed to deal with flow states. It could be ψ_{der}, but it could also be another one (for example: talking about integrals or built using infinitesimals as in [19]). We denote by Δ_ψ the associated infinitesimal generator.

We are actually discussing algorithms relative to some ψ, and to be more precise, we should be refering to ψ-algorithms. The point is that not every infinitesimal generator is appropriate and that appropriateness is actually relative to a time domain and to the class of allowed dynamics over this time domain. To see this, keep in mind that – when Δ_ψ corresponds to derivative – to be able to talk about derivatives, one implicitly restricts oneself to dynamics that are differentiable, hence non-arbitrary. In other words, one is restricting to a particular class of possible dynamics, and not all dynamics are allowed. Restricting to other classes of dynamics (for example, analytic ones) may lead to different notions of algorithm.

From the update generator Δ^+ and Δ_ψ, we build a generator also tagging states by the fact that they correspond to a jump or a flow:

Definition 11 (Generator of a State). *We define the tagged generator of a state X, denoted $\Delta_t(X)$, as a function that maps state X to $\{\mathcal{F}\} \times \Delta_\psi(X)$ when $Flow(X)$ and $\Delta_\psi(X)$ is defined and to $\{\mathcal{J}\} \times \Delta^+(X)$ when $Jump(X)$.*

Let T be a set of ground terms. We say that states X and Y *coincide* over T, if $[\![s]\!]_X = [\![s]\!]_Y$ for all $s \in T$. This will be abbreviated $X =_T Y$. The fact that X and Y coincide over T implies that X and Y necessarily share some common elements in their respective base sets, at least all the $[\![s]\!]_X$ for $s \in T$.

An algorithm should have a finite imperative description. Intuitively, the evolution of an algorithm from a given state is only determined by inspecting part of this state by means of the terms appearing in the algorithm description. The following corresponds to the *Bounded Exploration* postulate in [14].

Postulate III. *For any algorithm, there exists a finite set T of ground terms over vocabulary $\mathcal{V}$ such that for all states X and Y that coincide for T, $\Delta_t(X)$ and $\Delta_t(Y)$ both exist and $\Delta_t(X) = \Delta_t(Y)$.*

A ground term of T is a *critical term* and a *critical element* is the value (interpretation) of a critical term.

Definition 12 (Analog Algorithm). *An algorithm is an object satisfying Postulates I through III.*

3   Characterization Theorem

We now go on to define the rules of our programs (adding to those of ASM programs in [14]).

Definition 13.

- **Update Rule:** *An* update rule *of vocabulary $\mathcal{V}$ has the form $f(t_1, \ldots, t_j) := t_0$ where f is a j-ary function symbol in $\mathcal{V}$ and $t_1, \ldots, t_j$ are ground terms over $\mathcal{V}$.*
- **Parallel Update Rule:** *If $R_1, \ldots, R_k$ are update rules of vocabulary $\mathcal{V}$, then* par $R_1 R_2 \ldots R_k$ endpar *is a* parallel update rule *of vocabulary $\mathcal{V}$.*

$\Delta_t(R_i, X)$ denotes the interpretation of a rule R in state X and is defined as expected: If R is an update rule $f(t_1, \ldots, t_j) := t_0$ then $\Delta_t(R, X) = \{\mathcal{J}\} \times (f, ([\![t_i]\!]_X, \ldots, [\![t_j]\!]_X), [\![t_0]\!]_X)$ and when R is par $R_1, \ldots, R_k$ endpar then $\Delta_t(R, X) = \{\mathcal{J}\} \times (d_1 \cup \cdots \cup d_k)$ where $\Delta_t(R_i, X) = \{\mathcal{J}\} \times d_i$ for all i.

Next, we introduce rules to deal with *Flows*.

Definition 14.

- **Basic Continuous Rule:** *A basic continuous rule of vocabulary $\mathcal{V}$ has the form* DYNAMIC$(f(t_1, \ldots, t_j), t_0)$ *where f is a symbol of arity j and $t_0, t_1, \ldots, t_j$ are ground terms of vocabulary $\mathcal{V}$.*
- **Flow Rule:** *If $R_1, \ldots, R_k$ are basic continuous rules of vocabulary $\mathcal{V}$, then* flow R_1 R_2 $\ldots$ R_k endflow *is a flow rule of vocabulary $\mathcal{V}$.*

Their semantics are then defined as follows. If R is a basic continuous rule DYNAMIC$(f(t_1, \ldots, t_j), t_0)$ then $\Delta_t(R, X) = \{\mathcal{F}\} \times \{(f, (a_1, \ldots, a_j), a_0)\}$ where each $a_i = [\![t_i]\!]_X$. If R is a flow rule with constituents $R_1, \ldots, R_k$, then $\Delta_t(R, X) = \{\mathcal{F}\} \times (d_1 \cup \cdots \cup d_k)$ where $\Delta_t(R_i, X) = \{\mathcal{F}\} \times d_i$.

Finally, we allow conditionals:

Definition 15.

- **Selection Rule:** *If φ is a ground boolean term over vocabulary $\mathcal{V}$ and R_1 and R_2 are rules of vocabulary $\mathcal{V}$ then:* if φ then R_1 or else R_2 endif *is a rule of vocabulary $\mathcal{V}$.*

Given such a rule R and a state X, if φ evaluates to *true* (the interpretation of scalar function *true*) in X then $\Delta_t(R, X) = \Delta_t(R_1, X)$ else $\Delta_t(R, X) = \Delta_t(R_2, X)$.

An ASM program of vocabulary $\mathcal{V}$ is a rule of vocabulary $\mathcal{V}$. The first key result is the following, which can be seen as a completeness result.

Theorem 16 (Completeness). *For every algorithm of vocabulary $\mathcal{V}$, there is an ASM program Π over $\mathcal{V}$ with the identical behavior: $\Delta_t(\Pi, X) = \Delta_t(X)$ for all states X.*

4 Extended Statements

We are now very close to formulating our other theorems. First we define an abstract state machine relative to semantics ψ.

Definition 17. *A ψ-abstract state machine B comprises the following: (a) an ASM program Π; (b) a set $\mathcal{S}$ of (first-order) structures over some finite vocabulary $\mathcal{V}$ closed under isomorphisms, and a subset $\mathcal{S}_0 \subseteq \mathcal{S}$ closed under isomorphisms; (c) a map ι and a map φ such that $\langle \mathcal{S}, \mathcal{S}_0, \iota, \varphi \rangle$ is an algorithm, where Δ_ψ is fixed to be the infinitesimal generator associated with ψ, and for all states X in $\mathcal{S}$, $\Delta_t(\Pi, X) = \Delta_t(X)$.*

By definition, a ψ-abstract state machine B satisfies all the postulates and hence is an algorithm.

Definition 18. *An ASM program Π is ψ-solvable for a set S of (first-order) structures over some finite vocabulary $\mathcal{V}$ closed under isomorphisms and a subset $S_0 \subseteq S$ closed under isomorphisms if there exists a unique ι and φ such that $(\Pi, S, S_0, \iota, \varphi)$ is a ψ-abstract state machine.*

Definition 19. *A semantics ψ is unambiguous if for all sets S of (first-order) structures over some finite vocabulary $\mathcal{V}$ closed under isomorphisms, and for all subsets $S_0 \subseteq S$ closed under isomorphisms, whenever there exists some ι and φ such that $(\Pi, S, S_0, \iota, \varphi)$ is a ψ-abstract state machine, then ι and φ are unique.*

Our main results follow. (Proofs are relegated to the technical report [6].)

Theorem 20. *For every ψ-definable algorithm A, there exists an equivalent ψ-abstract state machine B.*

Theorem 21. *Assume that ψ is unambiguous. For every ψ-definable algorithm A, there exists a unique equivalent ψ-abstract state machine B with the same states and initial states.*

Corollary 22. *Assume that ψ is unambiguous. For every ψ-definable algorithm A, there exists an equivalent ψ-solvable ASM program.*

To any algorithm A that is ψ-definable there corresponds an equivalent ψ-abstract state machine B, and hence a ψ-solvable program Π. Conversely, a ψ-abstract state machine B corresponds to a ψ-definable algorithm. However, not every program Π is ψ-solvable.

When ψ-corresponds to ψ_{der}, unambiguity comes from (unicity in) the Cauchy-Lipschitz theorem. The fact that not every program Π is ψ-solvable is due to the fact that not all differential equations have a solution.

5   Examples

The examples in this section are for semantics ψ_{der}. Our settings cover, first of all, analog algorithms that are pure flow, in particular all systems that can be modeled as ordinary differential equations. A very simple, classical example is the pendulum: the motion of an idealized simple pendulum is governed by the second-order differential equation $\theta'' + \frac{g}{L}\theta = 0$, where θ is angular displacement, g is gravitational acceleration, and L is the length of the pendulum rod. This can indeed be modeled as the program

```
flow  DYNAMIC(θ, θ₁)  DYNAMIC(θ₁, -g/L · θ) endflow
```

using the fact that any ordinary differential equation can be put in the form of a vectorial first-order equation, here θ_1 corresponding to the derivative of θ.

As a consequence, our formalism covers very generic classes of continuous-time models of computation, including the GPAC, which corresponds to ordinary differential equations with polynomial right-hand sides [12,13]. Recall that the

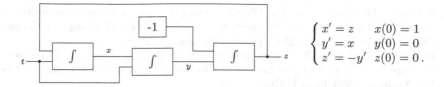

$$\begin{cases} x' = z & x(0) = 1 \\ y' = x & y(0) = 0 \\ z' = -y' & z(0) = 0. \end{cases}$$

Fig. 1. A GPAC for sine and cosine (left). Corresponding evolution (right).

GPAC was proposed as a mathematical model of differential analyzers (DAs), one of the most famous analog computer machines in history. Figure 1 (left) depicts a (non-minimal) GPAC that generates sine and cosine. In this picture, $\int$ signifies some integrator, and -1 denotes some constant block. This simple GPAC can be modeled by the program

```
flow DYNAMIC(x, z)  DYNAMIC(y, x)  DYNAMIC(z, −x) endflow
```

Our proposed model can also adequately describe hybrid systems, made of alternating sequences of continuous evolution and discrete transitions. This includes, for example, a simple model of a bouncing ball, the physics of which are given by the flow equations $x'' = -gm$, where g is the gravitational constant and $v = x'$ is the velocity, except that upon impact, each time $x = 0$, the velocity changes according to $v' = -k \cdot v'$, where k is the coefficient of impact. Every time the ball bounces, its speed is reduced by a factor k. This system can be described by a program like

```
if x = 0 then          v := −k · v
else flow DYNAMIC(x, v)  DYNAMIC(v, −g · m) endflow endif
```

Our setting is an extension of classical discrete-time algorithms; hence, all classical discrete-time algorithms can also be modeled.

As for examples with semantics other than ψ_{der}: Observe that one can consider timelines like $\mathbb{Q}$ instead of $\mathbb{R}$. (For such a timeline, we have $Flow(i)$ for all $i \in \mathbb{Q}$.) One can define a semantics on such a timeline where for every state X we have $Flow(X)$ by first extending the evolution function to $\mathbb{R}$ (for example by restricting to continuous dynamics) and then using the derivative. Constructions of [19] are also covered by our settings: In some sense, the example at the beginning of the paragraph is the spirit of the constructions from [19], where the timeline is the set of hyperreals obtained by multiplying some fixed infinitesimal by some hyperinteger (using hyperreals and infinitesimals). Notice that there is no need to consider derivatives or similar notions: we could also consider analytic dynamics, and consider a semantics related to the family of Taylor coefficients. Weaker notions of solution, like variational approaches, can also be considered.

References

1. Blum, L., Shub, M., Smale, S.: On a theory of computation and complexity over the real numbers; NP completeness, recursive functions and universal machines. Bull. Am. Math. Soc. **21**, 1–46 (1989)

2. Boker, U., Dershowitz, N.: The Church-Turing thesis over arbitrary domains. In: Avron, A., Dershowitz, N., Rabinovich, A. (eds.) Pillars of Computer Science. LNCS, vol. 4800, pp. 199–229. Springer, Heidelberg (2008)

3. Boker, U., Dershowitz, N.: Three paths to effectiveness. In: Blass, A., Dershowitz, N., Reisig, W. (eds.) Fields of Logic and Computation. LNCS, vol. 6300, pp. 135–146. Springer, Heidelberg (2010)

4. Bournez, O., Campagnolo, M.L.: A survey on continuous time computations. In: Cooper, S.B., Löwe, B., Sorbi, A. (eds.) Changing Conceptions of What is Computable, pp. 383–423. Springer, New York (2008)

5. Bournez, O., Dershowitz, N., Falkovich, E.: Towards an axiomatization of simple analog algorithms. In: Agrawal, M., Cooper, S.B., Li, A. (eds.) TAMC 2012. LNCS, vol. 7287, pp. 525–536. Springer, Heidelberg (2012)

6. Bournez, O., Dershowitz, N., Néron, P.: Axiomatizing analog algorithms. ArXiv e-prints (2016). http://arxiv.org/abs/1604.04295

7. Bush, V.: The differential analyser. J. Franklin Inst. **212**, 447–488 (1931)

8. Cohen, J., Slissenko, A.: On implementations of instantaneous actions real-time ASM by ASM with delays. In: Proceedings of 12th International Workshop on Abstract State Machines, Université de Paris, vol. 12, pp. 387–396 (2005)

9. Cohen, J., Slissenko, A.: Implementation of sturdy real-time abstract state machines by machines with delays. In: Proceedings of 6th International Conference on Computer Science and Information Technology, National Academy of Science of Armenia (2007)

10. Dershowitz, N., Gurevich, Y.: A natural axiomatization of computability and proof of Church's Thesis. Bull. Symbolic Logic **14**, 299–350 (2008)

11. Fu, M.Q., Zucker, J.: Models of computation for partial functions on the reals. J. Log. Algebraic Methods Program. **84**, 218–237 (2015)

12. Graça, D.S., Buescu, J., Campagnolo, M.L.: Computational bounds on polynomial differential equations. Appl. Math. Comput. **215**, 1375–1385 (2009)

13. Graça, D.S., Costa, J.F.: Analog computers and recursive functions over the reals. J. Complex. **19**, 644–664 (2003)

14. Gurevich, Y.: Sequential abstract-state machines capture sequential algorithms. ACM Trans. Comput. Log. **1**, 77–111 (2000)

15. Hasuo, I., Suenaga, K.: Exercises in *nonstandard static analysis* of hybrid systems. In: Madhusudan, P., Seshia, S.A. (eds.) CAV 2012. LNCS, vol. 7358, pp. 462–478. Springer, Heidelberg (2012)

16. Nyce, J.M.: Guest editor's introduction. IEEE Ann. Hist. Comput. **18**, 3–4 (1996)

17. Platzer, A.: Differential dynamic logic for hybrid systems. J. Autom. Reasoning **41**, 143–189 (2008)

18. Reisig, W.: On Gurevich's theorem on sequential algorithms. Acta Informatica **39**, 273–305 (2003)

19. Rust, H.: Hybrid abstract state machines: using the hyperreals for describing continuous changes in a discrete notation. In: International Workshop on Abstract State Machines, Swiss Federal Institute of Technology, pp. 341–356 (2000)

20. Shannon, C.E.: Mathematical theory of the differential analyser. J. Math. Phys. **20**, 337–354 (1941)

21. Suenaga, K., Hasuo, I.: Programming with infinitesimals: a WHILE-language for hybrid system modeling. In: Aceto, L., Henzinger, M., Sgall, J. (eds.) ICALP 2011, Part II. LNCS, vol. 6756, pp. 392–403. Springer, Heidelberg (2011)

22. Tucker, J.V., Zucker, J.I.: A network model of analogue computation over metric algebras. In: Cooper, S.B., Löwe, B., Torenvliet, L. (eds.) CiE 2005. LNCS, vol. 3526, pp. 515–529. Springer, Heidelberg (2005)

Generalized Effective Reducibility

Merlin Carl[✉]

Fachbereich für Mathematik und Statistik der Universität Konstanz,
Konstanz, Germany
merlin.carl@uni-konstanz.de

Abstract. We introduce two notions of effective reducibility for set-theoretical statements, based on computability with Ordinal Turing Machines (OTMs), one of which resembles Turing reducibility while the other is modelled after Weihrauch reducibility. We give sample applications by showing that certain (algebraic) constructions are not effective in the OTM-sense and considering the effective equivalence of various versions of the axiom of choice.

1 Introduction

From a sufficiently remote point of view, construction problems in mathematics can be seen as multi-valued, class-sized 'functions' from the set-theoretical universe V to itself. Examples of construction problems would be the problem assigning to fields their algebraic closures, to sets their well-orderings, to integrable functions their stem functions, to linear orderings their completions etc. Formally, this makes a construction problem a (class-sized) relation $R \subseteq V \times V$.

A 'solution' to or 'canonification' of a construction problem R is then a (class-sized) witness 'function' $F : V \to V$ such that, for all x in the domain of R, we have $R(x, F(x))$ and otherwise $F(x) = \emptyset$. Similarly, we can say that F witnesses the truth of a set-theoretical statement ϕ of the form $\forall x \exists y \psi$ if F is a solution for $\{(x, y) : \psi(x, y)\}$, the most natural candidates to consider being Π_2-statements, since ψ can be assumed to be absolute between transitive sets in that case.

Fixing an appropriate notion of effectiveness for set-theoretical constructions, we can now ask for specific construction problems R whether there exists an **effective** solution for R and similarly, whether some statement ϕ is 'effectively true'. Moreover, we can ask whether a construction or a statement 'effectively reduces' to another.

In the following, 'effectiveness' will be interpreted to mean computability by Ordinal Turing Machines (OTMs) without ordinal parameters. For the definition and basic results on OTMs, we refer the reader to [Ko1]. It was argued in [Ca] that OTM-computations are appropriate as a formalization of the intuitive notion of a 'transfinite effective procedure'.

For convenience, we assume that our machines work with three tapes, a 'miracle' tape (to be explained below), a scratch tape and an output tape.

Our set-theoretical notation is standard and can e.g. be looked up in [Je].

© Springer International Publishing Switzerland 2016
A. Beckmann et al. (Eds.): CiE 2016, LNCS 9709, pp. 225–233, 2016.
DOI: 10.1007/978-3-319-40189-8_23

2   Basic Methods and Notions

To apply OTMs to general mathematical constructions, we need to represent arbitrary sets as an input for OTMs, i.e. as sets of ordinals. This can be achieves as follows:

Definition 1. *Let x be a set, $t = tc(x)$ the transitive closure of x, $\alpha \in On$ and $f : \alpha \to tc(x)$ a well-ordering of $tc(x)$ in the order type α. We define $c_f(x)$, the f-code for x, recursively as the following set of ordinals: $c_f(x) := \{p(f^{-1}(y), \beta) : y \in x \wedge \beta \in c_f(y)\}$, where p denotes Cantor's ordinal pairing function. We say that $A \subseteq On$ 'is a code for' or 'codes' the set x if and only if there is some f for which $A = c_f(x)$. We write $rep(\tau, x)$ to indicate that τ codes x.*

Remark: By a certain abuse of notation, if x is a set, we will sometimes write $c(x)$ for an 'arbitrary' code for x. Note that $|c(x)| = |tc(x)|$ if $tc(x)$ is infinite.

We can now talk about OTM-computability of arbitrary functions from V to V:

Definition 2. *Let $F : V \to V$ be a functional class. We say that F is OTM-computable if and only if there is an OTM-program P such that, for every set x and every tape content τ, if $rep(\tau, x)$, then $P(\tau)$ converges to output σ such that $rep(\sigma, F(x))$, i.e. P takes representations of x to representations of $F(x)$.*

By this definition, the representation of a set x will depend on the choice of a well-ordering of $tc(x)$. The output of a computation on input x may hence depend on the choice of the representation of x. This is fine as long as only the output, but not the object coded by the output, depends on the choice of the input representation.

This allows us to make our notion of 'effectivity' precise:

Definition 3. *Let $R \subseteq V \times V$ be a construction problem. Then R is effectively solvable if and only if there is an OTM-computable solution F for R. Moreover, a set-theoretical Π_2-statement $\forall x \exists y \phi(x, y)$ (where ϕ is Δ_0) is effective if and only if the construction problem $\{(x, y) \in V \times V : \phi(x, y)\}$ is effectively solvable.[1] We write R_x for $\{y : (x, y) \in R\}$.*

One may now inquire whether various well-known construction problems and Π_2-statements are effective. Such questions were studied by Hodges in [Ho2], though with a different notion of effectivity based on Jensen and Karp's primitive recursive set functions. We note here that the two methods Hodges uses also work for our model, which allows us to carry over results.

The following lemma corresponds to Hodges' 'cardinality method', i.e. Lemma 3.2 of [Ho2]:

---

[1] In particular, this implies that an effective Π_2-statement must be true.

Lemma 1. *Let $\alpha \in On$, and let $R \subseteq V \times V$ be such that, for some cardinal $\kappa > \alpha$, there is $x \in V$ such that $|tc(x)| = \kappa$, $R_x \neq \emptyset$ and $\forall y \in R_x$ $|y| > \kappa$. Then no witness function for R is OTM-computable in the parameter α.*
Consequently, if R is such that there are such κ and x for every $\alpha \in On$, then no witness function for R is computable by a parameter-OTM (i.e. an OTM with a fixed tape cell marked with 1).
In particular, if, for some transitive x of infinite cardinality, $R_x \neq \emptyset$ and $\forall y \in R_x$ $|y| > |x|$ then no witness function for R is parameter-free OTM-computable.

Proof. Clearly, in less than κ^+ many steps, the machine cannot write a code of a structure of cardinality $> \kappa$.

It hence suffices to show that, when P is an OTM-program and P is given a (code c of a) set x of size $\kappa \geq \omega$ for input and the computation halts, then the output of the computation will be of size $\leq \kappa$. This follows if we can show that the computation will take less than κ^+ many steps, since P can write at most α many symbols in α many steps. Suppose for a contradiction that P takes $\lambda > \kappa$ many steps, and let δ be the smallest cardinal $> \lambda$. Let H be the Σ_1-Skolem hull of $\kappa \cup \{c\}$ in $L_\delta[c]$ and let M denote the transitive collapse of H. We may assume without loss of generality that $c \subseteq \kappa$, so that we have $c \in M$; as $L_\delta[c]$ contains the computation of P in the input c, so does H and hence there is $S \in M$ such that M believes that S is the computation of P with input c. By transitivity of M and absoluteness of computations, S is actually the computation of P with input c. Since S is contained in a transitive set of cardinality κ, $|S| \leq \kappa$, so the length of the computation is $< \kappa^+$, as desired.

There is also an analogue of the 'forcing method' (Lemma 3.7 of [Ho2]), which is given in Lemma 5 below.

Convention: For many of the following results, we will need the existence of generic filters for various partial orderings in L and some of its (symmetric) extensions. To avoid technical complications, we use as a shortcut an extra assumption that guarantees the existence of such filters. $0^\sharp$ is more than enough for our purposes, and we assume from now on that it exists.[2]

These lemmata can be seen as expressing the intuition that neither the power set operation on infinite sets nor the use of the axiom of choice are 'effective', not even in a very idealized sense. We note some sample applications.

Lemma 2. *None of the following construction problems is effectively solvable: (1) Field to its algebraic closures (2) Linear ordering to its completions (3) Set to its (constructible) power set (4) Set to its well-orderings*

Proof. (1) can be proved by an easy adaption of the proof of Theorem 4.1 of [Ho2]. There is only one point that requires a little care, namely the use of

---

[2] For some of the following results, this assumption is actually necessary: It is e.g. not hard to check that all choice principles considered in Sect. 4 are effective (and hence trivially reducible to each other) if $V = L$.

countable transitive models in that proof: For it might happen that an OTM-program P that halts in V does not halt in such a model M.[3] However, a check of Hodge's proof reveals that the countability of the ground model serves no purpose but to guarantee the existence of generic filters. We can hence circumvent this problem by doing the construction over L, using $0^\sharp$ to guarantee the existence of the required filters. (2) and (3) are easy applications of Lemma 1. (4) follows from Lemma 6 below.

It is, on the other hand, not hard to see that e.g. the construction problem of taking a ring to its quotient field is effectively solvable as in [Ho2]. The intuitions captured by Hodges' approach are hence preserved in our framework.

There are certainly various interesting questions to be asked about the effectivity, or otherwise, of various construction problems or Π_2-statements. However, we want to take the analogy with Turing computability a bit further: Instead of merely asking what problems are solvable, we want to consider what problems/statements are effectively reducible to which others in the sense that, given access to a solution to one as an 'oracle', one can effectively solve the other. A quite straightforward way to make this idea precise is the following:

Definition 4. *Assume that the OTM is equipped with an extra 'miracle tape'. Let F be a class function taking sets of ordinals to sets of ordinals. A miracle-OTM-program is defined like an OTM-program, but with an extra 'miracle' command. When this command is carried out, the set X of ordinals on the miracle tape is replaced (in one step) by $F(X)$. We write P^F to indicate that P is run and whenever the miracle command is applied to X, it is replaced by $F(X)$.*[4]

Definition 5. *Let C_1 and C_2 be construction problems. Then C_1 is reducible to C_2, written $C_1 \leq C_2$, if and only if there is some miracle-OTM-program P such that the following holds: Whenever F is a canonification of C_2 and whenever $G : V \to V$ is a class function taking each code for a set x to some code y for $F(x)$ and x is a set and c a code for x, we have $P^G(c) \downarrow = d$, where d is a code for a set z such that $(x, z) \in C_1$.*

Remark: Note that we do not demand in the conditions on G that $G(c)$ depends only on x when c is a code for x. By demanding that the same reduction works for every G, we rule out the possibility of coding extra information into the input representations.

Concerning this notion of reducibility, we observe that certainly a cardinality-raising construction is not reducible to one that is not:

---

[3] For example, suppose there is some minimal countable α such that $L_\alpha \models$ ZFC. Then the OTM-program that writes L on the tape until an L-level satisfying ZFC will halt in V, but not inside L_α.

[4] We thus make the implicit assumption that the miracle tape behaves deterministically, i.e. that, whenever the miracle command is applied to some X, the outcome will be the same. However, this property is not used anywhere in the arguments below. One may thus drop it, at the price of some extra formal complications.

Lemma 3. *Let C_1, C_2 be construction problems. Assume that there are some canonification F of C_2 and some infinite set x such that, for all sets y, (1) if $C_1(x,y)$, then $|y| > |tc(x)|$ and (2) if y is infinite, then $|F(y)| \leq |tc(y)|$. Then $C_1 \nleq C_2$.*

Proof. As in the proof of Lemma 1 above, OTM-computable functions cannot raise infinite cardinalities. By assumption, the miracle operation will also not raise the cardinality. Hence the output of a program P with a C_2-miracle will (for infinite input) always have at most the cardinality of the input and thus cannot in any case witness C_1.

Remark: In particular, the construction problem of taking a valued field to its linear compactifications (see [Ho2], Theorem 4.10) is not reducible to any of the following construction problems: Field to algebraic closure, formally real field to its real closure, field of characteristic p to its separable algebraic closure.

The above captures the idea that one construction 'helps' carrying out another. There is also a much more restrictive intuitive notion of reducibility between problems, namely that instances of one (construction) problem can be effectively 'translated' to particular instances of another: Given an instance of a problem C_1, we can first effectively turn it into an instance of a problem C_2 and then effectively turn the solution to C_2 into a solution to C_1. Another way to view this is that C_2 may only be used once in solving C_1. Thus, we define:

Definition 6. *Let C_1, C_2 be construction problems. Then C_1 is generalized Weihrauch reducible to C_2, written $C_1 \leq_{gW} C_2$, if and only if there are OTM-programs P and Q such that the following holds for all sets x in the domain of C_1, every code c for x and every canonification F of C_2:*

   (1) $Q(c)$ converges to output c', where c' is a code for a set y

   (2) For every code c'' of $F(y)$, $P(c'')$ converges to output c''', where c''' is a code for a set z

   (3) We have $C_1(x,z)$

   If these clauses hold, we say that (P,Q) witnesses the gW-reducibility of C_1 to C_2. Also, when F is a canonification, P and Q are OTM-programs and x is a set, we write $[P,F,Q](x)$ for the z obtained by the procedure just described.

   If $C_1 \leq_{gW} C_2$ and $C_2 \leq_{gW} C_1$, we write $C_1 \equiv_{gW} C_2$.

Remark: The name of the notion is due to its obvious resemblance with Weihrauch reducibility, which is an analogous notion for classical computability. For some results on classical Weihrauch reducibility, see e.g. [BGM]. Another notion of Weihrauch reducibility on higher cardinalities based on continuous transformations on generalized Baire spaces is introduced in [Ga]. We do not know how this approach connects to ours.[5]

---

[5] The generalization of uniform reducibility introduced in [HJ] based on (Turing-) computable strategies goes into a different direction, though it may be compatible with our setting. We thank one of our anonymous referees for pointing out this paper as well as the work of Galeotti to us.

We note that reducibility notions satisfy the general order-theoretic properties of reducibility relations:

Lemma 4. *Both $\leq$ and $\leq_{gW}$ are transitive and reflexive. Consequently, $\equiv_{gW}$ and $\equiv$ are equivalence relations.*

Definition 7. *Let C be a construction problem. Then $[C]$ denotes the $\equiv$-equivalence class of C and $[C]_{gW}$ denotes the $\equiv_{gW}$-equivalence class of C.*

3   A Method for Negative Results

We develop a method for showing that a construction problem is not gW-reducible to another. We will work with class-sized models of ZF^-, which denotes Zermelo-Fraenkel set theory without the axiom of powerset; more precisely, we take the formulation of ZF^- given in [GH].

Remark: Note that the following theorem is not trivial even when ZF^- is strengthened to full ZF, since a ZF model M may contain a set x without containing a suitable input format for x, so that the computation of an OTM cannot be simulated within M.

Lemma 5. *Let $M \models ZF^-$ be transitive and suppose that $x \in M$. Then $\mathbb{P}_x := \{f : \omega \to x : |f| < \omega \wedge f \text{ injective}\}$ is a set in M.*

Proof. Let $y := x \times \omega$. For each $n \in \omega$, we have $y^n \in M$ and the function $F : \omega \to M$ that maps n to y^n is definable in M. By replacement and union, $A := \bigcup\{y^n : n \in \omega\} \in M$. Now $\mathbb{P}_x$ can be obtained from A via separation.

Theorem 8. *Let F be a computable class function, $M \models ZF^-$ transitive such that $On^M = On$[6]. Assume moreover that $x \in M$ is such that there are (in V) two mutually generic $\mathbb{P}_x$-generic filters G_1 and G_2 over M. Then $F(x) \in M$.*

Proof. Let P be a program witnessing the computability of F. Let $x \in M$ be as in the assumption of the Theorem. By passing to $tc(x)$ if necessary, we may assume without loss of generality that x is transitive. Let G_1, G_2 be mutually M-generic filters over $\mathbb{P}_x$ which exist by assumption. In $M[G_1]$ and $M[G_2]$, x is well-ordered in order type α by $\bigcup G_1$ and $\bigcup G_2$, respectively. Hence both $M[G_1]$ and $M[G_2]$ contain tape contents coding x and thus contain the computations of P on these inputs. As ZF^- models, $M[G_1]$ and $M[G_2]$ contain the decoding of every tape content they contain. Thus $F(x) \in M[G_1] \cap M[G_2]$. As G_1 and G_2 are mutually generic, we have $M[G_1] \cap M[G_2] = M$, so $F(x) \in M$, as desired.

---

[6] Again, some condition on the height of M is required to ensure that the convergence of programs is absolute between V and M. In particular, a parameter-free OTM can run for more than α many steps, where α is minimal such that $L_\alpha \models ZF^-$.

This suggests a general method for proving, given constructions C_1 and C_2, that $C_1 \not\leq_{gW} C_2$. In general, find a class A sufficiently closed under OTM-computability and a canonification F of C_2 such that there is some $x \in A$ with the property that the closure of $F[A]$ under OTM-computability does not contain a C_1-solution for x. By Theorem 8, we can take for A a transitive class model M of ZF^-. We summarize the most important special case of this method in the following lemma:

Lemma 6. *Let C_1, C_2 be construction problems. Assume that there are a canonification F of C_2 and a transitive class-sized $M \models ZF^-$ and some $x \in M \cap dom(C_1)$ such that M is closed under F, but $\{y : C_1(x,y)\} \cap M = \emptyset$. Assume moreover that x is such that there are (in V) two mutually generic $\mathbb{P}_x$-generic filters G_1 and G_2 over M. Then $C_1 \not\leq_{gW} C_2$.*

Proof. Assume otherwise, and let P and Q be OTM-programs such that (P, Q) witnesses the gW-reducibility of C_1 to C_2. Pick F, M and x as in the statement of the Lemma. Then Q computes, for every code of x as an input, a code for some (unique) set y. By Theorem 8, we have $y \in M$. As M is closed under F, we have $F(y) \in M$. Now, for every code of $F(y)$ as an input, P computes a code for some (unique) set z. Again by Theorem 8, $z \in M$. Also, by the choice of P and Q, we have $C_1(x, z)$. So $z \in \{y : C_1(x, y)\} \cap M$ and so the latter is not empty, a contradiction.

4   Results on Generalized Effective Reducibility

As a sample application of the notions and methods developed above, we consider variants of the axiom of choice with respect to effective reducibility.

Definition 9. *Denote by AC the statement that for all sets x, there is a function f such that $f(\emptyset) = \emptyset$ and for $y \in x$, if $y \neq \emptyset$, then $f(y) \in y$. Denote by AC' the statement that for all sets x whose elements are non-empty and mutually disjoint, there is a set r such that $|r \cap y| = 1$ for all $y \in x$. Denote by WO the well-ordering principle, i.e. the statement that for every set x, there is an ordinal α and a bijection $f : \alpha \leftrightarrow x$. Denote by ZL Zorn's lemma, i.e. the statement that, for every partially ordered set $(X, \leq)$ in which every ascending chain has an upper bound, there is a $\leq$-maximal element in X. Finally, denote by HMP the Hausdorff maximality principle, i.e. the statement that, for every partially ordered set $(X, \leq)$, every totally ordered subset is contained in a $\subseteq$-maximal totally ordered subset.*

Remark: HMP can be seen as the combinatorial core behind the proofs of ZL from AC or WO. For effective reducibility, it is the more interesting formulation: For if ZL holds, one can obtain the set of maximal elements of a p.o.-set $(X, \leq)$ satisfying the conditions of ZL by merely searching through X.[7]

---

[7] This does not imply that ZL is effective, though, as there is no effective way to effectively choose one element from the set of all maximal elements.

It is not hard to see that all these principles are equivalent in the sense of reducibility: The usual equivalence proofs explain, modulo a transfinite version of Church's thesis, how each of these principles can be reduced to any other. However, for gW-reducibility, we can use Lemma 6 to show that (under $0^\sharp$) the well-ordering principle is not generalized Weihrauch reducible to the axiom of choice:

Theorem 10. *If $0^\sharp$ exists, then $WO \nleq_{gW} AC$.*

Proof. (Sketch) In Theorem D.-A.C. of [Z], it is shown how to construct a transitive model of $ZF^- + AC + \neg WO$ as a union of an ascending chain of symmetric extensions of a transitive ground model M of ZF^-. Starting with $M = L$, it is easily checked that, under the assumption that $0^\sharp$ exists, the construction leads to a definable transitive class model N of $ZF^- + AC$ such that some set $A \in N$ that is non-wellorderable in N is countable in V and moreover $\mathbb{P}_A$ is countable and thus has two mutually generic filters over N. Hence the assumptions of Lemma 6 are satisfied and the non-reducibility follows.

Many other relations between choice principles are effective, however:

Theorem 11. *We have (1) $AC' \equiv_{gW} AC \leq_{gW} ZL$ and (2) $ZL \leq_{gW} WO$*

Proof. The usual equivalence proofs over ZF (see e.g. [Je]) effectivize.

Remark: We do not know how HMP relates to AC in terms of $\leq_{gW}$. We suspect that HMP $\nleq_{gW}$ AC. Our current state of knowledge (combined with our remark on HMP above) hence gives some meaning to the humorous claim that 'The Axiom of Choice is obviously true, the well-ordering principle obviously false, and who can tell about Zorn's lemma?'[8].

5   Conclusion and Further Work

We have seen how our framework can be used to distinguish between various versions of set-theoretical principles usually regarded as equivalent.

One may now ask for many statements which are effectively reducible or gW-reducible to which others. This may be viewed as a cardinality-independent version of reverse mathematics (see [Sh]) and the theory of the Weihrauch lattice. Apart from that, it may be interesting to consider variants with OTM-computability replaced by other models like ITTMs [HL] or parameter-OTMs, along with variations of Weihrauch reducibility, such as strong Weihrauch reducibility.

Acknowledgements. We thank our three anonymous referees for suggesting various corrections and improvements to our work.

---

[8] See [Kr], where this quote is attributed to Jerry Bona.

References

[BGM] Brattka, V., Gherardi, G., Marcone, A.: The Bolzano-Weierstraß theorem is the jump of weak König's lemma. Ann. Pure Appl. Log. **163**(6), 623–655 (2012)

[Ca] Carl, M.: Approach to a church-turing-thesis for infinitary computations. Preprint. arXiv: 1307.6599

[Ga] Galeotti, L.: Computable analysis over the generalized Baire space. M.Sc. thesis, Amsterdam (2015)

[GH] Gitman, V., Hamkins, J., Johnstone, T.: What is the theory ZFC without power set? Math. Log. Q., Preprint (to appear). arXiv: 1110.2430

[HJ] Hirschfeldt, D., Jockusch, C.: On notions of computability theoretic reduction between Π_2^1-principles. Preprint. http://www.math.uchicago.edu/drh/Papers/Papers/upaper.pdf

[HL] Hamkins, J., Lewis, A.: Infinite time turing machines. J. Symb. Log. **65**(2), 567–604 (2000)

[Ho2] Hodges, W.: On the effectivity of some field constructions. Proc. London Math. Soc. **32**, 133–162 (1976)

[Je] Jech, T.: Set Theory, Third Millenium edn. Springer, Heidelberg (2003)

[Ko1] Koepke, P.: Turing computations on ordinals. Bull. Symb. Log. **11**, 377–397 (2005)

[Kr] Krantz, S.: The axiom of choice. In: Handbook of Logic and Proof Techniques for Computer Science, pp. 121–126. Birkhäuser, Boston (2002)

[Sh] Shore, R.A., Mathematics, R., Countable, U.: Effective mathematics of the uncountable. In: Greenberg, N., Hamkins, J., Hirschfeld, D., Miller, R. (eds.) Lecture Notes in Logic. Cambridge University Press, Cambridge (2013)

[Z] Zarach, A.: Unions of ZF$^-$-models that are themselves ZF$^-$-models. Stud. Logic Found. Math. **108**, 315–342 (1982)

Lightface Π_3^0-Completeness of Density Sets Under Effective Wadge Reducibility

Gemma Carotenuto[1(✉)] and André Nies[2]

[1] Dipartimento di Matematica, Universita di Salerno, Fisciano, Italy
gcarotenuto@unisa.it
[2] Department of Computer Science, University of Auckland, Auckland, New Zealand

Abstract. Let $\mathcal{A} \subseteq {}^\omega 2$ be measurable. The density set $D\mathcal{A}$ is the set of $Z \in {}^\omega 2$ such that the local measure of $\mathcal{A}$ along Z tends to 1. Suppose that $\mathcal{A}$ is a Π_1^0 set with empty interior and the uniform measure of $\mathcal{A}$ is a positive computable real. We show that $D\mathcal{A}$ is lightface Π_3^0 complete for effective Wadge reductions. This is an algorithmic version of a result in descriptive set theory by Andretta and Camerlo [1]. They show a completeness result for boldface Π_3^0 sets under plain Wadge reductions.

1 Introduction

We work in Cantor space ${}^\omega 2$ with the product measure μ. For a finite bit string s and a measurable set $\mathcal{A} \subseteq {}^\omega 2$, the local measure of $\mathcal{A}$ above s is

$$\mu_s(\mathcal{A}) = 2^{-|s|} \cdot \mu([s] \cap \mathcal{A}),$$

where $[s]$ is the clopen set in ${}^\omega 2$ consisting of all the extensions of s (this set is often denoted N_s in the literature). Let $D\mathcal{A}$ be the points Z of such that $\mathcal{A}$ has density 1 at Z, namely

$$D\mathcal{A} = \{Z \colon \lim_n \mu_{Z \restriction n}(\mathcal{A}) = 1\}.$$

We call *density set* every set of the form $D\mathcal{A}$, for some measurable A. We will be mainly interested in closed $\mathcal{A}$, in which case $D\mathcal{A} \subseteq \mathcal{A}$.

The Lebesgue density theorem for Cantor space states that if $\mathcal{A}$ is measurable, then almost every $Z \in \mathcal{A}$ is in $D\mathcal{A}$. This result has been the seed for recent investigations both in algorithmic randomness and in descriptive set theory.

In algorithmic randomness, density has been instrumental in solving the long-open "covering problem". Combining Bienvenu et al. [3] and Day and Miller [5] yielded the answer; for an overview see [2]. Khan [6] obtained a variety of results, in particular relating density for the reals with density for Cantor space. In a recent article, Myabe et al. [7] define density randomness of a point Z in Cantor space as the combination of Martin-Löf-randomness of Z and the property that $Z \in D\mathcal{P}$ for each Π_1^0 set $\mathcal{P}$ containing Z; they show the equivalence of density randomness with a number of notions stemming from effective analysis. Martin-Löf-randomness of Z does not necessarily imply that Z satisfies the effective

© Springer International Publishing Switzerland 2016
A. Beckmann et al. (Eds.): CiE 2016, LNCS 9709, pp. 234–239, 2016.
DOI: 10.1007/978-3-319-40189-8_24

version of Lebesgues's theorem; for instance, the least element of a non-empty Π_1^0 set $\mathcal{P}$ of ML-randoms is not in $D\mathcal{P}$.

In descriptive set theory Andretta and Camerlo [1] have analyzed the complexity of the density sets. It is easily seen that the $\mathbf{\Pi_3^0}$ pointclass is an upper bound for this study. For example, if $A = [s]$, then $DA = A$, which is closed. In [1, Sect. 7] the authors show that DA is $\mathbf{\Pi_3^0}$-complete with respect to Wadge reducibility in case that A has empty interior and positive measure. They also prove that density sets can have any complexity within $\mathbf{\Pi_3^0}$. The first author in [4] has conducted a similar study of the difference hierarchy over the closed sets in the setting of the real line with the Lebesgue measure.

The goal of this short paper is to connect the two approaches to Lebesgue's theorem. We give an algorithmic version of the result in [1]. We show that for any (lightface) Π_1^0 set $\mathcal{A} \subseteq {}^\omega 2$ with empty interior and measure $\mu\mathcal{A}$ a positive computable real, the density set $D\mathcal{A}$ is lightface Π_3^0 complete for algorithmic Wadge reductions.

To say that a real r is computable means that from a number $n \in \omega$ we can compute a rational that is within 2^{-n} of r. The algorithmic version of Wadge reducibility is as follows: for $\mathcal{C}, \mathcal{D} \subseteq {}^\omega 2$, we write $\mathcal{D} \leq_m \mathcal{C}$ if there is a total Turing functional Ψ such that $\mathcal{D} = \Psi^{-1}(\mathcal{C})$. Totality means that $\Psi(Y) \in {}^\omega 2$ for each Y; equivalently, $\Psi(Y;n)$ is obtained by evaluating a truth table computed from n on Y. Note that it is easy to construct such a set $\mathcal{A}$, for instance by a Cantor space version of the construction of a Cantor set of positive measure in the unit interval.

In our proof, while we import the basic combinatorics of the approach in [1], the details are more complicated because we need to build an algorithmic Wadge reduction. This is where we use the hypothesis that the measure of $\mathcal{A}$ is computable. It is not known at present whether this hypothesis is actually necessary. The other hypothesis (that the interior be empty) is of course necessary, as for instance shown by taking $\mathcal{A}$ to be the whole Cantor space.

2   Completeness for Lightface Point Classes

For the basics about arithmetical hierarchy see for example [9,10].

Definition 1. *Consider a lightface point class Γ in Cantor space, such as Π_2^0 or Π_3^0. We say that $\mathcal{C}$ is Γ-complete if $\mathcal{C}$ is in Γ, and for each $\mathcal{D} \in \Gamma$ we have $\mathcal{D} \leq_m \mathcal{C}$.*

The next result is folklore (see e.g. [8]). For the reader's benefit we provide the short proof.

Proposition 2. *The class $\mathcal{C}$ of all sequences with infinitely many 1's is Π_2^0-complete.*

Proof. The class $\mathcal{C}$ is clearly Π_2^0. Now suppose $\mathcal{D}$ is Π_2^0, so $\mathcal{D} = \bigcap_n \mathcal{G}_n$ with $\mathcal{G}_{n+1} \subseteq \mathcal{G}_n$ for a uniformly Σ_1^0 sequence $\langle \mathcal{G}_n \rangle_{n \in \omega}$. Define a total Turing functional

Ψ as follows. Given a sequence of bits Z, at stage t we use the first t bits of Z, and append one bit at the end of the output Ψ^Z. We append 0 unless we see at stage t that $[Z\restriction_t] \subseteq \mathcal{G}_n$ for the next n; in that case we append 1.

In the following we effectively identify $^{\omega \times \omega}2$ and $^\omega 2$ via the standard computable pairing function $\omega \times \omega \to \omega$. The following is presumably folklore.

Proposition 3. *The set $\mathcal{E} = \{Z \in {}^\omega 2 : \forall n \forall^\infty k\, Z(n,k) = 0\}$ is Π_3^0-complete.*

Proof. Clearly $\mathcal{E}$ is Π_3^0. Now suppose a given set $\mathcal{F}$ is Π_3^0, so $\mathcal{G} = \bigcap_n \mathcal{G}_n$ where $\mathcal{G}_n$ is uniformly Σ_2^0. By the previous proposition and the uniformity in $\mathcal{D}$ of its completeness part, for each n we effectively have a Turing functional Ψ_n such that $\mathcal{G}_n = \Psi_n^{-1}(^\omega 2 \backslash \mathcal{C})$. Define a total Turing functional $\Psi \colon {}^\omega 2 \to {}^{\omega \times \omega}2$ by

$$\Psi^Z(n,r) = \Psi_n^Z(r).$$

Clearly $Z \in \mathcal{F} \Leftrightarrow \forall n\, [Z \in \mathcal{G}_n] \Leftrightarrow \forall n\, [\Psi_n^Z \in {}^\omega 2 \backslash \mathcal{C}] \Leftrightarrow \Psi^Z \in \mathcal{E}$. $\quad\square$

3   Reaching the Maximal Complexity

Theorem 4. *Let $\mathcal{A} \subseteq {}^\omega 2$ be a Π_1^0 set with empty interior such that $\mu\mathcal{A} > 0$ and $\mu\mathcal{A}$ is a computable real. Then $D\mathcal{A}$ is Π_3^0-complete.*

We begin with some preliminaries. In the following s, t, u denote strings of bits. Note that if $\mu\mathcal{A}$ is a computable real then $\mu_s(\mathcal{A})$ is a computable real uniformly in s (see e.g. [9, 1.9.18]). Also, $\mu_s(\mathcal{A}) < 1$ for each s because $\mathcal{A}$ has empty interior.

Let $L(t) = \mu_t(\mathcal{A})$. We note that L is a computable martingale in the sense of algorithmic randomness. That is, $L \geq 0$, the "martingale equality" $L(s0) + L(s1) = 2L(s)$ holds for each string s, and $L(s)$ is a computable real uniformly in s; see e.g. [9, Chap. 7]. By hypothesis that $\mathcal{A}$ has empty interior, we have $L(s) < 1$ for each s.

We will show that the oscillation behaviour of L can be controlled sufficiently well in order to code the Π_3^0-complete set $\mathcal{E}$ from Proposition 3 into $D\mathcal{A}$. For $p \in \mathbb{N}$ let $\delta_p = 1 - 3^{-p}$. We write $\theta(s) = p$ if $\delta_{p-1} < L(s) < \delta_p$. We leave $\theta(s)$ undefined in case $L(s) = \delta_k$ for some k. In the following, when we write $\theta(s)$ we imply that $\theta(s)$ is defined. Observe that, if $\theta(X\restriction_n)$ is defined for each n and $\lim_n \theta(X \restriction n) = \infty$, then $X \in D(+\mathcal{A})$. Note that the binary relations $\{\langle s, p \rangle : \theta(s) = p\}$ and $\{\langle s, p \rangle : \theta(s) \geq p\}$ are Σ_1^0 because the martingale L is computable.

Lemma 5.

(i) (Increasing the value of L). Let $p < k$ and $\theta(s) = p$. There is $t \supset s$ such that $\theta(t) \geq k$ and $L(u) > \delta_{p-1}$ for each u with $s \subseteq u \subseteq t$.

(ii) (Decreasing the value of L). Let $p > q$ and $\theta(s) = p$. There is $t \supset s$ such that $\theta(t) = q$ and $L(u) > \delta_{q-1}$ for each u with $s \subseteq u \subseteq t$.

Proof. (i) By an application of the Lebesgue density theorem [1, Proposition 3.5] (where r there is δ_k), there exists a prefix minimal string $v \supset s$ such that $L(v) \geq \delta_k$ and $L(u) \geq L(s)$ for each u with $s \subseteq u \subseteq v$. If $\theta(v)$ is defined, the string $t = v$ is as required. Otherwise we have $L(v) = \delta_m$ for some $m \geq k$. By the Lebesgue density theorem, L is not constant on the set of extensions of v, so one can choose a prefix minimal string $w \supseteq v$ such that $L(w0) \neq L(w1)$. By the minimality of w we have $L(v') = \delta_m$ for each v' with $v \subseteq v' \subseteq w$.

First suppose that $L(w0) > L(w)$. Since $L(w0) < 1$, we have $L(w0) - L(w) < 3^{-m}$. Hence $L(w) - L(w1) = L(w0) - L(w) < 3^{-m}$ by the martingale equality, and thus $\delta_{m-1} < L(w1) < \delta_m$ because $\delta_{m-1} = \delta_m - 2 \cdot 3^{-m}$. Then $\theta(w1) = m \geq k$, so the string $t = w1$ is as required. If $L(w1) > L(w)$ instead, then $t = w0$ is as required by an analogous argument.

(ii) As $L(s) < 1$, by the Lebesgue density theorem there exists a prefix minimal $t \supset s$ such that $L(t) < \delta_q$. Let $t = vb$ where $b \in \{0, 1\}$. Since $L \leq 1$ the martingale equality for $1 - L$ implies that $2(1 - L(v)) \geq 1 - L(t)$. So $L(t) \leq \delta_{q-1}$ would imply $2(1 - L(v)) \geq 1 - L(t) \geq 3 \cdot 3^{-q} > 2 \cdot 3^{-q}$, and hence $L(v) < \delta_q$ contrary to the minimality of t. Hence $\theta(t) = q$ and $L(u) > \delta_{q-1}$ for each u with $s \subseteq u \subseteq t$, as required. This establishes the lemma.

Remark 6. By the remarks on θ above and since L is computable, all the conditions in the Lemma are Σ_1^0. Thus in (i) given a string s and numbers $p < k$, if $\theta(s) = p$ we can run a search for the string t. So the function $s, p, k \mapsto t$ is partial computable and defined whenever $\theta(s) = p < k$. A similar remark applies to (ii).

Proof (of Theorem 4). By Proposition 3, the set

$$\mathcal{E} = \{Z \in {}^{\omega \times \omega}2 \mid \forall q \forall^\infty n[Z(q, n) = 0]\}$$

is Π_3^0-complete; thus, it will be enough to show that $\mathcal{E} \leq_m D(\mathcal{A})$. In the following let a, b denote bit-valued square matrices. Given such a matrix a, we will compute a string $s = \psi(a) \in {}^{<\omega}2$ in such a way that $\theta(s)$ is defined. We ensure that $a \subseteq b \to \psi(a) \subseteq \psi(b)$. Then the function

$$\Psi : {}^{\omega \times \omega}2 \to {}^{\omega}2, \qquad \Psi(Z) = \bigcup_n \psi(Z \restriction n \times n)$$

is a total Turing functional.

Defining ψ. The definition of ψ is by recursion, using Lemma 5 and Remark 6. If $\mu\mathcal{A} > 1/2$, after removing from $\mathcal{A}$ either the elements of Cantor space starting with 0 or the elements starting with 1, we may assume that $\mu\mathcal{A} \leq 1/2$. So we may assume that $\theta(\emptyset)$ is defined. We set $\psi(\emptyset) = \emptyset$.

Now suppose that $\phi(a)$ has been defined for each $n \times n$ bit-valued matrix a in accordance with the conditions above. Given a matrix $b = \langle a(i, j) \mid i, j < n+1 \rangle$, let $s = \psi(a)$ where $a = b \restriction n \times n$. Then $p = \theta(s)$ is defined.

If there is a $q \leq n$ such that $b(q, n) = 1$, choose q least.

- If $q < p$, via (ii) of Lemma 5 compute a string $t \supset s$ such that $\theta(t) = q$ and $L(u) > \delta_{q-1}$ for each u with $s \subseteq u \subseteq t$, and define $\psi(b) = t$.

- If $q \geq p$, or there is no such $q \leq n$ at all, let $k = \max(p+1, n+1)$. Via (i) of Lemma 5 compute a string $t \supset s$ such that $\theta(t) \geq k$ and $L(u) > \delta_{p-1}$ for each u with $s \subseteq u \subseteq t$, and define $\psi(b) = t$. This completes the recursion step.

We verify that $\mathcal{E} \leq_m D\mathcal{A}$ via Ψ. First suppose that $Z \notin \mathcal{E}$. Let q be least such that $\exists^\infty n[Z(q,n) = 1]$. Then for arbitrarily large n, we have $\theta(\psi(Z \restriction n \times n)) = q$, hence $\theta(\Psi(Z) \restriction r) = q$ for infinitely many r. Therefore $\Psi(Z) \notin D\mathcal{A}$.

Now suppose that $Z \in \mathcal{E}$. Then for every $q \in \omega$, there is m_q such that $\forall m \geq m_q[Z(q,m) = 0]$. For $r \in \omega$, let $n_r = \max\{m_0, \ldots, m_r, r\}$. Note that for each $n \geq n_r$, if $Z(q,n) = 1$ then $q > r$.

The following claim will show that $\Psi(Z) \in D\mathcal{A}$.

Claim. Given $n > n_r$, let $s = \psi(Z \restriction n \times n)$ and $t = \psi(Z \restriction (n+1) \times (n+1))$.

For each string u with $s \subseteq u \subseteq t$, we have $L(u) \geq \delta_r$.

To see this, we prove inductively that $\theta(\psi(Z \restriction n \times n)) > r$ for each $n > n_r$. For the start of the induction at $n_r + 1$, suppose that $s = \psi(Z \restriction n_r \times n_r)$ has just been defined and consider the next step of the definition of ψ along Z. The least possible value of q is $r + 1$. Since $n_r \geq r$, no matter whether we apply (ii) or (i) of the lemma we ensure that $\theta(t) > r$ and hence $L(t) > \delta_r$, where $t = \psi(Z \restriction (n_r + 1) \times (n_r + 1))$.

For the inductive step, suppose that $n > n_r$ and for $s = \psi(Z \restriction n \times n)$ we have $\theta(s) > r$. Let $t = \psi(Z \restriction (n+1) \times (n+1))$. Again the least possible value of q is $r + 1$. If we apply (ii) of the lemma then $\theta(t) > r$ and $L(u) > \delta_r$ for each u with $s \subseteq u \subseteq t$. If we apply (i) then, where $\theta(s) = p > r$, we have $\theta(t) \geq p + 1$ and $L(u) > \delta_r$ for each u with $s \subseteq u \subseteq t$. This completes the claim.

We conclude that for each r, we have $L(\Psi(Z) \restriction m) \geq \delta_r$ for sufficiently large m. Hence $\Psi(Z) \in D\mathcal{A}$.

Theorem 4 leaves open the following question. Is there a Π_1^0 class with empty interior and non-computable measure such that the density set is not Π_3^0-complete?

References

1. Andretta, A., Camerlo, R.: The descriptive set theory of the Lebesgue density theorem. Adv. Math. **234**, 1–42 (2013)
2. Bienvenu, L., Day, A., Greenberg, N., Kučera, A., Miller, J., Nies, A., Turetsky, D.: Computing K-trivial sets by incomplete random sets. Bull. Symb. Logic **20**, 80–90 (2014)
3. Bienvenu, L., Greenberg, N., Kučera, A., Nies, A., Turetsky, D.: Coherent randomness tests and computing the K-trivial sets. J. Eur. Math. Soc. **18**, 773–812 (2016)
4. Carotenuto, C.: On the topological complexity of the density sets of the real line. Ph.D. thesis, Università di Salerno (2015). http://elea.unisa.it:8080/jspui/bitstream/10556/1972/1/tesi_G.Carotenuto.pdf
5. Day, A.R., Miller, J.S.: Density, forcing and the covering problem. Math. Res. Lett. **22**(3), 719–727 (2015)

6. Khan, M.: Lebesgue density and Π_1^0-classes. J. Symb. Logic, to appear
7. Miyabe, K., Nies, A., Zhang, J.: Using almost-everywhere theorems from analysis to study randomness. Bull. Symb. Logic, (2016, to appear)
8. Montagna, F., Sorbi, A.: Creativeness and completeness in recursion categories of partial recursive operators. J. Symb. Logic **54**, 1023–1041 (1989)
9. Nies, A.: Computability and randomness. Oxford Logic Guides, vol. 51. Oxford University Press, Oxford (2009). 444 pp. Paperback version 2011
10. Odifreddi, P.: Classical Recursion Theory, vol. 1. NorthHolland Publishing Co., Amsterdam (1989)

Program Size Complexity of Correction Grammars in the Ershov Hierarchy

John Case[1] and James S. Royer[2](✉)

[1] Department of Computer and Information Sciences,
University of Delaware, Newark, DE 19716-2586, USA
case@udel.edu
[2] Department of Electrical Engineering and Computer Science,
Syracuse University, Syracuse, NY 13244, USA
jsroyer@syr.edu

Abstract. A general correction grammar for a language L is a program g that, for each $(x,t) \in \mathbb{N}^2$, issues a yes or no (where when $t = 0$, the answer is always no) which is g's t-th approximation in answering "$x \in L$?"; moreover, g's sequence of approximations for a given x is required to converge after *finitely many* mind-changes. The set of correction grammars can be transfinitely stratified based on O, Kleene's system of notation for constructive ordinals. For $u \in O$, a u-correction grammar's mind changes have to fit a count-down process from ordinal notation u; these u-correction grammars capture precisely the Σ_u^{-1} sets in Ershov's hierarchy of sets for Δ_2^0. Herein we study the relative succinctness between these classes of correction grammars. *Example:* Given u and v, transfinite elements of O with $u <_o v$ (Kleene's ordering on O), for each $\emptyset^{(2)}$-computable $H \colon \mathbb{N} \to \mathbb{N}$, there is a v-correction grammar i_v for a finite (alternatively, a co-finite) set A such that the smallest u-correction grammar for A is $> H(i_v)$. We also exhibit relative succinctness progressions in these systems of grammars and study the "information-theoretic" underpinnings of relative succinctness. Along the way, we verify and improve slightly a 1972 conjecture of Meyer and Bagchi.

1 Correction Grammars and Relative Succinctness

Burgin [5] suggested that a person's knowledge of a language L may involve his/her storing a representation of L in terms of *two* grammars, say g_1 and g_2, where g_1's language over-generalizes L and g_2 is used to "edit" errors of (make corrections to) g_1. Thus, $L = (L_1 - L_2) = \{\, x \in L_1 \mid x \notin L_2 \,\}$, where L_i is the language generated by the grammar g_i. The pair $\langle g_1, g_2 \rangle$ can thus be seen as a single description of (or "grammar" for) the language L. Burgin called such pairs *grammars with prohibition*; we prefer the term *correction grammars*. These correction grammars may be seen as modeling self-correcting human behavior.[1]

---

[1] For more discussion, see [6].

© Springer International Publishing Switzerland 2016
A. Beckmann et al. (Eds.): CiE 2016, LNCS 9709, pp. 240–250, 2016.
DOI: 10.1007/978-3-319-40189-8_25

In the computability-theoretic context of the present paper, we first think of p as being a *correction grammar* for L if and only if $p = \langle i, j \rangle$ and $L = (W_i - W_j)$.[2] Hence, such a correction grammar $\langle i, j \rangle$ is an *index* from [18] for the *difference of c.e. sets*, i.e., for the *d.c.e. set*, $(W_i - W_j)$. Correction grammars are more powerful than ordinary c.e. grammars in the sense that there are languages (e.g., $L = \{ x \mid x \notin W_x \}$) that correction grammars can describe that ordinary c.e. grammars cannot.

This paper's focus is on a different advantage correction grammars have over c.e. grammars: *relative succinctness*. That is, correction grammars can provide *vastly* shorter descriptions of certain *c.e. languages* (in our cases, finite and co-finite languages) than can any c.e. grammar. Here is a sample result.

Theorem 1. *Suppose* $H \colon \mathbb{N} \to \mathbb{N}$ *is* $\emptyset^{(2)}$-*computable and* $W_{i_1} = \mathbb{N}$. *Then there is a c.e. grammar* i_2 *with* W_{i_2} *co-finite (hence,* $(W_{i_1} - W_{i_2})$ *is finite, and, thus, c.e.) such that* $\min\{ i \mid W_i = (W_{i_1} - W_{i_2}) \} > H(\langle i_1, i_2 \rangle)$; *that is, when one magnifies the numerical value of the correction grammar* $\langle i_1, i_2 \rangle$ *by* H, *the result is still smaller than the numerical value of any c.e. grammar for the finite language named by* $\langle i_1, i_2 \rangle$.

Scholium 2 (On Theorem 1).

(a) It is especially interesting to consider the cases where Theorem 1's function H is large valued and fast growing; then the value of $H(\langle i_1, i_2 \rangle)$ is a vast magnification of the value of $\langle i_1, i_2 \rangle$. Hence, thanks to the theorem, the minimum c.e. grammar for $(W_{i_1} - W_{i_2})$ is huge compared its correction grammar $\langle i_1, i_2 \rangle$.

(b) For the magnification function H, one can concretely take $H = \lambda x.10^{20^{30^x}}$ or $H = \lambda x. Ack(x + 100, x + 100)$ where Ack is Ackermann's function [20]. However, these are tame choices as there are $\emptyset^{(2)}$-computable functions that dominate every $\emptyset^{(1)}$-computable function. (f *dominates* $g \Leftrightarrow_{\mathrm{def}}$ $\{ x \mid g(x) < f(x) \}$ is co-finite.)

(c) In effect, the theorem identifies the *size* of the c.e. grammar i with the number i itself and the size of the correction grammar $\langle i_1, i_2 \rangle$ with the number $\langle i_1, i_2 \rangle$. As explained in Sect. 2, a consequence of Theorem 1 is that the analogous result holds for a wide choice of size-measurement schemes for grammars.

(d) Theorem 1's $\langle i_1, i_2 \rangle$ can be computed from a relativized program for H. Most of our relative succinctness results are uniformly algorithmic in this sense.

(e) For simplicity, Theorem 1 asserts that there is *a* grammar (e.g., $\langle i_1, i_2 \rangle$) witnessing the relative succinctness. In fact we can construct an infinite c.e. set of such witnesses for Theorem 1 (and analogously for Corollary 4 and Theorems 6 and 7 below).

---

[2] *A Dollop of Standard Terminology*: W_i is the i-th c.e. set, where i codes a program for generating or for accepting W_i [30]. $\langle \cdot, \cdot \rangle$ is a pairing function, i.e., a computable isomorphism from $\mathbb{N} \times \mathbb{N}$ to $\mathbb{N}$ [30], where $\mathbb{N} = $ the natural numbers. For $A \subseteq \mathbb{N}$, the A-*computable* (respectively, *partial A-computable*) functions are the total (respectively, partial) functions over $\mathbb{N}$ that are computable relative to oracle A. W_i^A is the i-th A-c.e. set, where i codes a relativized program that, with oracle A, generates or accepts W_i^A [30]. For each $k \in \mathbb{N}$, let $A^{(k)}$ be the k-th *jump* [30] of A, i.e.: $A^{(0)} = A$ and $A^{(i+1)} = \{ x \mid x \in W_x^{A^{(i)}} \}$. Thus, $\emptyset^{(1)} = $ the halting problem, $\emptyset^{(2)} = $ the jump of $\emptyset^{(1)}$, etc. For $A \subseteq \mathbb{N}$, $\overline{A} = \mathbb{N} - A$. Any unexplained terminology below is from [30].

(f) Theorem 1 turns out to *fail* if one attempts to strengthen it by allowing H to be an arbitrary $\emptyset^{(3)}$-computable function.

2   A Digression on Grammar Size Measures

Before discussing extensions of Theorem 1 we first consider how versions of these theorems still hold under a wide range of size-measurement schemes.

Definition 3. For a given oracle A, an *A-computable Blum program size measure* (or simply, *A-computable size measure*) is a finite-to-one, A-computable $s \colon \mathbb{N} \to \mathbb{N}$ such that $b_s = \lambda n. \max\{ i : s(i) \le n \}$ is also A-computable.

The $\emptyset$-computable size measures match Blum's original notion of program size measure [2]. These include $s(i) = i$, $s(i) = |i| =$ the length of the reduced binary representation of i (i.e., no redundant, leading 0s), and most standard ways of measuring program (or grammar) size. For some less-standard size measurements, we need a bit of terminology: For each $A \subseteq \mathbb{N}$, let U^A be an additively optimal universal partial A-computable function [22]. (Such an U^A exists by a straightforward relativization of Lemma 2.1.1 in [22].) Define $C^A = \lambda x. \min\{ |p| : U^A(p) = x \}$. $C^A(x)$ is the *A-Kolmogorov complexity of x*. Now, for each A, C^A is an $A^{(1)}$-computable size measure. E.g., $C^\emptyset$, the standard notion of (plain) Kolmogorov complexity, is an $\emptyset^{(1)}$-computable size measure.

It turns out that all of our $\emptyset^{(k)}$-relative-succinctness results in this paper (e.g., Theorem 1 for $k = 2$) still hold under *any* choice of $\emptyset^{(k)}$-computable size measures. So for example, Theorem 1 generalizes to:

Corollary 4. *Suppose $H \colon \mathbb{N} \to \mathbb{N}$ is $\emptyset^{(2)}$-computable, s_{ce} and s_{dce} are any $\emptyset^{(2)}$-computable size measures, and $W_{i_1} = \mathbb{N}$. Then there is a c.e. grammar i_2 with W_{i_2} co-finite and $\min\{ s_{ce}(i) : W_i = (W_{i_1} - W_{i_2}) \} > H(s_{dce}(\langle i_1, i_2 \rangle))$.*

Scholium 5 (On Corollary 4). Taking $s_{ce} = s_{dce} = C^{\emptyset^{(1)}}$ and $H = $ a large, fast-growing $\emptyset^{(2)}$-computable function in Corollary 4 provides some insight into the trade-offs behind relative succinctness. As measured by $C^{\emptyset^{(1)}}$, $\min\{ C^{\emptyset^{(1)}}(i) : W_i = (W_{i_1} - W_{i_2}) \}$ is the least amount of $\emptyset^{(1)}$-algorithmic information in any c.e. grammar for $(W_{i_1} - W_{i_2})$ which, by Corollary 4, is H-enormous compared to $s_{dce}(\langle i_1, i_2 \rangle)$.

(a) Thus, intuitively, a necessary price a then witnessing correction grammar $\langle i_1, i_2 \rangle$ pays for its small size is that $\langle i_1, i_2 \rangle$ is enormously deficient in $C^{\emptyset^{(1)}}$-information compared to that of any c.e. grammar for $(W_{i_1} - W_{i_2})$.

(b) Such a witness $\langle i_1, i_2 \rangle$ would have its own information minimality if $C^{\emptyset^{(1)}}(\langle i_1, i_2 \rangle)$ just happened to $= \min\{ C^{\emptyset^{(1)}}(\langle j_1, j_2 \rangle) : (W_{j_1} - W_{j_2}) = (W_{i_1} - W_{i_2}) \}$. However, as mentioned in Scholium 2(e), we can produce an infinite c.e. set, A, of witnesses, $\langle i_1, i_2 \rangle$, to Corollary 4, and, by straightforward relativization of well-known immunity properties of sets of minimal indices [25], all but finitely many elements of this A are *not* such $C^{\emptyset^{(1)}}$-minimal size d.c.e. grammars.

Section 5 below further discusses some information-theoretic aspects of relative succinctness.

3   The Ershov Hierarchy and Relative Succinctness

Versions of Theorem 1 hold for generalizations of the notion of correction grammar. One natural extension is to consider finite differences of c.e. languages. That is, for $k > 0$, define a k-fold correction grammar as being a number of the form $\langle i_1, \ldots, i_k \rangle$ that names the language $(W_{i_1} - (W_{i_2} - \cdots (W_{i_{k-1}} - W_{i_k}) \cdots))$. The languages named by k-fold correction grammars are called the k-c.e. sets [36]; they formalize the notion of a finite, fixed number of successive edits for errors. The Ershov hierarchy [12–14] includes and extends this basic idea into the transfinite by means of constructive ordinals as discussed below.

Ordinals are representations of well-orderings. A *constructive ordinal* is, roughly, an ordinal, α, for which there is a program, called a *notation*, that specifies how to build α algorithmically or lay α out end-to-end. Herein we use Kleene's well-known general ordinal notation system O (coded as a proper subset of $\mathbb{N}$) [21,30]. Each constructive ordinal has at least one O-notation. For each $k \in \mathbb{N}$, let $\underline{k}$ = the unique O-notation for the ordinal k. Kleene also introduced an order relation $<_o$ on O-notations that naturally embeds into the ordering of the corresponding constructive ordinals [21,30], e.g., $\underline{0} <_o \underline{1} <_o \underline{2} <_o \ldots <_o w$, where w is any O-notation for ω (the ordinal for $\mathbb{N}$ under the ordering: $0 < 1 < 2 < \ldots$).

Our most general formulation of correction grammars is based on the idea of using Kleene's O-notations to bound the number of corrections that such a grammar can make. To do this rigorously, we make use of the technique of algorithmic counting-down from O-notations. For a notation $u \in O$, our u-*correction grammars* are each algorithms for counting-down corrections from u.

For $u \in O$, here is *roughly* how level-u correction grammars work. Choose *uniform programs* c, d where c is for counting-down from u (along $<_o$), and where d, for each time t and input x, decides whether to exclude or include x at time t—but, initially, i.e., at $t = 0$, d excludes (independently of x). With each mind-change of d—about excluding or including a given x, c must walk/leap strictly further down u (along $<_o$). The correction grammar $p = \langle c, d \rangle$ accepts the set, W_p^u, of all x that d eventually includes—once c counts down from u no more. By the well-ordering property of ordinals, such a count-down sequence must be *finite* and, hence, such a grammar makes only finitely many corrections on excluding/including any given x.

Examples: For $k \in \mathbb{N}$, counting-down corrections from $\underline{k}$ is equivalent to an algorithm that initially excludes each item x and, then, the algorithm can change its mind about x's inclusion or exclusion up to k times on the way to giving its final, correct answer as to whether x is included or excluded. When $k = 1$, such algorithms are grammars for the c.e. sets; when $k = 2$, such algorithms are grammars for the d.c.e. sets; and in general, for a fixed k, they provide grammars for the k-c.e. sets.[3] It can be shown that counting-down corrections allowed from any notation for ω is equivalent to declaring algorithmically, at the time a first correction is made, a finite numerical bound on the number of *further* corrections to be allowed. This is more powerful than just initially setting a *fixed, finite* number of corrections allowed. As another example consider using a notation for the ordinal $\omega + \omega$ (two copies of ω laid end to end), also constructive and transfinite: in this case, when the first correction is made, the algorithm declares a finite bound on the number of further corrections it will make; this bound is, however, allowed to be changed once, at a later time. For a notation for the constructive ordinal $\omega + \omega + \omega$, the algorithm is allowed to update the bound twice. For at least natural notations for the constructive ordinal ω^2, the algorithm is allowed to make a finite number of changes to the bound, where the maximum number of changes allowed to the bound is announced when the algorithm makes the first correction!

Each W_p^u as sketched above is manifestly a u-c.e. set from the Σ_u^{-1} level in the general Ershov Hierarchy [12–14]. Hence, u-correction grammars provide a motivation for studying the u-c.e. sets.[4] It turns out that each u-c.e. set has a correction grammar as above, and, moreover, that $\langle W_p^u \rangle_{p \in \mathbb{N}}$ is an acceptable indexing of Σ_u^{-1}.[5] The details of the W_p^u's definition will appear in the full paper.

Here is our lift of Theorem 1 to our general formulation of correction grammars.

Theorem 6. *Suppose $\underline{0} <_o u <_o v \in O$, $H : \mathbb{N} \to \mathbb{N}$ is $\emptyset^{(2)}$-computable, and s_u and s_v are $\emptyset^{(2)}$-computable size measures. Then (a) and (b) just below each hold.*

(a) If v is transfinite or else finite and even, then there is an e with W_e^v finite such that $\min\{ s_u(p) : W_p^u = W_e^v \text{ or } W_p^u = \overline{W_e^v} \} > H(s_v(e))$.

(b) If v is transfinite or else finite and odd, then there is an e with W_e^v co-finite such that $\min\{ s_v(p) : W_p^u = W_e^v \text{ or } W_p^u = \overline{W_e^v} \} > H(s_v(e))$.

It turns out that if v is finite and odd, then the conclusion of Theorem 6(a) is *false*; and similarly with Theorem 6(b) for v finite and even. We note that these odd/even provisos go away if the theorem's H is restricted to be $\emptyset^{(1)}$-computable.

Our proof of Theorem 6 is a succinct and indirect combination of Ershov-Hierarchy refinements of index-set degree-calculations of Rogers [30] together with recursion theorems. The next section has a sample proof in this style, but where the refinement of Rogers' index-set degree-calculations is only a relativization.

---

[3] For each $k \in \mathbb{N}$, the system of $\underline{k}$-correction grammars turns out to be equivalent to the system of k-fold correction grammars introduced above.

[4] Algorithmic counting-down from constructive ordinals has been used, for example, in: computability theory [1,7,11], proof theory [29,38], term rewriting [4,40], and computational learning theory [15].

[5] $\langle W_p^u \rangle_{p \in \mathbb{N}}$'s acceptability provides s-m-n and recursion theorems that are needed in some of our proofs.

Let w be an O-*notation* for ω. While our w-c.e. sets are precisely Σ_w^{-1}, they are distinct from the well known *omega*-c.e. sets [37] (each of which has a computable approximation analogous to an w-c.e. set *except* the number of mind changes is bounded by some computable function). The *omega*-c.e. sets are in $\Delta_w^{-1} \subsetneq \Sigma_w^{-1}$.

4   Confirmation of the Meyer and Bagchi Conjecture

In 1972 Meyer and Bagchi [25] conjectured a (non-Ershov Hierarchy) succinctness result that, by adapting Theorem 6's proof, we can confirm (and improve[6]) in:

Theorem 7. *Let A and B be sets with $A^{(1)} \leq_T B$ and let s_A and s_B be $B^{(2)}$-computable size measures. Then, given any $B^{(2)}$-computable function H, one can compute an e (from a $\varphi^{B^{(2)}}$-program for H) such that W_e^B is co-finite and $\min\{ s_A(p) : W_p^A = W_e^B \} > H(s_B(e))$.*

Proof.[7] Since $A^{(1)} \leq_T B$, we have $\Sigma_1^{0,A} \cup \Pi_1^{0,A} \subseteq \Delta_1^{0,B}$. By standard results, $\{ \langle p,q \rangle : W_p^A \neq W_q^B \} \in \Sigma_2^{0,B}$. Theorem 14-XV's proof in [30] relativizes to give: $(B^{(3)}, \overline{B^{(3)}}) \leq_m (\{ i : W_i^B \text{ is co-finite} \}, \{ i : W_i^B \notin \Delta_1^{0,B} \})$. Let r witness this reduction. As s_A is a $B^{(2)}$-computable size measure, by Definition 3, $b_{s_A} = \lambda n. \max\{ i : s_A(i) \leq n \}$ is also $B^{(2)}$-computable. By the recursion theorem for $\varphi^{B^{(2)}}$, given a $\varphi^{B^{(2)}}$-program for H, we can compute an i_0 such that

$$\varphi_{i_0}^{B^{(2)}} = \lambda x. \begin{cases} 0, & \text{if } (\forall p \leq b_{s_A}(H(s_B(r(i_0)))))[W_p^A \neq W_{r(i_0)}^B]; \\ \uparrow, & \text{otherwise.} \end{cases}$$

Let $e = r(i_0)$. *Case 1:* $(\exists p \leq b_{s_A}(H(s_B(e))))[W_p^A = \overline{W_e^B}]$. Then $\varphi_{i_0}^{B^{(2)}}(i_0)\uparrow$. Thus, $W_e^B \notin \Delta_1^{0,B}$, and so, $W_e^B \notin \Pi_1^{0,A}$ contradicting Case 1. Hence, we must have *Case 2:* $(\forall p \leq b_{s_A}(H(s_B(e))))[W_p^A \neq \overline{W_e^B}]$. Then $\varphi_{i_0}^{\emptyset^{(2)}}(i_0)\downarrow$. Hence, W_e^B ($= W_{r(i_0)}^B$) is co-finite. Also, since $\max\{ p : s_A(p) \leq H(s_B(e)) \} = b_{s_A}(H(s_B(e)))$, we also have $(\forall p : s_A(p) \leq H(s_B(e)))[W_p^A \neq \overline{W_e^B}]$.   □

Theorem 7 can be shown by a *direct* (albeit involved) construction of a suitable W_e^B thus. By standard results, write H as the double limit of a B-computable function (with the inner limit also total). Double limits tend to *thrash*. To handle thrashing in W_e^B's construction, while crucially ensuring that $\overline{W_e^B}$ is *finite*, one can use a variant of Sacks' *restraint* trick [32] that allows a new number to be held out of W_e^B only when no higher priority concern holds a number out of W_e^B.

---

[6] Meyer and Bagchi's conjecture had s_A as a computable size measure and $s_B = \lambda p.p$.

[7] *Another dollop of terminology.* For partial function ψ, $\psi(x)\downarrow$ means that ψ is defined on x and $\psi(x)\uparrow$ means that ψ is not defined on x. Let $\langle \varphi_p^A \rangle_{p \in \mathbb{N}}$ be an acceptable programming system for the A-computable partial functions over $\mathbb{N}$. For this φ^A we have *(i) the s-1-1 theorem for φ^A:* There is a computable $s \colon \mathbb{N}^2 \to \mathbb{N}$ such that, for each $p, x, y \colon \varphi_{s(p,x)}^A(y) = \varphi_p^A(\langle x, y \rangle)$; and *(ii) the Kleene parametric recursion theorem for φ^A:* There is a computable $r \colon \mathbb{N}^2 \to \mathbb{N}$ such that, for each $p, x, y \colon \varphi_{r(p,x)}^A(y) = \varphi_p^A(\langle r(p,x), y \rangle)$. (A, B) *double m-reduces* [35] to (C, D) (written: $(A, B) \leq_m (C, D)$) if and only if there is an computable f such that, for each x, $[x \in A \Leftrightarrow f(x) \in C$ and $x \in B \Leftrightarrow f(x) \in D]$; such an f *witnesses* the reduction.

5   Why Are the Programs Succinct?

The succinct programs above are so, in part, because they are missing information. Scholium 5 above noted a link between relative succinctness and "information deficiency." Below is a *sample* of our results that explore this link. For *this* sample the notion of information deficiency is based on *unprovability*. Let T be a $\emptyset^{(2)}$-*computably axiomatized extension of Second Order Arithmetic (SOA)* [30,34] such that T's theorems expressible in SOA are true. E.g., $T =$ ZFC + all Δ_3^0-truths of *first*-order Arithmetic. T's being $\emptyset^{(2)}$-computably axiomatized makes its set of theorems a Σ_3^0-set by direct relativization of standard results [9,23,30]. T can thus be regarded as a $\emptyset^{(2)}$-*algorithmic extractor of SOA truth.*

Theorem 8. *Suppose (i) through (iv) just below.*

(i) T *is a $\emptyset^{(2)}$-computability axiomatized extension of SOA such that T's theorems expressible in SOA are true.*

(ii) $\underline{0} <_o u <_o v \in O$; $A \subseteq \Sigma_u^{-1}$; s_u *and* s_v *are $\emptyset^{(2)}$-computable size measures.*

(iii) *For each informal sentence E (translatable to SOA), we write $\langle\!\langle E \rangle\!\rangle$ for a naturally corresponding, fixed closed well-formed formula of SOA that semantically expresses E.*

(iv) *For each $\emptyset^{(2)}$-computable $H \colon \mathbb{N} \to \mathbb{N}$, there is an e_H with $W_{e_H}^v \in A$ and $\min\{\, s_u(p) \mathbin{\vdots} W_p^u = W_{e_H}^v \,\} > H(s_v(e_H))$.*

Then, for each sufficiently large, $\emptyset^{(2)}$-computable $H \colon \mathbb{N} \to \mathbb{N}$ and for each e_H as in (iv) just above, we have $T \nvdash \langle\!\langle u \, and \, v \in O \, and \, W_{e_H}^v \in \Sigma_u^{-1} \rangle\!\rangle$.

Theorem 8's conclusion just above provides an information deficiency of e_H, e.g., in many obvious cases where SOA $\vdash \langle\!\langle u$ and $v \in O \rangle\!\rangle$.

The base of T above can be first-order arithmetic [23,30]—*provided* that, roughly, we change $\langle\!\langle u$ and $v \in O \rangle\!\rangle$ above to $\langle\!\langle u$ and v settled$\rangle\!\rangle$, where *settled* is an approximation to O meaning that the limit, tied to our *uniform* count-down decision process, *converges.*[8]

6   Relative Succinctness Progressions

Fix a $\emptyset^{(2)}$-computable $H \colon \mathbb{N} \to \mathbb{N}$ and $\underline{0} <_o u_0 <_o u_1 <_o \dots <_o u_n \in O$. By Theorem 6 above, we know that, for each of $j = 1, \dots, n$, there is an e_j with $W_{e_j}^{u_j}$ finite or co-finite (subject to the even/odd conditions on the finite u_j's) with e_j H-more succinct than any $W^{u_{j-1}}$-program for $W_{e_j}^{u_j}$. It turns out that we can arrange for each of these $W_{e_j}^{u_j}$'s to be the same set or its complement. That is:

---

[8] One may wonder why $\langle\!\langle u$ and $v \in O \rangle\!\rangle$ (or the same thing but with 'settled' in place of '$\in O$') is in the *conclusion* of Theorem 8. The simple answer is that $\emptyset^{(2)}$ is not a strong enough oracle to remove either one from inside the $\langle\!\langle \cdots \rangle\!\rangle$ in the conclusion.

Theorem 9 (A sample succinctness progression). *Suppose H is $\emptyset^{(2)}$-computable and $\underline{0} <_o u_0 <_o u_1 <_o \cdots <_o u_n \in O$. Let $s_{u_0}, \ldots, s_{u_n}$ be $\emptyset^{(2)}$-computable size measures. For $j = 1, \ldots, n$, let op_j be an element of $\{\lambda A.A, \lambda A.\overline{A}\}$ that is chosen arbitrarily subject to two constraints: (i) $\mathsf{op}_j = \lambda A.A$ when u_j is finite and even, and (ii) $\mathsf{op}_j = \lambda A.\overline{A}$ when u_j is finite and odd. Then there are $e_1, \ldots, e_n$ and A, a finite set, such that (a) $\mathsf{op}_1(W_{e_1}^{u_1}) = \cdots = \mathsf{op}_n(W_{e_n}^{u_n}) = A$, and (b) for $j = 1, \ldots, n$: $\min\{s_{u_{j-1}}(p) \mid W_p^{u_{j-1}} = W_{e_j}^{u_j} \text{ or } W_p^{u_{j-1}} = \overline{W_{e_j}^{u_j}}\} > H(s_{u_j}(e_j))$.*

It is notable that a *single finite set* can witness the succinctness between each of the adjacent pairs in $\langle W_i^{u_0}\rangle_{i\in\mathbb{N}}, \ldots, \langle W_i^{u_n}\rangle_{i\in\mathbb{N}}$: *conceivably*, the sets witnessing the relative succinctness between distinct pairs must be very different—perhaps because the basis for relative succinctness between adjacent pairs are "orthogonal." Theorem 9 and related (but here not included) succinctness progression results show that this is not the case. We note that, as with Theorem 6, the theorem's odd/even provisos go away if the theorem's H is restricted to be $\emptyset^{(1)}$-computable. Theorem 9 is proved by a parameterized extension of the proof of Theorem 6.

7   Related Results

Blum in his seminal paper [2] provided a computability-theoretic, axiomatic treatment of *program size* and established the remarkable result that: for any computable function h, for *some* primitive recursive functions f, there are *general* recursive procedures for f which are h more succinct than *any* primitive recursive procedures for f. Hence, although one need not go beyond primitive recursions to compute such f, *judicious* use of more general recursion leads to programs for f which are *considerably*, i.e., h, smaller than any of those employing primitive recursions only! Drumm [10] and Constable [8] independently refined Blum's techniques to exhibit relative-succinctness results analogous to Blum's, but between successive levels of the LOOP-hierarchy [27].

The above results left open the possibility that the *functions* admitting more succinct programs might be pathological examples that no one would ever want to compute anyway. However, Meyer and Fischer [26] showed that push-down automata (PDAs) are computably more succinct than finite state automata for some *co-finite* sets, and Meyer [24] provided very general techniques for obtaining relative succinctness results for characteristic functions of *finite* sets! In particular, Meyer showed that *double* recursive procedures [28] are $\emptyset^{(1)}$-computably more succinct than primitive recursive procedures for some characteristic functions of finite sets. He was the first to notice that, in many cases, one can obtain programs that are more succinct by a $\emptyset^{(1)}$-computable amount.

Computable relative-succinctness results were shown by: Borodin [3] for context-sensitive grammars over context-free grammars (CFGs); Valiant [39] for ambiguous CFGs over deterministic CFGs; and Schmidt and Szymanski [33] for ambiguous CFGs over unambiguous CFGs. Hartmanis and Baker showed that, if P $\neq$ NP, then nondeterministic poly-time procedures are computably more succinct than deterministic poly-time procedures for some *finite* initial segments of a particular NP-complete set. By way of contrast, Hartmanis and Baker also showed that deterministic Turing Machines (TMs) that *provably* run in poly-time are computably more succinct than *nondeterministic* TMs that are explicitly clocked to run in poly-time. Hartmanis [16] proved that non-deterministic PDAs are computably more succinct than deterministic PDAs. Hay [19] showed how to improve from computable, to $\emptyset^{(1)}$-*computable*, the relative succinctness results mentioned above of [16,26,33,39]. She also showed that TMs are computably more succinct than finite state automata for *singleton* sets. Hartmanis [17] obtained a strong computability-theoretic, sufficient condition for computable relative succinctness between two programming systems. Royer and Case [31] constructed a systematic framework for relative succinctness between subrecursive programming systems, characterized when relative-succinctness phenomena occur between programming systems, and introduced the notion of relative-succinctness progressions (as in Theorem 9) and established a number of such progressions through a variety of subrecursive and complexity-bounded hierarchies.

Beyond the subrecursive context of the above results, Parikh [30, p. 216] showed that, for each computable $h\colon \mathbb{N} \to \mathbb{N}$, there is: (i) an i_1 with $W_{i_1} \in \Delta_1^0$ such that $\min\{\, p \mid W_p = \overline{W_{i_1}}\,\} > h(i_1)$, and (ii) an i_2 with $W_{i_2}^{\emptyset^{(1)}} \in \Sigma_1^0$ such that $\min\{\, p \mid W_p = W_{i_2}^{\emptyset^{(1)}}\,\} > h(i_2)$. Meyer and Bagchi [25] showed that, for sets A and B with $A^{(1)} \leq_T B$, for each $B^{(2)}$-computable H, there is an e such that $W_e^B \in \Sigma_1^A$, but $\min\{\, s_A(p) \mid W_p^A = W_e^B\,\} > H(s_B(e))$, where s_A and s_B are computable size measures; they conjectured that this W_e^B could be made co-finite. (Theorem 7 above confirms and improves slightly upon this conjecture.)

Acknowledgments. Thanks to Frank Stephan for alerting us to our earlier blunder of not noticing the need for even/odd cases in Theorem 6. Grant support was received by J. Case from NSF grant CCR-0208616, and by J. Royer from NSF grants CCR-0098198 and CCF-1319769.

References

1. Ash, C., Knight, J.: Recursive structures and Eshov's hierarchy. Math. Logic Q. **42**, 461–468 (1996)
2. Blum, M.: On the size of machines. Inf. Control **11**, 257–265 (1967)
3. Borodin, A.: Computational complexity: Theory and practice. In: Aho, A.V. (ed.) Currents in the Theory of Computing. Prentice-Hall, Englewood Cliffs (1973)
4. Buchholz, W.: Proof-theoretic analysis of termination proofs. Ann. Pure Appl. Logic **75**, 57–65 (1995)
5. Burgin, M.: Grammars with prohibition and human-computer interaction. In: Proceedings of the Business and Industry Simulation Symposium, pp. 143–147. Society for Modeling and Simulation International (2005)
6. Carlucci, L., Case, J., Jain, S.: Learning correction grammars. J. Symb. Logic **74**, 489–516 (2009)

7. Case, J., Jain, S.: Rice and Rice-Shapiro theorems for transfinite correction grammars. Math. Logic Q. **57**(5), 504–516 (2011)
8. Constable, R.: Subrecursive programming languages II: on program size. J. Comput. Syst. Sci. **5**, 315–334 (1971)
9. Craig, W.: On axiomatizability within a system. J. Symb. Logic **18**, 30–32 (1953)
10. Drumm, E.: Extensions to blum's size results in subrecursive formalisms. Master's thesis, University of Toronto (1970)
11. Epstein, R.L., Haas, R., Kramer, R.L.: Hierarchies of sets and degrees below $0'$. In: Lerman, M., Schmerl, J.H., Soare, R.I. (eds.) Logic Year 1979–80. Lecture Notes in Mathematics, vol. 859, pp. 32–48. Springer, Heidelberg (1981)
12. Ershov, Y.L.: A hierarchy of sets. I. Algebra Logic **7**, 25–43 (1968)
13. Ershov, Y.L.: A hierarchy of sets II. Algebra Logic **7**, 212–284 (1968)
14. Ershov, Y.L.: A hierarchy of sets III. Algebra Logic **9**, 20–31 (1970)
15. Freivalds, R., Smith, C.: On the role of procrastination in machine learning. Inf. Comput. **107**(2), 237–271 (1993)
16. Hartmanis, J.: On the succinctness of different representations of languages. SIAM J. Comput. **9**, 114–120 (1980)
17. Hartmanis, J.: On Gödel speed-up and succinctness of language representations. Theor. Comput. Sci. **26**, 335–342 (1983)
18. Hay, L.: Rice theorems for d.r.e. sets. Can. J. Math **27**, 352–365 (1975)
19. Hay, L.: On the recursion-theoretic complexity of relative succinctness of representations of languages. Inf. Control **52**, 2–7 (1982)
20. Hopcroft, J., Ullman, J.: Introduction to Automata Theory Languages and Computation. Addison-Wesley, Reading (1979)
21. Kleene, S.C.: On notation for ordinal numbers. J. Symb. Logic **3**, 150–155 (1938)
22. Li, M., Vitányi, P.: An Introduction to Kolmogorov Complexity and Its Applications, 3rd edn. Springer, New York (2008)
23. Mendelson, E.: Introduction to Mathematical Logic, 5th edn. Chapman & Hall, New York (2009)
24. Meyer, A.: Program size in restricted programming languages. Inf. Control **21**, 382–394 (1972)
25. Meyer, A., Bagchi, A.: Program size and economy of description. In: Proceedings of the 4th Annual ACM Symposium on Theory of Computing, pp. 183–186 (1972)
26. Meyer, A., Fischer, M.: Economy of description by automata, grammars and formal systems. In: Proceedings of the IEEE 12th Annual Symposium on Switching and Automata Theory, pp. 42–51 (1971)
27. Meyer, A., Ritchie, D.: The complexity of loop programs. In: Proceedings of the 22nd National ACM Conference, pp. 465–469. Thomas Book Co. (1967)
28. Péter, R.: Recursive Functions. Academic Press, New York (1967)
29. Rathjen, M.: The realm of ordinal analysis. In: Cooper, S., Truss, J. (eds.) Sets and Proofs. London Mathematical Society Lecture Note Series, vol. 258, pp. 219–279. Cambridge University Press, Cambridge (1999)
30. Rogers, H.: Theory of Recursive Functions and Effective Computability. McGraw-Hill, New York (1967). Reprinted by MIT Press, 1987
31. Royer, J., Case, J.: Subrecursive Programming Systems: Complexity & Succinctness. Monograph in Programming Theoretical Computer Science. Birkhäuser, Boston (1994). See www.cis.udel.edu/~case/RC94Errata.pdf for corrected errata
32. Sacks, G.: Degrees of Unsolvability. Princeton University Press, Princeton (1963)
33. Schmidt, E., Szymanski, T.: Succinctness of descriptions of ambiguous context-free languages. Inf. Control **32**, 547–553 (1976)

34. Simpson, S.: Subsystems of Second Order Arithmetic, 2nd edn. Cambridge University Press, Cambridge (2010)
35. Smullyan, R.: Theory of Formal Systems. Annals of Mathematics Studies, vol. 47. Princeton University Press, Princeton (1961)
36. Soare, R.: Recursively Enumerable Sets and Degrees. Springer, Heidelberg (1987)
37. Soare, R.: Turing oracle machines, online computing, and three displacements in computability theory. Ann. Pure Appl. Logic 160, 368–399 (2009)
38. Takeuti, G.: Proof Theory. Studies in Logic and the Foundations of Mathematics, vol. 81, 2nd edn. North-Holland, Amsterdam (1987)
39. Valiant, L.: Relative complexity of checking and evaluating. Inf. Process. Lett. 5, 20–23 (1976)
40. Weiermann, A.: Proving termination for term rewriting systems. In: Börger, E., Jäger, G., Büning, H., Richter, M. (eds.) CSL 1991. LNCS, vol. 626, pp. 419–428. Springer, Heidelberg (1992)

Automorphism Groups of Substructure Lattices of Vector Spaces in Computable Algebra

Rumen Dimitrov[1]([✉]), Valentina Harizanov[2], and Andrei Morozov[3]

[1] Department of Mathematics, Western Illinois University, Macomb, IL 61455, USA
rd-dimitrov@wiu.edu
[2] Department of Mathematics, George Washington University,
Washington, D.C. 20052, USA
harizanv@gwu.edu
[3] Sobolev Institute of Mathmematics, Novosibirsk 630090, Russia
morozov@math.nsc.ru

Abstract. For a Turing degree $\mathbf{x}$, we investigate the automorphisms of the lattice of $\mathbf{x}$-c.e. vector spaces. We establish the equivalence of the embedding relation for these automorphism groups with the order relation on the corresponding Turing degrees. By a result of Guichard the automorphisms of the lattice of $\mathbf{x}$-c.e. vector spaces are induced by $\mathbf{x}$-computable invertible semilinear transformations, $GSL_{\mathbf{x}}$. We prove that the Turing degree spectrum of the group $GSL_{\mathbf{x}}$ is the upper cone of Turing degrees $\geq \mathbf{x}''$.

1 Automorphisms of Effective Structures

The study of automorphisms on computable or computably enumerable structures connects computability theory and classical group theory. The set of all automorphisms of a computable structure forms a group under composition, and it is natural to ask questions about its complexity as well as the complexity of its subgroups. It is also interesting to connect the embedability of the subgroups with Turing reducibility.

The following notion is the focus of our investigation. Let $\mathbf{d}$ be a Turing degree. For an infinite computable structure $\mathcal{M}$, we define $Aut_{\mathbf{d}}(\mathcal{M})$ to be the set of all automorphisms of $\mathcal{M}$ computable in $\mathbf{d}$. The set $Aut_{\mathbf{d}}(\mathcal{M})$ under composition is a subgroup of $Aut(\mathcal{M})$. When the structure $\mathcal{M}$ is ω with equality, then its automorphism group $Aut(\mathcal{M})$ is usually denoted by $Sym(\omega)$, the symmetric group of ω. Hence we have

$$Sym_{\mathbf{d}}(\omega) = \{f \in Sym(\omega) : \deg(f) \leq \mathbf{d}\},$$

where $\deg(f)$ is the Turing degree of f. Our other computability theoretic notation is also standard and as in [18].

The three authors acknowledge partial support of the binational research grant DMS-1101123 from the National Science Foundation. The second author acknowledges support of the NSF grant DMS-1202328 and the Columbian College of Arts and Sciences of GWU.

© Springer International Publishing Switzerland 2016
A. Beckmann et al. (Eds.): CiE 2016, LNCS 9709, pp. 251–260, 2016.
DOI: 10.1007/978-3-319-40189-8_26

The *Turing degree spectrum* of a countable structure $\mathcal{A}$ is

$$DgSp(\mathcal{A}) = \{\deg(\mathcal{B}) : \mathcal{B} \cong \mathcal{A}\},$$

where $\deg(\mathcal{B})$ is the Turing degree of the atomic diagram of $\mathcal{B}$. Knight [8] proved that the degree spectrum of any structure is either a singleton or is upward closed. Only the degree spectrum of a so-called automorphically trivial structure is a singleton, and if the language is finite, that degree must be **0** (see [7]). Automorphically trivial structures include all finite structures, and also some special infinite structures, such as the complete graph on countably many vertices. Jockusch and Richter (see [15]) defined the *degree of the isomorphism type* of a structure, if it exists, to be the least Turing degree in its Turing degree spectrum. Richter [15,16] was first to systematically study such degrees. For these and more recent results about these degrees see [4]. In this paper we are especially interested in the following result by Morozov.

Theorem 1 [10]. *The degree of the isomorphism type of the group $Sym_{\mathbf{d}}(\omega)$ is $\mathbf{d}''$.*

We will establish a similar result in the context of effective vector spaces.

Let V_∞ be a canonical fully effective $\aleph_0$-dimensional vector space over a computable field F. We can think of the vectors in V_∞ as (the codes of) the finitely non-zero ω-sequences of elements of F. By $\mathcal{L}$ we denote the lattice of all subspaces of V_∞. For a Turing degree $\mathbf{d}$, by $\mathcal{L}_{\mathbf{d}}(V_\infty)$ we denote the following sublattice of $\mathcal{L}$:

$$\mathcal{L}_{\mathbf{d}}(V_\infty) = \{V \in \mathcal{L} : V \text{ is } \mathbf{d}\text{-computably enumerable}\}.$$

Note that in the literature $\mathcal{L}_{\mathbf{0}}(V_\infty)$ is usually denoted by $\mathcal{L}(V_\infty)$. Guichard [6] established that there are countably many automorphisms of $\mathcal{L}_{\mathbf{0}}(V_\infty)$ by showing that each computable automorphism is generated by a $1-1$ and onto computable semilinear transformation of V_∞. Recall that a map $\mu : V_\infty \to V_\infty$ is called a *semilinear transformation* of V_∞ if there is an automorphism σ of F such that

$$\mu(\alpha u + \beta v) = \sigma(\alpha)\mu(u) + \sigma(\beta)\mu(v)$$

for every $u, v \in V_\infty$ and every $\alpha, \beta \in F$.

Notation 2. *By $GSL_{\mathbf{d}}$ we denote the group of $1-1$ and onto semilinear transformations $\langle \mu, \sigma \rangle$ such that $\deg(\mu) \leq \mathbf{d}$ and $\deg(\sigma) \leq \mathbf{d}$.*

Hence Guichard proved that every element of $Aut(\mathcal{L}_{\mathbf{0}}(V_\infty))$ is generated by an element of $GSL_{\mathbf{0}}$. This result can be relativized to an arbitrary Turing degree $\mathbf{d}$. The proof of Theorem 3 below is essentially identical to the proof in [6].

Theorem 3 [6]. *Every $\Phi \in Aut(\mathcal{L}_{\mathbf{d}}(V_\infty))$ is generated by some $\langle \mu, \sigma \rangle \in GSL_{\mathbf{d}}$. Moreover if Φ is also generated by some other $\langle \mu_1, \sigma_1 \rangle \in GSL_{\mathbf{d}}$, then there is $\gamma \in F$ such that*

$$(\forall v \in V_\infty)\,[\mu(v) = \gamma\mu_1(v)].$$

There are two main results in this paper. The first is Theorem 4 in Sect. 2, which establishes that for every pair $\mathbf{a}$, $\mathbf{b}$ of Turing degrees, we have $Aut(\mathcal{L}_{\mathbf{a}}(V_\infty)) \hookrightarrow Aut(\mathcal{L}_{\mathbf{b}}(V_\infty))$ if and only if $\mathbf{a} \leq \mathbf{b}$. The second main result is Theorem 7 in Sect. 3, which establishes that the isomorphism degree type of the group $GSL_{\mathbf{d}}$ is $\mathbf{d}''$.

2   Group Embeddings and Turing Reducibility

Morozov showed that the correspondence $\mathbf{a} \to Sym_{\mathbf{a}}(\omega)$ can be used to substitute Turing reducibility with group-theoretic embedding. More precisely, Morozov [11] established that

$$(Sym_{\mathbf{a}}(\omega) \hookrightarrow Sym_{\mathbf{b}}(\omega)) \Leftrightarrow (\mathbf{a} \leq \mathbf{b})$$

for every pair $\mathbf{a}$, $\mathbf{b}$ of Turing degrees. It follows from this result that $\mathbf{a} = \mathbf{b}$ if and only if $Sym_{\mathbf{a}}(\omega) \cong Sym_{\mathbf{b}}(\omega)$. Here, we establish an analogous result for the subgroups of the group of automorphisms of the corresponding sublattices of $\mathcal{L}$. In the proof of the next, main theorem we will use the standard notation: $[x, y] = x^{-1}y^{-1}xy$ and $x^y = y^{-1}xy$.

Theorem 4. *For any pair of Turing degrees* $\mathbf{a}, \mathbf{b}$ *we have*

$$(Aut(\mathcal{L}_{\mathbf{a}}(V_\infty)) \hookrightarrow Aut(\mathcal{L}_{\mathbf{b}}(V_\infty))) \Leftrightarrow \mathbf{a} \leq \mathbf{b}.$$

Proof. Obviously, if $\mathbf{a} \leq \mathbf{b}$, then $Aut(\mathcal{L}_{\mathbf{a}}(V_\infty)) \hookrightarrow Aut(\mathcal{L}_{\mathbf{b}}(V_\infty))$.

Now, assume that $Aut(\mathcal{L}_{\mathbf{a}}(V_\infty)) \hookrightarrow Aut(\mathcal{L}_{\mathbf{b}}(V_\infty))$. Let $\{e_0, e_1, \ldots\}$ be a fixed computable basis of V_∞. For $\langle \mu_1, \sigma_1 \rangle, \langle \mu_2, \sigma_2 \rangle \in GSL_{\mathbf{a}}$, we define $\langle \mu_1, \sigma_1 \rangle \sim \langle \mu_2, \sigma_2 \rangle$ iff:

(1) $\sigma_1 = \sigma_2$, and
(2) there is $\alpha \in F$ such that $\alpha \neq 0$ and $(\forall v \in V_\infty) [\mu_1(v) = \alpha \mu_2(v)]$.

Note that $Aut(\mathcal{L}_{\mathbf{a}}(V_\infty)) \cong GSL_{\mathbf{a}}/\sim$. We can define a group embedding $\delta : Sym_{\mathbf{a}}(\omega) \hookrightarrow GSL_{\mathbf{a}}/\sim$ as follows. For any $f \in Sym_{\mathbf{a}}(\omega)$, we let $\delta(f)$ be the $\sim$-equivalence class of a linear transformation $\langle \widetilde{f}, id \rangle$ such that

$$\widetilde{f}(e_i) = e_{f(i)}.$$

Note that if $\delta(f_1) = \delta(f_2)$, then $\widetilde{f}_1 = c\widetilde{f}_2$ for some $c \in F$, and thus

$$(\forall i \in \omega) [e_{f_1(i)} = \widetilde{f}_1(e_i) = c\widetilde{f}_2(e_i) = ce_{f_2(i)}].$$

Since the vectors e_i, $i \in \omega$, are independent, we must have

$$(\forall i \in \omega) [f_1(i) = f_2(i)].$$

Therefore, there exists a map

$$K : Sym_{\mathbf{a}}(\omega) \hookrightarrow GSL_{\mathbf{b}}/\sim$$

such that if $f \in Sym_\mathbf{a}(\omega)$, then $K(f)$ is a $\mathbf{b}$-computable linear transformation of V_∞ modulo scalar multiplication.

We claim that if a set A is c.e. in $\mathbf{a}$, then A is c.e. in $\mathbf{b}$. Fix $A \subseteq \omega$ such that A is c.e. in $\mathbf{a}$, and let $h : \omega \to \omega$ be an $\mathbf{a}$-computable enumeration of A. Hence $rng(h) = A$. Fix a partition of the natural numbers into uniformly computable infinite sets R_i for $i \in \mathbb{Z}$ with enumerations $R_i = \{c_i^0 < c_i^1 < \cdots \}$. Let the permutations $g_0, g_1, w, b \in Sym_\mathbf{a}(\omega)$ be defined as follows:

$w(c_i^j) = c_{i+1}^j$ for each $i \in \mathbb{Z}$ and $j \in \omega$,

$g_0 = \prod_{j \in \omega} (c_0^{2j}, c_0^{2j+1})$,

$g_1 = \prod_{j \in \omega} (c_0^{2j+1}, c_0^{2j+2})$, and

$b = \prod_{n, t \in \omega \,\wedge\, h(t) = n} (c_n^t, c_n^{t+1})$.

We will also use the following abbreviation: $w^n = \underbrace{w \cdots w}_{n \text{ times}}$. Then we have

$$n \notin A \Leftrightarrow \left([g_0, b^{w^n}] = 1 \wedge [g_1, b^{w^n}] = 1 \right).$$

This is because g_0 and b^{w^n} commute iff n is not enumerated into A at an odd stage t, and, similarly, g_1 and b^{w^n} commute iff n is not enumerated into A at an even stage t. Hence for $\widetilde{g_0} = K(g_0)$, $\widetilde{g_1} = K(g_1)$, $\widetilde{w} = K(w)$, and $\widetilde{b} = K(b)$, we have

$$n \notin A \Leftrightarrow \left([K(g_0), K(b)^{(K(w))^n}] = 1 \wedge [K(g_1), K(b)^{(K(w))^n}] = 1 \right)$$

$$\Leftrightarrow \left([\widetilde{g_0}, \widetilde{b}^{\widetilde{w}^n}]_{/\sim} = 1 \wedge [\widetilde{g_1}, \widetilde{b}^{\widetilde{w}^n}]_{/\sim} = 1 \right)$$

We will now show that $[\widetilde{g_0}, \widetilde{b}^{\widetilde{w}^n}] \nsim 1$ is c.e. relative to $\mathbf{b}$. Let $\tau_n =_{def} [\widetilde{g_0}, \widetilde{b}^{\widetilde{w}^n}]$. Then

$$\tau_n \nsim 1 \Leftrightarrow \tau_n(e_0) \text{ and } e_0 \text{ are linearly independent, or}$$

$$(\exists m \in \omega)\, (\exists \alpha \neq 0)\, [\tau_n(e_0) = \alpha e_0 \wedge \tau_n(e_m) \neq \alpha e_m].$$

Let $A \in \mathbf{a}$. Then A and $\overline{A}$ are both c.e. in $\mathbf{b}$, and, therefore, A is computable in $\mathbf{b}$. Hence $\mathbf{a} \leq \mathbf{b}$. $\qquad\square$

3   Complexity of $GSL_\mathbf{d}$

In this section we will determine the Turing degree spectrum of $GSL_\mathbf{d}$. For the statement of the main theorem we will use terminology and notation from the following definition.

Definition 1. *A permutation p on a set M is:*

(i) $1_{\inf}2_{\inf}$ on M if it is a product of infinitely many 1 -cycles and infinitely many 2-cycles;

(ii) $1_{\inf}2_{fin}$ on M if it is a product of infinitely many 1-cycles and finitely many 2-cycles.

The main theorem about the degree spectrum of $GSL_{\mathbf{d}}$ will be derived from the following embeddability theorem.

Theorem 5. *Let G be an X-computable group, and let $H : Sym_0(\omega) \hookrightarrow G$ be an embedding. Suppose that for every $1_{\inf}2_{\inf}$ permutation $p \in Sym_0(\omega)$, the image $H(p)$ is not a conjugate of the image of any $1_{\inf}2_{fin}$ permutation in $Sym_0(\omega)$.*
Then $\mathbf{0}'' \leq \deg(X)$.

Proof. Let A be a Π_2^0-complete set and let $R(x,t)$ be a computable predicate such that
$$n \in A \Leftrightarrow (\exists^\infty t)\, R(n,t).$$

We will prove that $A \leq_T X$. Fix a partition of the natural numbers into uniformly computable infinite sets $S_{i,j}$ for $i \in \mathbb{Z}$ and $j \in \{1,2\}$ with enumerations $S_{i,j} = \{c_{i,j}^0 < c_{i,j}^1 < \cdots\}$. The sets $S_{i,1}$ and $S_{i,2}$ will be referred to as the left and the right parts of the i-th column, $S_i = S_{i,1} \cup S_{i,2}$. This reference will be useful when we define certain maps below. We can graphically present this partition as follows:

$$
\begin{array}{cccccccc}
\vdots & \vdots & & \vdots & \vdots & & \vdots & \vdots \\
c_{-1,1}^2 & c_{-1,2}^2 & & c_{0,1}^2 & c_{0,2}^2 & & c_{1,1}^2 & c_{1,2}^2 \\
\cdots \quad c_{-1,1}^1 & c_{-1,2}^1 & & c_{0,1}^1 & c_{0,2}^1 & & c_{1,1}^1 & c_{1,2}^1 \quad \cdots \\
c_{-1,1}^0 & c_{-1,2}^0 & & c_{0,1}^0 & c_{0,2}^0 & & c_{1,1}^0 & c_{1,2}^0 \\
\hline
S_{-1,1} & S_{-1,2} & & S_{0,1} & S_{0,2} & & S_{1,1} & S_{1,2}
\end{array}
$$

Column S_{-1} Column S_0 Column S_1

We will now define the following maps.

(i) $w(c_{i+1,j}^k) =_{def} c_{i,j}^k$ for each $i \in \mathbb{Z}$, $k \in \omega$ and $j = 1,2$.

Clearly, the map w is such that $w(S_{i+1,1}) = S_{i,1}$ and $w(S_{i+1,2}) = S_{i,2}$. It maps the left (right) part of the $(i+1)$-st column to the left (right) part of the i-th column for each i.

(ii) $p_0 =_{def} \prod_{k \in \omega} (c_{0,1}^k, c_{0,2}^k)$.

It is easy to see that the map p_0 switches the left and right parts of the 0-th column (i.e., $p_0(S_{0,1}) = S_{0,2}$ and $p_0(S_{0,2}) = S_{0,1}$), and is identity on all other elements of ω.

(iii) $p_n =_{def} p_0^{w^n} = w^{-n} p_0 w^n$.

Note that the map p_n switches the left and right parts of the n-th column (i.e., $p_n(S_{n,1}) = S_{n,2}$ and $p_n(S_{n,2}) = S_{n,1}$), and is identity on all other elements of ω.

$$\text{(iv) } z(k) =_{def} \begin{cases} 0 & \text{if } k = 0, \\ 1 & \text{if } k = 2, \\ k - 2 & \text{if } k = 2t \geq 4, \\ k + 2 & \text{if } k = 2t + 1. \end{cases}$$

Note that the map z is a permutation of ω, which contains only one infinite cycle and (0).

(v) $\tau =_{def} (0, 1)$.

For $k \in \mathbb{Z}$ we have

$$\tau^{z^k} = \begin{cases} (0, 2k), & \text{if } k \geq 1, \\ (0, 2\,|k| + 1), & \text{if } k \leq 0, \end{cases}$$

so

$$(\forall n, m \in \omega)\,(\exists n_1, m_1 \in \mathbb{Z})\,\Big[\Big(\tau^{z^{n_1}}\Big)^{\tau^{z^{m_1}}} = (n, m)\Big]. \tag{1}$$

Note that property (1) guarantees that any $1_{\inf}2_{\text{fin}}$ permutation on ω can be represented as a finite product of the permutations τ and z.

(vi) We will now construct a permutation b on ω with the following properties:

$$b \upharpoonright_{S_{n,1}} = id \upharpoonright_{S_{n,1}}$$

$$b \upharpoonright_{S_{n,2}} \text{ is } \begin{cases} 1_{\inf}2_{\inf}, & \text{if } n \in A, \\ 1_{\inf}2_{\text{fin}}, & \text{if } n \notin A. \end{cases}$$

We will define b in stages. At each stage s we will have $E^s =_{def} dom(b^s) = rng(b^s)$.

Construction

Stage 0.

Let $b^0 \upharpoonright S_i =_{def} id$ for $i \leq -1$, and $E^0 = \bigcup_{i \leq -1} S_i$.

Stage $s + 1 = \langle n, t \rangle$.

Case 1. If $R(n, t)$, then find the least elements $p, q, r \in S_{n,2}$ such that $p, q, r \notin E^s$. Let $b^{s+1} = b^s \cdot (p, q)$ and assume that $b^{s+1}(r) = r$. Thus, we have $E^{s+1} = E^s \cup \{p, q, r\}$ and $b^{s+1} \upharpoonright E^s = b^s$.

Case 2. If $\neg R(n, t)$, then find the least elements $p, q, r \in S_{n,2}$ such that $p, q, r \notin E^s$. Let $b^{s+1} \upharpoonright E^s = b^s$ and $b^{s+1}(p) = p$, $b^{s+1}(q) = q$, $b^{s+1}(r) = r$. Then $E^{s+1} = E^s \cup \{p, q, r\}$.

End of construction.

By construction, $dom(b) = rng(b) = \omega$.

It follows that if $n \in A$, then $(\exists^\infty t)\, R(n, t)$, so Case 1 applies infinitely often for this n, and hence the map b switches infinitely many pairs in the right part of the n-th column. Therefore, $b \upharpoonright S_{n,2}$ is $1_{\inf}2_{\inf}$ and $b \upharpoonright S_{n,1} = id$.

If $n \notin A$, then $(\exists^{<\infty} t)\, R(n, t)$, so Case 1 applies finitely often for this n, and hence the map b switches only finitely many pairs in the right part of the n-th column. Therefore, $b \upharpoonright S_{n,2}$ is $1_{\inf}2_{\text{fin}}$ and $b \upharpoonright S_{n,1} = id$.

In both cases, the map b^{p_n} reverses the action of b on the left and right parts of the n-th column S_n, while for $k \neq n$, we have $b^{p_n} \upharpoonright S_k = b \upharpoonright S_k$.

$$\text{Then } b \cdot b^{p_n} \text{ is } \begin{cases} 1_{\inf}2_{\inf} \text{ on } S_n, & \text{if } n \in A, \\ 1_{\inf}2_{\text{fin}} \text{ on } S_n, & \text{if } n \notin A, \\ id \text{ on } S_k, & \text{if } n \neq k. \end{cases}$$

Therefore, $b \cdot b^{p_n}$ is $\begin{cases} 1_{\inf}2_{\inf} \text{ on } \omega, \text{ if } n \in A, \\ 1_{\inf}2_{\text{fin}} \text{ on } \omega, \text{ if } n \notin A. \end{cases}$

Finally, note that on ω, every computable $1_{\inf}2_{\inf}$ permutation is the conjugate of a fixed computable $1_{\inf}2_{\inf}$ permutation and some other computable permutation. Therefore, assume that f is a fixed computable $1_{\inf}2_{\inf}$ permutation such that:

$$(\forall z_1 \in Sym_0(\omega))\,(\exists h \in Sym_0(\omega))\,[z_1 = f^h].$$

Hence for every n, we have

$$
\begin{aligned}
n \in A &\Leftrightarrow b \cdot b^{p_n} \text{ is a } 1_{\inf}2_{\inf} \text{ permutation on } \omega \\
&\Leftrightarrow (\exists h \in Sym_0(\omega))\,[b \cdot b^{p_n} = f^h] \\
&\Leftrightarrow (\exists u \in H(Sym_0(\omega)))\,[H(b) \cdot H(b)^{H(p_n)} = H(f)^u], \text{ and}
\end{aligned}
\tag{2}
$$

$$
\begin{aligned}
n \notin A &\Leftrightarrow b \cdot b^{p_n} \text{ is a } 1_{\inf}2_{\text{fin}} \text{ permutation on } \omega \\
&\Leftrightarrow b \cdot b^{p_n} = \prod_{(i,j)\in F} \left(\tau^{z^i}\right)^{\tau^{z^j}} \\
&\Leftrightarrow H(b) \cdot H(b)^{H(p_n)} = \prod_{(i,j)\in F} \left(H(\tau)^{H(z)^i}\right)^{H(\tau)^{H(z)^j}}.
\end{aligned}
\tag{3}
$$

The set F in the last line of (3) denotes some finite set of pairwise disjoint cycles and the maps referenced in (2) and (3) are those that we defined in (i)–(vi) above. For the map $H : Sym_0(\omega) \hookrightarrow G$ note that $H(p_n) = H(w)^{-n} \cdot H(p_0) \cdot H(w)^n$.

We claim that the last equivalence in (2) can be strengthened so that we have:

$$n \in A \Leftrightarrow (\exists u \in G)\,[H(z) = H(f)^u] \tag{4}$$

That is, if $n \in A$, then

$$(\exists h \in Sym_0(\omega))\,[b \cdot b^{p_n} = f^h] \text{ and, therefore,}$$
$$(\exists u \in H(Sym_0(\omega)))\,[H(b) \cdot H(b)^{H(p_n)} = H(f)^u], \text{ and}$$
$$(\exists u \in G)\,[H(b) \cdot H(b)^{H(p_n)} = H(f)^u].$$

For the proof of the other direction of (4) suppose that for some fixed $h \in G$ we have

$$H(b) \cdot H(b)^{H(p_n)} = H(f)^h, \text{ but } n \notin A.$$

Then, because of (3), we have the following:
(i) $b \cdot b^{p_n}$ is a $1_{\inf}2_{\text{fin}}$ permutation on ω,
(ii) $H(b) \cdot H(b)^{H(p_n)}$ is the image of the $1_{\inf}2_{\text{fin}}$ permutation $b \cdot b^{p_n}$, while
(iii) $H(f)$ is the image of the $1_{\inf}2_{\inf}$ permutation f.
This contradicts our assumption that the image under H of the $1_{\inf}2_{\text{fin}}$ permutation $b \cdot b^{p_n}$ cannot be the conjugate of the image of the $1_{\inf}2_{\inf}$ permutation f.

Theorem 6. *The degree of the isomorphisms type of the group GSL_0 is $\mathbf{0}''$.*

Proof. Let $V = \{v_0, v_1, \ldots\}$ be a computable basis of V_∞. Define

$$H : Sym_0(\omega) \hookrightarrow GSL_0$$

so that for any $p \in Sym_0(\omega)$ the image $H(p) = \langle L, id \rangle$ is a semilinear map such that

$$L(v_i) = v_{p(i)} \text{ for every } i \in \omega.$$

We claim that under H the image of a $1_{\inf}2_{\inf}$ permutation from $Sym_0(\omega)$ cannot be a conjugate of the image of a $1_{\inf}2_{\fin}$ permutation from $Sym_0(\omega)$. To establish this fact, suppose that $\langle f, id \rangle$, $\langle f_1, id \rangle \in GSL_0$ are the images of some $1_{\inf}2_{\inf}$ and $1_{\inf}2_{\fin}$ computable permutations on ω, respectively. Suppose that $\langle f, id \rangle$ and $\langle f_1, id \rangle$ are conjugates, and let $\langle h, \sigma \rangle \in GSL_0$ be such that $\langle f, id \rangle^{\langle h, \sigma \rangle} = \langle f_1, id \rangle$. Note that the map $h : V_\infty \to V_\infty$ is $1-1$ and onto. The associated field automorphism $\sigma : F \to F$ from $\langle h, \sigma \rangle$ is used to indicate that $h(av + bw) = \sigma(a) h(v) + \sigma(b) h(w)$. To simplify the notation, we will refer to the semilinear maps $\langle f, id \rangle$, $\langle f_1, id \rangle$, and $\langle h, \sigma \rangle$ simply as f, f_1, and h, respectively.

Note that the definition of the map H allows us to view $f \restriction V$ and $f_1 \restriction V$ as $1_{\inf}2_{\inf}$ and $1_{\inf}2_{\fin}$ permutations on V, respectively. We now claim that f_1 satisfies the following property:

$$(\exists W \subset_{fin} V_\infty)(\forall v \in V_\infty)\left[(v - f_1(v)) \in W\right]. \tag{5}$$

Here, $W \subset_{fin} V_\infty$ stands for W being a finite-dimensional subspace of V_∞. For a set $U \subseteq V_\infty$, by $cl(U)$ we will denote the closure of U, which is the set of all linear combinations of the vectors in U. To prove (5), assume that $B = \{x_1, \ldots, x_k, y_1, \ldots, y_k\} \subseteq V$ is such that $f_1 \restriction V = \prod_{1 \leq i \leq k}(x_i, y_i)$. Note that for every $v \in V_\infty$, there are $v_1 \in cl(V - B)$ and $v_2 \in cl(B)$ such that $v = v_1 + v_2$. Then

$$f_1(v) = f_1(v_1) + f_1(v_2) = v_1 + f_1(v_2),$$

and so

$$v - f_1(v) = v_1 + v_2 - v_1 - f_1(v_2) = v_2 - f_1(v_2) \in cl(B),$$

because $f_1(v_2) \in cl(B)$. Therefore, $W = cl(B)$ is a finite-dimensional subspace of V_∞ for which property (D) holds.

We will now prove that f^h does not satisfy property (5), which will contradict the assumption that $f^h = f_1$. Thus, assume that W is a finite-dimensional subspace of V_∞ such that

$$(\forall x \in V_\infty)\left[(x - f^h(x)) \in W\right]. \tag{6}$$

The *support* of a vector x with respect to a basis $Z = \{z_j : j \in J\}$, denoted by $supp_Z(x)$, is the set $\{z_{j_l} : l \in \{0, \ldots, t\}\}$ such that

$$x = \sum_{l=0}^{t} \lambda_l z_{j_l}$$

and $(\forall l \in \{0, ..., t\})[\lambda_l \neq 0]$.

Let $W_1 = h(W)$ and note that W_1 is finite-dimensional. Let B_1 be a finite subset of the basis V such that

$$(\forall x \in W_1)\,[supp_V(x) \subseteq B_1].$$

We will now find $u_1 \in V_\infty$ such that $u_1 - f(u_1) \notin W_1$. Since $f \upharpoonright V$ is a $1_{\inf}2_{\inf}$ permutation on V, there are infinitely many pairs $(u, v) \in V \times V$ such that

$$u \neq v, \; f(u) = v \text{ and } f(v) = u. \tag{7}$$

Since B_1 is finite, we can also find $u_1, v_1 \in V - B_1$, which have property (7). Then:

(i) $u_1 - f(u_1) = u_1 - v_1 \neq 0$, and
(ii) $u_1 - f(u_1) = (u_1 - v_1) \notin cl(B_1)$ because $B_1 \cup \{u_1, v_1\} \subseteq V$.

Since $W_1 \subseteq cl(B_1)$, we have that $u_1 - f(u_1) \notin W_1$. Therefore,

$$\left(h^{-1}(u_1) - h^{-1}(f(u_1))\right) \notin h^{-1}(W_1), \text{ and so}$$
$$\left(h^{-1}(u_1) - h^{-1}fhh^{-1}(u_1)\right) \notin W.$$

If we let $x_1 = h^{-1}(u_1)$, we obtain

$$x_1 - f^h(x_1) \notin W,$$

which contradicts that f^h satisfies (6).

We constructed an embedding $H : Sym_0(\omega) \hookrightarrow GSL_0$ such that the images of any $1_{\inf}2_{\inf}$ and $1_{\inf}2_{\text{fin}}$ permutations from $Sym_0(\omega)$ cannot be conjugates in GSL_0. We use Theorem 5 to conclude that $\mathbf{0}''$ is computable in any copy of GSL_0. We can construct a copy of GSL_0, which is computable in $\mathbf{0}''$. Therefore, the degree of the isomorphisms type of GSL_0 is $\mathbf{0}''$.

Note that the result of the previous theorem can be easily relativized to any Turing degree $\mathbf{d}$.

Theorem 7. *The degree of the isomorphisms type of the group $GSL_{\mathbf{d}}$ is $\mathbf{d}''$.*

References

1. Dimitrov, R., Harizanov, V., Morozov, A.S.: Dependence relations in computably rigid computable vector spaces. Ann. Pure Appl. Logic **132**, 97–108 (2005)
2. Downey, R.G., Remmel, J.B.: Computable algebras, closure systems: coding properties. In: Ershov, Y., Goncharov, S.S., Nerode, A., Remmel, J.B. (eds.) Handbook of Recursive Mathematics, vol. 2. Studies in Logic and the Foundations of Mathematics, vol. 139, pp. 997–1039. North-Holland, Amsterdam (1998)
3. Ershov, Y.L., Goncharov, S.S.: Constructive Models. Siberian School of Algebra and Logic. Kluwer Academic/Plenum Publishers, New York (2000). (English translation)

4. Fokina, E., Harizanov, V., Melnikov, A.: Computable model theory. In: Downey, R. (ed.) Turing's Legacy: Developments from Turing Ideas in Logic, pp. 124–194. Cambridge University Press/ASL, Cambridge (2014)

5. Goncharov, S., Harizanov, V., Knight, J., Morozov, A., Romina, A.: On automorphic tuples of elements in computable models. Siberian Math. J. **46**, 405–412 (2005). (English translation)

6. Guichard, D.R.: Automorphisms of substructure lattices in recursive algebra. Ann. Pure Appl. Logic **25**, 47–58 (1983)

7. Harizanov, V., Miller, R.: Spectra of structures and relations. J. Sym. Logic **72**, 324–348 (2007)

8. Knight, J.F.: Degrees coded in jumps of orderings. J. Sym. Logic **51**, 1034–1042 (1986)

9. Metakides, G., Nerode, A.: Recursively enumerable vector spaces. Ann. Pure Appl. Logic **11**, 147–171 (1977)

10. Morozov, A.S.: Permutations and implicit definability. Algebra Logic **27**, 12–24 (1988). (English translation)

11. Morozov, A.S.: Turing reducibility as algebraic embeddability. Siberian Math. J. **38**, 312–313 (1997). (English translation)

12. Morozov, A.S.: Groups of computable automorphisms. In: Ershov, Y.L., Goncharov, S.S., Nerode, A., Remmel, J.B. (eds.) Handbook of Recursive Mathematics, vol. 1. Studies in Logic and the Foundations of Mathematics, vol. 139, pp. 311–345. North-Holland, Amsterdam (1998)

13. Morozov, A.S., On theories of classes of groups of recursive permutations. Tr. Inst. Matematiki (Novosibirsk) **12** (1989). Mat. Logika i Algoritm. Probl. 91–104 (Russian). (English translation. Siberian Adv. Math. **1**, 138–153 (1991))

14. Morozov, A.S.: Computable groups of automorphisms of models. Algebra Logic **25**, 261–266 (1986). (English translation)

15. Richter, L.J.: Degrees of unsolvability of models. Ph.D. dissertation, University of Illinois at Urbana-Champaign (1977)

16. Richter, L.J.: Degrees of structures. J. Sym. Logic **46**, 723–731 (1981)

17. Rogers, H.: Theory of Recursive Functions and Effective Computability. McGraw-Hill, New York (1967)

18. Soare, R.I.: Recursively Enumerable Sets and Degrees. Springer, Heidelberg (1987)

Parameterized Complexity and Approximation Issues for the Colorful Components Problems

Riccardo Dondi[1] and Florian Sikora[2(✉)]

[1] Dipartimento di Scienze umane e sociali,
Università degli Studi di Bergamo, Bergamo, Italy
`riccardo.dondi@unibg.it`
[2] Université Paris-Dauphine, PSL, CNRS UMR 7243, LAMSADE, Paris, France
`florian.sikora@dauphine.fr`

Abstract. The quest for colorful components (connected components where each color is associated with at most one vertex) inside a vertex-colored graph has been widely considered in the last ten years. Here we consider two variants, Minimum Colorful Components (MCC) and Maximum Edges in transitive Closure (MEC), introduced in the context of orthology gene identification in bioinformatics. The input of both MCC and MEC is a vertex-colored graph. MCC asks for the removal of a subset of edges, so that the resulting graph is partitioned in the minimum number of colorful connected components; MEC asks for the removal of a subset of edges, so that the resulting graph is partitioned in colorful connected components and the number of edges in the transitive closure of such a graph is maximized. We study the parameterized and approximation complexity of MCC and MEC, for general and restricted instances.

1 Introduction

The quest for colorful components inside a vertex colored graph has been a widely investigated problem in the last years, with application for example in bioinformatics [5,8,12]. Roughly speaking, given a vertex-colored graph, the problem asks to find the colorful components of the graph, that is connected components that contain at most one vertex of each color. While most of the approaches have focused on the identification of a single connected colorful component, the identification of the minimum number of colorful connected components that match a given motif has been considered in [4,7].

Here we consider a similar framework, where instead of looking for a single colorful component inside a vertex-colored graph, we ask for a partition of the graph vertices in colorful components. This approach stems from a problem in bioinformatics, and more specifically in comparative genomics. In this context, a fundamental task is to infer the relations between genes in different genomes and, more precisely, to infer which genes are orthologous, that is those genes that originate via a speciation event from a gene of an ancestral genome.

A graph approach has been proposed aiming to identify disjoint orthology sets, where each of such sets corresponds to colorful disjoint component in the given graph [13].

A. Beckmann et al. (Eds.): CiE 2016, LNCS 9709, pp. 261–270, 2016.
DOI: 10.1007/978-3-319-40189-8_27

Different combinatorial problem formulations, based on different objective functions, have been proposed and studied in this direction [2,13]. Here, we considered two such approaches, MINIMUM COLORFUL COMPONENTS (MCC) and MAXIMUM EDGES IN TRANSITIVE CLOSURE (MEC). Given a vertex-colored graph, both problems ask for the removal of some edges so that the resulting graph is partitioned in colorful components but with different objective functions. The former aims to minimize the number of connected colorful components, while the latter aims to maximize the transitive closure of the resulting graph. A related but different problem has been considered in [5], where the objective function is the minimization of edge removal, so that the computed graph consists only of colorful components.

Previous Results. Given a graph on n vertices, MCC is known not only to be NP-hard, but also not approximable within factor $O(n^{1/14-\varepsilon})$ unless P = NP [2]. It is easy to see that the reduction leading to this inapproximability result implies also that MCC cannot be solved in time $n^{f(k)}$ for any function f, where k is the number of colorful components.

MEC is known to be APX-hard even when colored by at most three colors (while it is solvable in polynomial time for two colors), and, unless P = NP, it is not approximable within factor $O(n^{1/3-\varepsilon})$ when the number of colors is arbitrary, even when the input graph is a tree where each color appears at most twice [1]. A heuristic to solve MEC is presented in [13], while in [1], the authors present a polynomial-time $\sqrt{2 \cdot OPT}$ approximation algorithm.

Contributions and Organization of the Paper. In this paper we investigate more deeply the complexity of MCC and MEC. More precisely, we show in Sect. 3 that MCC on trees is essentially equivalent to MINIMUM MULTICUT on Trees, thus MCC is not approximable within factor $1.36 - \varepsilon$ unless P = NP for any $\varepsilon > 0$, but 2-approximable, it is fixed-parameter tractable and it admits a poly-kernel (when the parameter is the number of colorful components). Moreover, in Sect. 4 we show that MCC is easily solvable in polynomial time on paths, while it is not in XP class when parameterized by the structural parameter Distance to Disjoint Paths.

Then we consider the parameterized complexity of MEC with respect to the number k of edges in the transitive closure of a solution. For this parameter we give in Sect. 5 a parameterized algorithm, by reducing the problem to an exponential kernel. We use a similar idea in Sect. 6, to improve it to a poly-kernel for MEC when the input graph is a tree. Finally, we show in Sect. 7 that results similar to those of Sect. 4, hold also for MEC. Due to space constraints, some proofs (marked with a ★) are deferred to the full version of the paper.

2   Definitions

In this section we introduce some preliminary definitions. Consider a set of colors $C = \{c_1, \ldots, c_q\}$. A C–colored graph $G = (V, E, C)$ is a graph where every vertex in V is associated with a color in C; the color associated with a vertex $v \in V$ is denoted by $c(v)$. A connected component induced by a vertex set $V' \subseteq V$ is called *a colorful component*, if it does not contain two vertices having the same color. If a graph has t connected components where each component $i \in [t]$ has exactly n_i vertices, the number of edges in its transitive closure is defined by $\sum_{i=1}^{t} \frac{n_i(n_i-1)}{2}$.

Next, we introduce the formal definitions of the optimization problems we deal with.

MINIMUM COLORFUL COMPONENTS (MCC)

- **Input**: a C-colored graph $G = (V, E, C)$.
- **Output**: remove a set of edges $E' \subseteq E$ such that each connected component in $G' = (V, E \setminus E', C)$ is colorful, and the number of connected components of G' is minimized.

MAXIMUM EDGES IN TRANSITIVE CLOSURE (MEC)

- **Input**: a C-colored graph $G = (V, E, C)$.
- **Output**: remove a set of edges $E' \subseteq E$ such that each connected component in $G' = (V, E \setminus E', C)$ is colorful, and the number of edges in the transitive closure of G' is maximum.

The parametorized versions of MCC and MEC are defined analogously (and abusively denoted with the same names), with the addition in the input of an integer k, that denotes the number of connected components in G' for MCC and the number of edges in the transitive closure of G' for MEC.

Notice that, when considering an instance of MCC and MEC, we assume that E contains no edge $\{u, v\}$ with $c(u) = c(v)$, otherwise such an edge can be deleted from E as u and v will not be part of the same colorful component in any feasible solution of MCC or MEC.

Complexity. A parameterized problem (I, k) is said *fixed-parameter tractable* (or in the class FPT) with respect to a parameter k if it can be solved in $f(k) \cdot |I|^c$ time (in *fpt-time*), where f is any computable function and c is a constant (see [9] for more details about fixed-parameter tractability). The class XP contains problems solvable in time $|I|^{f(k)}$, where f is an unrestricted function.

A powerful technique to design parameterized algorithms is *kernelization*. In short, kernelization is a polynomial-time self-reduction algorithm that takes an instance (I, k) of a parameterized problem P as input and computes an equivalent instance (I', k') of P such that $|I'| \leqslant h(k)$ for some computable function h and $k' \leqslant k$. The instance (I', k') is called a *kernel* in this case. If the function h is polynomial, we say that (I', k') is a polynomial kernel.

Concerning approximation definitions, we refer the reader to some reference textbook like [3].

3   MCC for Trees: Parameterized Complexity and Approximability

In this section, we show that MCC on trees is essentially equivalent to the MINI-MUM MULTI-CUT problem on Trees (M-CUT-T), thus the positive and negative results of (M-CUT-T) for parameterized complexity and approximability transfer to MCC. We recall that M-CUT-T, given a tree T_M and a set S_M of pairs of terminals, asks if there exist a minimum cut (that is a set of removed edges) such that, for each pair $(x, y) \in S_M$, x and y are disconnected through that cut.

3.1   Positive Results

We show that MCC on trees admits an FPT algorithm (and a polynomial kernel) and a 2-approximation algorithm by reducing MCC to M-CUT-T. We first describe the reduction. Given a colored tree $G_T = (V, E, C)$ as an instance of MCC, we define an instance (T_M, S_M) of M-CUT-T as follows: T_M is exactly G_T (except for the colors of the vertices); for each pair (x, y) of vertices in G_T such that $c(x) = c(y)$, we define a pair (x, y) in S_M.

Now, we prove the main lemma of this section.

Lemma 1. *Consider an instance G_T of MCC and the corresponding instance (T_M, S_M) of M-CUT-T. Then: (1) given a solution of MCC on G_T consisting of $k + 1$ connected components, a solution of M-CUT-T on (T_M, S_M) consisting of k edges cut can be computed in polynomial time; (2) given a solution of M-CUT-T on (T_M, S_M) consisting of k edges, a solution of MCC on G_T consisting of $k + 1$ connected components can be computed in polynomial time.*

Proof. Consider a solution of MCC consisting of $k + 1$ components obtained by removing a set E' of k edges. Then, E' is a solution of M-CUT-T over instance (T_M, S_M). Indeed, for each pair $(x, y) \in S_M$, $c(x) = c(y)$, hence the two vertices belong to different connected components after the removal of edges in E'.

Conversely, consider a solution E' of M-CUT-T over instance (T_M, S_M), with $|E'| = k$. Then, remove the edges in E' from G_T and consider the $k + 1$ connected components induced by this removal in G_T. Since each pair $(x, y) \in S_M$ is disconnected after the removal of E', it follows that each connected component of G_T after the removal of E' is colorful.   □

We can now easily give the main result of this section:

Theorem 2 (★). *If the input graph of MCC is a tree, MCC can be solved in time $O^*(1.554^k)$[1] where k is the natural parameter and also admits a 2-approximation algorithm.*

Lemma 1 implies also a poly-kernel for MCC on trees.

Theorem 3 (★). *If the input graph of MCC is a tree, it is possible to compute in polynomial time a kernel of size $O(k^3)$ where k is the natural parameter.*

---

[1] The O^* notation suppresses polynomial factors.

3.2 Approximation Lower Bound of MCC on Trees

Let us now prove a lower bound for the approximation of MCC on trees, by giving a reduction from M-CUT-T. Starting from an instance (T_M, S_M) of M-CUT-T, we compute a colored tree $G_T = (V, E, C)$, input of MCC, as follows. First, G_T is isomorphic to T_M, and we color each vertex v of G_T as c_v. Denote by E_1 the edge set of such a tree. Then for each pair $(u, v) \in S_M$, we define a leaf u_v adjacent to v and colored $c_{u,v}$ and a leaf v_u adjacent to u and colored $c_{u,v}$ (see Fig. 1). Denote by E_2 the edge set introduced by adding these edges.

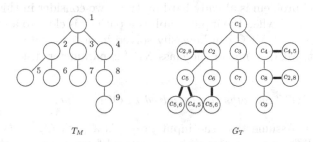

$$T_M \qquad\qquad\qquad G_T$$

Fig. 1. Sample construction of G_T from T_M with $S_M = \{(2, 8), (5, 6), (4, 5)\}$. Edge set E_2 of T is drawn thick. For ease, colors of G_T are drawn inside the nodes. On possible solution for this instance of M-CUT-T cuts edges $\{\{2, 6\}, \{1, 4\}\}$ and implies 3 colorful connected components in the corresponding instance of MCC.

Lemma 4. *Given a solution of MCC on $G_T = (V, E, C)$ consisting of k colorful components, we can compute in polynomial time a solution of MCC on $G_T = (V, E, C)$ consisting of at most k colorful components such that the edges cut belong only to set E_1.*

Proof. Consider the case that an edge $\{u, v\}$ has been deleted, where v is a leaf introduced in G_T. Then, notice that the removal of edge $\{u, v\}$ makes v an isolated vertex. By construction u and v (and each leaf adjacent to u) have different colors. Hence there are two possible cases: either the colorful component H that contains u does not include vertices colored by c_v, hence we can add v to H, thus we can avoid removing edge $\{u, v\}$, or there is a vertex w colored by c_v in H. In this case we can remove an edge of E_1, which separates w from u without removing edge $\{u, v\}$; such an edge must exist, since v and w are leaves incident in different internal vertices. $\qquad\square$

Lemma 5 (★). *Consider an instance (T_M, S_M) of M-CUT-T and the corresponding instance $G_T = (V, E, C)$ of MCC. Then: (1) given a solution of M-CUT-T over instance (T_M, S_M) that cuts k edges, we can compute in polynomial time a solution of MCC over instance $G_T = (V, E, C)$ consisting of at most $k + 1$ colorful components; (2) given a solution of MCC over instance $G_T = (V, E, C)$ consisting of at most $k + 1$ colorful components, we can compute in polynomial time a solution of M-CUT-T over instance (T_M, S_M) that cuts at most k edges.*

Since M-CUT-T cannot be approximated within factor 1.36 (since it is as hard as MINIMUM VERTEX COVER to approximate [10]), Lemmas 4 and 5 allow to extend the result to MCC.

Theorem 6 (★). *MCC on trees cannot be approximated within factor $1.36-\varepsilon$, for any constant $\varepsilon > 0$ unless $P = NP$.*

4   Structural Parameterization of MCC

Since the MCC problem is already hard on trees, we consider in this section the complexity of MCC when the input graph is a path or is close to a set of disjoint paths. We show that MCC can be easily solved in polynomial time, while, as a sharp contrast, MCC is not in the class XP for parameter distance to disjoint paths.

Theorem 7. *MCC on paths can be solved in $O(n^3)$-time.*

Proof (Sketch). Assume that the input graph is a path $G_P = (V, E, C)$, and assume that the vertices on the path are ordered from v_1 to v_n. Define $M[j]$ as the minimum number of colorful components of a solution of MCC over instance G_P restricted to vertices $\{v_1, \ldots, v_j\}$. $M[j]$, with $j > 1$, can be computed as follows:

$$M[j] = \min_{0 \leqslant t < j} M[t] + 1, \text{ such that } v_{t+1}, \ldots, v_j \text{ induce a colorful component.}$$

In the base cases, it holds $M[1] = 1$, and $M[0] = 0$. Next, we prove the correctness of the dynamic programming recurrence.

We claim that given a path $G_P = (V, E, C)$ instance of MCC, there exists a solution of MCC on instance G_P restricted to vertices $\{v_1, \ldots, v_j\}$ consisting of h colorful components if and only if $M[j] = h$.

It is then easy to see that the value of an optimal solution of MCC on path $G_P = (V, E, C)$ is stored in $M[n]$. Since the table $M[j]$ consists of n entries and each entry can be computed in time $O(n^2)$, it follows that MCC on paths can be computed in time $O(n^3)$.   □

Let us now prove that MCC is not in XP when parameterized by the *Distance to Disjoint Paths* number d (the minimum number of vertices to remove from the input graph to have disjoint paths), even when the input graph is a tree. We prove this result by giving a reduction from MINIMUM VERTEX COVER (MinVC) to MCC on trees.

Consider an instance $G = (V, E)$ of MinVC, and let $G_C = (V_C, E_C)$ be the corresponding instance of MCC. G_C is a rooted tree, defined as follows. First, we define $|V|$ paths, one for each vertex in G. Path P_i contains vertex $v_{c,i}$, colored by c_i, and vertices $e_{c,i,j}$, for each $\{v_i, v_j\} \in E$, colored by c_{ij}. Notice that vertices $e_{c,i,j}$ appears in P_i based on the lexicographic order of the corresponding edges. Moreover, there exist two vertices associated with edge $\{v_i, v_j\} \in E$, namely

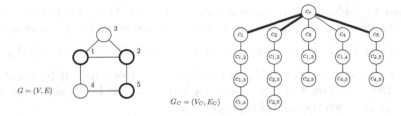

Fig. 2. Sample construction of an instance of MCC from an instance of MinVC. A possible solution for MinVC is given in thick while edges to be cut for the instance of MCC are also in thick.

$e_{c,i,j}$ (in P_i) and $e_{c,j,i}$ (in P_j), which are both colored by c_{ij}. The tree G_C is obtained by connecting the paths $P_1, \ldots P_{|V|}$ to a root r, which is colored by c_r, where c_r is a fresh new color (see Fig. 2).

Lemma 8 (★). *Let $G = (V, E)$ be an instance of MinVC, and let $G_C = (V_C, E_C)$ be the corresponding instance of MCC. Then: (1) given a vertex cover of G of size k, we can compute in polynomial time a solution of MCC over instance G_C consisting of $k + 1$ colorful components; (2) given a solution of MCC over instance G_C consisting of $k+1$ colorful components, we can compute in polynomial time a vertex cover of G of size k.*

By the previous lemma, the following result holds.

Theorem 9 (★). *MCC is NP-hard even when the input graph is at distance 1 to Disjoint Paths.*

It is worth noticing that this result extends to parameter pathwidth or distance to interval graph, as these last parameters are "stronger" than distance to disjoint path in the sense of [11].

5   An FPT Algorithm for MEC Parameterized by k

We present a parameterized algorithm for MEC with respect to the natural parameter k. To do so, we will show that the problem admits an exponential size kernel.

Given a colored graph G, we first compute a Depth-First-Search (DFS) $D = (V, E_D, E_B)$ of G. Recall that D consists of a tree induced by $D' = (V, E_D)$ (hence not considering edges in E_B), while $E_B = E \setminus E_D$ are called *backward edges* and have the following well-known property (see [6] for details).

Lemma 10. *Consider a graph G and the corresponding DFS $D = (V, E_D, E_B)$. Let $\{u, v\}$ be a backward edge. Then there exists a path p in $D' = (V, E_D)$ that starts in the root of D' and contains both u and v.*

We will first show some easy cases where there is a solution of MEC of size at least k. Let V_A be the set of vertices of V which are parent of a leaf in D'.

Lemma 11 (★). *If there exists a path in D' from the root $r(D')$ to a leaf of D' of length at least $2k$, then there exists a solution of MEC of size at least k.*

Lemma 12 (★). *There exists a solution of MEC of size at least k if $|V_A| \geqslant k$.*

Now, for each vertex $v \in V_A$ we consider the leaves adjacent to v and their colors. Define the set $C_x(v)$ as the set of leaves colored by c_x and adjacent to $v \in V_A$ in D. Then the following property holds.

Lemma 13 (★). *Given a vertex $v \in V_A$, if there exist $\sqrt{2k}$ non-empty sets $C_x(v)$ associated with distinct colors c_x, then there exists a solution of MEC of size at least k.*

Given a vertex $v \in V_A$ and a set $C_x(v)$, consider the sets of vertices connected with backward edges to a vertex $u \in C_x(v)$. Define $Adj(C_x(v)) = \{V'_A \subseteq V_A : \{u, w\} \in E, u \in V'_A, w \in C_x(v)\}$.

The following property holds.

Lemma 14 (★). *Given a vertex-colored graph G such that the hypothesis of Lemma 11 does not hold, consider a vertex v in V_A and a set $C_x(v)$. of possible sets of adjacent vertices to a node u of $C_x(v)$.*

Based on Lemma 14, we can partition the vertices of each $C_x(v)$ into sets $C_{x,1}(v), \ldots, C_{x,p}(v)$, with $p \leqslant 2^{2k+1}$, depending on their set of adjacent vertices (that is two vertices of $C_x(v)$ belong to the same set $C_{x,t}(v)$ if they have the same set of adjacent vertices).

Now, assume that the hypotheses of Lemmas 11–13 do not hold. Consider an algorithm that, for each set $C_{x,i}(v)$, computes a set $C'_{x,i}(v)$ by picking at most k vertices of $C_{x,i}(v)$ and removing the other vertices of $C_{x,i}(v)$. Let G' be the resulting graph. We claim that G' contains at most $O(k^2 2^{2k+1})$ vertices. First, notice that each $C'_{x,t}(v)$ contains at most k vertices and that, for each vertex v, there exists at most 2^{2k+1} sets $C'_{x,t}(v)$. Since, there exist at most $O(k\sqrt{k})$ sets $C_x(v)$ (at most $\sqrt{2k}$ colors c_x and at most k vertices $v \in V_A$), we can conclude that G' contains at most $O(k^2\sqrt{k}2^{2k+1})$ vertices in sets $C'_{x,i}(v)$.

Now, consider the vertices G' which are not contained in some set $C'_{x,i}(v)$. These vertices correspond to internal vertices of D'. Since the hypothesis of Lemma 11 does not hold, D' is a tree of depth at most $2k$, and there exist at most k vertices adjacent to leaves, as $|V_A| < k$. Hence there exist at most k paths of length $2k$ in D' from the root to vertices adjacent to leaves, thus we can conclude that there exist at most $2k^2$ internal vertices in D'. Hence there exists at most $2k^2$ vertices in G' which are not contained in some set $C'_{x,i}(v)$.

Now, we prove that (G', k) is a kernel for MEC.

Lemma 15 (★). *There exists a collection of disjoint colorful components $V_1, \ldots, V_h$ of size at least 2 in G if and only if there exists a collection of disjoint colorful components $V'_1, \ldots, V'_h$ in G', with $|V_i| = |V'_i|$, $1 \leqslant i \leqslant h$.*

Hence we have the following result.

Theorem 16 (★). *There exists a kernel of size $O(k^2\sqrt{k}2^{2k+1})$ for MEC.*

6   A Poly-Kernel for MEC on Trees

In this section, we show that in the special case of MEC where the input graph is a tree, the kernel size can be quadratic. The algorithm is similar to the one of Sect. 5. Consider a colored tree $G_T = (V, E, C)$, and let $r(G_T)$ denote the root of G_T. Lemmata 11–13 hold for G_T. Hence, we focus only on the leaves of G_T.

Since G_T is a tree, it follows that a leaf u having ancestor v belongs to a component of size at least 2 only if u and v belongs to the same component. It follows that among the leaves having color c_x and adjacent to a vertex u, only one can belong to a colorful component of size at least 2. Hence, given $v \in V_A$, let $C_x(v)$ be the set of leaves adjacent to v and colored by c_x. We remove all but one vertex from $C_x(v)$. Let G'_T be the resulting tree. We have the following property for G'_T.

Lemma 17 ($\bigstar$).   *There exists a collection of disjoint colorful components* $V_1, \ldots, V_h$ *of size at least 2 in* G_T *if and only if there exists a collection of disjoint colorful components* $V'_1, \ldots, V'_h$ *in* G'_T, *with* $|V_i| = |V'_i|$, $1 \leqslant i \leqslant h$.

Theorem 18.   *There exists a kernel of size* $O(k^2)$ *for MEC on trees.*

Proof. The result follows from Lemma 17 and from the fact that tree G'_T contains at most k^2 internal vertices (by Lemma 11 and by Lemma 12) and there exist at most $O(k\sqrt{k})$ sets $C_{x,i}(v)$ (by Lemma 13), each of size bounded by 1. $\square$

7   Structural Parameterization of MEC

It is easy to see that the results on structural parameterization for MCC hold also for MEC (after appropriate modifications).

Theorem 19 ($\bigstar$).   *MEC on paths can be solved in* $O(n^3)$-*time.*

Similarly to MCC, MEC is NP-hard even if we restrict the instance to graphs having distance 1 to Disjoint Paths. As for MCC, it is worth noticing that this hardness result extends to other stronger parameters like pathwidth [11].

Theorem 20 ($\bigstar$).   *MEC is NP-hard even when the input graph has distance* 1 *to Disjoint Paths.*

8   Conclusion

In the future, we aim at refining the parameterized complexity analysis, for example deepen the structural results for MCC and MEC. Moreover, it would be interesting to study the parameterized complexity of the two problems under other meaningful parameters in the direction of parameterizing above a guaranteed value.

References

1. Adamaszek, A., Blin, G., Popa, A.: Approximation and hardness results for the maximum edges in transitive closure problem. In: Jan, K., Miller, M., Froncek, D. (eds.) IWOCA 2014. LNCS, vol. 8986, pp. 13–23. Springer, Heidelberg (2015)
2. Adamaszek, A., Popa, A.: Algorithmic and hardness results for the colorful components problems. Algorithmica **73**(2), 371–388 (2015)
3. Ausiello, G., Crescenzi, P., Gambosi, G., Kann, V., Marchetti-Spaccamela, A., Protasi, M.: Complexity and Approximation: Combinatorial Optimization Problems and Their Approximability Properties. Springer, Heidelberg (1999)
4. Betzler, N., van Bevern, R., Fellows, M.R., Komusiewicz, C., Niedermeier, R.: Parameterized algorithmics for finding connected motifs in biological networks. IEEE/ACM Trans. Comput. Biol. Bioinform. **8**(5), 1296–1308 (2011)
5. Bruckner, S., Hüffner, F., Komusiewicz, C., Niedermeier, R., Thiel, S., Uhlmann, J.: Partitioning into colorful components by minimum edge deletions. In: Kärkkäinen, J., Stoye, J. (eds.) CPM 2012. LNCS, vol. 7354, pp. 56–69. Springer, Heidelberg (2012)
6. Cormen, T.H., Leiserson, C.E., Rivest, R.L., Stein, C.: Introduction to Algorithms, 3rd edn. MIT Press, Cambridge (2009)
7. Dondi, R., Fertin, G., Vialette, S.: Complexity issues in vertex-colored graph pattern matching. J. Discrete Algorithms **9**(1), 82–99 (2011)
8. Dondi, R., Fertin, G., Vialette, S.: Finding approximate and constrained motifs in graphs. Theor. Comput. Sci. **483**, 10–21 (2013)
9. Downey, R.G., Fellows, M.R.: Fundamentals of Parameterized Complexity. Springer, New York (2013)
10. Garg, N., Vazirani, V.V., Yannakakis, M.: Primal-dual approximation algorithms for integral flow and multicut in trees. Algorithmica **18**(1), 3–20 (1997)
11. Komusiewicz, C., Niedermeier, R.: New races in parameterized algorithmics. In: Rovan, B., Sassone, V., Widmayer, P. (eds.) MFCS 2012. LNCS, vol. 7464, pp. 19–30. Springer, Heidelberg (2012)
12. Lacroix, V., Fernandes, C.G., Sagot, M.: Motif search in graphs: application to metabolic networks. IEEE/ACM Trans. Comput. Biol. Bioinform. **3**(4), 360–368 (2006)
13. Zheng, C., Swenson, K., Lyons, E., Sankoff, D.: OMG! Orthologs in multiple genomes – competing graph-theoretical formulations. In: Przytycka, T.M., Sagot, M.-F. (eds.) WABI 2011. LNCS, vol. 6833, pp. 364–375. Springer, Heidelberg (2011)

A Candidate for the Generalised Real Line

Lorenzo Galeotti[(✉)]

Universität Hamburg, Hamburg, Germany
lorenzo.galeotti@gmail.com

Abstract. Let κ be an uncountable cardinal with $\kappa^{<\kappa} = \kappa$. In this paper we introduce $\mathbb{R}_\kappa$, a Cauchy-complete real closed field of cardinality 2^κ. We will prove that $\mathbb{R}_\kappa$ shares many features with $\mathbb{R}$ which have a key role in real analysis and computable analysis. In particular, we will prove that the Intermediate Value Theorem holds for a non-trivial subclass of continuous functions over $\mathbb{R}_\kappa$. We propose $\mathbb{R}_\kappa$ as a candidate for extending computable analysis to generalised Baire spaces.

1 Introduction

Computable analysis is the study of the computational properties of real analysis. We refer the reader to [21] for an introduction to computable analysis. In classical computability theory one studies the computational properties of functions over natural numbers and transfers these properties to arbitrary countable spaces via coding. The same approach is taken in computable analysis. By using coding, in fact, one can transfer topological and computational results from the Baire space ω^ω to sets of cardinality $2^{\aleph_0}$. In particular, by encoding the real numbers, one can use the Baire space to study computability in the context of real analysis.

Of particular interest in computable analysis is the study of the computational content of theorems from classical analysis. The idea is that of formalizing the complexity of theorems by means similar to those used in computability theory to classify functions over the natural numbers. In this context, the *Weihrauch* theory of reducibility plays an important role. For an introduction to the theory of Weihrauch reductions, see [4]. Weihrauch reductions can be used to classify functions over the Baire space ω^ω. By using this concept it is possible to arrange many theorems from classical real analysis in a complexity hierarchy called the Weihrauch hierarchy. A study of the Weihrauch degrees of some of the most important theorems from real analysis can be found in [3,4].

Recently, the study of the descriptive set theory of the generalised Baire spaces κ^κ for cardinals $\kappa > \omega$ has been catching the interest of set theorists (see [12] for an overview on the subject). This fact is also witnessed by the increasing

L. Galeotti—The author would like to thank the Isaac Newton Institute for Mathematical Sciences for the hospitality during the research programme *Mathematical, Foundational and Computational Aspects of the Higher Infinite* and the Royal Society for financial support during his stay in Cambridge (via the project *Infinite games in logic and Weihrauch degrees*; IE141198). This paper reports on results from the author's Master's thesis [13] supervised by Benedikt Löwe and Hugo Nobrega.

A. Beckmann et al. (Eds.): CiE 2016, LNCS 9709, pp. 271–281, 2016.
DOI: 10.1007/978-3-319-40189-8_28

number of workshops dedicated to generalised Baire spaces organized in the last two years (AST 2014 in Amsterdam and a satellite workshop to DMV 2015 in Hamburg). Even though generalised Baire spaces are not a new concept in set theory, many aspects of this theory are still unknown. In particular there has been no attempt to generalise computable analysis to spaces of cardinality 2^κ.

This paper provides the foundational basis for the study of *generalised computable analysis*, namely the generalisation of computable analysis to generalised Baire spaces. Since in classical computable analysis and classical Weihrauch theory the field of real numbers has a central role, a question arises naturally in this context: what is the right generalisation of $\mathbb{R}$ in the context of generalised computable analysis?

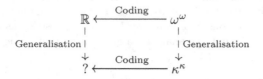

In this paper we answer this question. In particular, we will introduce $\mathbb{R}_\kappa$, a generalisation of the real line, which provides a well behaved environment for generalising classical results from real analysis to uncountable cardinals. We propose $\mathbb{R}_\kappa$ as the starting point for the study of generalised computable analysis.

The problem of generalising the real line is not new in mathematics. Different approaches have been tried for very different purposes. A good introduction to these number systems can be found in [9]. Among the most influential contributions to this field particularly important are the works of Sikorski [20] and Klaua [15] on the *real ordinal numbers* and that of Conway [6] on the *surreal numbers*. Sikorski's idea was to repeat the classical Dedekind construction of the real numbers starting from an ordinal equipped with the Hessenberg operations (i.e., commutative operations over the ordinal numbers). Unfortunately, one can prove that these fields do not have the density properties that, as we will see, will have a central role in the context of real analysis. The surreal numbers were introduced by Conway in order to generalise both the Dedekind construction of real numbers and the Cantor construction of ordinal numbers. In his introduction to surreal numbers, Conway proved that they form a (class) real closed field. Later, Ehrlich [15] proved that every real closed field is isomorphic to a subfield of the surreal numbers, showing that they behave like a universal (class) model for real closed fields. It is then natural for us to use this framework in the developing of $\mathbb{R}_\kappa$.

As we will see, doing analysis over field extensions of $\mathbb{R}$ is not an easy task. This is due to the fact that no proper ordered field extension of $\mathbb{R}$ is connected. However, many of the basic theorems of real analysis are linked to the fact that $\mathbb{R}$ is a connected space. To overcome this problem, instead of using standard topological tools, we will use a different mathematical framework which, under specific conditions over the density of $\mathbb{R}_\kappa$, will allow us to see our field extension of $\mathbb{R}$ as a linear continuum.

In this paper, we shall give necessary requirements for a space $\mathbb{R}_\kappa$ to be the generalisation of the real numbers. Considering the Intermediate Value Theorem (IVT) as one of the pillars of real analysis, we place particular emphasis on its validity in the generalised case, and we develop the requirements in such a way that they will allow us to prove it.

2   The Surreal Numbers

In this paper κ will refer to a fixed cardinal larger than ω. As usual in generalised descriptive set theory, let κ be an uncountable cardinal with $\kappa^{<\kappa} = \kappa$. Note in particular that this assumption implies that κ is a regular cardinal.

We will assume basic knowledge of topology, field theory and computable analysis. A good introduction to these subjects can be found in [5,18] and [21], respectively.

The following definition as well as most of the results in this section, are due to Conway [6] and have also been deeply studied by Gonshor in [14].

Definition 1 (Surreal numbers). *A surreal number is a function from an ordinal $\alpha \in \mathrm{On}$ to $\{+, -\}$, i.e., a sequence of pluses and minuses of ordinal length. We will denote the class of surreal numbers by* No. *The length of a surreal number $x \in$ No is the smallest ordinal $\ell(x) \in \mathrm{On}$ for which x is not defined. Moreover, we will use* $\mathrm{No}_{<\alpha}$ *and* $\mathrm{No}_{\leq\alpha}$ *to denote the set of $x \in$ No with $\ell(x) < \alpha$ and the set of $x \in$ No with $\ell(x) \leq \alpha$ respectively.*

We can define a total order over No as follows:

Definition 2. *Let $x, y \in$ No be two surreal numbers. We say that x is smaller than y in symbol $x < y$ iff $x(\alpha) < y(\alpha)$, where α is the smallest ordinal s.t. $x(\alpha) \neq y(\alpha)$ using the order $- < 0 < +$ where $x(\alpha) = 0$ if x is not defined at α.*

According to Conway's original idea, every surreal number is generated by filling some cut between shorter numbers. The following theorem gives us a connection between this intuition and the surreal numbers as we have defined them.

Theorem 3 (Simplicity theorem). *Let L and R be two sets of surreal numbers such that $L < R$ (i.e. $\forall \ell \in L \forall r \in R\, \ell < r$). Then there is a unique surreal z, denoted by $[L|R]$, of minimal length such that $L < \{z\} < R$. We will call $[L|R]$ a representation of z.*

By using the Simplicity Theorem Conway defined the field operations over No and proved that these operations satisfy the axioms of real closed fields. Later Ehrlich proved that the class field No behaves like a universal model for the theory of real closed fields, this means that every set-like model of the theory of real closed fields is isomorphic to a subfield of No. In particular Conway proved that the real numbers are a subfield of $\mathrm{No}_{\leq\omega}$.

The last theorem we want to mention in this section is due to van der Dries and Ehrlich [8]:

Theorem 4 (van der Dries and Ehrlich). *The set of surreal numbers* $No_{<\kappa}$ *is a real closed field.*

3   Super Dense κ-real Extensions of $\mathbb{R}$

In this paper we will have a quasi-axiomatic approach. In particular, we will first determine the properties that we need on $\mathbb{R}_\kappa$ in order to prove some basic facts from classical analysis. Then we will show how it is possible to define $\mathbb{R}_\kappa$ as a subfield of the surreal numbers.

Let us consider some of the basic properties that we expect from $\mathbb{R}_\kappa$. First of all we want $\mathbb{R}_\kappa$ to be a generalisation of $\mathbb{R}$ to the uncountable cardinal κ. Therefore we require that $\mathbb{R}_\kappa$ is a proper ordered field extension of $\mathbb{R}$. As we said, we want to use $\mathbb{R}_\kappa$ to do analysis. For this reason, we expect $\mathbb{R}_\kappa$ to behave as much as possible like $\mathbb{R}$. Formally we will require that $\mathbb{R}_\kappa$ is a real closed field extending $\mathbb{R}$. Since the theory of real closed fields is complete [17, Corollary 3.3.16], this implies that $\mathbb{R}_\kappa$ has the same first order properties as $\mathbb{R}$.[1]

> REQUIREMENT R1: $\mathbb{R}_\kappa$ is a real closed field extending $\mathbb{R}$.

Now, since we want to use $\mathbb{R}_\kappa$ to do computable analysis over sets of cardinality 2^κ, we require that $|\mathbb{R}_\kappa| = 2^\kappa$.

> REQUIREMENT R2: $\mathbb{R}_\kappa$ has cardinality 2^κ.

Finally, since the set of rational numbers $\mathbb{Q}$ has a central role in the representation theory of $\mathbb{R}$ (the interested reader is referred to [21]), we want $\mathbb{R}_\kappa$ to have a dense subset which can play the same role as $\mathbb{Q}$.

> REQUIREMENT R3: $\mathbb{R}_\kappa$ has a dense subset of cardinality κ.

In general we define:

Definition 5 (κ-real extension of $\mathbb{R}$). *Let K be a field satisfying* R1, R2, R3. *Then we will call K a κ-real extension of $\mathbb{R}$.*

Now we focus on those properties that are needed to extend theorems from classical analysis to $\mathbb{R}_\kappa$. Many of these classical results depend on the order over $\mathbb{R}$ and on its interval topology. So we will start considering interval topologies over κ-real extensions of $\mathbb{R}$ and their properties. First we recall few facts from field theory and classical analysis. It is a well known result from classical analysis that $\mathbb{R}$ has no Dedekind complete ordered field extensions (see [5, Corollary 8.7.4]). Therefore $\mathbb{R}_\kappa$ will not be Dedekind complete. More generally we have:

---

[1] In this paper we will use gray boxes for Requirements. Requirements are properties that the definition of $\mathbb{R}_\kappa$ will have to satisfy.

Corollary 6. *Let K be a κ-real extension of $\mathbb{R}$. Then K is not Dedekind complete.*

As usual, given an ordered field K, one can define Cauchy sequences over K.

Definition 7 (Cauchy sequences). *Let α be an ordinal and K^+ the positive part of K. Then a sequence $(x_i)_{i\in\alpha}$ of elements of K is Cauchy iff*

$$\forall \varepsilon \in K^+ \exists \beta < \alpha \forall \gamma, \gamma' \geq \beta |x_{\gamma'} - x_\gamma| < \varepsilon.$$

The sequence is convergent *if there is $x \in K$ such that*

$$\forall \varepsilon \in K^+ \exists \beta < \alpha \forall \gamma \geq \beta |x_\gamma - x| < \varepsilon.$$

We will call x the limit *of $(x_i)_{i\in\alpha}$. Given an ordered field K it is said to be Cauchy complete iff every Cauchy sequence whose length is equal to the degree $\mathrm{Deg}(K) = \mathrm{Coi}(K^+)$ of K has a limit in K.*

It is an easy exercise to see that Cauchy and Dedekind completeness coincide on Archimedean fields, while on non-Archimedean fields Cauchy completeness is weaker than Dedekind completeness.

Another property which is central in mathematical analysis is connectedness. It turns out that connectedness and Dedekind completeness are equivalent properties. Therefore, it is easy to see that we will not be able to define $\mathbb{R}_\kappa$ in such a way that its interval topology is connected.

As we said, our main purpose is that of proving basic facts from analysis over $\mathbb{R}_\kappa$. In particular we want to be able to prove the Intermediate Value Theorem (IVT). It turns out that if the IVT holds on an ordered field K then K is connected. This means that we cannot aim to prove the IVT over κ-real extensions of $\mathbb{R}$ in all its strength.

3.1 κ-topologies

Given what we have proved in the previous section, it is hard to do analysis over κ-real extensions of $\mathbb{R}$ by using standard topological tools. To overcome this problem we will use a tool introduced by Alling called κ-topology. A similar approach to do analysis over the surreal numbers was taken in [19].

Definition 8 (κ-topology). *A κ-topology τ over a set X is a collection of subsets of X such that:*

1. *$\emptyset, X \in \tau$.*
2. *$\forall \alpha < \kappa$ if $\{A_i\}_{i\in\alpha}$ is a collection of sets in τ, then $\bigcup_{i<\alpha} A_i \in \tau$.*
3. *$\forall A, B \in \tau A \cap B \in \tau$.*

The elements of τ are called κ-open sets.

Intuitively, the reason why we use κ-topologies is that, as we have seen in the previous section, interval topologies over κ-real extensions of $\mathbb{R}$ are too fine. As we will see κ-topologies will be coarser than topologies and will allow us to prove a weaker version of the Intermediate Value Theorem over particularly well-behaved κ-real extensions of $\mathbb{R}$.

Theorem 9 (Alling). *Let X be a set and B be a topological base over X. Then the set τ_κ defined as follows: $\emptyset, X \in \tau_\kappa$ and union of less than κ elements of B is in τ_κ, is a κ-topology. We will call τ_κ the κ-topology generated by B. Moreover we will call B a base for the κ-topology.*

Obviously many topological definitions can be relativized to κ-topologies. In particular we have the following:

Definition 10 (κ-continuity). *Let X and Y be two sets and τ, τ' be two κ-topologies respectively on X and on Y. Then $f : X \to Y$ is a κ-continuous function iff $\forall U \in \tau' f^{-1}[U] \in \tau$.*

Definition 11 (κ-connectedness). *Let X be a set and τ be a κ-topology over X. Then X is κ-connected iff $\forall U, V \in \tau X = U \cup V \wedge U \cap V = \emptyset \Rightarrow U = \emptyset \vee V = \emptyset$.*

All these definitions behave quite well with respect to their topological counterparts. Indeed, many classical theorems from topology hold for κ-topologies (see [1]). However, there are theorems from topology that are not valid on κ-topologies. Typically for κ-topologies local properties do not transfer to global properties (e.g. in κ-topologies openness is not implied by local openness).

Now we will introduce a κ-topological analogue of the interval topology over an ordered set.

Definition 12 (Interval κ-topology). *Let X be an ordered set and B be the set of open intervals with end points in $X \cup \{+\infty, -\infty\}$. We will call interval κ-topology over X the κ-topology generated by B.*

From now on we will consider the interval κ-topology as the standard κ-topology over κ-real extensions of $\mathbb{R}$.

As we have seen, in order to be able to prove some basic theorems from analysis we need to work within a connected space. However, as we have already pointed out, we can not aim for connectedness of κ-real extensions of $\mathbb{R}$. The next result is due to Alling [1] and it makes precise the connection between the density of an ordered set and the connectedness of its interval κ-topology.

Definition 13 (Hausdorff η_κ-set). *Let X be an ordered set and κ be a cardinal. We say that X is an η_κ-set iff given $L, R \subseteq X$, such that $L < R$ and $|L| + |R| < \kappa$ then there is $x \in X$ such that $L < \{x\} < R$.*

Theorem 14 (Alling). *Let X be an η_κ-set endowed with the interval κ-topology and X' a subset of X. Then X' is κ-connected iff X' is an interval in X.*

In view of Theorem 14, it is natural to require:

REQUIREMENT R4: $\mathbb{R}_\kappa$ is an η_κ-set.

Definition 15. *A field $K \supseteq \mathbb{R}$ is called a super dense κ-real extension of $\mathbb{R}$ if it satisfies requirements R1, R2, R3, and R4.*

As in classical topology κ-continuous functions preserve κ-connectedness.

Theorem 16. *Let $f : X \to Y$ be a κ-continuous function. If X is κ-connected then $f(X)$ is κ-connected.*

3.2   Analysis over Super Dense κ-real Extensions of $\mathbb{R}$

Using the results from the previous section we can modify the standard topological proof of the IVT to show that its restriction to κ-continuous functions holds over super dense κ-real extensions of $\mathbb{R}$.

Theorem 17 (IVT^K_κ). *Let K be a super dense κ-real extension of $\mathbb{R}$, the set $[a,b] \subset K$ be a closed subinterval of K and $f : [a,b] \to K$ be a κ-continuous function. Then for every $r \in K$ such that r is in between $f(a)$ and $f(b)$, there is $c \in [a,b]$ such that $f(c) = r$.*

It is a well-known fact that in every real closed field the IVT holds for polynomials in one variable (see [17, Theorem 3.3.9]), therefore it is natural to ask if polynomials over super dense κ-real extensions of $\mathbb{R}$ are κ-continuous.

Theorem 18. *Let K be a super dense κ-real extension of $\mathbb{R}$ and p be a polynomial in one variable with coefficients in K. Then p is κ-continuous.*

4   The Generalised Real Line $\mathbb{R}_\kappa$

We are now ready to define $\mathbb{R}_\kappa$. A naïve attempt to define such extension would be that of starting from κ endowed with the surreal operations (i.e., the Hessenberg operations) and try to repeat the standard construction of $\mathbb{Z}^\kappa$ and $\mathbb{Q}^\kappa$. Then, we could define $\mathbb{R}^\kappa$ as the Cauchy completion of $\mathbb{Q}^\kappa$ obtaining a Cauchy complete field. Unfortunately this approach does not work. This is due to the fact that, as Sikorski proved, the field $\mathbb{Q}^\kappa$ is Cauchy complete and then $\mathbb{R}^\kappa = \mathbb{Q}^\kappa$. Recall that $\mathbb{Q}^\kappa$ is a set of equivalence classes of pairs of elements in $\mathbb{Z}^\kappa$, hence it has cardinality at most κ. Therefore $\mathbb{R}^\kappa$ violates R2 and is not a good candidate for our purposes. This construction appeared for the first time in a paper from Sikorski in 1948 [20] (see also [2,15] for a complete study of this approach). For this reason we have to take a different approach in defining $\mathbb{R}_\kappa$.

We will follow the work done by Dales and Woodin in [7] for $\kappa = \aleph_1$ [2]. By Theorem 4 we know that $\mathrm{No}_{<\kappa}$ is a real closed field. Moreover, since $\kappa > \omega$, it is easy to see that $\mathbb{R} \subset \mathrm{No}_{<\kappa}$. In particular this means that R1 holds for $\mathrm{No}_{<\kappa}$. Furthermore, it is not hard to prove that $\mathrm{No}_{<\kappa}$ also satisfies R4. Then we have:

Proposition 19. *The field $\mathrm{No}_{<\kappa}$ has the following properties:*

1. $|\mathrm{No}_{<\kappa}| = \kappa$ *and* $\mathrm{Deg}(\mathrm{No}_{<\kappa}) = \kappa$.
2. $\mathrm{Cof}(\mathrm{No}_{<\kappa}) = \mathrm{Coi}(\mathrm{No}_{<\kappa}) = \kappa$ *and* $\mathrm{No}_{<\kappa}$ *has weight* κ.

Proposition 19 tells us that $\mathrm{No}_{<\kappa}$ has almost all the properties that we want from $\mathbb{R}_\kappa$ but is still too small. Moreover, it is not hard to see that $\mathrm{No}_{<\kappa}$ is not Cauchy complete in the sense of Definition 7 this fact is particularly problematic

---

[2] Note that in [7] Dales and Woodin do not make use of surreal numbers giving a different construction of $\mathbb{R}_{\aleph_1}$.

in the context of computable analysis, where most of the classical representations of $\mathbb{R}$ rely on the fact that $\mathbb{R}$ is the Cauchy completion of $\mathbb{Q}$.

It is therefore natural to consider $\mathrm{No}_{<\kappa}$ as generalised rational numbers, and to define $\mathbb{R}_\kappa$ as the Cauchy completion of $\mathrm{No}_{<\kappa}$ as in classical analysis[3]. Since we are working within the surreal numbers, this can be done in a natural way.

Definition 20 (Veronese cuts). *Let K be an ordered field. We call $\langle L, R \rangle$ a cut over K iff $L, R \subseteq K$ and $L < R$. Moreover we will say that $\langle L, R \rangle$ is a Veronese cut iff it is a cut such that, L has no maximum, R has no minimum and for each $\varepsilon \in K^+$ there are $\ell \in L$ and $r \in R$ such that $r < \ell + \varepsilon$. We will say that K is Veronese complete iff for each Veronese cut $\langle L, R \rangle$, there is $x \in K$ such that $L < \{x\} < R$.*

It is a well known fact that Cauchy and Veronese completeness are equivalent notions (see [7,10]). For this reason we can define the Cauchy completion of $\mathrm{No}_{<\kappa}$ by using the Simplicity Theorem as follows:

Definition 21 ($\mathbb{R}_\kappa$). *We define $\mathbb{R}_\kappa$ as follows:*

$$\mathbb{R}_\kappa = \mathrm{No}_{<\kappa} \cup \{x \mid x = [L|R] \text{ where } \langle L, R \rangle \text{ is a Veronese cut on } \mathrm{No}_{\leq\kappa}\}.$$

Now we will show that $\mathbb{R}_\kappa$ is a super dense κ-real extension of $\mathbb{R}$. First of all we have that $\mathrm{No}_{<\kappa}$ is a dense subfield of $\mathbb{R}_\kappa$ and that $\mathbb{R}_\kappa$ is Cauchy complete (i.e., $\mathbb{R}_\kappa$ the Cauchy completion of $\mathrm{No}_{<\kappa}$).

Lemma 22. *The field $\mathrm{No}_{<\kappa}$ is dense in $\mathbb{R}_\kappa$. Moreover the set $\mathbb{R}_\kappa$ is Cauchy complete.*

In view of the previous lemma from now on we will call $\mathrm{No}_{<\kappa}$ the κ-rational numbers and we will use the symbol $\mathbb{Q}_\kappa$ instead of $\mathrm{No}_{<\kappa}$.

Since we have showed that $\mathbb{R}_\kappa$ is the Cauchy completion of a real closed field, by a standard model theoretical argument we have:

Corollary 23. *The set $\mathbb{R}_\kappa$ is a real closed field extending $\mathbb{R}$.*

Now that we have shown that $\mathbb{R}_\kappa$ is a real closed field extending $\mathbb{R}$ we want to check that all the other properties of super dense κ-real extensions of $\mathbb{R}$ hold for $\mathbb{R}_\kappa$.

Theorem 24. *The field $\mathbb{R}_\kappa$ is the unique Cauchy complete real closed field of cardinality 2^κ, with degree and weight κ which is an η_κ-set.*

Proof. We will only prove $|\mathbb{R}_\kappa| = 2^\kappa$ the rest follows from the fact that $\mathbb{Q}_\kappa$ is dense in $\mathbb{R}_\kappa$. We want to prove $2^\kappa \leq |\mathbb{R}_\kappa| \leq 2^\kappa$. On the one hand we have that $\mathbb{R}_\kappa \subset \mathrm{No}_{\leq\kappa}$. Indeed, $\mathrm{No}_{\leq\kappa}$ contains the Dedekind completion of $\mathrm{No}_{<\kappa}$, hence also its Cauchy completion $\mathbb{R}_\kappa$. Then, since $|\mathrm{No}_{\leq\kappa}| = 2^\kappa$, we have that $|\mathbb{R}_\kappa| \leq 2^\kappa$.

---

[3] Note that this also reflects the fact that $\mathrm{No}_{<\omega}$ are the dyadic numbers and $\mathbb{R}$ is the Cauchy completion of $\mathrm{No}_{<\omega}$.

On the other hand let $\{0,1\}^{<\kappa}$ be the full binary tree of height κ, we define a tree T which is in bijection with $\{0,1\}^{<\kappa}$ and whose nodes are subintervals of $\mathbb{R}_\kappa$ and whose branches corresponds to different elements of $\mathbb{R}_\kappa$. We define the tree by recursion as follows: set $T_\lambda = (0,1)$ as the root of the tree. Now assume that for $p \in 2^{<\kappa}$ and that the element $T_p \neq \emptyset$ is already defined. We define T_{p0} and T_{p1} as two non-empty disjoint subintervals of T_p such that $T_{p0} = (a_{p0}, b_{p0})$ and $T_{p1} = (a_{p1}, b_{p1})$, where $a_{p0}, b_{p0}, a_{p1}, b_{p1} \in \mathbb{Q}_\kappa$ with $|a_{p0} - b_{p0}| \leq \frac{1}{\ell(p)+1}$ and $|a_{p1} - b_{p1}| \leq \frac{1}{\ell(p)+1}$, where $\ell(p)$ is the length of p. Finally if $p \in 2^{<\kappa}$ is of limit length γ and $T_{p\restriction\alpha}$ has already been defined for every $\alpha < \gamma$, we define $T'_p = \bigcap_{\alpha<\gamma} T_{p\restriction\alpha}$. Note that by the fact that $\mathbb{R}_\kappa$ is an η_κ-set, the set T'_p non-empty, moreover T'_p is trivially convex (i.e., if $x,y \in T'_p$ and $x \leq z \leq y$, then $z \in T'_p$). Therefore we can define T_p as we have done for the successor stage starting from T'_p. It follows trivially by the way in which we have defined the tree that for $p \in 2^\kappa$ the set $\bigcap_{\alpha\in\kappa} T_{p\restriction\alpha}$ contains a single element of $\mathbb{R}_\kappa$. Indeed by the properties of the tree we have that $[\{a_{p\restriction\alpha} \mid \alpha \in \kappa\} \mid \{b_{p\restriction\alpha} \mid \alpha \in \kappa\}]$ is a Veronese cut in $\mathbb{Q}_\kappa$. Therefore $\bigcap_{\alpha\in\kappa} T_{p\restriction\alpha}$ is a singleton in $\mathbb{R}_\kappa$ as desired. Moreover given $p, p' \in 2^\kappa$ such that $p \neq p'$ we have $\bigcap_{\alpha\in\kappa} T_{p\restriction\alpha} \neq \bigcap_{\alpha\in\kappa} T_{p'\restriction\alpha}$. Therefore we trivially have $2^\kappa \leq |\mathbb{R}_\kappa|$ as desired.

5   Conclusions and Future Work

In this paper we have introduced a real closed field extending $\mathbb{R}$ suitable for doing real analysis over the generalised Baire space κ^κ. We have showed that, although it has some limitations intrinsic to the problem, $\mathbb{R}_\kappa$ preserves many interesting properties of the real numbers. In particular we showed:

1. $\mathbb{R}_\kappa$ is a Cauchy complete super dense κ-real extension of $\mathbb{R}$ of cardinality 2^κ.
2. $\mathbb{R}_\kappa$ has a dense subset of cardinality κ and $\mathrm{Coi}(\mathbb{R}_\kappa^+) = \kappa$.
3. The IVT holds for κ-continuous functions.

As we have seen, most of these properties are motivated by computable analysis. For this reason we propose $\mathbb{R}_\kappa$ as the generalised real line in the context of computable analysis. An example of how $\mathbb{R}_\kappa$ can be used to study the topological Weihrauch complexity of theorems of analysis can be found in [13].

There are two natural continuations of this paper. On one hand it is natural to ask for a study of the computational strength of generalisations of theorems from real analysis. To accomplish this, a theory of generalised type two computability is needed. As we shall show in a paper soon to appear, it is possible to modify the notions of Ordinal Turing Machine introduced by Koepke in [16] to define a generalised version of Type Two Turing Machine (T2TM). The intuition behind this notion is that generalised T2TMs should run classical programs for Turing machines for κ steps instead of just ω. These machines lead to a very natural notion of computability, in which, because of the properties of $\mathbb{R}_\kappa$ and $\mathbb{Q}_\kappa$, the field operations restricted to $\mathbb{Q}_\kappa$ are computable in less than κ steps, while one may need to run forever (i.e., up to κ) to compute the same operations over $\mathbb{R}_\kappa$.

Moreover, this notion of computability preserves the correspondence between continuous functions and functions which are computable with an oracle.

A second natural continuation of this paper is the systematic study of the real analysis of $\mathbb{R}_\kappa$. Particularly interesting would be the study of a notion of integral. This problem is not new in the theory of surreal numbers and partial solutions have been proposed in the last decades (see [11, pp. 2–3]). Recently a solution to the problem of integration over the surreal numbers has been proposed in [11]. We are currently working on the problem of integration over $\mathbb{R}_\kappa$.

References

1. Alling, N.: Foundations of Analysis over Surreal Number Fields. North-Holland Mathematics Studies, Elsevier Science, Holland (1987)
2. Asperó, D., Tsaprounis, K.: Long reals (2015). https://archive.uea.ac.uk/bfe12ncu/long_reals3.pdf
3. Brattka, V.: Computable versions of Baire's category theorem. In: Sgall, J., Pultr, A., Kolman, P. (eds.) MFCS 2001. LNCS, vol. 2136, pp. 224–235. Springer, Heidelberg (2001)
4. Brattka, V., Gherardi, G.: Effective choice and boundedness principles in computable analysis. Bull. Symbolic Logic **17**(1), 73–117 (2011)
5. Cohn, P.: Basic Algebra: Groups, Rings, and Fields. Springer, Heidelberg (2003)
6. Conway, J.: On Numbers and Games. Ak Peters Series. Taylor & Francis, London (2000)
7. Dales, H., Woodin, W.: Super-real Fields: Totally Ordered Fields with Additional Structure. London Mathematical Society Monographs. Clarendon Press, Oxford (1996)
8. Dries, L., Ehrlich, P.: Fields of surreal numbers and exponentiation. Fundamenta Mathematicae **167**, 173–188 (2011)
9. Ehrlich, P.: Real Numbers, Generalizations of the Reals, and Theories of Continua. Synthese Library. Springer, Heidelberg (1994)
10. Ehrlich, P.: Dedekind cuts of Archimedean complete ordered abelian groups. Algebra universalis **37**(2), 223–234 (1997)
11. Ehrlich, P., Costin, O., Friedman, H.: Integration on the surreals: a conjecture of Conway, Kruskal and Norton (2015). arXiv:1505.02478
12. Friedman, S., Hyttinen, T., Kulikov, V.: Generalized Descriptive Set Theory and Classification Theory. Memoirs of the American Mathematical Society. American Mathematical Society, Providence (2014)
13. Galeotti, L.: Computable Analysis Over the Generalized Baire Space. Master's thesis, ILLC Master of Logic Thesis Series MoL-2015-13, Universiteit van Amsterdam (2015)
14. Gonshor, H.: An Introduction to the Theory of Surreal Numbers. London Mathematical Society Lecture Note Series. Cambridge University Press, Cambridge (1986)
15. Klaua, D.: Rational and real ordinal numbers. In: Ehrlich, P. (ed.) Real Numbers, Generalizations of the Reals, and Theories of Continua. Synthese Library, vol. 242, pp. 259–276. Springer, Amsterdam (1994)
16. Koepke, P.: Turing computations on ordinals. Bull. Symbolic Logic **11**(3), 377–397 (2005)

17. Marker, D.: Model Theory: An Introduction. Graduate Texts in Mathematics. Springer, New York (2006)
18. Munkres, J.: Topology. Featured Titles for Topology Series. Prentice Hall, Upper Saddle River (2000)
19. Rubinstein-Salzedo, S., Swaminathan, A.: Analysis on surreal numbers. J. Logic Anal. **6**, 1–39 (2014)
20. Sikorski, R.: On an ordered algebraic field. Soc. Sci. Litt. Varsovic Sci. Math. Phys. **41**, 69–96 (1948)
21. Weihrauch, K.: Computable Analysis: An Introduction. Texts in Theoretical Computer Science. An EATCS Series. Springer, Heidelberg (2012)

Finitely Generated Semiautomatic Groups

Sanjay Jain[1], Bakhadyr Khoussainov[2], and Frank Stephan[1,3(✉)]

[1] Department of Computer Science, National University of Singapore,
13 Computing Drive, COM1, Singapore 117417, Republic of Singapore
sanjay@comp.nus.edu.sg
[2] Department of Computer Science, University of Auckland,
Private Bag 92019, Auckland, New Zealand
bmk@cs.auckland.ac.nz
[3] Department of Mathematics, National University of Singapore,
10 Lower Kent Ridge Road, S17, Singapore 119076, Republic of Singapore
fstephan@comp.nus.edu.sg

Abstract. The present work shows that Cayley automatic groups are semiautomatic and exhibits some further constructions of semiautomatic groups and in particular shows that every finitely generated group of nilpotency class 3 is semiautomatic.

1 Introduction

Hodgson [4,5] as well as Khoussainov and Nerode [8] initiated the study of automatic structures, including that of groups. In their approach, such a group is given by a regular set A as the domain (denoting the representatives of the group) such that both, the group operation $\circ$ and the equality $=$, are automatic; that is, there is an automaton which reads the convoluted tuples (x, y, z) or (x, y) and decides whether such a tuple satisfies $x \circ y = z$ or $x = y$, respectively. Here, for $x = x_0 x_1 \ldots x_m$ and $y = y_0 y_1 \ldots y_n$ with $x_i, y_i \in \Sigma$, the convolution of the pair (x, y) is the string $z_0 z_1 \ldots z_{\max\{m,n\}}$, over the new alphabet $(\Sigma \cup \{\#\})^2$, where $z_i = (x_i, y_i)$ and x_i (respectively y_i) is taken to be $\#$ in case of $i > m$ (respectively, $i > n$). The convolution over triples or tuples in general is defined similarly. The advantage of this setting is that every function and relation definable in the language of group theory using parameters from the group is again automatic. Furthermore, automata providing the mappings can be found algorithmically. This also leads to the conclusion that for every fixed automatic group, the first-order theory is decidable [8]. Furthermore, automatic functions are precisely those which can be computed in linear time by a position-faithful one-tape Turing machine [2], thus the automatic functions coincide with the smallest reasonable time complexity class for functions.

S. Jain is supported in part by NUS grants C252-000-087-001, R146-000-181-112 and R146-000-184-112; B. Khoussainov is supported in part by the Marsden Fund grant of the Royal Society of New Zealand; F. Stephan is supported in part by NUS grants R146-000-181-112 and R146-000-184-112.

A. Beckmann et al. (Eds.): CiE 2016, LNCS 9709, pp. 282–291, 2016.
DOI: 10.1007/978-3-319-40189-8_29

Epstein et al. [3] argued that in the above formalisation, automaticity is, at least from the viewpoint of finitely generated groups, too restrictive. They furthermore wanted that the representatives of the group elements are given as words over the generators, leading to more meaningful representatives than arbitrary strings. Their concept of automatic groups led, for finitely generated groups, to a larger class of groups, though, by definition, it of course does not include groups which require infinitely many generators; groups with infinitely many generators, to some extent, were covered in the notion of automaticity by Hodgson, Khoussainov and Nerode. Nies and Thomas [10,11] provide results which contrast and compare these two notions of automaticity and give an overview on results for groups which are automatic in the sense of Hodgson, Khoussainov and Nerode.

Kharlampovich et al. [7] generalised the notion further to Cayley automatic groups. Here a finitely generated group $(A, \circ)$ is Cayley automatic iff the domain A is a regular set, for every group element there is a unique representative in A and, for every $a \in A$, the mapping $x \mapsto x \circ a$ is automatic. Note that the above requires multiplication by constants to be automatic only from one side; when multiplication by a constant from both sides are automatic, then the group is called Cayley biautomatic.

Finitely generated Cayley automatic groups have word problem decidable in quadratic time, carrying over the corresponding result from the two previous versions of automaticity. As opposed to the case of automatic groups (in the original sense of Hodgson), Miasnikov and Šunić [9] showed that several natural problems like the conjugacy problem can be undecidable for some Cayley automatic groups.

Jain et al. [6] investigated the general approach where, in a structure for some relations and functions, it is only required that the versions of the functions or relations with all but one variable fixed to constants is automatic. Here the convention is to put the automatic domains, functions and relations before a semicolon and the semiautomatic relations after the semicolon. For a group, the semiautomatic group $(A, \circ; =)$ would be a structure where the domain A is regular, the group operation (with both inputs) is automatic and for each fixed element $a \in A$ the set $\{b \in A : b = a\}$ is regular — note that group elements might have several representatives in semiautomatic groups.

In the present work, for any group, ε represents the neutral element. One of the basic results obtained is that the notion $(A, \circ; =)$ collapses to an automatic group (in the sense of Hodgson, Khoussainov and Nerode), as

$$a = b \Leftrightarrow \exists c \, [a \circ c = \varepsilon \text{ and } b \circ c = \varepsilon].$$

For semiautomatic groups, the two interesting group structures are $(A, =; \circ)$ and $(A; \circ, =)$. In the first one, the equality is automatic, while in the second one, both the group operation and the equality are only semiautomatic. If a group is finitely generated, then the definition of being Cayley biautomatic is the same as having a presentation of the form $(A, =; \circ)$.

Finitely generated semiautomatic groups share with the other types of automatic groups one important property: The word problem can be decided in quadratic time and the algorithm is the same as known for the Cayley automatic groups [7]. Thus finitely generated groups with an undecidable or very complex word problem are not semiautomatic.

2   Constructions of Semiautomatic Groups

Recall that Cayley automatic groups are finitely generated group $(A, \circ)$ iff the domain A is regular, every group element has a unique representative in A and for every $a \in A$, the mapping $x \mapsto x \circ a$ is automatic. This notion is equivalent to allowing multiple representatives in A for the group elements, but additionally requiring that equality is automatic.

Miasnikov and Šunić [9] showed that there are Cayley automatic groups which are not Cayley biautomatic, that is, which have no semiautomatic representation of the form $(A, =; \circ)$; furthermore, there are Cayley automatic groups for which the conjugacy problem is undecidable and that the isomorphism problem is also undecidable for the class of Cayley automatic groups. Berdinsky and Khoussainov [1] have shown that every Baumslag Solitar group is Cayley automatic and Jain et al. [6] announced that every Baumslag Solitar group is semiautomatic. This and other results follow now from a general transfer theorem which shows that every Cayley automatic group is semiautomatic.

Theorem 1. *If $(A, =; \{x \mapsto x \circ a : a \in A\})$ is a Cayley automatic group, then the group has a semiautomatic presentation $(B, x \mapsto x^{-1}; \circ, =)$; in this presentation the domain is regular and the inversion is an automatic function, whereas the equality and the group operation are semiautomatic.*

Proof. Given the Cayley automatic group A as in the statement of the theorem, let $B = \{(v, w) : v, w \in A\}$ be the set of convoluted pairs (v, w), where the pair (v, w) stands for the element $v^{-1} \circ w$ of A. Now $(v, w) \circ (\varepsilon, u) = v^{-1} \circ w \circ \varepsilon^{-1} \circ u = v^{-1} \circ w \circ u = (v, w \circ u)$, $(u, \varepsilon) \circ (v, w) = u^{-1} \circ \varepsilon \circ v^{-1} \circ w = (v \circ u)^{-1} \circ w = (v \circ u, w)$ and $(v, w) \circ (w, v) = v^{-1} \circ w \circ w^{-1} \circ v = \varepsilon$. As every fixed element $a \in A$ can be represented by either (ε, a) or (a^{-1}, ε), multiplication with a fixed group element from either side is automatic. Furthermore, the mapping $(v, w) \mapsto (w, v)$ is automatic and computes the inverse. The set of representations of a fixed element $a \in A$ is the set $\{(v, w) : (a, \varepsilon) \circ (v, w) = (\varepsilon, \varepsilon)\} = \{(v, w) : v \circ a = w\}$, where the latter set is easily seen to be regular.   □

The above result shows that the undecidability results for Cayley automatic groups by Miasnikov et al. [9,12] generalise to finitely generated semiautomatic groups.

Corollary 2. *There is a semiautomatic group $(A; \circ, =)$ for which the conjugacy problem is undecidable. Furthermore, the isomorphism problem for semiautomatic groups is undecidable.*

The next result shows that all semiautomatic groups can have an automatic inversion.

Proposition 3. *Every semiautomatic group* $(A; \circ, =)$ *has a further semiautomatic presentation* $(B, x \mapsto x^{-1}; \circ, =)$.

Proof. The proposition is proven by introducing two new symbols, $+$ and $-$, such that $B = \{+, -\} \cdot A$ consists of all $+x$ representing $x \in A$ and $-x$ representing x^{-1} for $x \in A$. The inverse is now computed by the function interchanging $+$ and $-$. For fixed $a \in A$, $x \mapsto x \circ a$ becomes $+x \mapsto +(x \circ a)$ and $-x \mapsto -(a^{-1} \circ x)$, $x \mapsto a \circ x$ is implemented similarly and $\{+x : x = a\} \cup \{-x : x = a^{-1}\}$ is the regular set of representatives of a in B. $\qquad\square$

Theorem 4. *Assume that* $(A, \circ, =)$ *is an automatic group (in the sense of Hodgson, Khoussainov and Nerode),* $(B; \circ, =)$ *is a semiautomatic group and* $\{\varphi_b : b \in B\}$ *is a family of homomorphisms from* A *to* A *such that for each* $b, b' \in B$, $\varphi_{b \circ b'}(a) = \varphi_b(\varphi_{b'}(a))$ *and each* φ_b *is an automatic mapping, then the semidirect product* $A \rtimes_{\varphi} B$ *is also a semiautomatic group. Here the group operation in* $A \rtimes_{\varphi} B$ *is defined by* $(a, b) \circ (a', b') = (a \circ \varphi_b(a'), b \circ b')$.

Proof. Consider the representation set $C = \{(a, b, \tilde{a}) : a, \tilde{a} \in A \text{ and } b \in B\}$, where $(a, b, \tilde{a}) \in C$ stands for $(a, \varepsilon) \circ (\varepsilon, b) \circ (\tilde{a}, \varepsilon)$ in the group $A \rtimes_{\varphi} B$. Now, for a fixed $a' \in A$, $b' \in B$ and arbitrary $(a, b, \tilde{a}) \in C$, the multiplications are defined as follows:

$$(a, b, \tilde{a}) \circ (a', \varepsilon) \mapsto (a, b, \tilde{a} \circ a');$$
$$(a', \varepsilon) \circ (a, b, \tilde{a}) \mapsto (a' \circ a, b, \tilde{a});$$
$$(a, b, \tilde{a}) \circ (\varepsilon, b') \mapsto (a, b \circ b', \varphi_{b'^{-1}}(\tilde{a}));$$
$$(\varepsilon, b') \circ (a, b, \tilde{a}) \mapsto (\varphi_{b'}(a), b' \circ b, \tilde{a}).$$

Note that multiplying with $(a', b', \tilde{a}')$ in C can be defined using the above as $(a', b', \tilde{a}') = (a', \varepsilon) \circ (\varepsilon, b') \circ (\tilde{a}', \varepsilon)$. Now, all the four mappings above are automatic as they only use homomorphisms from B, which are automatic, and multiplication with fixed group elements in the basic groups A and B, which are automatic. It follows that $\circ$ is semiautomatic in C.

For equality, note that $(a, b, \tilde{a}) = (a', b', \tilde{a}')$ iff $b = b'$ (in group B) and $a \circ \varphi_b(\tilde{a}) = a' \circ \varphi_{b'}(\tilde{a}')$ (in group A). Thus, for a fixed $(a', b', \tilde{a}') \in C$ and any $(a, b, \tilde{a}) \in C$, $(a, b, \tilde{a}) = (a', b', \tilde{a}')$ iff $b = b'$ (in group B) and $a \circ \varphi_{b'}(\tilde{a}) = a' \circ \varphi_{b'}(\tilde{a}')$. As $\varphi_{b'}$, $\circ$ restricted to A and equality in A are automatic, it follows that equality is semiautomatic in C. $\qquad\square$

It can also be shown that the free product of finitely many semiautomatic groups is semiautomatic. The construction is very much inline with the one of Kharlampovich et al. [7] for Cayley automatic groups with some adjustments for semiautomaticity.

Theorem 5. *The free product of finitely many semiautomatic groups is semiautomatic.*

Proof. Let $(A_1; \circ, =), \ldots, (A_n; \circ, =)$ be semiautomatic groups which all share the empty word ε as neutral element and use disjoint alphabets to represent the other elements. Note that, for each fixed $a \in A_k$, there is an automatic mapping $x \mapsto a \circ x$ (for $x \in A_k$) such that the length of $a \circ x$ is at most a constant more than the length of x. Let $\#$ be a symbol not appearing in the members of $A_1, \ldots, A_k$. Now each member of the free product B of $A_1, \ldots, A_k$ is a word of the form $\#^+ u_1 \#^+ u_2 \#^+ \ldots \#^+ u_m \#^+$ with $u_1, \ldots, u_m$ representing elements different from ε and no two subsequent u_h, u_{h+1} are from the same group A_k. Any word from $\#^+$ denotes the neutral element of the group. It is sufficient to show that the multiplication with any fixed element from $A_1 \cup A_2 \cup \ldots \cup A_n - \{\varepsilon\}$ is automatic, multiplying with ε can be realised by the identity function. Consider $a \in A_k - \{\varepsilon\}$.

Now $x \mapsto a \circ x$ is given as follows: x is mapped to $\#a\#$ in the case that $x \in \#^+$; x is mapped to $\#a\#x$ in the case that the first component u_1 from x is not from A_k; x is mapped to the word x', where u_1 is replaced by $\#^{|u_1|}$ in the case that $u_1 = a^{-1}$; otherwise x is mapped to the word x', where u_1 is replaced by a word from $(a \circ u_1)\#^*$. To ensure automaticity of the mapping, in the last two cases above, enough $\#$'s are filled in to make sure that length of x and x' do not differ by more than a constant (independent of x).

The mapping $x \mapsto x \circ a$ is given as follows: x is mapped to $\#a\#$ in the case that $x \in \#^+$; x is mapped to $x\#a\#$ in the case that the last component u_m of x is not from A_k; the last part of the form $u_m\#^+$ is erased from x by the mapping in the case that $u_m = a^{-1}$ and the last part $u_m\#^+$ is replaced by $(u_m \circ a)\#$ in the case that $u_m \in A_k - \{a^{-1}\}$.

Furthermore, for comparing whether x of the form $\#^+ u_1 \#^+ \ldots \#^+ u_n \#^+$ represents a fixed element $\#a_1\#a_2\# \ldots \#a_m\#$, consider the automaton consisting of m distinct subautomatons: the h-th subautomaton checks whether the component u_h of x is from the same A_k as a_h and has the same value; the automaton accepts iff all these checks succeed and the number of components n of x is exactly m. $\qquad\square$

3 Nilpotent Groups

Kharlampovich et al. [7] showed that finitely generated groups of nilpotency class 1 or 2 are Cayley automatic. The next theorem uses semiautomatic groups in place of Cayley automatic groups and pushes the above result one step further. As it is open whether all the finitely generated groups of nilpotency class 3 are Cayley automatic, this result provides a candidate for separating the two notions within the finitely generated groups.

Theorem 6. *Every finitely generated group of nilpotency class 3 can be represented as a semiautomatic group $(A; \circ, =)$.*

Proof. Let $a_1, \ldots, a_n$ be the finitely many generators in the original nilpotent group.

Consider the factor group of the given group over the quotient group generated by all elements of the form $x \circ y \circ x^{-1} \circ y^{-1}$. This group is Abelian and is isomorphic to

$$\mathbb{Z}^r \times \{0, 1, \ldots, p_{r+1} - 1\} \times \ldots \times \{0, 1, \ldots, p_n - 1\}$$

for some $r \leqslant n$ and natural numbers $p_{r+1}, \ldots, p_n \geqslant 2$.

Let $b_1, \ldots, b_{n'}$ be all the group elements of the form $a_i^{-1} \circ a_{i'}^{-1} \circ a_i \circ a_{i'}$ and let $c_1, \ldots, c_{n''}$ be all the group elements of the form $a_i^{-1} \circ b_j^{-1} \circ a_i \circ b_j$ or $b_j^{-1} \circ a_i^{-1} \circ b_j \circ a_i$. Note that the $c_1, \ldots, c_{n''}$ commute with all group elements, that for each i, j there is a k with

$$a_i \circ b_j = b_j \circ a_i \circ c_k, \; a_i \circ b_j^{-1} = b_j^{-1} \circ a_i \circ c_k^{-1}$$

and that for each i, i' there are j, k with

$$a_i \circ a_{i'} = a_{i'} \circ a_i \circ b_j, \; a_i \circ a_{i'}^{-1} = a_{i'}^{-1} \circ a_i \circ b_j^{-1} \circ c_k.$$

Similar rules allow to move a_i^{-1} over $a_{i'}, b_j$. Note that the group elements $b_j, b_{j'}$ also commute with each other, as when, for example, $b_{j'} = a_i^{-1} \circ a_{i'}^{-1} \circ a_i \circ a_{i'}$ then $b_j \circ b_{j'} = b_j \circ a_i^{-1} \circ a_{i'}^{-1} \circ a_i \circ a_{i'} = a_i^{-1} \circ a_{i'}^{-1} \circ a_i \circ a_{i'} \circ b_j = b_{j'} \circ b_j$. The reason is that the $c_k, c_{k'}$ produced by moving $a_i^{-1}, a_{i'}^{-1}$, respectively, over b_j, are cancelled out when moving $a_i, a_{i'}$ over b_j. Now, each group element is given by a convoluted tuple of integers

$$(m_1, \ldots, m_n, m_1', \ldots, m_{n'}', m_1'', \ldots, m_{n''}'')$$

where, for $i = r + 1, \ldots, n$, $m_i \in \{0, 1, \ldots, p_i - 1\}$. The above member of A represents

$$a_1^{m_1} \circ \ldots \circ a_n^{m_n} \circ b_1^{m_1'} \circ \ldots \circ b_{n'}^{m_{n'}'} \circ c_1^{m_1''} \circ \ldots \circ c_{n''}^{m_{n''}''}.$$

Note that several tuples of this type may represent the same group element due to products of some b_j and c_k being ε.

In the representation set A, the integers m_i and m_j' in the above are represented in binary, with the reverse ordering of the bits to allow automatic addition of components. Furthermore, each m_k'' is represented as a convoluted tuple $(h_0, h_1, \ldots, h_n)$ of integers (in binary using reverse ordering of the bits) satisfying

$$m_k'' = h_0 + h_1 \cdot m_1 + \ldots + h_n \cdot m_n,$$

The number of integers used in the overall representation described above is $n + n' + (n + 1) \cdot n''$, which is a constant independent of the group element; therefore convolution can indeed be used to represent the group element.

Now it will be shown that multiplication with a fixed a_i is automatic and that equality is semiautomatic.

First, for automaticity of the multiplication with a fixed element, note that it is sufficient to show that multiplication with a fixed generator from

$a_1, a_1^{-1}, \ldots, a_n, a_n^{-1}$ is automatic, as every other group element is a fixed product of these. This is shown in several steps, the number of steps is constant and each step is automatic. For showing that the mapping $x \mapsto a_i \circ x$ is automatic, it is now described how, $a_i \circ a_{i'}^{m_{i'}} a_{i'+1}^{m_{i'+1}} \ldots a_n^{m_n} b_1^{m_1'} \ldots b_{n'}^{m_{n'}'} c_1^{m_1''} \ldots c_{n''}^{m_{n''}''}$
is updated to $a_{i'}^{m_{i'}} \circ a_i \circ a_{i'+1}^{m_{i'+1}} \ldots a_n^{m_n} b_1^{s_1'} \ldots b_{n'}^{s_{n'}'} c_1^{s_1''} \ldots c_{n''}^{s_{n''}''}$, where $i' < i$ and $m_k'' = (h_0^k, h_1^k, \ldots, h_n^k)$. Repeatedly using this mechanism to shift a_i over $a_1^{m_1} a_2^{m_2} \ldots a_{i-1}^{m_{i-1}}$ and then showing how to handle the increase of m_i by 1, gives the multiplication by a_i for any member of the group as represented in A. Now suppose $1 \leqslant i' < i$, $1 \leqslant q \leqslant n$, and $m_{i'} > 0$. There are $j, k_1, \ldots, k_n$ such that $a_i a_{i'} = a_{i'} a_i b_j$ and $b_j a_q = a_q b_j c_{k_q}$. Then, the following operations are done to update m_j' and various m_k'' to obtain the corresponding m_j' and s_j'' (values not updated are unchanged).

(a) $m_{i'}$ is added to m_j' (to handle the increase in the b_j).
(b) $m_{i'}(m_{i'} - 1)/2$ is added to $m_{k_{i'}}''$ (to handle the increase in $c_{k_{i'}}$ due to moving of b_j generated in (a) over $a_{i'}^{m_{i'}}$). If $m_{i'}$ is odd, then the above addition can be achieved by adding $m_{i'}/2$ to $h_0^{k_{i'}}$ and adding $(m_{i'} - 2)/2$ to $h_{i'}^{k_{i'}}$. If $m_{i'}$ is even, $m_{i'}(m_{i'} - 1)/2 = m_{i'}(m_{i'} - 2)/2 + m_{i'}/2$. Thus, the above addition can be achieved by adding $m_{i'}(m_{i'} - 2)/2$ to $h_0^{k_{i'}}$ and adding $m_{i'}/2$ to $h_{i'}^{k_{i'}}$.
(c) $m_{i'} * m_q$ is added to m_{k_q}'', for $q = i' + 1, \ldots, n$ (to handle the increase in c_{k_q} due to moving of $b_j^{m_{i'}}$ over $a_q^{m_q}$). This can be done by adding $m_{i'}$ to $h_q^{k_q}$.

Note that $a_i \circ a_{i'}^{-1} = a_{i'}^{-1} \circ a_i \circ b_j^{-1} \circ c_{k'}$ for some k' which permits to handle the case when $m_{i'} < 0$ in a similar manner. For the multiplication

$$a_i \circ a_i^{m_i} a_{i+1}^{m_{i+1}} \ldots a_n^{m_n} b_1^{m_1'} \ldots b_{n'}^{m_{n'}'} c_1^{m_1''} \ldots c_{n''}^{m_{n''}''},$$

one updates m_i to $m_i + 1$ and, as a chain reaction, for $k = 1, \ldots, n''$, update h_0^k to $h_0^k - h_i^k$, for the tuple $(h_0^k, h_1^k, \ldots, h_n^k)$ representing m_k'' (so that the new value of m_i is used rather than the older value in the computation of m_k'').

Similarly it can be shown that also the mappings $x \mapsto a_i^{-1} \circ x$, $x \mapsto x \circ a_i$ and $x \mapsto x \circ a_i^{-1}$ are automatic.

The above handles all multiplications by a_i on the left except for the case of $i > r$ and $m_i + 1 = p_i$. To handle this, an additional multiplication by $a_i^{-p_i}$ can be done using the above mechanism to bring the power of a_i to 0.

Now, for showing semiautomaticity of equality, for any fixed element

$$a_1^{m_1} \circ \ldots \circ a_n^{m_n} \circ b_1^{m_1'} \circ \ldots \circ b_{n'}^{m_{n'}'} \circ c_1^{m_1''} \circ \ldots \circ c_{n''}^{m_{n''}''} \in A$$

it is shown that the set of its representatives in A is regular. Note that in the vector of the exponents, for each further representative of the group element, the first n coordinates must also be $m_1, m_2, \ldots, m_n$, which can be easily checked. In the case that these n coordinates are equal, one can tailormake an automaton to check for equality. The automaton can, for each k and for the coordinates $(h_0, h_1, \ldots, h_n)$ representing m_k'', use the formula

$$h_0 + h_1 \cdot m_1 + \ldots + h_n \cdot m_n$$

to get the explicit value corresponding to m_k'' in the other representative, in binary notation, as multiplication of integers by constants can be done automatically. However, the m'-coordinates and m''-coordinates can be different for the two representatives. The difference of these coordinates must, however, give a vector representing ε. Thus, it suffices to give a test for neutrality in the m'-coordinates and m''-coordinates in order to be able to decide equality to the fixed given element. Call a set $\{v_1, \ldots, v_r\}$ of vectors representing these coordinates to be independent over $\mathbb{Z}$ iff no v_h can be obtained from a linear combination of the others using coefficients from $\mathbb{Z}$. If one of the sets $\{v_1, v_2, \ldots, v_r\}$, $\{-v_1, v_2, \ldots, v_r\}$, $\{v_1 - v_2, v_2, \ldots, v_r\}$, $\{v_2 - v_1, v_2, \ldots, v_r\}$ is independent, then all of them are. So Euclid's algorithm can be run on the vectors until all but one vector have a 0 in the first coordinate; then one can run the algorithm until, among all those vectors with a 0 in the first coordinate, all but at most one have a 0 in the second coordinate and so on. This implies that the number of independent vectors is not larger than the number of coordinates. Thus there are fixed vectors $\{v_1, \ldots, v_\ell\}$ such that two vectors

$$(m_1, \ldots, m_n, m_1', \ldots, m_{n'}', m_1'', \ldots, m_{n''}'') \text{ and}$$
$$(m_1, \ldots, m_n, \tilde{m}_1', \ldots, \tilde{m}_{n'}', \tilde{m}_1'', \ldots, \tilde{m}_{n''}'')$$

represent the same element iff the difference

$$(0, \ldots, 0, m_1' - \tilde{m}_1', \ldots, m_{n'}' - \tilde{m}_{n'}', m_1'' - \tilde{m}_1'', \ldots, m_{n''}'' - \tilde{m}_{n''}'')$$

is of the form $r_1 \cdot v_1 + r_2 \cdot v_2 + \ldots r_\ell \cdot v_\ell$ for some $r_1, \ldots, r_\ell \in \mathbb{Z}$. This is an existentially quantified formula, where the multiplication with fixed vectors (represented as convoluted tuples) can be done by an automatic function and their adding and comparing with the target as well. Thus the predicate whether the two vectors from above representing the two group elements are the same is automatic. Thus for each group element

$$a_1^{m_1} \circ \ldots \circ a_n^{m_n} \circ b_1^{m_1'} \circ \ldots \circ b_{n'}^{m_{n'}'} \circ c_1^{m_1''} \circ \ldots \circ c_{n''}^{m_{n''}''}$$

there is a finite automaton which decides whether another group element is equal to it. So the group $(A; \circ, =)$ is semiautomatic. $\square$

For the representation used in the above theorem, by using the natural subgroup B of all elements in A generated by the b_j and c_k, the following Theorem 7 below can be shown. The key idea is to represent each group element in the form $b \circ a \circ \tilde{b}$ where $b, \tilde{b}$ are in B and a is either $a_1^{m_1} \circ \ldots \circ a_n^{m_n}$ or $a_n^{m_n} \circ \ldots \circ a_1^{m_1}$; these two orderings are used in order to facilitate inversion. Item **(b)** in the theorem below is proven by coding a computationally difficult problem into the theory of the structure of the group and then conclude that this problem would be solvable in the case that the given structure is semiautomatic.

Theorem 7. *In the following, $(A, \circ)$ denotes a finitely generated group of nilpotency class 3, B denotes the commutator subgroup generated by all elements of the form $x \circ y \circ x^{-1} \circ y^{-1}$ with $x, y \in A$ and $\bullet$ denotes the restriction of $\circ$ to the domain $(A \times B) \cup (B \times A)$.*

(a) *For every A as above, the structure $(A, B, x \mapsto x^{-1}, \bullet; \circ, =)$ is semiautomatic.*

(b) *For some A as above, the structure $(A, B, \bullet, =; \circ)$ is not semiautomatic.*

Proof. (a): The notation from the proof of Theorem 6 is carried over for this proof. The group elements are represented as products $b \circ a \circ \tilde{b}$ where (i) $b, \tilde{b}$ are products of b_j's and c_k's represented in the same format as they are represented in Theorem 6 and (ii) a is a member of $(a_1^* a_2^* \ldots a_n^* \cup a_n^* \ldots a_2^* a_1^*)$ represented as a convoluted tuple $(m_0, m_1, \ldots, m_n)$, where $m_0 \in \{-1, 1\}$; the tuple $(m_0, m_1, \ldots, m_n)$ represents $a_1^{m_1} \circ \ldots \circ a_n^{m_n}$, if $m_0 = 1$, and $a_n^{m_n} \circ \ldots \circ a_1^{m_1}$, if $m_0 = -1$.

The mappings $b \mapsto b^{-1}$, $a \mapsto a^{-1}$ and $\tilde{b} \mapsto \tilde{b}^{-1}$ are realised by negating all entries in the corresponding tuples; for mapping $(b \circ a \circ \tilde{b})$ to $(b \circ a \circ \tilde{b})^{-1}$, one has to exchange the entries of b and $\tilde{b}$ as well, as $(b \circ a \circ \tilde{b})^{-1} = \tilde{b}^{-1} \circ a^{-1} \circ b^{-1}$. Thus the mapping $x \mapsto x^{-1}$ (in the chosen representation) is automatic.

Note that, in the representation for $\hat{b} \in B$, all the m-coordinates are 0. Thus for the component m_k'' in the representation of $\hat{b}$, the integers $h_1, h_2, \ldots$ (as in Theorem 6) can be ignored. Hence, the multiplication $\hat{b} \bullet (b \circ a \circ \tilde{b})$, can be done by adding m_j' coordinate of the representation of $\hat{b}$ to the corresponding m_j' coordinate of b and the h_0-component of the m_k'' coordinate of $\hat{b}$ to the corresponding h_0-component of the m_k'' coordinates of b. Similarly, when computing $(b \circ a \circ \tilde{b}) \bullet \hat{b}$, the coordinates of $\hat{b}$ are added as above to those of $\tilde{b}$.

Note that, in the representation chosen for this proof, $\hat{b} \in B$ is actually a product of the form: $b' \circ \varepsilon \circ \tilde{b}'$, where $b', \tilde{b}'$ are represented as in Theorem 6. The coordinates of b' and $\tilde{b}'$ as above can be contracted to the coordinates of one member of B by simply adding, component-wise, prior to carrying out the multiplication $\bullet$ as described above. These arguments explain why $\bullet$ is an automatic function.

Multiplication of an element x with generators a_i from either side as done in Theorem 6, can easily be adjusted to the representation chosen here.

Now, assume a fixed group element $x = a_1^{m_1} a_2^{m_2} \ldots a_n^{m_n} \circ b'$, where $b' \in B$, is given. Let $a = a_1^{m_1} a_2^{m_2} \ldots a_n^{m_n}$. Note that for all representatives $y = b \circ a' \circ \tilde{b}$ of x in the representation chosen, the coordinates $m_1, m_2, \ldots, m_n$ corresponding to a' must match to that of a as above. Furthermore, there is a fixed element $\hat{b} \in B$ such that

$$a_n^{m_n} \circ \ldots \circ a_1^{m_1} = a_1^{m_1} \circ \ldots \circ a_n^{m_n} \circ \hat{b}.$$

Now, for any given representative $y = b \circ a' \circ \tilde{b}$ with the coordinates for a' being $m_0 = -1, m_1, \ldots, m_n$, the coordinate m_0 can be converted to $+1$, by replacing $\tilde{b}$ by $\hat{b} \circ \tilde{b}$. Furthermore, using the arguments given in the proof of Theorem 6, there is a fixed automatic homomorphism $\varphi_a : B \to B$ such that $b \circ a = a \circ \varphi_a(b)$. The products $\varphi_a(b) \circ \tilde{b}$ or $\varphi_a(b) \circ \hat{b} \circ \tilde{b}$ can be carried out by componentwise addition of the vectors involved. Once this is done, the algorithm from Theorem 6 can be used to compare y in the resulting representation with that of x. Thus equality is semiautomatic in the representation chosen.

The full proof of part (b) is omitted due to space constraints. The idea is to code a hard problem into a system of equations for a specific semiautomatic structure of the form $(A, B, \bullet, =; \circ)$. Each of the equations involves multiplication of an existentially quantified variable x from A with constants from A and existentially quantified variables from B, the latter is automatic as it falls into the domain of $\bullet$. Then it is shown that solving such equations permits to solve the original computationally hard problem which cannot be solved by an automaton, implying that the structure cannot be semiautomatic. The coding mentioned above is either of an NP-hard problem (which is more direct to do) or in a more involved way of a Diophantine unsolvable problem, which then gives the undecidability result. □

References

1. Berdinsky, D., Khoussainov, B.: On automatic transitive graphs. In: Shur, A.M., Volkov, M.V. (eds.) DLT 2014. LNCS, vol. 8633, pp. 1–12. Springer, Heidelberg (2014)
2. Case, J., Jain, S., Seah, S., Stephan, F.: Automatic functions, linear time and learning. Logical Meth. Comput. Sci. **9**(3), 1–26 (2013)
3. Epstein, D.B.A., Cannon, J.W., Holt, D.F., Levy, S.V.F., Paterson, M.S., Thurston, W.P.: Word Processing in Groups. Jones and Bartlett Publishers, Boston (1992)
4. Hodgson, B.R.: Théories décidables par automate fini. Ph.D. thesis, Département de mathématiques et de statistique, Université de Montréal (1976)
5. Hodgson, B.R.: Décidabilité par automate fini. Annales des sciences mathématiques du Québec **7**(1), 39–57 (1983)
6. Jain, S., Khoussainov, B., Stephan, F., Teng, D., Zou, S.: Semiautomatic structures. In: Hirsch, E.A., Kuznetsov, S.O., Pin, J.É., Vereshchagin, N.K. (eds.) CSR 2014. LNCS, vol. 8476, pp. 204–217. Springer, Heidelberg (2014)
7. Kharlampovich, O., Khoussainov, B., Miasnikov, A.: From automatic structures to automatic groups. Groups Geom. Dyn. Syst. **8**(1), 157–198 (2014)
8. Khoussainov, B., Nerode, A.: Logic and computational complexity. In: Leivant, D. (ed.) Automatic Presentations of Structures. LNCS, vol. 960, pp. 367–392. Springer, Heidelberg (1995)
9. Miasnikov, A., Šunić, Z.: Cayley graph automatic groups are not necessarily Cayley graph biautomatic. In: Dediu, A.-H., Martín-Vide, C. (eds.) LATA 2012. LNCS, vol. 7183, pp. 401–407. Springer, Heidelberg (2012)
10. Nies, A.: Describing groups. Bull. Symbolic Logic **13**(3), 305–339 (2007)
11. Nies, A., Thomas, R.: FA-presentable groups and rings. J. Algebra **320**, 569–585 (2008)
12. Šunić, Z., Ventura, E.: The conjugacy problem in automaton groups is not solvable. J. Algebra **364**, 148–154 (2012)

The Boolean Algebra of Piecewise Testable Languages

Anton Konovalov and Victor Selivanov$^{(\boxtimes)}$

A.P. Ershov Institute of Informatics Systems,
Siberian Division Russian Academy of Sciences,
Novosibirsk State University, Novosibirsk, Russia
Jack@sibmail.ru, vseliv@iis.nsk.su

Abstract. We characterize up to isomorphism the Boolean algebra (BA, for short) of regular piecewise testable languages and show the decidability of classes of regular languages related to this characterization. This BA turns out isomorphic to several other natural BAs of regular languages, in particular to the BA of regular aperiodic languages.

Keywords: Boolean algebra · Frechét ideal · Regular language · Aperiodic language · Piecewise testable language

1 Introduction

Boolean algebras are of principal importance for several branches of mathematics. Accordingly, characterization of naturally arising BA's attracts attention of many researchers. As examples we mention characterizations of natural BA's in logic and computability theory [3,7,12–14].

In automata theory, people consider many natural classes of languages which form BA's whose characterizations could be of some interest because they provide new information on well-known classes of regular languages. Due to the Stone duality, this contributes to understanding of the corresponding Stone spaces which are closely related to the profinite topology [1,10].

The first papers in this direction appeared relatively recently [8,16]. In [5,6,16] some fundamental BA's of regular languages and ω-languages were characterized up to isomorphism. A surprising discovery was that those BAs are either quite simple (informally, very similar to the countable atomless BA) or else they are isomorphic to a distinguished BA $\mathbb{A}$ described as follows (we use some well-known terminology on BA's which may be found e.g. in [2], see also the next section).

If $\mathbb{B}$ is a BA and α an ordinal, let $F_\alpha(\mathbb{B})$ be the α-th iterated Frechét ideal of $\mathbb{B}$, and $\mathbb{B}^{(\alpha)} = \mathbb{B}/F_\alpha(\mathbb{B})$ is the α-th Frechét derivative of $\mathbb{B}$. Frechét ideals are central for a useful classification of isomorphism types of countable BA's [4]. A very particular case of this is the following assertion: There is a unique,

A. Konovalov and V. Selivanov supported by RFBR project 13-01-00015a.

A. Beckmann et al. (Eds.): CiE 2016, LNCS 9709, pp. 292–301, 2016.
DOI: 10.1007/978-3-319-40189-8_30

up to isomorphism, countable BA $\mathbb{A}$ such that $F_0(\mathbb{A}) \subset F_1(\mathbb{A}) \subset \cdots \subset F_\omega(\mathbb{A}) = F_{\omega+1}(\mathbb{A})$, for each $n < \omega$ the BA $\mathbb{A}^{(n)}$ is atomic with infinitely many atoms, and $\mathbb{A}^{(\omega)}$ is a countable atomless BA.

The BA's considered in [5,6,16] are large in the sense that any of them extends the BA $\mathbb{A}_\Sigma$ of regular aperiodic languages over an alphabet Σ (or the corresponding BA of ω-languages). In this note, we start to investigate "small" BA's of regular languages like the self-dual levels of Straubing-Thérien or Brzozowski hierarchies [18,20]. Probably, the most popular of those is the BA of piecewise testable languages $\mathbb{P}_\Sigma$ over Σ. Recall that a language L is piecewise testable if there exists a number k (depending only on L) such that membership of a word in the language depends only on the subwords (also called pieces) of length less than k (we recall some characterizations of piecewise testable languages in the next section).

The main result of this paper may be formulated as follows: If Σ contains at least two letters then $\mathbb{P}_\Sigma$ is isomorphic to $\mathbb{A}$.

Therefore, the BA $\mathbb{A}$ appears also among the "small" BA's of regular languages. Note that for the unary alphabet Σ the BA $\mathbb{P}_\Sigma$ coincides with $\mathbb{A}_\Sigma$ of regular aperiodic languages and is much simpler [16].

In Sect. 2 we provide the necessary background on BA's and regular languages, in Sect. 3 we give some important examples of piecewise testable languages, and in Sect. 4 we prove the main result.

2   Preliminaries

2.1   Preliminaries on Boolean Algebras

Here we briefly recall some relevant notions about BA's used in the sequel. We assume the reader to be familiar with basic notions related to BA's like ideal of a BA, quotient-algebra of a BA modulo a given ideal, and canonical homomorphism of a BA onto its quotient-algebra (for a detailed treatment of countable BA's see e.g. [2]). BA's are considered in the signature $\{\cup, \cap, ^{-}, 0, 1\}$.

Recall that an element a of a BA $\mathbb{B}$ is an *atom* if $a \neq 0$ and $x < a$ implies $x = 0$. A BA $\mathbb{B}$ is called *atomless* if it has no atom, and it is called *atomic* if below any non-zero element there is an atom. The ideal of a BA $\mathbb{B}$ generated by atoms is called *Frechét ideal* of $\mathbb{B}$. It consists of all finite unions (including the empty union which coincides with 0) of atoms and is denoted by $F(\mathbb{B})$. The quotient-algebra $\mathbb{B}/F(\mathbb{B})$ is called *Frechét derivative* of $\mathbb{B}$ and is also denoted by $\mathbb{B}'$.

Define the transfinite sequence $\{F_\alpha(\mathbb{B})\}$ of *iterated Frechét ideals* of a BA $\mathbb{B}$ as follows: $F_0(\mathbb{B}) = \{0\}$, $F_{\beta+1}(\mathbb{B}) = \{x \mid x/F_\beta(\mathbb{B}) \in F(\mathbb{B}^{(\beta)})\}$ where $\mathbb{B}^{(\beta)} = \mathbb{B}/F_\beta(\mathbb{B})$, and $F_\alpha(\mathbb{B}) = \bigcup_{\beta < \alpha} F_\beta(\mathbb{B})$ for a limit ordinal α. This sequence is ascending under inclusion, and $F_\alpha(\mathbb{B}) = F_{\alpha+1}(\mathbb{B})$ for some ordinal α.

2.2   Preliminaries on Regular Languages

Here we briefly recall some notation, notions and facts on regular languages used in the sequel. Let Σ^* denote the set of words over Σ including the empty word

ε, and let $\Sigma^+ := \Sigma^* \backslash \{\varepsilon\}$. For any $n < \omega$, let Σ^n be the set of words over Σ of length n. The length of a word w is denoted by $|w|$. Let $\sqsubseteq$ be the prefix partial order on Σ^*, i.e. $u \sqsubseteq v$ iff $ux = v$ for some $x \in \Sigma^*$.

We freely use some standard definitions and notation on words, regular languages and finite automata like regular expressions or deterministic finite automata (DFA) denoted as $\mathcal{A} = (Q, \Sigma, f, q_0, F)$ (for a more systematic treatment see e.g. [18,20,21]). The automaton $\mathcal{A}$ recognizes the language $L_{\mathcal{A}} = \{u \in \Sigma^* \mid q_0 \cdot u \in F\}$ where $q \cdot u = f(q, u)$ is the state reached by $\mathcal{A}$ when it reads the word u started from a state q. Let $\mathcal{R}_{\Sigma}$ denote the class of regular languages over Σ (i.e., languages recognized by DFA's or, equivalently, denoted by regular expressions). For $L \in \mathcal{R}_{\Sigma}$, $\mathcal{M}_L$ denotes the minimal DFA that recognizes L; the minimal DFA is unique up to isomorphism.

For $k < \omega$, a language $L \in \mathcal{R}_{\Sigma}$ is called k-*sparse* if the function $p_L(n) = |\Sigma^n \cap L|$ is $O(n^k)$ (cf. [19,21]). Let $\mathcal{S}_k$ denote the class of regular k-sparse languages over Σ. Note that $\mathcal{S}_0 \subset \mathcal{S}_1 \subset \cdots$. Languages from $\mathcal{S}_\omega = \bigcup_{k<\omega} \mathcal{S}_k$ are called *sparse languages*. Note that 0-sparse languages are also known as *slender languages*. We will use the following characterizations of k-sparse languages in terms of regular expressions (cf. [19,21]).

Proposition 1. *For any $L \in \mathcal{R}_{\Sigma}$ and $k < \omega$, $L \in \mathcal{S}_k$ iff L is a finite union of languages $xy_0^* z_0 \cdots y_k^* z_k$ where $x, y_i, z_i \in \Sigma^*$.*

There is also a useful characterization of sparse languages in terms of (graphs of) their minimal automata (see e.g. [9,16]). We say that a DFA $\mathcal{A}$ has an ω-*pattern* if there are $u, v_1, v_2, w \in \Sigma^*$ such that v_1, v_2 are $\sqsubseteq$-incomparable (i.e., $v_1(i) \neq v_2(i)$ for some $i < |v_1|, |v_2|$), $q_0 \cdot u = q_0 \cdot uv_1 = q_0 \cdot uv_2$ and $q_0 \cdot uv_1 w \in F$.

Proposition 2. *For any $L \in \mathcal{R}_{\Sigma}$, $L \in \mathcal{S}_\omega$ iff the minimal DFA of L has no ω-pattern.*

For $q \in Q$ and $u \in \Sigma^*$, let (q, u) denote the path in $\mathcal{A}$ along u started at q, and let $Q(q, u) = \{q \cdot v | v \sqsubseteq u\}$. We say that the path (q, u) *meets* a set of states $G \subseteq Q$ if $Q(q, u) \cap G \neq \emptyset$. The path (q, u) is a *cycle of* $\mathcal{A}$ if u is nonempty and $q \cdot u = q$. A cycle is *simple* if it has no repeated vertices other than the starting and ending vertices. Let $C_{\mathcal{A}}$ be the set of all $Q(q, u)$ where (q, u) is a cycle of $\mathcal{A}$. Define the preorder $\leq$ on $C_{\mathcal{A}}$ as follows: $G \leq H$ if there is a path of $\mathcal{A}$ starting in G and ending in H. Let $\equiv$ denote the equivalence relation on $C_{\mathcal{A}}$ induced by $\leq$.

Note that if $G \equiv H$ then $K \equiv G$ for some $K \supseteq G \cup H$, i.e. any element $[G]$ of the quotient-set $C_{\mathcal{A}}/\equiv$ has a largest set under inclusion; these largest sets are called *strongly connected components* (SCC's) of $\mathcal{A}$ and they may serve as canonical representatives for the equivalence classes $[G]$. The next result is known and easily follows from the previous proposition:

Proposition 3. *For any sparse language $L = L_{\mathcal{A}} \in \mathcal{R}_{\Sigma}$ and any $u \in L$, if the path (q_0, u) meets a SCC G of $\mathcal{A}$ then $[G] = \{G\}$ and $G = Q(s, v)$ for some simple cycle of $\mathcal{A}$.*

We also recall the following classical fact of automata theory. A DFA $\mathcal{A}$ is called *counter-free* if $q \cdot u^n = q$ implies $q \cdot u = q$, for all $q \in Q$, nonempty word $u \in \Sigma^*$ and $n > 0$.

Proposition 4. *For any $L \in \mathcal{R}_\Sigma$ the following conditions are equivalent:*

1. *There is $n < \omega$ such that $xy^n z \in L$ iff $xy^{n+1}z \in L$ for all $x, y, z \in \Sigma^*$.*
2. *The minimal DFA of L is counter-free.*

Languages satisfying the conditions in the last proposition are called *aperiodic*. There are several other important characterizations of the class $\mathcal{A}_\Sigma$ of aperiodic languages, in particular $L \subseteq \Sigma^+$ is aperiodic iff L is defined by a first order sentence of signature $\sigma = \{<, Q_a\}_{a \in \Sigma}$ (see e.g. [18,20]) for additional details). The class $\mathcal{A}_\Sigma$ is closed under the Boolean operations, let $\mathbb{A}_\Sigma$ be the corresponding BA.

A popular subclass of the aperiodic languages is the class $\mathcal{P}_\Sigma$ of *piecewise testable languages* (see e.g. [11,17]). It has several nice characterizations, in particular $L \subseteq \Sigma^+$ is piecewise testable iff L is defined by a boolean combination of existential sentences of signature σ. Below we use the following characterization [11,17] in terms of forbidden patterns T_0, T_1 (where $x, z \in \Sigma^*$, $u, v \in \Sigma^+$) shown on Figs. 1 and 2 below.

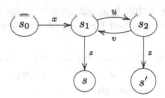

Fig. 1. Pattern T_0, where s_0 is the initial state and $s \in F \Leftrightarrow s' \notin F$.

Proposition 5. *For any regular language L, $L \in \mathcal{P}_\Sigma$ iff any DFA recognizing L does not contain any of the patterns T_0 and T_1.*

Restricting to minimal automata we immediately obtain:

Corollary 1. *For any regular language L, $L \in \mathcal{P}_\Sigma$ iff the minimal DFA of L contains neither non-singleton cycles nor the patterns T_1.*

Remark. In the literature one often meets two versions of piecewise testable languages, namely the languages which can or cannot contain the empty word. Usually, proofs for the two cases are slightly different but formulations essentially coincide. This also applies to our case. In the sequel we stick for simplicity to the case of languages $L \subseteq \Sigma^+$ of non-empty words; the other version is considered in a similar fashion.

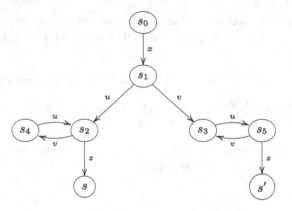

Fig. 2. Pattern T_1, where s_0 is the initial state and $s \in F \Leftrightarrow s' \notin F$.

3   A Family of Piecewise Testable Languages

Here we show that for all $x_0, \ldots, x_n \in \Sigma^*$ with $x_0 \cdots x_n \neq \varepsilon$, and for all $a_1, \ldots, a_n \in \Sigma$, the language $x_0 a^*_1 x_1 \cdots a^*_n x_n$ is piecewise testable. This easy fact is crucial for the characterization of BA $\mathbb{P}_\Sigma$. Since we have not seen it in the literature, we provide a detailed proof using the automata-theoretic characterization of piecewise testable languages from the previous section. From a discussion with J.-E. Pin we learned that a natural semigroup-theoretic proof is also possible.

By *simple expression* we mean a regular expression $x_0 a^*_1 x_1 \cdots a^*_n x_n$ where $x_1, \ldots, x_{n-1} \in \Sigma^+$, $x_0, x_n \in \Sigma^*$, $x_0 \cdots x_n \neq \varepsilon$, $a_1, \ldots, a_n \in \Sigma$, and the first letter of x_i is distinct from a_i for all $0 < i < n$, and also for $i = n$ if x_n is non-empty.

Lemma 1. *Let $L \subseteq \Sigma^+$ be the regular language defined by a simple expression as above. Then the minimal DFA of L does not contain non-singleton cycles and patterns T_1 from the previous section.*

Proof. For $n = 0$ the assertion is clear. Let $n > 0$ and let m_i be the length of the word x_i for $i \leq n$, so $x_i = x_{i,1} \cdots x_{i,m_i}$ for unique letters $x_{i,1}, \ldots, x_{i,m_i} \in \Sigma$. Note that $m_i > 0$ for all $0 < i < n$. For $n = 1$ the minimal DFA for $L = x_0 a^*_1 x_1$ looks as shown on Fig. 3 where s_0 is the initial state, s_2 is the unique accepting state, and the arrows to the rejecting sink state s carry all letters required by the definition of DFA. Note that in the case $x_0 = \varepsilon$ we have $x_1 \neq \varepsilon$, $s_0 = s_1$ and the nodes $q_1, \ldots, q_{m_0-1}$ and the corresponding arrows disappear, while in the case $x_1 = \varepsilon$ we have $x_0 \neq \varepsilon$, $s_1 = s_2$ is the accepting state, and the nodes $p_1, \ldots, p_{m_1-1}$ and the corresponding arrows disappear. It is easy to see that the minimal DFA has the desired property.

For $n = 2$ the minimal DFA for $L = x_0 a^*_1 x_1 a^*_2 x_2$ looks similar: x_0 leads from s_0 to s_1, x_1 leads from s_1 to s_2, x_2 leads from s_2 to the unique accepting state s_3, and all other arrows lead to the rejecting sink state s.

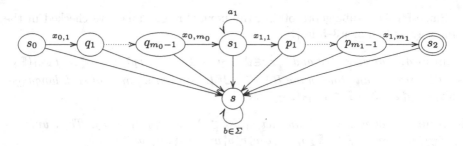

Fig. 3. Minimal DFA for $n = 1$.

The argument works in a similar way for any $n \geq 3$. $\qquad\qquad\square$

Lemma 2. *For any $n > 0$, $x_0, \ldots x_n \in \Sigma^*$ and $a_1, \ldots a_n \in \Sigma$, the language $L = x_0 a_1^+ x_1 \cdots a_n^+ x_n$ may be represented by a simple expression $y_0 b_1^* y_1 \cdots b_m^* y_m$ such that $m \leq n$, $\{b_1, \ldots, b_m\} = \{a_1, \ldots, a_n\}$ and y_{i-1} has the letter b_i for all $1 \leq i \leq m$.*

Proof. By induction on n. For $n = 1$, write x_1 as $x_1 = a_1^l y_1$ where $l \geq 0$ and y_1 is empty or starts with a letter distinct from a_1. Then a desired representation is $L = (x_0 a_1^{l+1}) a_1^* y_1$.

For $n > 1$, set $K = x_1 a_2^+ x_2 \ldots x_{n-1} a_n^+ x_n$, then $L = x_0 a_1^+ K$. By induction, K has a desired representation $y_1 b_2^* y_2 \ldots b_m^* y_m$, so $L = x_0 a_1^+ y_1 b_2^* y_2 \ldots b_m^* y_m$. If $a_1 \neq b_2$, represent $x_0 a_1^+ y_1$ as $(x_0 a_1^{l+1}) a^* z_1$ similar to the case $n = 1$, then $L = (x_0 a_1^{l+1}) a^* z_1 b_2^* y_2 \ldots b_m^* y_m$ is a desired representation. The same argument works for the case when $a_1 = b_2$ and $y_1 \notin b_2^*$. Finally, let $a_1 = b_2$ and $y_1 = b_2^l$ for some $l \geq 0$. Then $L = (x_0 b_2^{l+1}) b_2^* y_2 \ldots b_m^* y_m$ is a desired representation. $\quad\square$

Lemma 3. *For all $x_0, \ldots, x_n \in \Sigma^*$ with $x_0 \cdots x_n \neq \varepsilon$, and for all $a_1, \ldots, a_n \in \Sigma$, the language $L = x_0 a^*_1 x_1 \cdots a_n^* x_n$ is piecewise testable.*

Proof. It is easy to see that L is the union of languages $L(S)$ for all $S \subseteq \{1, \ldots, n\}$ where $L(S)$ is obtained from $x_0 a_1^+ x_1 \cdots a_n^+ x_n$ by deleting the expressions a_i^+ with $i \in S$ (e.g., for $n = 2$ we have $x_0 a_1^* x_1 a_2^* x_2 = x_0 x_1 x_2 \cup x_0 a_1^+ x_1 x_2 \cup x_0 x_1 a_2^+ x_2 \cup x_0 a_1^+ x_1 a_2^+ x_2$).

By Lemmas 1, 2 and Corollary 1, any $L(S)$ is piecewise testable. Since $\mathbb{P}_\Sigma$ is closed under finite union, L is piecewise testable. $\qquad\square$

Remark. As is well known (see e.g. [15] and references therein), the class of piecewise testable languages is stratified to the ω levels of the difference hierarchy over the class of languages defined by existential sentences of signature σ. It is not hard to show that no fixed level of this hierarchy contains all languages from the previous lemma.

4   Proof of the Main Result

Now it is easy to adjust the proof in [16] to the case of piecewise testable languages, so we give only short comments.

Since $\mathbb{P}_\Sigma$ is a subalgebra of $\mathbb{A}_\Sigma$, the next three lemmas are checked in the same way as Lemmas 3–5 in [16].

Lemma 4. *For all $k < \omega$ and $x, z_i \in \Sigma^*$, $y_i \in \Sigma \cup \{\varepsilon\}$, $xy_1^*z_1 \cdots y_k^*z_k/F_{k+1}(\mathbb{P}_\Sigma)$ is either zero or an atom in the BA $\mathbb{P}_\Sigma^{(k)}$. If L is a finite union of such languages $xy_1^*z_1 \cdots y_k^*z_k$ then $L \in F_{k+1}(\mathbb{P}_\Sigma)$.*

Lemma 5. *Let $u, w \in \Sigma^*$ and let $v_1, v_2 \in \Sigma$ satisfy $v_1 \neq v_2$. Then $uv_1^*w \notin F_1(\mathbb{P}_\Sigma)$, $uv_1^*v_2v_1^*w \notin F_2(\mathbb{P}_\Sigma)$, $uv_1^*v_2v_1^*v_2v_1^*w \notin F_3(\mathbb{P}_\Sigma)$ and so on.*

Lemma 6. *For any $L \in F_\omega(\mathbb{P}_\Sigma)$ the minimal DFA of L has no ω-pattern.*

The next two theorems and corollaries are also proved in the same way as the corresponding facts in [16].

Theorem 1. *For any $L \in \mathcal{R}_\Sigma$ the following conditions are equivalent:*

1. $L \in F_\omega(\mathbb{P}_\Sigma)$.
2. *The minimal DFA of L has neither non-singleton cycles nor ω-patterns.*
3. $L \in \mathcal{P}_\Sigma \cap \mathcal{S}_\omega$.
4. L *is a finite union of languages $xy_0^*z_0 \cdots y_k^*z_k$, where $k < \omega$, $x, z_i \in \Sigma^*$, and $y_0, \ldots, y_k \in \Sigma$.*

Corollary 2. *The class of regular languages $F_\omega(\mathbb{P}_\Sigma)$ is decidable.*

Next we characterize the ideals $F_k(\mathbb{P}_\Sigma)$ for $k < \omega$. Observe that $F_0(\mathbb{P}_\Sigma) = \{\emptyset\}$ and $F_1(\mathbb{P}_\Sigma)$ is the class of finite languages over Σ.

Theorem 2. *For any $k < \omega$ and $L \in \mathcal{R}_\Sigma$ the following conditions are equivalent:*

1. $L \in F_{k+2}(\mathbb{P}_\Sigma)$.
2. *The minimal DFA of L has neither non-singleton cycles nor ω-patterns, nor chains $F_0 < \cdots < F_{k+1}$ of SCC's of $\mathcal{M}_L$ such that some accepting state of $\mathcal{M}_L$ is reachable from F_{k+1}.*
3. $L \in \mathcal{P}_\Sigma \cap \mathcal{S}_k$.
4. L *is a finite union of languages $xy_0^*z_0 \cdots y_k^*z_k$, where $x, z_i \in \Sigma^*$, $y_0, \ldots, y_k \in \Sigma$.*

Corollary 3. *For any $k < \omega$, the class of regular languages $F_k(\mathbb{P}_\Sigma)$ is decidable uniformly on k.*

We are ready to complete the proof of the main result:

Theorem 3. *The BA $\mathbb{P}_\Sigma$ is a unique, up to isomorphism, countable BA $\mathbb{A}$ such that $F_0(\mathbb{A}) \subset F_1(\mathbb{A}) \subset \cdots \subset F_\omega(\mathbb{A}) = F_{\omega+1}(\mathbb{A})$, for each $n < \omega$ the BA $\mathbb{A}^{(n)}$ is atomic with infinitely many atoms, and $\mathbb{A}^{(\omega)}$ is a countable atomless BA.*

Proof. First we check that $F_k(\mathbb{P}_\Sigma) \subset F_{k+1}(\mathbb{P}_\Sigma)$ for each $k < \omega$. For $k = 0$ this is clear. Choose distinct $a, b \in \Sigma$, it suffices to show that

$$a^* \in F_2(\mathbb{P}_\Sigma) \backslash F_1(\mathbb{P}_\Sigma), \ a^*ba^* \in F_3(\mathbb{P}_\Sigma) \backslash F_2(\mathbb{P}_\Sigma), \ a^*ba^*ba^* \in F_4(\mathbb{P}_\Sigma) \backslash F_3(\mathbb{P}_\Sigma),$$

so on. By Theorem 1 and Lemma 4,

$$a^* \in F_2(\mathbb{P}_\Sigma), \ a^*ba^* \in F_3(\mathbb{P}_\Sigma), \ a^*ba^*ba^* \in F_4(\mathbb{P}_\Sigma),$$

so on. By Lemma 5,

$$a^* \notin F_1(\mathbb{P}_\Sigma), \ a^*ba^* \notin F_2(\mathbb{P}_\Sigma), \ a^*ba^*ba^* \notin F_3(\mathbb{P}_\Sigma),$$

so on. Note that precise proof needs induction on k, as in [16].

By Lemmas 4–6 the elements $a^*/F_1(\mathbb{P}_\Sigma), a^*ba^*/F_2(\mathbb{P}_\Sigma), \ldots$ are atoms respectively in the BA's $\mathbb{P}_\Sigma^{(1)}, \mathbb{P}_\Sigma^{(2)}, \ldots$, and this also holds for the elements

$$b^n a^*/F_1(\mathbb{P}_\Sigma), b^n a^*ba^*/F_2(\mathbb{P}_\Sigma), \ldots.$$

As the languages $b^n a^*$ (and also the languages $b^n a^*ba^*$ and so on.) are pairwise disjoint for distinct n, the BA's $\mathbb{P}_\Sigma^{(1)}, \mathbb{P}_\Sigma^{(2)}, \ldots$ have infinitely many atoms (as also the BA $\mathbb{P}_\Sigma^{(0)} = \mathbb{P}_\Sigma$).

Next we check that the BA $\mathbb{P}_\Sigma^{(k)}$ is atomic for each $k < \omega$. For $k < 2$ this is clear, it remains to check that for any $k < \omega$ and $L \in \mathbb{P}_\Sigma \backslash F_{k+2}(\mathbb{P}_\Sigma)$ there is $A \in \mathbb{P}_\Sigma$ such that $A \subseteq L$ and $A/F_{k+2}(\mathbb{P}_\Sigma)$ is an atom in $\mathbb{P}_\Sigma^{(k+2)}$. We distinguish the cases $L \notin F_\omega(\mathbb{P}_\Sigma)$ and $L \in F_\omega(\mathbb{P}_\Sigma)$.

In the first case Proposition 2 and Corollary 1 imply that the minimal DFA of L has an ω-pattern with the words u, v_1, v_1, w such that v_1, v_2 are distinct letters from Σ, so A exists by Theorem 1 and Lemma 5. In the second case, by Theorem 2 there are SCC's $F_0 < \cdots < F_{k+1}$ (having by Corollary 1 the form $F_i = \{s_i\}$, $s_i y_i = s_i$ for some $y_0, \ldots, y_{k+1} \in \Sigma$) and words $z_0, \ldots, z_k \in \Sigma^+$, $x, z_{k+1} \in \Sigma^*$, such that the first letter of z_i is distinct from y_i for each $i \le k$, $q_0 x = s_0$ and $x z_0 \cdots z_{k+1} \in L$. Then the language $A = x y_0^* z_0 \cdots y_{k+1}^* z_{k+1}$ has the desired properties.

It remains to show that for any $L \in \mathbb{P}_\Sigma \backslash F_\omega(\mathbb{P}_\Sigma)$ there is $M \in \mathbb{P}_\Sigma \backslash F_\omega(\mathbb{P}_\Sigma)$ such that $M \subset L$ and $M, L \backslash M \notin F_\omega(\mathbb{P}_\Sigma)$. The minimal DFA $\mathcal{M}_L$ of L has again an ω-pattern for some u, v_1, v_2, w as above. We can thus take $M = u v_1 (v_1 + v_2)^* w$. $\qquad\square$

The main result and the corresponding result in [16] immediately imply:

Corollary 4. *Over any alphabet* Σ, $\mathbb{P}_\Sigma$ *is isomorphic to* $\mathbb{A}_\Sigma$.

5 Conclusion

We see that the BA $\mathbb{A}$ from Introduction is astonishingly robust in characterising BAs of regular languages: all such non-degenerate BAs considered so far are isomorphic to $\mathbb{A}$. At the same time, the structure of many BA's of regular languages remains unclear. In particular, this is the case for higher self-dual levels of the Straubing-Thérien, Brzozowski and other similar hierarchies of regular languages [15]. The same applies to the corresponding BA's of regular ω-languages. An intriguing open question is whether any such non-degenerate BA is isomorphic to $\mathbb{A}$.

References

1. Gehrke, M., Grigorieff, S., Pin, J.É.: Duality and equational theory of regular languages. In: Aceto, L., Damgård, I., Goldberg, L.A., Halldórsson, M.M., Ingólfsdóttir, A., Walukiewicz, I. (eds.) ICALP 2008, Part II. LNCS, vol. 5126, pp. 246–257. Springer, Heidelberg (2008)
2. Goncharov, S.S.: Countable Boolean Algebras and Decidability. Plenum, New York (1996)
3. Hanf, W.: The boolean algebra of logic. Bull. Amer. Math. Soc. **20**(4), 456–502 (1975)
4. Ketonen, J.: The structure of countable Boolean algebras. Ann. Math. **108**, 41–89 (1978)
5. Konovalov, A.: Boolean algebras of regular quasi-aperiodic languages. In: Brattka, V., Diener, H., Spreen, D. (eds.) Logic, Computation, Hierarchies, pp. 191–204. Ontos Publishing de Gruiter, Boston (2014)
6. Selivanov, V., Konovalov, A.: Boolean algebras of regular ω-languages. In: Dediu, A.-H., Martín-Vide, C., Truthe, B. (eds.) LATA 2013. LNCS, vol. 7810, pp. 504–515. Springer, Heidelberg (2013)
7. Lempp, S., Peretyat'kin, M., Solomon, R.: The Lindenbaum algebra of the theory of the class of all finite models. J. Math. Log. **2**(2), 145–225 (2002)
8. Marini, C., Sorbi, A., Simi, G., Sorrentino, M.: A note on algebras of languages. Theor. Comput. Sci. **412**, 6531–6536 (2011)
9. Pin, J.-E.: Private communication
10. Pippenger, N.: Regular languages and stone duality. Theor. Comput. Syst. **30**(2), 121–134 (1997)
11. Schmitz, H.: Some forbidden patterns in automata for dot-depth one languages. Ph.D. thesis, Technical report, 220, University of Würzburg, Department of Computer Science (1999)
12. Selivanov, V.L.: Universal Boolean algebras with applications. In: Abstracts of International Conference in Algebra, Novosibirsk, p. 127 (1991) (in Russian)
13. Selivanov, V.L.: Hierarchies, numerations, index sets. In: Handwritten Notes, p. 290 (1992)
14. Selivanov, V.L.: Positive structures. In: Cooper, S.B., Goncharov, S.S. (eds.) Computability and Models, Perspectives East and West, pp. 321–350. Kluwer Academic/Plenum Publishers, New York (2003)
15. Selivanov, V.L.: Hierarchies and reducibilities on regular languages related to modulo counting. RAIRO Theor. Inform. Appl. **41**, 95–132 (2009)
16. Selivanov, V., Konovalov, A.: Boolean algebras of regular languages. In: Mauri, G., Leporati, A. (eds.) DLT 2011. LNCS, vol. 6795, pp. 386–396. Springer, Heidelberg (2011)
17. Stern, J.: Characterizations of some classes of regular events. Theoret. Comput. Sci. **35**, 17–42 (1985)

18. Straubing, H.: Finite Automata, Formal Logic and Circuit Complexity. Birkhäuser, Boston (1994)
19. Szilard, A., Yu, S., Zhang, K., Shallit, J.: Characterizing regular languages with polynomial densities. In: Havel, I.M., Koubek, V. (eds.) MFCS 1992. LNCS, vol. 629, pp. 494–503. Springer, Heidelberg (1992)
20. Thomas, W.: Languages, automata and logic. In: Rozenberg, G., Salomaa, A. (eds.) Handbook of Formal Language theory, vol. B, pp. 133–191. Springer, Heidelberg (1996)
21. Yu, S.: Regular languages. In: Rozenberg, G., Salomaa, A. (eds.) A Chapter of Handbook of Formal Languages. Springer, Heidelberg (1997)

On the Lattices of Effectively Open Sets

Oleg V. Kudinov[1] and Victor L. Selivanov[2(✉)]

[1] S.L. Sobolev Institute of Mathematics SB RAS, Novosibirsk, Russia
kud@math.nsc.ru
[2] A.P. Ershov Institute of Informatics Systems SB RAS,
Novosibirsk State University, Novosibirsk, Russia
vseliv@iis.nsk.su

Abstract. We show that for many natural computable metric spaces and computable domains the first order theory of the lattice of effectively open sets is hereditarily undecidable. Moreover, for several important spaces (e.g., finite-dimensional Euclidean spaces and the domain $P\omega$) this theory is m-equivalent to the first-order arithmetic.

Keywords: Lattice · Effectively open set · First-order theory · Interpretation

1 Introduction

The lattice $\mathcal{E}$ of computably enumerable (c.e.) sets is a popular object of study in Computability Theory [12]. A principal fact about this lattice is the undecidability of its first-order theory $Th(\mathcal{E})$ [5,6]. Moreover, $Th(\mathcal{E})$ is known [7] to be m-equivalent to the first-order arithmetic $Th(\mathbb{N})$ where $\mathbb{N} := (\omega; +, \times)$, i.e., to the ω'th iteration $\emptyset^{(\omega)}$ of Turing jump starting with the empty set.

The lattice $\mathcal{E}$ may be considered as the lattice $\Sigma_1^0(\omega)$ of effectively open subsets of the discrete topological space ω. The lattices of effectively open sets $\Sigma_1^0(X)$ of various "effective" topological spaces X are of interest in Computable Analysis [13] and Effective Descriptive Set Theory [9,11], hence it is natural to ask which results about $\mathcal{E}$ hold true for the lattices of effectively open sets of "natural" effective topological spaces. It seems that not much is known about this question. To our knowledge, only the case of Cantor space $\mathcal{C}$ and Baire space $\mathcal{N}$ was studied to some extent, in the context of the theory of Π_1^0-classes. In [10] (see remarks after Main Theorem 3.1 in Introduction) it is shown that $Th(\Pi_1^0(\mathcal{C}))$ is m-equivalent to $\emptyset^{(\omega)}$. Since the lattice $\Pi_1^0(\mathcal{C})$ is anti-isomorphic to $\Sigma_1^0(\mathcal{C})$, this settles the question for $\mathcal{C}$. To our knowledge, similar questions for $\mathcal{N}$ (even the decidability of $Th(\Sigma_1^0(\mathcal{N}))$) were open.

In this paper, we attempt to make next steps in this direction. First, in Sect. 2 we show that for many natural effective spaces X the theory $Th(\Sigma_1^0(X))$ (including Baire space) is undecidable. This includes the computable metric

---

O.V. Kudinov—Supported by RFBR projects 13-01-00015a and 14-01-00376.

V.L. Selivanov—Supported by RFBR project 13-01-00015a.

A. Beckmann et al. (Eds.): CiE 2016, LNCS 9709, pp. 302–311, 2016.
DOI: 10.1007/978-3-319-40189-8_31

spaces without isolated points, and many natural classes of domains (as well as many non-complete partial orders with Scott topology).

In Sect. 3, we try to precisely estimate the m-degree of $Th(\Sigma_1^0(X))$. We prove that for Euclidean finite-dimensional spaces X and for some natural domains, $Th(\Sigma_1^0(X))$ is m-equivalent to $\emptyset^{(\omega)}$. The methods of this section apply mainly to locally compact spaces. The precise estimation of the complexity of $Th(\Sigma_1^0(X))$ turns out to be very subtle and strongly depending on the topology of X. For many natural spaces X we still have a big gap between the known (to us) lower and the upper bound for $Th(\Sigma_1^0(X))$. In particular, for Baire space we currently only know the estimate $\emptyset' \leq_m Th(\Sigma_1^0(X)) \leq_m O^{(\omega)}$ where O is the Kleene ordinal notation system which is a Π_1^1-complete set.

2 Undecidability of $Th(\Sigma_1^0(X))$

Here we show that for many natural effective spaces X the theory $Th(\Sigma_1^0(X))$ is undecidable. First we explain what we mean by effectively open sets. For any countably based topological space X and any numbering $\beta : \omega \to P(X)$ of a basis of X, define a numbering $\pi : \omega \to P(X)$ by $\pi(n) = \bigcup \beta[W_n]$ where $\{W_n\}$ is the standard numbering of c.e. sets [12] and $\beta[W_n]$ is the image $\{\beta(a) \mid a \in W_n\}$. The sets in $\pi[\omega]$ are called *effectively open sets in X*. Thus, the set of effectively open sets is always equipped with the induced numbering π, hence it makes sense to speak about computable sequences of effectively open sets.

We define some particular classes of effective spaces relevant to this paper. A *computable metric space* [13] is a triple (X, d, ν), where (X, d) is a metric space and $\nu : \omega \to X$ is a numbering of a dense subset $rng(\delta)$ of X such that the set $\{(i, j, k, l) \mid \varkappa(k) < d(\nu(i), \nu(j)) < \varkappa(l)\}$ is c.e. Here $\varkappa$ is a computable numbering of the set $\mathbb{Q}$ of rationals. Any computable metric space (X, d, ν) gives rise to a numbering β of the standard basis $\beta_{\langle m, n \rangle} = B(\nu_m, \varkappa_n)$ wcre $\langle m, n \rangle$ is the Cantor pairing and $B(\nu_m, \varkappa_n)$ is the basic open ball with center ν_m and radius $\varkappa_n$. By a *strongly computable metric space* (SCMS) we mean a computable metric space such that there exists an infinite computable sequence $\{B_n\}$ of pairwise disjoint basic open balls. The metric spaces ω, $\mathbb{Q}$, $\mathcal{C} := 2^\omega$, $\mathcal{N} := \omega^\omega$, $\mathbb{R}^n$, and $\mathbb{R}^\omega$ equipped with the standard metrics and with natural numberings of dense subsets are SCMS. Any computable metric space without isolated points is a SCMS. Recall that a theory of signature σ is *hereditarily undecidable* if any its subtheory of signature σ is undecidable.

Theorem 1. *Let X be a SCMS. Then $Th(\Sigma_1^0(X))$ is hereditarily undecidable.*

Proof. Since the theory $Th(\mathcal{E})$ is hereditarily undecidable [5,6], it suffices to define the structure $\mathcal{E}$ in the structure $\Sigma_1^0(X)$ by first-order formulas with parameters; in our case two parameters v, w will suffice. Consider the formulas $\phi_U(x, v, w) := v \subseteq x \wedge x \subseteq w$ and $\phi_\subseteq(x, y, v, w) := \phi_U(x, v, w) \wedge \phi_U(x, v, w) \wedge x \subseteq y$. For any values $V, W \in \Sigma_1^0(X)$ for parameters v, w with $V \subseteq W$, the formulas define the sublattice $\mathcal{D}$ of $\Sigma_1^0(X)$ formed by the sets lying between V and W.

Therefore, it suffices to find effectively open sets V, W such that the lattice $\mathcal{D}$ is isomorphic to $\mathcal{E}$.

Let $\{B_n\}$ be the sequence from the definition of SCMS and let B'_n be obtained from the ball B_n by removing its center c_n. Then $\{B'_n\}$ is a computable sequence of effectively open sets, hence $V := \bigcup_n B'_n$ and $W := \bigcup_n B_n$ are effectively open and $W \backslash V = \{c_0, c_1, \ldots\}$. From the definition of SCMS it is easy to see that the function $D \mapsto \{n \mid c_n \in D\}$ is a desired isomorphism between $\mathcal{D}$ and $\mathcal{E}$.   □

Next we recall some definitions related to domain theory (for more details see e.g. [1,3,11]). Let X be a T_0-space. For $x, y \in X$, let $x \leq y$ denote that $x \in U$ implies $y \in U$, for all open sets U. The relation $\leq$ is a partial order known as the *specialization order*. Let $F(X)$ be the set of *finitary elements* of X (known also as *compact elements*), i.e. elements $p \in X$ such that the upper cone $\uparrow p = \{x \mid p \leq x\}$ is open. Such open cones are called *f-sets*. The space X is called a *φ-space* if every open set is a union of f-sets. The φ-space X is an *f-space* if any compatible elements $c, d \in F(X)$ have a least upper bound w.r.t. $\leq$ (compatibility means that c, d have an upper bound in $F(X)$). An f-space X is an *f_0-space* if $F(X)$ has a least element.

By a *computable φ-space* we mean a pair (X, δ) consisting of a φ-space X and a numbering $\delta : \omega \to F(X)$ of all the finitary elements such that the specialization order is c.e. on the finitary elements (i.e., the relation $\delta_x \leq \delta_y$ is c.e.). Setting $\beta(n) := \uparrow \delta_n$ we obtain a numbering of a topological basis of X. Thus, we have a notion of an effective open set in every computable φ-space.

By a *strongly computable φ-space* (SCPS) we mean a computable φ-space X such that the specialization order is computable on the finitary elements, and there is a computable sequence $\{c_n\}$ of pairwise incomparable finitary elements. Although the restrictions on SCPSs are rather strong, many popular domains are SCPSs. In particular, this applies to all φ-spaces mentioned below in this section. A SCPS X is a *strongly computable f_0-space (SCFS)* if it is an f_0-space, the relation of compatibility is computable on $F(X)$, and the supremum of compatible finitary elements is computable.

Let $\omega^{\leq \omega}$ be the completion of the partial ordering $(\omega^*; \sqsubseteq)$ where ω^* is the set of finite strings of naturals and $\sqsubseteq$ is the prefix relation. Of course, $\omega^{\leq \omega} = \omega^* \cup \omega^\omega$ consists of all finite and infinite strings of natural numbers. For every $2 \leq n < \omega$, let $n^{\leq \omega}$ be obtained in the same way from $(n^*; \sqsubseteq)$. Thus, $n^{\leq \omega} = n^* \cup n^\omega$ consists of all finite and infinite words over the alphabet $\{0, \ldots, n-1\}$. Let $P\omega$ be the powerset of ω with the Scott topology for the subset relation, hence the finitary elements of $P\omega$ are exactly the finite subsets of ω.

Let $\omega_\perp$ be the "flat" domain obtained from the discrete space of ω by adjoining a new bottom element $\perp$. Let $\omega_\perp^\omega$ be the space of partial functions $g : \omega \rightharpoonup \omega$ with the usual structure of a φ-space given by the Scott topology for the subgraph relation (as usual, we identify the partial function g with the total function $\tilde{g} : \omega \to \omega_\perp = \omega \cup \{\perp\}$ where $g(x)$ is undefined iff $\tilde{g}(x) = \perp$, for some "bottom" element $\perp \notin \omega$). For each n, $2 \leq n < \omega$, let $n_\perp^\omega$ be the space of partial functions $g : \omega \rightharpoonup \{0, \ldots, n-1\}$ defined similarly to $\omega_\perp^\omega$. Obviously, $\omega^{\leq \omega}, n^{\leq \omega}, P\omega, \omega_\perp^\omega, n_\perp^\omega$ are complete SCFSs. As is well known (see e.g. [2]), for any (complete) f_0-spaces

X, Y the space Y^X of continuous functions from X to Y with the topology of pointwise convergence is again a (complete) f_0-space. An inspection of the corresponding proofs shows that if X, Y are (complete) SCFSs then so is also Y^X. Therefore, any space of continuous partial functionals over ω of a finite type is a complete SCFS. In particular, this applies to the spaces $\mathbb{F}_n$ defined by induction $\mathbb{F}_0 := \omega_\perp$, $\mathbb{F}_{n+1} := \omega_\perp^{\mathbb{F}_n}$.

Theorem 2. *Let X be an SCPS. Then $Th(\Sigma_1^0(X))$ is hereditarily undecidable.*

Proof. We use the same interpretation scheme as in the previous proof. Let $\{c_n\}$ be the sequence of finitary elements from the definition of SCPS. This time we take as the parameters values $W := \bigcup_n \uparrow c_n$ and $V := \bigcup_n (\uparrow c_n \setminus \{c_n\})$. From the definition of SCPS it is easy to see that $\{\uparrow c_n \setminus \{c_n\}\}$ is a computable sequence of effectively open sets, hence again V and W are effectively open and $W \setminus V = \{c_0, c_1, \ldots\})$. Moreover, the function $D \mapsto \{n \mid c_n \in D\}$ is again an isomorphism between $\mathcal{D}$ and $\mathcal{E}$. □

3 m-Degree of $Th(\Sigma_1^0(X))$

Here we give precise estimate of the algorithmic complexity of $Th\Sigma_1^0(X)$ for some spaces X. First we establish a natural upper bound that applies to many locally compact spaces. We need a technical notion related to local compactness. By *arithmetically locally compact space* (ALCS) we mean a triple (X, β, κ) consisting of a topological space X, a numbering β of a basis in X, and a numbering κ of compact sets in X such that any set β_n is a union of some sets in $\{\kappa_i \mid i < \omega\}$, and the relation $\kappa_i \subseteq \bigcup \beta[D_n]$ is arithmetical where $\{D_n\}$ is the canonical numbering of finite subsets of ω. Note that although ALCSs are not automatically locally compact, many popular locally compact spaces may be considered as ALCSs. In particular, the computable φ-spaces, the finite dimensional Euclidean spaces, and Cantor space are ALCSs (say, for a computable φ-space (X, δ) we can set $\kappa_n := \uparrow \delta_n$ which is compact; the relation $\kappa_i \subseteq \bigcup \beta[D_n]$ in this case is c.e.).

Proposition 1. *If (X, β, κ) is an ALCS then $Th(\Sigma_1^0(X)) \leq_m \emptyset^{(\omega)}$.*

Proof. It suffices to show that the relation $\pi_i \subseteq \pi_j$ is arithmetical because then the elementary diagram of the numbered structure $(\Sigma_1^0(X); \subseteq, \pi)$, and hence also $Th(\Sigma_1^0(X))$, is m-reducible to $\emptyset^{(\omega)}$.

Obviously, $\pi_i \subseteq \pi_j$ is equivalent to $\forall n (\kappa_n \subseteq \pi_i \rightarrow \kappa_n \subseteq \pi_j)$, hence it suffices to show that the relation $\kappa_n \subseteq \pi_i$ is arithmetical. We have $\kappa_n \subseteq \pi_i$ iff $\kappa_n \subseteq \bigcup \beta[W_i]$ iff $\exists m (D_m \subseteq W_i \wedge \kappa_n \subseteq \bigcup \beta[D_m])$, by compactness of κ_n. The last relation is arithmetical by the definition of ALCS. □

We turn to the precise estimation of m-degrees of $Th(\Sigma_1^0(\mathbb{R}^n))$ and start with the following lemma.

Lemma 1. *Any connected component of an effectively open set in $\mathbb{R}^n$ is effectively open.*

Proof. Let U be a connected component of $V \in \Sigma_1^0(\mathbb{R}^n)$ and let a be a rational point in U. Then

$$U = \bigcup_m \{B(b,r) \mid b \in \mathbb{Q}^n \wedge r \in \mathbb{Q}^+ \wedge \exists a_1 \cdots a_{m-1} \in \mathbb{Q}^n \exists r_1 \cdots r_{m-1} \in \mathbb{Q}^+$$
$$(\bigwedge_{i=1}^m \overline{B(a_i, r_i)} \subseteq V \wedge \bigwedge_{i=1}^{m-1} (B(a_i, r_i) \cap B(a_{i+1}, r_{i+1}) \neq \emptyset) \wedge a \in B(a_1, r_1))\}$$

where $b = a_m$ and $r = r_m$. Since we can computably enumerate basic closed balls $\overline{B(b,r)} \subseteq V$ [8] with $b \in \mathbb{Q}^n, r \in \mathbb{Q}^+$, U is effectively open. □

Now we prove the main technical result of this paper:

Theorem 3. *For any $n \geq 1$, $Th(\Sigma_1^0(\mathbb{R}^n)) \equiv_m \emptyset^{(\omega)}$.*

Since the upper bound holds by the previous proposition, we only have to prove the lower bound. Since $\Sigma_1^0(\mathbb{R}^n)$ is a distributive lattice, we can use in the definitions not only the symbol of inclusion but also the symbols of Boolean operations and the constants $\emptyset, \mathbb{R}^n$. Note that for any $x \in \mathbb{R}^n$, x is computable iff $\mathbb{R}^n \setminus \{x\}$ is effectively open, hence we can use (to simplify notation) in our defining formulas the computable points (as complements of sets maximal w.r.t. inclusion among the sets below $\mathbb{R}^n$). More precisely, we will use the variable x to range over the computable points, the atomic formulas $x \in u$ (where u range as usual through $\Sigma_1^0(\mathbb{R}^n)$), and we can quantify over x.

Proof for $n = 1$. Since $Th(\mathcal{E}) \equiv_m \emptyset^{(\omega)}$ by [7], it suffices to m-reduce $Th(\mathcal{E})$ to $Th(\Sigma_1^0(\mathbb{R}))$. For this we again find a definition of the lattice $\mathcal{E}$ in $\Sigma_1^0(\mathbb{R})$ with parameters but this time the set of parameters itself should be definable.

First we show that the property of being a connected effectively open set is definable without parameters. We start with some auxiliary formulas. Let $\widetilde{Con}(u)$ be the formula $u \neq \emptyset \wedge \neg \exists v, w(u \subseteq v \cup w \wedge v \cap w = \emptyset \wedge u \cap v \neq \emptyset \wedge u \cap w \neq \emptyset)$ saying that u is "effectively connected" i.e. it cannot be partitioned into two disjoint non-empty effectively open sets. In particular, the connected effectively open sets satisfy this formula. Let $\xi(u,v)$ be the formula

$$u \neq \emptyset \wedge v \neq \emptyset \wedge u \cap v = \emptyset \wedge \forall u'(u' \cap v = \emptyset \rightarrow u' \subseteq u) \wedge \forall v'(v' \cap u = \emptyset \rightarrow v' \subseteq v)$$

saying that u, v are disjoint non-empty effectively open sets such that $u = Int(\mathbb{R} \setminus v)$ (where Int is the interior operator) and $v = Int(\mathbb{R} \setminus u)$.

Let now $CInt(u) := \widetilde{Con}(u) \wedge \exists v \neq \emptyset \xi(u,v)$. Then $\Sigma_1^0(\mathbb{R}) \models CInt(U)$ iff U is a computable interval (i.e. the endpoints of U are computable or infinite) distinct from $\emptyset, \mathbb{R}$. For the nontrivial direction, suppose $\Sigma_1^0(\mathbb{R}) \models CInt(U)$, V is a satisfying value for v and let $a := inf(U)$, $b := sup(U)$. Then $a < b$ and at least one of a, b is finite. For each computable point $p \in V$, sets $U_1 := \{x \in U \mid x < p\}$ and $U_2 := \{x \in U \mid p < x\}$ are effectively open and partition U, hence one of them is empty. From this one easily deduces that $V = (b, +\infty)$ if $a = -\infty$, $V = (-\infty, a)$ if $b = +\infty$, and $V = (-\infty, a) \cup (b, +\infty)$ if $a \neq -\infty, b \neq +\infty$. Since $U, V \in \Sigma_1^0(\mathbb{R})$, a and b are computable. It is now clear that $U = (a, b)$, as desired.

Let $\psi(u,x) := x \in u \wedge \forall y \in u \exists w (CInt(w) \wedge x, y \in w \subseteq u)$ and $\widetilde{Cmp}(u,v,x) := u \subseteq v \wedge \psi(u,x) \wedge \forall u'(x \in u' \subseteq v \wedge CInt(u') \to u' \subseteq u)$. Using Lemma 1 one easily checks that $\Sigma_1^0(\mathbb{R}) \models \widetilde{Cmp}(U,V,x)$ iff U is the connected component of V containing the computable point x. Let $Con(u) := (\exists x \in u)\widetilde{Cmp}(u,u,x)$. Then $\Sigma_1^0(\mathbb{R}) \models Con(U)$ iff U is connected, hence the property of being connected is definable.

Let $Cmp(u,v) := u \subseteq v \wedge Con(u) \wedge \forall u'(u' \subseteq v \wedge Con(u') \wedge u \cap u' \neq \emptyset \to u' \subseteq u)$. Then $\Sigma_1^0(\mathbb{R}) \models Cmp(U,V)$ iff U is a connected component of V. Let $Cmp^*(u,v)$ be the formula $u \subseteq v \wedge \forall u'(Cmp(u',v) \to (u' \subseteq u \vee u \cap u' = \emptyset))$ saying that u is the union of some connected components of v.

Let $ICmp(v)$ be the formula $\exists u(Cmp^*(u,v) \wedge \neg \exists w(w = v \backslash u \wedge Com^*(w,v)))$. Then $\Sigma_1^0(\mathbb{R}) \models ICmp(V)$ implies that V has infinitely many connected components. Suppose the contrary, then $V = V_0 \cup \cdots \cup V_n$ for some $n \geq 0$ and pairwise disjoint components $V_0, \ldots, V_n \in \Sigma_1^0(\mathbb{R})$. Then any U with $\Sigma_1^0(\mathbb{R}) \models Cmp^*(U,V)$ is a union of some of $V_0, \ldots, V_n$, hence $V \backslash U$ is the finite union of the remaining V_i, a contradiction.

Finally, let $\theta(u,v)$ be the formula

$$\xi(u,v) \wedge ICmp(u) \wedge ICmp(v) \wedge \forall w(Cmp(w, u \cup v) \to CInt(w)).$$

Then $\Sigma_1^0(\mathbb{R}) \models \theta(U,V)$ iff both U, V have infinitely many connected components which are computable intervals, $U = Int(\mathbb{R} \backslash V)$, $V = Int(\mathbb{R} \backslash U)$, between any U-components there is a V-component (this means that for any U-components $U_1 < U_2$, i.e. $\forall x \in U_0 \forall y \in U_1(x < y)$, there is a V-component V_1 with $U_1 < V_1 < U_2$ and between any V-components there is a U-component. Indeed, from right to left this is obvious. Conversely, if $\Sigma_1^0(\mathbb{R}) \models \theta(U,V)$ then the only non-trivial fact to check is that between any U-components there is a V-component. Suppose the contrary, then the interval $W := (inf(U_1), sup(U_2))$ is disjoint with V, hence $W \subseteq U$. This is a contradiction because $sup(U_1) \in W \backslash U$.

Let now $\Sigma_1^0(\mathbb{R}) \models \theta(U,V)$ and let $\{q_0, q_1, \cdots\}$ be a computable enumeration of the set $U \cap \mathbb{Q}$ without repetitions. Define the equivalence relation $\sim$ on ω by: $m \sim n$ iff q_m, q_n are in the same U-component. We claim that this relation is c.e. Indeed, since $m \sim n$ is equivalent to the disjunction of $q_m \leq q_n \wedge [q_m, q_n] \subseteq U$ and $q_n \leq q_m \wedge [q_n, q_m] \subseteq U$, so it suffices to check that the relation $q_m \leq q_n \wedge [q_m, q_n] \subseteq U$ is c.e. We have $U = \bigcup_i B_i$ for a computable sequence $\{B_i\}$ of basic open balls (i.e., intervals with rational endpoints). Since closed intervals are compact, the relation $q_m \leq q_n \wedge [q_m, q_n] \subseteq U$ is equivalent to

$$\exists l \exists i_0, \ldots, i_l (q_m \in B_{i_0} \wedge q_n \in B_{i_l} \wedge B_{i_0} \cap B_{i_1} \neq \emptyset \wedge \cdots \wedge B_{i_{l-1}} \cap B_{i_l} \neq \emptyset).$$

Alternately, the last assertion follows from the results in [8]. Therefore, $\sim$ is c.e. It is also co-c.e. because, by the previous paragraph, $m \not\sim n$ is equivalent to the disjunction of predicates $q_m < q_n \wedge \exists r \in V \cap \mathbb{Q}(q_m < r < q_n)$ and $q_n < q_m \wedge \exists r \in V \cap \mathbb{Q}(q_n < r < q_m)$.

Therefore, the relation $\sim$ is computable with infinitely many equivalence classes, hence the lattice $\mathcal{E}$ is isomorphic to the lattice $\mathcal{F}$ of c.e. sets closed under

$\sim$. The lattice $\mathcal{F}$ is in turn isomorphic to the lattice $(\mathcal{G}; \subseteq)$ where $\mathcal{G} := \{G \mid \Sigma_1^0(\mathbb{R}) \models Cmp^*(G, U)\}$ consists of effective open subsets of U closed under components, for each U as above.

Now it is easy to define a computable transformation of $\subseteq$-sentences $\phi \mapsto \phi^*$ such that $\mathcal{E} \models \phi$ iff $\Sigma_1^0(\mathbb{R}) \models \phi^*$ which gives a desired m-reduction of $Th(\mathcal{E})$ to $Th(\Sigma_1^0(\mathbb{R}))$. According to the method of interpretation, there is a computable transformation of $\subseteq$-formulas $\phi(\ldots) \mapsto \bar{\phi}(u, v, \ldots)$ that adds two new free variables u, v to those of ϕ and has the property that $\mathcal{E} \models \phi(\ldots)$ iff $\Sigma_1^0(\mathbb{R}) \models \bar{\phi}(U, V, \ldots)$ for any values U, V of u, v with $\Sigma_1^0(\mathbb{R}) \models \theta(U, V)$. Now it suffices to set $\phi^* := \forall u, v(\theta(u, v) \rightarrow \bar{\phi}(u, v))$. $\square$

Proof for $n \geq 2$. In this case we m-reduce $Th(\mathbb{N})$ to $Th(\Sigma_1^0(\mathbb{R}^n))$. Let Cof be the unary relation on $\Sigma_1^0(\mathbb{R}^n)$ which is true precisely on the cofinite subsets of $\mathbb{R}^n$, then $\Sigma_1^0(\mathbb{R}^n) \models Cof(U)$ iff $\mathbb{R}^n \backslash U$ is a finite set of computable points. We will first use in our reduction of $Th(\mathbb{N})$ to $Th(\Sigma_1^0(\mathbb{R}^n))$ this relation Cof (or, equivalently, the relation Fin on $\Pi_1^0(\mathbb{R}^n)$ such that $Fin(A)$ iff A is a finite sets of computable points). At the end of the proof we will show that the relation Cof is definable in $\Sigma_1^0(\mathbb{R}^n)$.

Our proof is closely related to that in [4] where a similar reduction was established for the lattice of all closed sets in $\mathbb{R}^n$. To keep our notation as close as possible to that of [4], we let our formulas to use, along with the usual variables U, V, W, G, H, K, T ranging over $\Sigma_1^0(\mathbb{R}^n)$, the variable x ranging over the computable points of $\mathbb{R}^n$, and variables A, B, C, D, E, F ranging over $\Pi_1^0(\mathbb{R}^n)$. Distinctions with the proof in [4] are mainly caused by the fact that we currently do not know whether the relation of being connected is definable in $\Sigma_1^0(\mathbb{R}^n)$ for $n \geq 2$. Instead, we use effective versions of some notions from the previous proof. Note that formulas $\widetilde{Con}$ and ξ have the same meaning as above in any reasonable effective space. Obviously, if effectively open sets U, V are effectively connected (i.e., satisfy $\widetilde{Con}$) and are not disjoint then $U \cup V$ is also effectively connected.

Let $\widetilde{Cmp}(u, v)$ be the formula

$$u \subseteq v \wedge \widetilde{Con}(u) \wedge \forall u'(u' \subseteq v \wedge \widetilde{Con}(u') \wedge u \cap u' \neq \emptyset \rightarrow u' \subseteq u)$$

saying that u is an effectively connected component (we say "effective component" for brevity) of v. Let $\Phi(V)$ be the formula $\forall x \in V \exists U \subseteq V(x \in U \wedge \widetilde{Cmp}(U, V))$ saying that any computable point of V belongs to some effective component of V. From Lemma 1 and remarks above it follows that if $\Sigma_1^0(\mathbb{R}^n) \models \Phi(V)$ then V is the union of its effective components.

Similar to axiom A6 in [4], for any finite disjoint sets A, B of computable points the following formula is easily seen to be true in $\Sigma_1^0(\mathbb{R}^n)$:

$$(A \cup B \subseteq U \wedge \widetilde{Con}(U)) \rightarrow \exists V \exists W(A \subseteq V \wedge B \subseteq W \wedge$$
$$V \cap W = \emptyset \wedge V \cup W \subseteq U \wedge \widetilde{Con}(V) \wedge \widetilde{Con}(W)).$$

Let now $A \approx_G B$ be the ternary relation meaning that A, B are finite disjoint sets of computable points and G is an effectively open set such that

$$A \cup B \subseteq G \wedge \Phi(G) \wedge \forall H(\widetilde{Cmp}(H, G) \rightarrow (|H \cap A| = 1 \wedge |H \cap B| = 1)).$$

Then $A \approx_G B$ implies that A, B are of the same cardinality, $|A| = |B|$. Note that for any finite sets A, B of computable points we have: $|A| = |B|$ iff $\Sigma_1^0(\mathbb{R}^n) \models Eq(A, B)$ where $Eq(A, B)$ is the formula

$$Fin(A) \wedge Fin(B) \wedge \exists C, D, G(Fin(C) \wedge Fin(D) \wedge C \approx_G D \wedge C = A \backslash B \wedge D = B \backslash A).$$

As in [4], it is now straightforward to interpret the structure $\mathbb{N}$ in the structure $(\Sigma_1^0(\mathbb{R}^n); Cof)$ without parameters by interpreting natural numbers as the cardinalities of finite sets A, B of computable points (i.e., by taking the quotient-set of all such A under the equivalence relation Eq) and interpreting $+, \times$ as follows: $|A| + |B| = |C|$ iff $\exists A', B'(Eq(A', A) \wedge Eq(B', B) \wedge A' \cap B' = \emptyset \wedge Eq(A' \cup B', C)$, and $|A| \times |B| = |C|$ iff

$$\exists U \exists F(Eq(B, F) \wedge F, C \subseteq U \wedge \forall V(Cmp(U, V) \rightarrow (|V \cap F| = 1 \wedge |V \cap C| = |A|))).$$

It remains to define the relation Fin in $\Pi_1^0(\mathbb{R}^n)$ with parameters ranging through a definable set (similar to the case $n = 1$). Let $W \sqsubseteq U$ abbreviates the formula $\forall D(\widetilde{Cmp}(D, W) \rightarrow \widetilde{Cmp}(D, U))$ meaning that any effective W-component is an effective U-component. Consider the following formulas (denoted below by (a), (b), (c), (d) respectively) with free variables U, V, W, W_1, x:

$$x \in U \wedge U \cap V = \emptyset \wedge \Phi(U) \wedge \Phi(V) \wedge \Phi(W) \wedge \Phi(W_1),$$
$$U \cup V \subseteq W \wedge V \subseteq W_1 \wedge \forall G((\widetilde{Cmp}(G, U) \wedge x \notin G) \rightarrow G \subseteq W_1),$$
$$(\forall G)_{\widetilde{Cmp}(G, W)}(\exists! H)_{\widetilde{Cmp}(H, U)}(\exists! K)_{\widetilde{Cmp}(K, V)}(H \subseteq G \wedge K \subseteq G),$$
$$(\forall G)_{\widetilde{Cmp}(G, W_1)}(\exists! H)_{\widetilde{Cmp}(H, U)}(\exists! K)_{\widetilde{Cmp}(K, V)}(H \subseteq G \wedge K \subseteq G \wedge x \notin H).$$

Informally, the effective W-components induce a bijection (denoted by f) between the effective U-components and the effective V-components while W_1 induces a bijection (denoted by g) between the effective V-components and the effective U-components which do not contain the point x. Note that the graphs of these bijections are definable with parameters from the tuple $\mathcal{A} = (U, V, W, W_1, x)$, hence we can use the new function symbols $f_\mathcal{A}, g_\mathcal{A}$ to denote the bijections. For $T \sqsubseteq U$, we denote by $f_\mathcal{A}[T]$ the union of $f_\mathcal{A}$-images of effective T-components. Set now $s_\mathcal{A} := g_\mathcal{A} \circ f_\mathcal{A}$ (this is a definable function on the effective U-components), and consider the following formulas (the second of which is not formalized completely)

(e) $s_\mathcal{A}(D) = s_\mathcal{A}(E) \rightarrow D = E$,
(f) $\neg \exists T \sqsubseteq U$ ($s_\mathcal{A}$ is a bijection on the effective T-components).

In particular, formula (f) forbids the finite $s_\mathcal{A}$-cicles, hence all the iterates $G, s_\mathcal{A}(G), s_\mathcal{A}^2(G), s_\mathcal{A}^3(G), \ldots$ of the function $s_\mathcal{A}$ started with the "initial" effective U-component (i.e., the unique effective component G containing the point x), are distinct. Let $\phi(U, V, W, W_1, x)$ be the conjunction of the formulas (a)–(f).

Define the unary relation $\widetilde{Iso}$ on $\Pi_1^0(\mathbb{R}^n)$ (an effective analogue of the relation Iso from [4]) as follows: $\widetilde{Iso}(A)$ iff $\exists V(A \subseteq V \wedge \Phi(V) \wedge \forall U(\widetilde{Cmp}(U, V) \rightarrow \exists! x(x \in$

$A \cap U)))$ and $\forall U(U \cap A \neq \emptyset \rightarrow \exists x(x \in U \cap A))$. This relation is definable, $\widetilde{Iso}(A)$ implies that any point in A is computable and isolated, any infinite set satisfying $\widetilde{Iso}$ is not bounded, and any finite set of computable points satisfies $\widetilde{Iso}$.

Now it suffices to check that for any $A \in \Pi_1^0(\mathbb{R}^n)$, $Fin(A)$ is equivalent to

$$\widetilde{Iso}(A) \wedge \forall U, V, W, W_1, x((\phi(U, V, W, W_1, x) \wedge A \subseteq U \cup V) \rightarrow$$
$$\exists T \sqsubset U(x \in T \wedge A \subseteq T \cup f_A[T]) \wedge$$
$$(\forall G)_{\widetilde{Cmp}(G,T)}(x \notin G \rightarrow (\exists H)_{\widetilde{Cmp}(H,T)}(s_A[H] = G))).$$

Indeed, the direction from left to right is obvious. Conversely, if A is infinite and satisfies $\widetilde{Iso}(A)$, take $x := 0$ to be the zero-vector in $\mathbb{R}^n$ and find a strictly increasing computable unbounded sequence $\{r_k\}$ of positive reals such that the boundaries of all balls $B(0, r_k)$ are disjoint with A (this is possible since the relation "the boundary of $B(0, r)$ is contained in $\mathbb{R}^n \backslash A$" is effectively open in $\mathbb{R}$). The assignments $U := B(0, r_0) \cup \bigcup_{k>0}(B(0, r_{2k}) \backslash \overline{B(0, r_{2k-1})})$, $V := \bigcup_k(B(0, r_{2k+1}) \backslash \overline{B(0, r_{2k})})$, $W := B(0, r_1) \cup \bigcup_{k>0}(B(0, r_{2k+1}) \backslash \overline{B(0, r_{2k-1})})$, $W_1 := \bigcup_k(B(0, r_{2k+2}) \backslash \overline{B(x, r_{2k})})$ make the defining formula false because $A \subseteq U \cup V$ and $\phi(U, V, W, W_1, x)$ are true but there is no corresponding $T \sqsubset U$ (such bounded sets T do not satisfy the condition $A \subseteq T \cup f_A[T]$). $\square$

For many domains we have even better definability result:

Theorem 4. *For any $X \in \{\omega^{\leq\omega}, n^{\leq\omega}, P\omega, \omega_\perp^\omega, n_\perp^\omega\}$, the lattice $\mathcal{E}$ is definable without parameters in the lattice $\Sigma_1^0(X)$.*

Proof. We give the proof only for the space $P\omega$ but the argument works for the other spaces as well. First we check that the set elements $\uparrow F$, for all finite $F \subseteq \omega$, is definable in $\Sigma_1^0(P\omega)$ (without parameters). Indeed, the defining formula is $ir(v) := v \neq 0 \wedge \forall x, y(v \subseteq x \cup y \rightarrow v \subseteq x \vee v \subseteq y)$ which says that v is a non-smallest join-irreducible element. Obviously, any set $\uparrow F$ satisfies this formula. Conversely, let an element $V \in \Sigma_1^0(P\omega)$ satisfy the formula. Since $V \neq \emptyset$, for some canonically enumerable sequence $\{F_n\}$ of finite sets we have $V = \bigcup_n \uparrow F_n$. Since the partial order $(\{F_n \mid n < \omega\}; \subseteq)$ is well founded, it has a minimal element F. Then of course $\uparrow F \subseteq V$, so it suffices to show that $V \subseteq \uparrow F$. Let $S := \{\uparrow F_n \mid F \not\subseteq F_n\}$. Since $S \in \Sigma_1^0(P\omega)$, $V \subseteq \uparrow F \cup S$ and V is join-irreducible, it suffices to show that $V \not\subseteq S$. Suppose the contrary, then $F \in S$, so $F_n \subseteq F$ for some n with $F \not\subseteq F_n$. By the minimality of F, $F_n = F$ which is a contradiction.

Now we check that the set $V_n := \{\uparrow F : |F| = n\}$ is definable in $\Sigma_1^0(P\omega)$ for each $n < \omega$. The sequence $\{\phi_n(v)\}$ of defining formulas is given by induction on n as follows: $\phi_0 := \forall x(x \subseteq v)$ (saying that v is the largest element of a lattice), and

$$\phi_{n+1} := ir(v) \wedge \neg\phi_0(v) \wedge \cdots \wedge \neg\phi_n(v) \wedge \forall x(v \subset x \rightarrow \phi_0(v) \vee \cdots \vee \phi_n(v))$$

(saying that v is a maximal join-irreducible element among those not in the set defined by the formula $\phi_0(v) \vee \cdots \vee \phi_n(v)$). By the previous paragraph and induction on n, ϕ_n defines V_n for each n.

Now let $U_n := \{S \subseteq \omega : n \leq |S|\}$, so $U_0 = P\omega$ and $U_n = \bigcup V_n$. Then the singleton set $\{U_{n+1}\}$ is clearly defined by the formula

$$\psi_{n+1}(u) := \forall x(\phi_{n+1}(x) \to x \subseteq u) \land \neg \exists x(\phi_n(x) \land x \subseteq u).$$

By the proof of Theorem 2, the lattice $\mathcal{E}$ is isomorphic to the sublattice $\{S \mid U_1 \subseteq S \subseteq U_2\}$ of $\Sigma_1^0(P\omega)$, and is thus definable without parameters. $\square$

Corollary 1. *For any* $X \in \{\omega^{\leq \omega}, n^{\leq \omega}, P\omega, \omega_\perp^\omega, n_\perp^\omega\}$, $Th(\Sigma_1^0(X)) \equiv_m \emptyset^{(\omega)}$.

Proof. The upper bound holds by Proposition 1, the lower bound by the previous theorem. $\square$

Acknowledgement. We thank André Nies for a discussion of the lower bound problem for $Th(\Sigma_1^0(\mathcal{N}))$, and the referees for valuable comments.

References

1. Abramsky, S., Jung, A.: Domain theory. In: Abramsky, S., Gabbay, D., Maibaum, T.S.E. (eds.) Handbook of Logic in Computer Science, vol. 3, pp. 1–168. Oxford University Press, Oxford (1994)
2. Ershov, Y.: Computable functionals of finite types. Algebra Log. **11**(4), 367–433 (1972)
3. Ershov, Y.: Theory of domains and nearby. In: Bjørner, D., Broy, M., Pottosin, I.V. (eds.) Formal Methods in Programming and Their Applications. LNCS, vol. 735, pp. 1–7. Springer, Heidelberg (1993)
4. Grzegorczyk, A.: Undecidability of some topological theories. Fundamenta Mathematicae **38**, 137–152 (1951)
5. Herrmann, E.: Definable boolean pairs in the lattice of recursively enumerable sets. In: Proceedings of 1st Easter Conference in Model Theory, Diedrichshagen, pp. 42–67 (1983)
6. Herrmann, E.: The undecidability of the elementary theory of the lattice of recursively enumerable sets. In: Proceedings of 2nd Frege Conference at Schwerin, GDR, vol. 20, pp. 66–72 (1984)
7. Harrington, L., Nies, A.: Coding in the lattice of enumerable sets. Adv. Math. **133**, 133–162 (1998)
8. Korovina, M.V., Kudinov, O.V.: The uniformity principle for sigma-definability. J. Log. Comput. **19**(1), 159–174 (2009)
9. Moschovakis, Y.N.: Descriptive Set Theory. North Holland, Amsterdam (2009)
10. Nies, A.: Effectively dense Boolean algebras and their applications. Trans. Am. Math. Soc. **352**(11), 4989–5012 (2000)
11. Selivanov, V.L.: Towards a descriptive set theory for domain-like structures. Theoret. Comput. Sci. **365**, 258–282 (2006)
12. Soare, R.I.: Recursively Enumerable Sets and Degrees. Springer, Berlin (1987)
13. Weihrauch, K.: Computable Analysis. Springer, Berlin (2000)

On the Executability of Interactive Computation

Bas Luttik and Fei Yang[(⊠)]

Eindhoven University of Technology, Eindhoven, The Netherlands
{s.p.luttik,f.yang}@tue.nl

Abstract. The model of *interactive Turing machines* (ITMs) has been proposed to characterise which stream translations are *interactively computable*; the model of *reactive Turing machines* (RTMs) has been proposed to characterise which behaviours are *reactively executable*. In this article we provide a comparison of the two models. We show, on the one hand, that the behaviour exhibited by ITMs is reactively executable, and, on the other hand, that the stream translations naturally associated with RTMs are interactively computable. We conclude from these results that the theory of reactive executability subsumes the theory of interactive computability. Inspired by the existing model of ITMs with advice, which provides a model of evolving computation, we also consider RTMs with advice and we establish that a facility of advice considerably upgrades the behavioural expressiveness of RTMs: every countable transition system can be simulated by some RTM with advice up to a fine notion of behavioural equivalence.

1 Introduction

According to the Church-Turing thesis, the classical Turing machine model adequately formalises which functions from natural numbers to natural numbers are effectively computable. There is, however, a considerable semantic gap between computing the result of a function applied to a natural number and the way computing systems operate nowadays. Modern computing systems are reactive, they are in continuous interaction with their environment, and their operation is not supposed to terminate. Quite a number of extended models of computation have been proposed in recent decades to study the combination of computation and interaction (see, e.g., the collection in [8]). In this paper we compare *interactive Turing machines* and *reactive Turing machines*.

Van Leeuwen and Wiedermann have developed a theory of interactive computation from the stance that an interactive computation can be viewed as a never-ending exchange of symbols between a component and its unpredictable interactive environment [9]. Semantically, this amounts to studying the recognition, generation and translation of infinite streams of symbols. In [10], the notion of interactive Turing machine (ITM) is put forward as a tool to formally characterise which stream translations are interactively computable. The notion is subsequently extended with an (non-computable) advice mechanism in order

---

F. Yang—This author is sponsored by the China Scholarship Council (CSC).

© Springer International Publishing Switzerland 2016
A. Beckmann et al. (Eds.): CiE 2016, LNCS 9709, pp. 312–322, 2016.
DOI: 10.1007/978-3-319-40189-8_32

to obtain a non-uniform machine model. Van Leeuwen and Wiedermann argue that the resulting model of *interactive Turing machines with advice* is as powerful as their model of evolving finite automata, and they conclude from this, on intuitive grounds, that ITMs with advice are adequate to model evolving system such as the Internet [17]. Moreover, in a recent article by Cabessa and Villa it is shown that ITMs with advice are as powerful as the model of interactive evolving recurrent neural networks for computing stream translations [4].

The model of interactive Turing machines focusses on capturing the computational content of sequential interactive behaviour. The included mechanism of interaction is therefore limited to achieving this goal, and does not easily generalise to more than one distributed component, nor does it allow for more fine-grained considerations of the behaviour of reactive systems. The behavioural theory of reactive systems, on the other hand, has focussed on aspects of modelling, specification and verification (see, e.g., [1]).

To integrate computability theory and the behavioural theory of reactive systems, the notion of reactive Turing machine (RTM) has been proposed in [2,3]. It extends Turing machines with concurrency-style interaction. Semantically, the operational behaviour of an RTM is given by a transition system. From this transition system one may extract a set of computations, or stream translations, but a more refined analysis is also possible. In fact, to study the effect of interaction of multiple components many refined notions of behavioural equivalence have been developed in the concurrency theory literature [7]. The notion of RTM gives rise to a general theory of *executability*: a transition system is executable (usually up to some preferred notion of behavioural equivalence) if there exists an RTM that has the transition system as its semantics. (We refer to [3] for more elaborate motivation of the notion of RTM.)

The aim of this paper is to make a connection between the theory of interactive computability and the theory of reactive systems, providing a comparison of the models of ITMs and RTMs in both their semantic domains. We shall first, in Sect. 2, recapitulate both models. Then, in Sect. 3 we present a transition-system semantics for ITMs; the transition system associated with an ITM is executable up to a fine notion of behavioural equivalence. In Sect. 4 we shall identify a subclass of RTMs that can be considered suitable for stream translation, and prove that the stream translation associated with an RTM in this subclass is interactively computable. In Sect. 5 we consider an extension of RTMs with an advice mechanism adapted from the advice mechanism considered for ITMs. RTMs with advice can execute every countable transition system, at the cost of introducing divergence in the computation. The paper ends with a conclusion in Sect. 6.

A full version of this paper is available as [14].

2   Preliminaries

2.1   The Theory of Interactive Computation

In [12], van Leeuwen and Wiedermann present an analysis of interactive computation on the basis of a *component C* (thought to behave according to a deterministic program) interacting with an unpredictable *environment E*. They discuss the consequences of a few general postulates pertaining to the behaviour and interaction of C and E for interactive recognition, interactive generation and interactive translation. In their analysis, the component C acts as a stream transducer, transforming an infinite input stream of data symbols from $\Sigma = \{0,1\}$ presented by E at its input port into an infinite output stream of symbols from Σ produced at its output port. Henceforth, by an *ω-translation* we mean a mapping $\phi : \Sigma^\omega \to \Sigma^\omega$ (with Σ^ω denoting the set of streams, i.e., infinite sequences, over Σ).

Interactive computation is a step-wise process. It is not required that the environment offers a symbol in every step, nor that the component produces a symbol in every step. For the purpose of modelling components, however, it is convenient to record that nothing is offered or produced. The symbol λ is used to indicate the situation that no symbol is offered at the input port or produced at the output port, and we let $\Sigma_\lambda = \Sigma \cup \{\lambda\}$. It is assumed that when E offers a non-λ symbol in some step, then the component C produces a non-λ symbol at its output port within finitely many steps, and vice versa; this assumption is referred to as the *interactiveness* (or *finite delay*) condition in the work of van Leeuwen and Wiedermann.

In order to formally define which ω-translations are interactively computable by a computational device, van Leeuwen and Wiedermann proposed the notion of *interactive Turing machine* [10,11]. It extends the classical notion of Turing machine with an input port and an output port, through which it exchanges an infinite, never ending stream of data symbols with its environment. Interactive Turing machines use a two-way infinite tape as memory on which they can write symbols from some presupposed set $\mathcal{D}_\square$ of *tape symbols*, not necessarily disjoint from Σ and including the special $\square$ symbol to denote an empty tape cell. Our formal definition below is adapted from [16] (but we leave out the distinction between internal and external states).

Definition 1. *A (deterministic) interactive Turing machine (ITM) with a single work tape is a triple $\mathcal{I} = (Q, \longrightarrow_\mathcal{I}, q_{in})$, where*

1. *Q is its set of* states;
2. *$\longrightarrow_\mathcal{I}: Q \times \mathcal{D}_\square \times \Sigma_\lambda \to Q \times \mathcal{D}_\square \times \{L,R\} \times \Sigma_\lambda$ is a transition function; and*
3. *$q_{in} \in Q$ is its initial state.*

The contents of the tape of an ITM may be represented by an element of $(\mathcal{D}_\square)^*$. We denote by $\breve{\mathcal{D}}_\square = \{\breve{d} \mid d \in \mathcal{D}_\square\}$ the set of *marked* symbols; a *tape instance* is a sequence $\delta \in (\mathcal{D}_\square \cup \breve{\mathcal{D}}_\square)^*$ such that δ contains exactly one element of $\breve{\mathcal{D}}_\square$. The marker indicates the position of the tape head.

A *computation* of an ITM $\mathcal{I} = (Q, \longrightarrow_\mathcal{I}, q_{in})$ is an infinite sequence of transitions

$$(q_{in}, \breve{\square}) = (q_0, \delta_0) \xrightarrow{i_0/o_0}_\mathcal{I} (q_1, \delta_1) \xrightarrow{i_1/o_1}_\mathcal{I} \cdots (q_k, \delta_k) \xrightarrow{i_k/o_k}_\mathcal{I} \cdots . \qquad (1)$$

The *input stream* associated with the computation in (1) is obtained from $i_0, i_1, \ldots$ by omitting all occurrences of λ, and the *output stream* associated with the computation in (1) is obtained from $o_0, o_1, \ldots$ by omitting all occurrences of λ. A pair $(\boldsymbol{x}, \boldsymbol{y}) \in \Sigma^\omega \times \Sigma^\omega$ is an *interaction pair* associated with $\mathcal{I}$ if there exists a computation of $\mathcal{I}$ with $\boldsymbol{x}$ as input stream and $\boldsymbol{y}$ as output stream. The set of all interaction pairs associated with an ITM $\mathcal{I}$ is called its *interactive behaviour*. (In Sect. 3 we shall present a more refined view on its behaviour when we associate with every ITM a transition system.) The computation in (1) is *interactive* if, for all $k \in \mathbb{N}$, if $i_k \neq \lambda$, then there exists $\ell \geq k$ such that $o_\ell \neq \lambda$. The computation in (1) is *input-active* if $i_k \neq \lambda$ for all $k \in \mathbb{N}$.

An ITM satisfies the *interactiveness* condition if all its computations are interactive. Clearly, if a deterministic ITM $\mathcal{I}$ satisfies the interactiveness condition, then its interactive behaviour is total, in the sense that for every $\boldsymbol{x} \in \Sigma^\omega$ there is at least one $\boldsymbol{y} \in \Sigma^\omega$ such that $(\boldsymbol{x}, \boldsymbol{y})$ is an interaction pair of $\mathcal{I}$. By confining our attention to the input-active computations—which, in the terminology of [12], corresponds to adopting the full environmental activity postulate—, we may then associate with every such ITM an ω-translation: we say that ITM $\mathcal{I}$ produces $\boldsymbol{y}$ on input $\boldsymbol{x}$ if $(\boldsymbol{x}, \boldsymbol{y})$ is the interaction pair associated with an input-active computation of $\mathcal{I}$.

Definition 2. *An ω-translation $\phi : \Sigma^\omega \to \Sigma^\omega$ is* interactively computable *if there exists a deterministic ITM that satisfying the interactiveness condition that produces $\phi(\boldsymbol{x})$ on input $\boldsymbol{x}$ for all $\boldsymbol{x} \in \Sigma^\omega$.*

In [12] a criterion of the interactively computable ω-translations is presented by using limit-continuous functions (a formal definition is included in [14]).

Theorem 1. *A total ω-translation is interactively computable iff it is limit-continuous.*

2.2 The Theory of Executability

The theory of executability combines computation and concurrency-style interaction in such a way that both are treated on equal footing; thus, an integration of computability and concurrency theory is realised.

The transition system is the central notion in the mathematical theory of discrete-event behaviour. It is parameterised by a set $\mathcal{A}$ of *action symbols*, denoting the observable events of a system. We extend $\mathcal{A}$ with a special symbol τ, which intuitively denotes unobservable internal activity. We shall abbreviate $\mathcal{A} \cup \{\tau\}$ by $\mathcal{A}_\tau$.

Definition 3. *An $\mathcal{A}_\tau$-labelled transition system $\mathcal{T}$ is a triple $(\mathcal{S}, \longrightarrow, \uparrow)$, where,*

1. *$\mathcal{S}$ is a set of states,*
2. *$\longrightarrow \subseteq \mathcal{S} \times \mathcal{A}_\tau \times \mathcal{S}$ is an $\mathcal{A}_\tau$-labelled transition relation,*
3. *$\uparrow \in \mathcal{S}$ is the initial state.*

Transition systems can be used to give semantics to programming languages and process calculi. The standard method is to first associate with every program or process expression a transition system (its operational semantics), and then consider programs and process expressions modulo one of the many behavioural equivalences on transition systems that have been studied in the literature. In this paper, we shall use the notion of (divergence-preserving) branching bisimilarity [5,6], which is the finest behavioural equivalence in van Glabbeek's linear time - branching time spectrum [7] that abstracts from internal computation steps (represented in the transition system by transitions labelled with τ). We adopt the notation $\underline{\leftrightarrow}_b^{\Delta}$ for divergence-preserving branching bisimilarity, and $\underline{\leftrightarrow}_b$ for the divergence-insensitive version (see [14] for a formal definition).

The notion of reactive Turing machine (RTM) was put forward in [3] to mathematically characterise which behaviour is executable by a conventional computing system. We recall the definition of RTMs and the ensued notion of executable transition system.

Definition 4. *A reactive Turing machine (RTM) $\mathcal{M}$ is a triple $(\mathcal{S}, \longrightarrow, \uparrow)$, where*

1. *$\mathcal{S}$ is a finite set of states,*
2. *$\longrightarrow \subseteq \mathcal{S} \times \mathcal{D}_{\square} \times \mathcal{A}_{\tau} \times \mathcal{D}_{\square} \times \{L, R\} \times \mathcal{S}$ is a $(\mathcal{D}_{\square} \times \mathcal{A}_{\tau} \times \mathcal{D}_{\square} \times \{L, R\})$-labelled transition relation (we write $s \xrightarrow{a[d/e]M} t$ for $(s, d, a, e, M, t) \in \longrightarrow$),*
3. *$\uparrow \in \mathcal{S}$ is a distinguished initial state.*

Intuitively, the meaning of a transition $s \xrightarrow{a[d/e]M} t$ is that whenever $\mathcal{M}$ is in state s, and d is the symbol currently read by the tape head, then it may execute the action a, write symbol e on the tape (replacing d), move the read/write head one position to the left or the right on the tape, and then end up in state t.

To formalise the intuitive understanding of the operational behaviour of RTMs, we associate with every RTM $\mathcal{M}$ an $\mathcal{A}_{\tau}$-labelled transition system $\mathcal{T}(\mathcal{M})$. The states of $\mathcal{T}(\mathcal{M})$ are the configurations of $\mathcal{M}$, pairs consisting of a state and a tape instance.

Definition 5. *Let $\mathcal{M} = (\mathcal{S}, \longrightarrow, \uparrow)$ be an RTM. The transition system $\mathcal{T}(\mathcal{M})$ associated with $\mathcal{M}$ is defined as follows:*

1. *its set of states consists of the set of all configurations of $\mathcal{M}$;*
2. *its transition relation $\longrightarrow$ is the least relation satisfying, for all $a \in \mathcal{A}_{\tau}$, $d, e \in \mathcal{D}_{\square}$ and $\delta_L, \delta_R \in \mathcal{D}_{\square}^*$:*
   - *$(s, \delta_L \check{d} \delta_R) \xrightarrow{a} (t, \delta_L^< e \delta_R)$ iff $s \xrightarrow{a[d/e]L} t$, and*
   - *$(s, \delta_L \check{d} \delta_R) \xrightarrow{a} (t, \delta_L e ^> \delta_R)$ iff $s \xrightarrow{a[d/e]R} t$*
   ($\delta_L^<$ is obtained from δ_L by placing the tape head marker on the right-most symbol in δ_L, and $^>\delta_R$ is obtained analogously from δ_R);
3. *its initial state is the configuration $(\uparrow, \check{\square})$.*

Turing introduced his machines to define the notion of *effectively computable function* in [15]. By analogy, we have a notion of *effectively executable behaviour* [3].

Definition 6. *A transition system is executable if it is the transition system associated with some RTM.*

3   Executability of Interactive Turing Machines

In this section we associate a transition system with every ITM, and then prove that it is executable modulo divergence-preserving branching bisimilarity. It is convenient to consider input and output as separate actions in the transition system associated with an ITM. We denote by $?i$ the action of inputting the symbol $i \in \Sigma$, and by $!o$ the action of outputting the symbol $o \in \Sigma$.

Definition 7. *Let* $\mathcal{I} = (Q, \longrightarrow_{\mathcal{I}}, q_{in})$ *be an ITM. The transition system* $\mathcal{T}(\mathcal{I})$ *associated with* $\mathcal{I}$ *is defined as follows:*

1. *its set of states is the set* $\{(s, \delta) \mid s \in Q \cup \{s_o \mid o \in \Sigma_\lambda, s \in Q\}, \delta$ *is a tape instance}*;
2. *its transition relation* $\longrightarrow$ *is the least relation satisfying, for all* $i, o \in \Sigma_\lambda$, $d, e \in \mathcal{D}_\square$, *and* $\delta_L, \delta_R \in \mathcal{D}_\square^*$:
   - $(s, \delta_L \check{d} \delta_R) \xrightarrow{?i} (t_o, \delta_L^< e \delta_R)$ *iff* $(s, d, i) \longrightarrow_{\mathcal{I}} (t, e, L, o)$ *and* $i \in \Sigma$,
   - $(s, \delta_L \check{d} \delta_R) \xrightarrow{?i} (t_o, \delta_L e {}^> \delta_R)$ *iff* $(s, d, i) \longrightarrow_{\mathcal{I}} (t, e, R, o)$ *and* $i \in \Sigma$,
   - $(s, \delta_L \check{d} \delta_R) \xrightarrow{\tau} (t_o, \delta_L^< e \delta_R)$ *iff* $(s, d, i) \longrightarrow_{\mathcal{I}} (t, e, L, o)$ *and* $i = \lambda$,
   - $(s, \delta_L \check{d} \delta_R) \xrightarrow{\tau} (t_o, \delta_L e {}^> \delta_R)$ *iff* $(s, d, i) \longrightarrow_{\mathcal{I}} (t, e, R, o)$ *and* $i = \lambda$,
   - $(s_o, \delta) \xrightarrow{!o} (s, \delta)$ *iff* $o \in \Sigma$, *and* $(s_o, \delta) \xrightarrow{\tau} (s, \delta)$ *iff* $o = \lambda$.
3. *its initial state is the configuration* $(q_{in}, \check{\square})$.

The following theorem shows that every transition systems associated with an ITM can be simulated by an RTM. (A proof of the theorem is included in [14].)

Theorem 2. *For every ITM* $\mathcal{I}$ *there exists an RTM* $\mathcal{M}$, *such that* $\mathcal{T}(\mathcal{I}) \underleftrightarrow{}_b^\Delta \mathcal{T}(\mathcal{M})$.

As a consequence we have the following corollary.

Corollary 1. *The transition system associated with an ITM is executable modulo divergence-preserving branching bisimilarity.*

4   Executable ω-Translations

Recall that an ω-translation is defined to be interactively computable if, and only if, it can be realised by an ITM. RTMs are designed for exhibiting the expressive power of executable transition systems, rather than ω-translations, and not every RTM naturally has an ω-translation associated with it. Imposing some restrictions on the formalism of RTMs, however, we shall define a subclass of RTMs with which an ω-translation is naturally associated. The ω-translation realised by such an RTM is then called *executable*, and we shall establish that an ω-translation is interactively computable if, and only if, it is executable.

By analogy to the systems described in the theory of interactive computation, we let the RTMs for ω-translations execute in steps, in such a way that with every step a pair of input and output actions can be associated. With every infinite computation of the RTM we can then associate a interaction pair, and the RTM will thus give rise to an ω-translation.

Definition 8. *Let* $\mathcal{A}_\tau = \{?i, !o \mid i, o \in \{0, 1\}\} \cup \{\tau\}$, *and let* $\mathcal{M} = (\mathcal{S}, \longrightarrow, \uparrow)$ *be an RTM with* $\mathcal{A}_\tau$ *as its set of labels. Then* $\mathcal{M}$ *is an* RTM *for* ω-*translation if it satisfies the following properties:*

1. *the set of states* $\mathcal{S}$ *is partitioned into disjoint sets* I *of input states and* E *of execution states, i.e.,* $\mathcal{S} = I \cup E$ *and* $I \cap E = \emptyset$;
2. *the initial state* $\uparrow$ *is an input state, i.e.,* $\uparrow \in I$;
3. *for a transition* $s \xrightarrow{a[d/e]M} t$, *if* $s \in I$, *then* $a \in \{?0, ?1\}$ *and* $t \in E$; *if* $s \in E$, *then* $a \in \{!0, !1, \tau\}$ *and* $t \in I$; *and*
4. *for all* $(s, d) \in E \times \mathcal{D}_\square$, *there is at most one transition of the form* $s \xrightarrow{a[d/e]M} t$; *and*
5. *for all* $(s, d) \in I \times \mathcal{D}_\square$, *there are exactly two transitions of the form* $s \xrightarrow{a[d/e]M} t$, *one with* $a = ?0$ *and one with* $a = ?1$.

In the following lemma we establish some properties of the transition system associated with an RTM for ω-translation. (See [14] for a proof of the lemma.)

Lemma 1. *Let* $\mathcal{M}$ *be an* RTM *for* ω-*translation. Then* $T(\mathcal{M}) = (\mathcal{S}_\mathcal{M}, \longrightarrow_\mathcal{M}, \uparrow_\mathcal{M})$ *satisfies the following properties:*

1. (Alternation). *The set of states* $\mathcal{S}_\mathcal{M}$ *is partitioned into a set of input states* $I_\mathcal{M}$ *and a set of output states* $E_\mathcal{M}$, *i.e.,* $\mathcal{S}_\mathcal{M} = I_\mathcal{M} \cup E_\mathcal{M}$ *and* $I_\mathcal{M} \cap E_\mathcal{M} = \emptyset$. *For every transition* $s \xrightarrow{a} s'$, *if* $s \in I_\mathcal{M}$, *then* $a \in \{?0, ?1\}$ *and* $s' \in E_\mathcal{M}$; *if* $s \in E_\mathcal{M}$, *then* $a \in \{!0, !1, \tau\}$ *and* $s' \in I_\mathcal{M}$.
2. (Unambiguity). *For every* $s \in E_\mathcal{M}$, *there is exactly one outgoing transition* $s \xrightarrow{a} s'$ *with* $a \in \{!0, !1, \tau\}$.
3. (Totality). *For every* $s \in I_\mathcal{M}$, *there are exactly two outgoing transitions, labelled with* ?0 *and* ?1, *respectively.*

We call a transition system that satisfies the conditions of Lemma 1 an *i/o transition system*. Moreover, by analogy to the interactiveness condition for ITMs, we impose an interactiveness condition on RTMs for ω-translation.

Definition 9. *An* i/o *transition system is* interactive, *if for every* $s \in S$ *and* $s \xrightarrow{?i} s_0$ *with* $i \in \{0, 1\}$, *and for every sequence* $s_0 \longrightarrow s_1 \longrightarrow \cdots$, *there exists a natural number* k, *such that* $s_k \xrightarrow{!o} s_{k+1}$ *with* $o \in \{0, 1\}$.
An RTM *for* ω-*translation is* interactive *if the associated* i/o *transition system is.*

We define the ω-translation realized by an RTM by defining the ω-translation realized by the i/o transition system associated with it. Let $T = (\mathcal{S}, \longrightarrow, \uparrow)$ be an i/o transition system, let $s \in \mathcal{S}$, and let $\sigma \in \mathcal{A}^\omega$, say $\sigma = a_0, a_1, \ldots$; we write $s \xrightarrow{\sigma}$ if there exist $s_0, s'_0, s_1, s'_1, \ldots \in \mathcal{S}$ such that $s = s_0$, and $s_i \longrightarrow^* s'_i \xrightarrow{a_i} s_{i+1}$ for all $i \geq 0$. (By $\longrightarrow^*$ we denote the reflexive-transitive closure of the relation $\xrightarrow{\tau}$.) If $\sigma \in \mathcal{A}^\omega$ and $s \xrightarrow{\sigma}$, then σ is a *weak infinite trace* from s. We denote by $Tr_w^\infty(s)$ the set of weak infinite traces from s.

Definition 10. *Let $\mathcal{T}$ be an i/o transition system, and s_0 be the initial state. For $\sigma \in Tr_w^{\infty}(s_0)$, the input stream realised by σ is the stream $\boldsymbol{x} \in \Sigma^{\omega}$ such that $\boldsymbol{x} = x_1 x_2 \ldots$, where $x_j = i$ if $?i$ is the j-th input action in σ, and similarly for the output stream realized by σ. We say that $\mathcal{T}$ realizes ω-translation $\phi : \Sigma^{\omega} \to \Sigma^{\omega}$ iff, for every $\boldsymbol{x} \in \Sigma^{\omega}$, there exists a trace $\sigma \in Tr_w^{\infty}(s_0)$ with $\boldsymbol{x}$ as its input stream, and for every such trace, its output stream is $\boldsymbol{y} = \phi(\boldsymbol{x})$.*

We can now define when an ω-translation is executable.

Definition 11. *An ω-translation is executable if it can be realized by an executable i/o transition system.*

The following lemma establishes that an ω-translation can be associated with every interactive i/o transition system.

Lemma 2. *If an i/o transition system is interactive, then it realises an ω-translation.*

Moreover, we have the following theorem. (A proof can be found in [14].)

Theorem 3. *An ω-translation is executable iff it is a limit-continuous total function.*

By Theorem 1, we have the following corollary.

Corollary 2. *An ω-translation is executable iff it is interactively computable.*

Therefore, the classes of computable limit-continuous functions, interactively computable ω-translations and executable ω-translations coincide.

5   Advice

In [10], the computational power of evolving interactive systems is studied using ITMs. Particularly, a mechanism called *advice function* is introduced to enhance the computational power of an ITM. In this way, the insertion of external information into the course of a computation is allowed, which leads to a non-uniform operation. In this section, we introduce the notion of advice as a process in parallel composition with an RTM, and show that advice processes indeed give the systems more expressive power.

In this section, we consider advices as functions over natural numbers. In order to record a number on the tape, a natural number n is encoded by a sequence n "1"s ending with a "0". In [10], the notion of ITM with advice is defined as follows.

Definition 12. *An advice function is a function $f : \mathbb{N} \to \mathbb{N}$. An ITM with advice (ITM/A) is equipped with a separate advice tape and a distinguished advice state. By writing the value of the argument x on the advice tape and by entering into the advice state, the value of $f(x)$ will appear on the advice tape in a single step. By this action, the original contents of the advice tape is completely overwritten.*

Here we do not put the restriction on the length of the advice function as in [12], since it does not make a difference in the issue of computability, and we are not yet interested in the issue of complexity. It is obvious that ITMs with uncomputable advice functions cannot be simulated by any RTM, as uncomputable advice function cannot be evaluated by the mechanism of RTMs. As an extension, we equip RTMs with advice processes which enable the simulation of ITM/As.

An advice process A_f is designed to compute the function f, and can only interact with a certain RTM $\mathcal{M}$. As an advice function is not necessarily computable, we cannot associate with every advice process an executable transition system. An RTM $\mathcal{M}$ communicates with A_f as follows: when it needs to get the result of $f(i)$, it enters a special control state a_f, and starts to send a sequence of i "1" s and a "0", which is already written on the tape, to the channel $\overline{in}$, and then, it receives the result sequence $f(i)$ "1"s and a "0" from out channel, and write them on the tape. This procedure ends up with another control state. We can model an advice process as follows.

Definition 13. *Let $f : \mathbb{N} \to \mathbb{N}$ be a function, A_f is an advice process for f with transition system $\mathcal{T}(A_f) = (\mathcal{S}, \to, \uparrow)$, where*

1. $\mathcal{S} = \{s_i \mid i = 0, 1, 2, \ldots\} \cup \{t_i \mid i = 0, 1, 2, \ldots\}$, *and*
2. $s_i \xrightarrow{in?1} s_{i+1}$, $i = 0, 1, 2 \ldots$ $s_i \xrightarrow{in?0} t_{f(i)}$, $i = 1, 2 \ldots$
   $t_i \xrightarrow{out!1} t_{i-1}$, $i = 1, 2 \ldots$ $t_0 \xrightarrow{out!0} s_0$
3. $\uparrow = s_0$.

The behaviour of A_f is deterministic. It receives a sequence of i "1"s from the channel in, followed by a "0" symbol, indicating the end of the sequence, and then, it produces $f(i)$ "1"s to the channel out, also followed by a "0" symbol. This procedure is repeated indefinitely.

The parallel composition of an RTM $\mathcal{M}$ and an advice process A_f, we write as $[\mathcal{M} \parallel A_f]_C$. The parallel composition is defined in the same way as the parallel composition of two RTMs in [3], where $C = \{in, out\}$ is the set of restricted names for communication. If $\mathcal{M}$ is an RTM and A_f is an advice process, then we call $[\mathcal{M} \parallel A_f]_C$ a reactive Turing machine with advice (RTM/A).

Note that, since advice functions and advice processes have the same computational power, by Corollary 2, an ω-translation is realisable by an ITM/A if, and only if, it is realisable by an RTM/A.

Let $\mathcal{T}$ be any bounded branching transition system (not necessarily effective). Based on a presupposed encoding of its sets of states and actions and its transition relation, let the advice function $f_{\mathcal{T}}$ be such that for the code of a state it yields the code of the set of all outgoing transitions of that state. It is straightforward to define an RTM that simulates $\mathcal{T}$ with the help of $f_{\mathcal{T}}$. Then we obtain the following result.

Theorem 4. *If $\mathcal{T}$ is a boundedly branching labelled transition system, then there exists an RTM/A $[\mathcal{M} \parallel A_f]_C$ such that $\mathcal{T}([\mathcal{M} \parallel A_f]_C) \leftrightarroweq_b^\Delta \mathcal{T}$.*

If we, instead, let the advice function f_T be such that on the code of a pair of a state s and a natural number i yields the code of the ith outgoing transition of s, then we can extend the simulation to transition systems with countable many states and transitions.

Theorem 5. *If T is a countable labelled transition system, then there exists an RTM/A $[\mathcal{M} \parallel A_f]_C$ such that $\mathcal{T}([\mathcal{M} \parallel A_f]_C) \underset{b}{\leftrightarrow} T$.*

Note that the transition system associated with an RTM/A is boundedly branching. Hence, by Theorem 2 in [13], if a transition system has no divergence up to $\underset{b}{\leftrightarrow}^{\Delta}$ and is unboundedly branching up to $\underset{b}{\leftrightarrow}^{\Delta}$, then it is not executable modulo $\underset{b}{\leftrightarrow}^{\Delta}$. It follows that there exist countable unboundedly branching transition systems that cannot be simulated by an RTM/A modulo $\underset{b}{\leftrightarrow}^{\Delta}$.

6   Conclusion

We have discussed the relationship between two models of computation that take interaction into account. We have established that the model of RTMs subsumes and is more expressive than the model of ITMs when it comes specifying behaviour, and coincides with the model of ITMs when it comes defining ω-translations.

Furthermore, we have shown that RTMs admit an extension with advice that facilitates modelling non-uniform behaviour. In [3] it was established that every effective transition system can be simulated by an RTM. Our result that every countable transition system can be simulated by an RTM with advice further confirms the universal expressiveness of the notion of RTM.

In [16], a complexity theory for interactive computation has been defined on the basis of ITMs and ω-translations. Clearly, such a complexity theory could also be based on the restricted class of RTMs for ω-translation. Such a complexity theory could then further be generalised towards a complexity theory for general executable behaviour.

References

1. Aceto, L., Ingólfsdóttir, A., Larsen, K.G., Srba, J.: Reactive Systems-Modelling, Specification and Verification. Cambridge University Press, Cambridge (2007)
2. Baeten, J.C.M., Cuijpers, P.J.L., Luttik, B., van Tilburg, P.J.A.: A process-theoretic look at automata. In: Arbab, F., Sirjani, M. (eds.) FSEN 2009. LNCS, vol. 5961, pp. 1–33. Springer, Heidelberg (2010)
3. Baeten, J.C.M., Luttik, B., van Tilburg, P.: Reactive Turing machines. Inf. Comput. **231**, 143–166 (2013)
4. Cabessa, J., Villa, A.E.P.: The super-Turing computational power of interactive evolving recurrent neural networks. In: Mladenov, V., Koprinkova-Hristova, P., Palm, G., Villa, A.E.P., Appollini, B., Kasabov, N. (eds.) ICANN 2013. LNCS, vol. 8131, pp. 58–65. Springer, Heidelberg (2013)
5. van Glabbeek, R.J., Weijland, W.P.: Branching time and abstraction in bisimulation semantics. J. ACM (JACM) **43**(3), 555–600 (1996)

6. van Glabbeek, R., Luttik, B., Trčka, N.: Branching bisimilarity with explicit divergence. Fundam. Informaticae **93**(4), 371–392 (2009)
7. van Glabbeek, R.J.: The linear time — branching time spectrum II. In: Best, E. (ed.) CONCUR 1993. LNCS, vol. 715, pp. 66–81. Springer, Heidelberg (1993)
8. Goldin, D., Smolka, S.A., Wegner, P.: Interactive Computation: The New Paradigm. Springer, Heidelberg (2006)
9. van Leeuwen, J., Wiedermann, J.: On algorithms and interaction. In: Nielsen, M., Rovan, B. (eds.) MFCS 2000. LNCS, vol. 1893, pp. 99–113. Springer, Heidelberg (2000)
10. van Leeuwen, J., Wiedermann, J.: Beyond the Turing limit: evolving interactive systems. In: Pacholski, L., Ružička, P. (eds.) SOFSEM 2001. LNCS, vol. 2234, pp. 90–109. Springer, Heidelberg (2001)
11. van Leeuwen, J., Wiedermann, J.: The Turing machine paradigm in contemporary computing. In: Enquist, B., Schmidt, W. (eds.) Mathematics Unlimited-2001 and Beyond. LNCS, vol. 2001, pp. 1139–1155. Springer, Heidelberg (2001)
12. van Leeuwen, J., Wiedermann, J.: A theory of interactive computation. In: Goldin, D., Smolka, S.A., Wegner, P. (eds.) Interactive Computation, pp. 119–142. Springer, Heidelberg (2006)
13. Luttik, B., Yang, F.: Executability and the π-calculus (extended abstract). In: Proceedings of ICE 2015, pp. 37–52 (2015)
14. Luttik, B., Yang, F.: On the Executability of Interactive Computation. CoRR abs/1601.01546 (2016). http://arxiv.org/abs/1601.01546
15. Turing, A.M.: On computable numbers, with an application to the Entscheidungs-problem. J. Math. **58**, 345–363 (1936)
16. Verbaan, P.R.A.: The computational complexity of evolving systems. Ph.D. thesis, Utrecht University (2006)
17. Wiedermann, J., van Leeuwen, J.: How we think of computing today. In: Beckmann, A., Dimitracopoulos, C., Löwe, B. (eds.) CiE 2008. LNCS, vol. 5028, pp. 579–593. Springer, Heidelberg (2008)

Circuit Satisfiability and Constraint Satisfaction Around Skolem Arithmetic

Christian Glaßer[1], Peter Jonsson[2], and Barnaby Martin[3(✉)]

[1] Theoretische Informatik, Julius-Maximilians-Universität, Würzburg, Germany
[2] Department of Computer and Information Science,
Linköpings Universitet, 581 83 Linköping, Sweden
[3] School of Science and Technology, Middlesex University,
The Burroughs, Hendon, London NW4 4BT, UK
barnabymartin@gmail.com

Abstract. We study interactions between Skolem Arithmetic and certain classes of Circuit Satisfiability and Constraint Satisfaction Problems (CSPs). We revisit results of Glaßer et al. [16] in the context of CSPs and settle the major open question from that paper, finding a certain satisfiability problem on circuits—involving complement, intersection, union and multiplication—to be decidable. This we prove using the decidability of Skolem Arithmetic. Then we solve a second question left open in [16] by proving a tight upper bound for the similar circuit satisfiability problem involving just intersection, union and multiplication. We continue by studying first-order expansions of Skolem Arithmetic without constants, $(\mathbb{N}; \times)$, as CSPs. We find already here a rich landscape of problems with non-trivial instances that are in P as well as those that are NP-complete.

1 Introduction

Skolem Arithmetic is the weak fragment of first-order arithmetic involving only multiplication. Thoralf Skolem gave a quantifier-elimination technique and argued for decidability of the theory in [27]. However, his proof was rather vague and a robust demonstration was not given of this result until Mostowski [22]. Skolem Arithmetic is somewhat less fashionable than *Presburger Arithmetic*, which involves only addition, and was proved decidable by Presburger in [25]. Indeed, Mostowski's proof made use of a reduction from Skolem Arithmetic to Presburger Arithmetic through the notion of weak direct powers (an excellent survey on these topics is [3]). The central thread of this paper is putting to work results about Skolem Arithmetic from the past, to solve open and naturally arising problems from today. Many of our results, like that of Mostowski, will rely on the interplay between Skolem and Presburger Arithmetic.

A *constraint satisfaction problem* (CSP) is a computational problem in which the input consists of a finite set of variables and a finite set of constraints, and

P. Jonsson—was partially supported by the *Swedish Research Council* (VR) under grant 621-2012-3239.

B. Martin—was supported by EPSRC grant EP/L005654/1.

A. Beckmann et al. (Eds.): CiE 2016, LNCS 9709, pp. 323–332, 2016.
DOI: 10.1007/978-3-319-40189-8_33

where the question is whether there exists a mapping from the variables to some fixed domain such that all the constraints are satisfied. When the domain is finite, and arbitrary constraints are permitted in the input, the CSP is NP-complete. However, when only constraints from a restricted set of relations are allowed in the input, it can be possible to solve the CSP in polynomial time. The set of relations that is allowed to formulate the constraints in the input is often called the *constraint language*. The question which constraint languages give rise to polynomial-time solvable CSPs has been the topic of intensive research over the past years. It has been conjectured by Feder and Vardi [13] that CSPs for constraint languages over finite domains have a complexity dichotomy: they are either in P or NP-complete. This conjecture remains unsettled, although dichotomy is now known on substantial classes (e.g. structures with domains of size ≤ 3 [9,26] and smooth digraphs [2,17]). Various methods, combinatorial (graph-theoretic), logical and universal-algebraic have been brought to bear on this classification project, with many remarkable consequences. A conjectured delineation for the dichotomy was given in the algebraic language in [10].

By now the literature on infinite-domain CSPs is also beginning to mature. Here the complexity can be much higher (e.g. undecidable) but on natural classes there is often the potential for structured classifications, and this has proved to be the case for reducts of, e.g. the rationals with order [5], the random (Rado) graph [7] and the integers with successor [6]; as well as first-order (fo) expansions of linear program feasibility [4]. Skolem and Presburger Arithmetic represent perfect candidates for continuation in this vein. These natural classes around Skolem and Presburger Arithmetic have the property that their CSPs sit in NP and a topic of recent interest for the second and third authors has been natural CSPs sitting in higher complexity classes.

Meanwhile, a literature existed on satisfiability of circuit problems over sets of integers involving work of the first author [16], itself continuing a line of investigation begun in [29] and pursued in [21,31,32]. The problems in [16] can be seen as variants of certain functional CSPs whose domain is all singleton sets of the non-negative integers and whose relations are set operations of the form: complement, intersection, union, addition and multiplication (the latter two are defined set-wise, e.g. $A \times B := \{ab : a \in A \wedge b \in B\}$). An open problem was the complexity of the problem when the permitted set operators were precisely complement, intersection, union and multiplication. In this paper we resolve that this problem is in fact decidable, indeed in triple exponential space. We prove this result by using the decidability of the theory of Skolem Arithmetic with constants. We take here Skolem Arithmetic to be the non-negative integers with multiplication (and possibly constants). In studying this problem we are able to bring to light existing results of [16] as results about their related CSPs, providing natural examples with interesting super-NP complexities. In addition, we improve one of the upper bounds of [16] to a tight upper bound. This is the circuit satisfiability problem where the permitted set operators are just intersection, union and multiplication, and where we improve the bound from NEXP to PSPACE. Interestingly, this result does not immediately translate to a similar upper bound for the corresponding functional CSP.

In the second part of the paper, Skolem Arithmetic takes centre stage as we initiate the study of the computational complexity of the CSPs of its reducts, i.e. those constraint languages whose relations have a fo-definition in $(\mathbb{N}; \times)$. $\mathrm{CSP}(\mathbb{N}; \times)$ is in P, indeed it is trivial. The object therefore of our early study is its fo-expansions. We show that $\mathrm{CSP}(\mathbb{N}; +, \neq)$ is NP-complete, as is $\mathrm{CSP}(\mathbb{N}; \times, c)$ for each $c > 1$. We further show that $\mathrm{CSP}(\mathbb{N}; \times, U)$ is NP-complete when U is any non-empty set of integers greater than 1 such that each has a prime factor p, for some prime p, but omits the factor p^2. Clearly, $\mathrm{CSP}(\mathbb{N}; \times, U)$ is in P (and is trivial) if U contains 0 or 1. As a counterpoint to our NP-hardness results, we prove that $\mathrm{CSP}(\mathbb{N}; \times, U)$ is in P whenever there exists $m > 1$ so that $U \supseteq \{m, m^2, m^3, \ldots\}$.

Related Work. Apart from the research on circuit problems mentioned above there has been work on other variants like circuits over integers [30] and positive natural numbers [8], equivalence problems for circuits [15], functions computed by circuits [24], and equations over sets of natural numbers [18,19].

2 Preliminaries

Let $\mathbb{N}$ be the set of non-negative integers, and let $\mathbb{N}^+$ be the set of positive integers. For $m \in \mathbb{N}$, let Div_m be the set of factors of m. Finally, let $\{\mathbb{N}\}$ be the set of singletons $\{\{x\} : n \in \mathbb{N}\}$. In this paper we use a version of the CSP permitting both relations and functions (and constants). Thus, a *constraint language* consists of a domain together with functions, relations and constants over that domain. One may thus consider a constraint language to be a first-order structure. A *homomorphism* from a constraint language Γ to a constraint language Δ, over the same signature, is a function f from the domain of Γ to the domain of Δ that preserves the relations, i.e. if $(x_1, \ldots, x_k) \in R^\Gamma$, then also $(f(x_1), \ldots, f(x_k)) \in R^\Delta$. A homomorphism from a constraint language to itself is an *endomorphism*. An endomorphism that also preserves the negations of relations is termed an *embedding* and a bijective embedding is an *automorphism*.

A constraint language is a *core* if all of its endomorphisms are embeddings (equivalently, if the domain is finite, automorphisms). The functional version of the CSP has previously been seen in, e.g., [12]. For a purely functional constraint language, a *primitive positive* (pp) sentence is the existential quantification of a conjunction of term equalities. More generally, and when relations present, we may have positive atoms in this conjunction. The problem $\mathrm{CSP}(\Gamma)$ takes as input a primitive positive sentence ϕ, and asks whether it is true on Γ. The problem $\mathrm{CSP}^c(\Gamma)$ is similar but allows input constants naming the domain elements. We will allow that the functions involved on ϕ be defined on a larger domain than the domain of Γ. This is rather *unheimlich*[1] but it allows the problems of [16] to

---

[1] Weird. Thus spake Lindemann about Hilbert's non-constructive methods in the resolution of Gordon's problem (see [28]).

be more readily realised in the vicinity of CSPs. For example, one such typical domain is $\{\mathbb{N}\}$, but we will allow functions such as $^-$ (complement), $\cup$ (union) and $\cap$ (intersection) whose domain and range is the set of all subsets of $\mathbb{N}$. We will also employ the operations of set-wise addition $A + B := \{a + b : a \in A \wedge b \in B\}$ and multiplication $A \times B := \{ab : a \in A \wedge b \in B\}$.

Σ_i^P, Π_i^P, and Δ_i^P are levels of the polynomial-time hierarchy, while Σ_i, Π_i, and Δ_i are levels of the arithmetical hierarchy. Moreover, we use the classes $NP = \Sigma_1^P$, $PSPACE = \bigcup_{k \geq 1} DSPACE(n^k)$, and $3EXPSPACE = \bigcup_{k \geq 1} DSPACE(2^{2^{2^{n^k}}})$. Where no SPACE is written explicitly, the complexity classes may be assumed to refer to time. For more on these complexity classes we refer the reader to [23].

For sets A and B we say that A is *polynomial-time many-one reducible* to B, in symbols $A \leq_m^P B$, if there exists a polynomial-time computable function f such that for all x it holds that $(x \in A \Longleftrightarrow f(x) \in B)$. If f is even computable in logarithmic space, then A is *logspace many-one reducible* to B, in symbols $A \leq_m^{\log} B$. A is *nondeterministic polynomial-time many-one reducible* to B, in symbols $A \leq_m^{NP} B$, if there is a nondeterministic Turing transducer M that runs in polynomial time such that for all x it holds that $x \in A$ if and only if there exists a y computed by M on input x with $y \in B$. The reducibility notions $\leq_m^P$, $\leq_m^{\log}$, and $\leq_m^{NP}$ are transitive and NP is closed under these reducibilities.

A *circuit* $C = (V, E, g_C)$ is a finite, non-empty, directed, acyclic multi-graph (V, E) with a specified node $g_C \in V$. The graph does not need to be connected and only has multiple edges between two nodes when a binary operator is applied on both sides to a single set (e.g. $A \times A$). Let $V = \{1, 2, \ldots, n\}$ for some $n \in \mathbb{N}$. The nodes in the graph (V, E) are topologically ordered, i.e., for all $v_1, v_2 \in V$, if $v_1 < v_2$, then there is no path from v_2 to v_1. Nodes are also called *gates*. Nodes with indegree 0 are called *input gates* and g_C is called the *output gate*. If there is an edge from gate u to gate v, then we say that u is a *predecessor* of v and v is a *successor* of u.

Let $\mathcal{O} \subseteq \{\cup, \cap, ^-, +, \times\}$. An $\mathcal{O}$-*circuit with unassigned input gates* $C = (V, E, g_C, \alpha)$ is a circuit (V, E, g_C) whose gates are labeled by the labeling function $\alpha : V \to \mathcal{O} \cup \mathbb{N} \cup \{\star\}$ such that the following holds: Each gate has an indegree in $\{0, 1, 2\}$, gates with indegree 0 have labels from $\mathbb{N} \cup \{\star\}$, gates with indegree 1 have label $^-$, and gates with indegree 2 have labels from $\{\cup, \cap, +, \times\}$. Input gates with a label from $\mathbb{N}$ are called *assigned* (or constant) input gates; input gates with label $\star$ are called *unassigned* (or variable) input gates. An $\mathcal{O}$-*formula* is an $\mathcal{O}$-circuit that only contains nodes with outdegree one.

Let $u_1 < \cdots < u_n$ be the unassigned inputs in C and $x_1, \ldots, x_n \in \mathbb{N}$. By assigning value x_i to the input u_i, we obtain an $\mathcal{O}$-*circuit* $C(x_1, \ldots, x_n)$ whose input gates are all assigned. In this circuit, each gate g computes the following set $I(g)$: If g is an assigned input gate where $\alpha(g) \neq \star$, then $I(g) = \{\alpha(g)\}$. If $g = u_k$ is an unassigned input gate, then $I(g) = \{x_k\}$. If g has label $^-$ and predecessor g_1, then $I(g) = \mathbb{N} \setminus I(g_1)$. If g has label $\circ \in \{\cup, \cap, +, \times\}$ and predecessors g_1 and g_2, then $I(g) = I(g_1) \circ I(g_2)$. Finally, let $I(C(x_1, \ldots, x_n)) = I(g_C)$ be the set computed by the circuit $C(x_1, \ldots, x_n)$.

Definition 1 (membership, equivalence, and satisfiability problems of circuits and formulas).

Let $\mathcal{O} \subseteq \{\cup, \cap, {}^{-}, +, \times\}$.

$\mathrm{MC_N}(\mathcal{O}) = \{(C, b) \mid C \text{ is an } \mathcal{O}\text{-circuit without unassigned inputs and } b \in I(C)\}$

$\mathrm{EC_N}(\mathcal{O}) = \{(C_1, C_2) \mid C_1 \text{ and } C_2 \text{ are } \mathcal{O}\text{-circuits without unassigned inputs and}$
$\qquad\qquad\qquad we \text{ have } I(C_1) = I(C_2)\}$

$\mathrm{SC_N}(\mathcal{O}) = \{(C, b) \mid C \text{ is an } \mathcal{O}\text{-circuit with unassigned inputs } u_1 < \cdots < u_n \text{ and}$
$\qquad\qquad\qquad there \text{ exist } x_1, \ldots, x_n \in \mathbb{N} \text{ such that } b \in I\big(C(x_1, \ldots, x_n)\big)\}$

$\mathrm{MF_N}(\mathcal{O})$, $\mathrm{EF_N}(\mathcal{O})$, *and* $\mathrm{SF_N}(\mathcal{O})$ *are the variants that deal with $\mathcal{O}$-formulas instead of $\mathcal{O}$-circuits.*

When an $\mathcal{O}$-circuit is used as input for an algorithm, then we use a suitable encoding such that it is possible to verify in deterministic logarithmic space whether a given string encodes a valid circuit.

In Sect. 3, for $i \in \mathbb{N}$, we often identify $\{i\}$ with i, where this can not cause a harmful confusion.

3   Circuit Satisfiability and Functional CSPs

We investigate the computational complexity of functional CSPs. In many cases we can translate known lower and upper bounds for membership, equivalence, and satisfiability problems of arithmetic circuits [15,16,21] to CSPs. Our main result is the decidability of $\mathrm{SC_N}({}^{-}, \cup, \cap, \times)$ and $\mathrm{CSP}^c(\{\mathbb{N}\}; {}^{-}, \cup, \cap, \times)$, which solves the main open question of the paper [16]. We emphasise that the domain of $\mathrm{CSP}^c(\{\mathbb{N}\}; {}^{-}, \cup, \cap, \times)$ is the set of singletons that we defined as $\{\mathbb{N}\}$ and not, e.g., the set of subsets of all natural numbers. This would be a different CSP. Our unusual definition is motivated by the circuit problems whose relationship to CSPs we wish to formalise.

We start with the observation that the equivalence of arithmetic terms reduces to functional CSPs. This yields several lower bounds for the CSPs.

Proposition 1. *For* $\mathcal{O} \subseteq \{{}^{-}, \cup, \cap, +, \times\}$ *it holds that* $\mathrm{EF_N}(\mathcal{O}) \leq_{\mathrm{m}}^{\log}$ $\mathrm{CSP}^c(\{\mathbb{N}\}; \mathcal{O})$.

Corollary 1.

1. $\mathrm{CSP}^c(\{\mathbb{N}\}; {}^{-}, \cup, \cap, +)$ *and* $\mathrm{CSP}^c(\{\mathbb{N}\}; {}^{-}, \cup, \cap, \times)$ *are* $\leq_{\mathrm{m}}^{\log}$*-hard for* PSPACE.
2. $\mathrm{CSP}^c(\{\mathbb{N}\}; \cup, \cap, +)$, $\mathrm{CSP}^c(\{\mathbb{N}\}; \cup, \cap, \times)$, $\mathrm{CSP}^c(\{\mathbb{N}\}; \cup, +)$, *and* $\mathrm{CSP}^c(\{\mathbb{N}\}; \cup, \times)$ *are* $\leq_{\mathrm{m}}^{\log}$*-hard for* Π_2^{P}.

CSPs with $+$ and $\times$ can express diophantine equations, which implies the Turing-hardness of such CSPs.

Proposition 2. $\mathrm{CSP}^c(\{\mathbb{N}\}; +, \times)$ *is* $\leq_m^{\log}$*-hard for* Σ_1, $\mathrm{CSP}^c(\{\mathbb{N}\}; \cup, \cap, +, \times) \in \Sigma_1$ *and* $\mathrm{CSP}^c(\{\mathbb{N}\}; ^-, \cup, \cap, +, \times) \in \Sigma_2$.

We now show that the decidability of Skolem arithmetic [14] can be used to decide the satisfiability of arithmetic circuits without $+$. From this we obtain the decidability of CSPs where exactly one arithmetic operation is forbidden.

Theorem 1. $\mathrm{SC}_{\mathbb{N}}(^-, \cup, \cap, \times)$, $\mathrm{CSP}^c(\{\mathbb{N}\}; ^-, \cup, \cap, \times)$ *and* $\mathrm{CSP}^c(\{\mathbb{N}\}; ^-, \cup, \cap, +)$ *are in* 3EXPSPACE.

The following proposition transfers the NP-hardness from satisfiability problems for arithmetic circuits to $\mathrm{CSP}^c(\{\mathbb{N}\}; \times)$ and $\mathrm{CSP}^c(\{\mathbb{N}\}; +)$.

Proposition 3. $\mathrm{CSP}^c(\{\mathbb{N}\}; \times)$ *and* $\mathrm{CSP}^c(\{\mathbb{N}\}; +)$ *are* $\leq_m^{\log}$*-hard for* NP.

The remaining results in this section show that certain functional CSPs belong to NP. This needs non-trivial arguments of the form: If a CSP can be satisfied, then it can be satisfied even with small values. These arguments are provided by the known results that integer programs, existential Presburger arithmetic, and existential Skolem arithmetic are decidable in NP.

Proposition 4. $\mathrm{CSP}^c(\{\mathbb{N}\}; ^-, \cap, \cup)$ *is* $\leq_m^{\log}$*-complete for* NP.

Proposition 5. $\mathrm{CSP}^c(\{\mathbb{N}\}; +) \in$ NP.

Proposition 6. $\mathrm{CSP}^c(\{\mathbb{N}\}; \cap, +) \leq_m^{\mathrm{NP}} \mathrm{CSP}^c(\{\mathbb{N}\}; +, =, \neq)$ *and* $\mathrm{CSP}^c(\{\mathbb{N}\}; \cap, \times)$ $\leq_m^{\mathrm{NP}} \mathrm{CSP}^c(\{\mathbb{N}\}; \times, =, \neq)$. *Therefore,* $\mathrm{CSP}^c(\{\mathbb{N}\}; \cap, +), \mathrm{CSP}^c(\{\mathbb{N}\}; \cap, \times) \in$ NP.

A Second Open Problem from [16]**.** We now improve another of the upper bounds of [16] to a tight upper bound. Here we have the circuit satisfiability problem where the permitted set operators are just intersection, union and multiplication, where we improve the bound from NEXP to PSPACE.

Theorem 2. $\mathrm{SC}_{\mathbb{N}}(\cup, \cap, \times) \in$ PSPACE.

Table 1 summarizes the results obtained in Sect. 3 and shows open questions. In particular, we would like to improve the gap between the lower and upper bounds for $\mathrm{CSP}^c(\{\mathbb{N}\}; \mathcal{O})$, where $\mathcal{O}$ contains $\cup$ and exactly one arithmetic operation ($+$ or $\times$).

4 CSPs over Fo-expansions of Skolem Arithmetic

We now commence our exploration of the complexity of CSPs generated from the simplest expansions of $(\mathbb{N}; \times)$. Abandoning our set-wise definitions, we henceforth use $\times$ to refer to the syntactic multiplication of Skolem Arithmetic (which may additionally carry semantic content). When we wish to refer to multiplication in a purely semantic way, we prefer $\cdot$s or $\prod$. We will consider $\times$ as a ternary relation rather than a binary function. We will never use syntactic $\times$ in a non-standard way, i.e. holding on a triple of integers for which it does not already hold in natural arithmetic.

Table 1. Upper and lower bounds for $\mathrm{CSP}^c(\{\mathbb{N}\}; \mathcal{O})$. All lower bounds are with respect to $\leq_{\mathrm{m}}^{\log}$-reductions.

$\mathcal{O}$				$\mathrm{CSP}^c(\{\mathbb{N}\}; \mathcal{O})$	
				Lower bound	Upper bound
$^-$ $\cup$ $\cap$ $+$ $\times$				Σ_1	Σ_2
$^-$ $\cup$ $\cap$ $+$				PSPACE	3EXPSPACE
$^-$ $\cup$ $\cap$ $\times$				PSPACE	3EXPSPACE
$^-$ $\cup$ $\cap$				NP	NP
$\cup$ $\cap$ $+$ $\times$				Σ_1	Σ_1
$\cup$ $\cap$ $+$				Π_2^P	3EXPSPACE
$\cup$ $\cap$ $\times$				Π_2^P	3EXPSPACE
$\cup$ $+$ $\times$				Σ_1	Σ_1
$\cup$ $+$				Π_2^P	3EXPSPACE
$\cup$ $\times$				Π_2^P	3EXPSPACE
$\cap$ $+$ $\times$				Σ_1	Σ_1
$\cap$ $+$				NP	NP
$\cap$ $\times$				NP	NP
$+$ $\times$				Σ_1	Σ_1
$+$				NP	NP
$\times$				NP	NP

Proposition 7. Let Γ be a finite signature reduct of $(\mathbb{N}; \times, 1, 2, \ldots)$. Then $\mathrm{CSP}(\Gamma)$ is in NP.

Upper Bounds. We continue with polynomial upper bounds. Note that constants are no longer assumed to necessary exist in our structures (in contrast to the situation in Proposition 7).

Lemma 1. Let $U \subseteq N$ be non-empty and $U \cap \{0, 1\} = \emptyset$. Then $\mathrm{CSP}(\mathbb{N}; \times, U)$ is polynomial-time reducible to $\mathrm{CSP}(\mathbb{N}^+; \times, U)$.

We now borrow the following slight simplification of Lemma 6 from [20].

Lemma 2 (Scalability [20]). Let Γ be a finite signature constraint language with domain $\mathbb{R}$, whose relations are quantifier-free definable in $+, \leq$ and $<$, such that the following holds.

- Every satisfiable instance of $\mathrm{CSP}(\Gamma)$ is satisfied by some rational point.
- For each relation $R \in \Gamma$, it holds that if $\bar{x} := (x_1, x_2, \ldots, x_k) \in R$, then $(ax_1, ax_2, \ldots, ax_k) \in R$ for all $a \in \{y : y \in \mathbb{R}, y \geq 1\}$.
- $\mathrm{CSP}(\Gamma)$ is in P.

Then $\mathrm{CSP}(\Delta)$ is in P, where Δ is obtained from Γ by substituting the domain $\mathbb{R}$ by $\mathbb{Z}$.

Lemma 3. *Arbitrarily choose $m > 1$ and $U \subseteq \mathbb{N}^+$ such that $\{m, m^2, m^3, \ldots\} \subseteq U$. Then, $\mathrm{CSP}(\mathbb{N}^+; \times, U)$ is in P.*

Proposition 8. *Arbitrarily choose $m > 1$ and $U \subseteq \mathbb{N}$ such that $\{m, m^2, m^3, \ldots\} \subseteq U$. Then, $\mathrm{CSP}(\mathbb{N}; \times, U)$ is in P.*

Cores. We say that an integer $m > 1$ has a *degree-one factor p* if and only if p is a prime such that $p|m$ and $p^2 \nmid m$. Let Div_m be the set of divisors of m, pp-definable in $(\mathbb{N}; \times, m)$ by $\exists y\ x \times y = m$. We can pp-define the relation $\{1\}$ in $(\mathrm{Div}_m; \times, m)$ since $x = 1$ iff $x \times x = x$ (recalling $0 \notin \mathrm{Div}_m$). It follows that $\{1, m\}$ are contained in the core of $(\mathrm{Div}_m; \times, m)$.

Lemma 4. *Let $m > 1$ be an integer that has a degree-one factor p. Then $(\mathrm{Div}_m; \times, m)$ has a two-element core.*

Lemma 5. *Let m be an integer that does not have a degree-one factor. Then $(\mathrm{Div}_m; \times, m)$ does not have a two-element core.*

Lower Bounds. We now move to lower bounds of NP-completeness.

Proposition 9. *$\mathrm{CSP}(\mathbb{N}; \neq, \times)$ is NP-complete.*

An operation $t : D^k \to D$ is a *weak near-unanimity* operation if t is idempotent and satisfies $t(y, x, \ldots, x) = t(x, y, x, \ldots, x) = \cdots = t(x, \ldots, x, y)$.

Theorem 3 [1]. *Let Γ be a constraint language over a finite set D. If Γ is a core and does not have a weak near-unanimity polymorphism, then $\mathrm{CSP}(\Gamma)$ is NP-hard.*

Lemma 6. *Arbitrarily choose an $m > 1$ such that $m \neq k^n$ for all $k, n > 1$ together with a finite set $\{1, m\} \subseteq S \subseteq \mathbb{N}\backslash\{0\}$. If $(S; \times, m)$ is a core, then $\mathrm{CSP}(S; \times, m)$ is NP-hard.*

Note that the proof of this last lemma is made easier by our assumption that $\times$ is a relation and not a function. Were it a function we would need to prove the domain S is closed under it.

Theorem 4. *$\mathrm{CSP}(\mathbb{N}; \times, m)$ is NP-hard for every integer $m > 1$.*

Theorem 5. *Let U be any subset of $\mathbb{N}\backslash\{0, 1\}$ so that every $x \in U$ has a degree-one factor. Then $\mathrm{CSP}(\mathbb{N}; \times, U)$ is NP-hard.*

For $x \in \mathbb{N}\backslash\{0, 1\}$, define its *minimal exponent*, $\mathrm{min\text{-}exp}(x)$, to be the smallest j such that x has a factor of p^j, for some prime p, but not a factor of p^{j+1}. Thus an integer with a degree-one factor has minimal exponent 1. Call $x \in \mathbb{N}\backslash\{0, 1\}$ *square-free* if it omits all repeated prime factors. For a set $U \subseteq \mathbb{N}\backslash\{0, 1\}$, define its *basis*, $\mathrm{basis}(U)$ to be the set $\{\mathrm{min\text{-}exp}(x) : x \in U\}$.

Lemma 7. *Let $U \subseteq \mathbb{N}\backslash\{0, 1\}$, so that $\mathrm{basis}(U)$ is finite and $\mathrm{basis}(U) \neq \{1\}$. There is some set X pp-definable in $(\mathbb{N}; \times, U)$ so that $\mathrm{basis}(X) = \{1\}$.*

Theorem 6. *Let $U \subseteq \mathbb{N}\backslash\{0, 1\}$ be so that $\mathrm{basis}(U)$ is finite. Then $\mathrm{CSP}(\mathbb{N}; \times, U)$ is NP-complete.*

5   Final Remarks

There are two major directions in which more work is necessary.

A perfunctory glance at the results of Sect. 3 shows that some of our bounds are not tight, and it would be great to see some natural CSPs in this region manifesting complexities such as PSPACE-complete. It is informative to compare our Table 1 with Table 1 in [16]. Our weird formulation of these CSPs belies the fact there are more natural versions where, for $\mathcal{O} \subseteq \{^-, \cap, \cup, +, \times\}$, we ask about $CSP(\mathcal{P}(\mathbb{N}); \mathcal{O})$, where $\mathcal{P}(\mathbb{N})$ is the power set of $\mathbb{N}$, rather than the somewhat esoteric $CSP(\{\mathbb{N}\}; \mathcal{O})$. Indeed, if we replace complement "$-$" by set difference "\", these questions could also be phrased for just the finite sets of $\mathcal{P}(\mathbb{N})$ (see recent work [11]).

Meanwhile, the results of Sect. 4 need to be extended to a classification of complexity for all $CSP(\Gamma)$, where Γ is a reduct of Skolem Arithmetic $(\mathbb{N}; \times)$. We anticipate the first stage is to complete the classification for $CSP(\mathbb{N}; \times, U)$ where U is fo-definable in $(\mathbb{N}; \times)$.

References

1. Barto, L., Kozik, M.: Constraint satisfaction problems of bounded width. In: FOCS, pp. 595–603 (2009)
2. Barto, L., Kozik, M., Niven, T.: The CSP dichotomy holds for digraphs with no sources and no sinks (a positive answer to a conjecture of Bang-Jensen and Hell). SIAM J. Comput. **38**(5), 1782–1802 (2009)
3. Bès, A.: A tribute to Maurice Boffa. Soc. Math. Belgique, 1–54 (2002)
4. Bodirsky, M., Jonsson, P., von Oertzen, T.: Essential convexity and complexity of semi-algebraic constraints. Log. Methods Comput. Sci. **8**, 4 (2012). Extended abstract titled Semilinear Program Feasibility at ICALP 2010
5. Bodirsky, M., Kára, J.: The complexity of temporal constraint satisfaction problems. J. ACM **57**, 2 (2010)
6. Bodirsky, M., Martin, B., Mottet, A.: Constraint satisfaction problems over the integers with successor. In: Halldórsson, M.M., Iwama, K., Kobayashi, N., Speckmann, B. (eds.) ICALP 2015. LNCS, vol. 9134, pp. 256–267. Springer, Heidelberg (2015)
7. Bodirsky, M., Pinsker, M.: Schaefer's theorem for graphs. In: Proceedings of STOC 2011, pp. 655–664 (2011). Preprint of the long version available at arxiv.org/abs/1011.2894
8. Breunig, H.-G.: The complexity of membership problems for circuits over sets of positive numbers. In: Csuhaj-Varjú, E., Ésik, Z. (eds.) FCT 2007. LNCS, vol. 4639, pp. 125–136. Springer, Heidelberg (2007)
9. Bulatov, A.: A dichotomy theorem for constraint satisfaction problems on a 3-element set. J. ACM **53**(1), 66–120 (2006)
10. Bulatov, A., Krokhin, A., Jeavons, P.G.: Classifying the complexity of constraints using finite algebras. SIAM J. Comput. **34**, 720–742 (2005)
11. Dose, T.: Complexity of constraint satisfaction problems over finite subsets of natural numbers. In: ECCC (2016)
12. Feder, T., Madelaine, F.R., Stewart, I.A.: Dichotomies for classes of homomorphism problems involving unary functions. Theor. Comput. Sci. **314**(1–2), 1–43 (2004)

13. Feder, T., Vardi, M.: The computational structure of monotone monadic SNP and constraint satisfaction: a study through datalog and group theory. SIAM J. Comput. **28**, 57–104 (1999)

14. Ferrante, J., Rackoff, C.W.: The Computational Complexity of Logical Theories. Lecture Notes in Mathematics. Springer, Heidelberg (1979)

15. Glaßer, C., Herr, K., Reitwießner, C., Travers, S.D., Waldherr, M.: Equivalence problems for circuits over sets of natural numbers. Theor. Comput. Syst. **46**(1), 80–103 (2010)

16. Glaßer, C., Reitwießner, C., Travers, S.D., Waldherr, M.: Satisfiability of algebraic circuits over sets of natural numbers. Discrete Appl. Math. **158**(13), 1394–1403 (2010)

17. Hell, P., Nešetřil, J.: On the complexity of H-coloring. J. Comb. Theor. Ser. B **48**, 92–110 (1990)

18. Jez, A., Okhotin, A.: Complexity of equations over sets of natural numbers. Theor. Comput. Sci. **48**(2), 319–342 (2011)

19. Jez, A., Okhotin, A.: Computational completeness of equations over sets of natural numbers. Inform. Comput. **237**, 56–94 (2014)

20. Jonsson, P., Lööw, T.: Computational complexity of linear constraints over the integers. Artif. Intell. **195**, 44–62 (2013). An extended abstract appeared at IJCAI 2011

21. McKenzie, P., Wagner, K.W.: The complexity of membership problems for circuits over sets of natural numbers. Comput. Complex. **16**(3), 211–244 (2007). Extended abstract appeared at STACS 2003

22. Mostowski, A.: On direct products of theories. J. Symb. Log. **17**(3), 1–31 (1952)

23. Papadimitriou, C.H.: Computational Complexity. Addison-Wesley, Reading (1994)

24. Pratt-Hartmann, I., Düntsch, I.: Functions definable by arithmetic circuits. In: Ambos-Spies, K., Löwe, B., Merkle, W. (eds.) CiE 2009. LNCS, vol. 5635, pp. 409–418. Springer, Heidelberg (2009)

25. Presburger, M.: Über die Vollständigkeit eines gewissen Systems der Arithmetik ganzer Zahlen. In: welchem die Addition als einzige Operation hervortritt, Comptes Rendus du I congres de Mathématiciens des Pays Slaves, pp. 92–101 (1929)

26. Schaefer, T.J.: The complexity of satisfiability problems. In: Proceedings of STOC 1978, pp. 216–226 (1978)

27. Skolem, T.: Über gewisse satzfunktionen in der arithmetik. Skr, Norske Videnskaps-Akademie i Oslo (1930)

28. Smorynski, C.: The incompleteness theorems. In: Barwise, J. (ed.) Handbook of Mathematical Logic, pp. 821–865. North-Holland, Amsterdam (1977)

29. Stockmeyer, L.J., Meyer, A.R.: Word problems requiring exponential time: preliminary report. In: Proceedings of the 5th Annual ACM Symposium on Theory of Computing, (STOC), pp. 1–9 (1973)

30. Travers, S.D.: The complexity of membership problems for circuits over sets of integers. Theor. Comput. Sci. **369**(1–3), 211–229 (2006)

31. Wagner, K.: The complexity of problems concerning graphs with regularities. In: MFCS, pp. 544–552 (1984)

32. Yang, K.: Integer circuit evaluation is Pspace-complete. J. Comput. Syst. Sci. **63**(2), 288–303 (2001). An extended abstract of appeared at CCC 2000

The Complexity of Counting Quantifiers on Equality Languages

Barnaby Martin[1(✉)], András Pongrácz[2], and Michał Wrona[3]

[1] School of Science and Technology, Middlesex University,
The Burroughs, Hendon, London NW4 4BT, UK
barnabymartin@gmail.com
[2] Debreceni Egyetem TTK, Algebra és Számelmélet Tanszék,
Debrecen PF. 18, 4010, Hungary
[3] Faculty of Mathematics and Computer Science,
Theoretical Computer Science Department,
Jagiellonian University, Kraków, Poland

Abstract. An equality language is a relational structure with infinite domain whose relations are first-order definable in equality. We classify the extensions of the quantified constraint satisfaction problem over equality languages in which the native existential and universal quantifiers are augmented by some subset of counting quantifiers. In doing this, we find ourselves in various worlds in which dichotomies or trichotomies subsist.

1 Introduction

The *constraint satisfaction problem* CSP(Γ), much studied in artificial intelligence, is known to admit several equivalent formulations, two of the best known of which are the query evaluation of primitive positive (pp) sentences – those involving only existential quantification and conjunction – on Γ, and the homomorphism problem to Γ (see, e.g., [14]). For finite Γ the problem CSP(Γ) is NP-complete in general, and a great deal of effort has been expended in classifying its complexity for certain restricted cases. Notably it is conjectured [9,12] that for all fixed finite Γ, the problem CSP(Γ) is in P or NP-complete. While this has not been settled in general, a number of partial results are known – e.g. over structures of size at most three [8,17] and over smooth digraphs [2,13].

A popular generalisation of the CSP involves considering the query evaluation problem for *positive Horn* logic – involving only the two quantifiers, $\exists$ and $\forall$, together with conjunction. The resulting *quantified constraint satisfaction problems* QCSP(Γ) allow for a broader class, used in artificial intelligence to capture non-monotonic reasoning, whose complexities rise to Pspace-completeness.

B. Martin and A. Pongrácz were supported by EPSRC grant EP/L005654/1.

A. Pongrácz—Supported also by the Hungarian Scientific Research Fund (OTKA) grant no. K109185.

M. Wrona—Partially supported by NCN grant number 2014/14/A/ST6/00138.

A. Beckmann et al. (Eds.): CiE 2016, LNCS 9709, pp. 333–342, 2016.
DOI: 10.1007/978-3-319-40189-8_34

Once upon a time, Bodirsky and Kára gave a systematic classification for $\mathrm{CSP}(\Gamma)$, where Γ consists of relations first-order (fo) definable in equality, over some countably infinite domain [7]. These so-called *equality languages* Γ displayed dichotomy between those for which $\mathrm{CSP}(\Gamma)$ was in P and those for which it was NP-complete. Pursuing this line of investigation, Bodirsky and Chen gave a trichotomy for $\mathrm{QCSP}(\Gamma)$, where Γ is an equality language – each problem being either in P, NP-complete or co-NP-hard [4]. In the conference version of that paper, the trichotomy was claimed to be across P, NP-complete or Pspace-complete [3], but the proof in the tricky case of $x = y \rightarrow y = z$ was flawed, and so in the journal version this became the weaker co-NP-hard (and in Pspace). The trichotomy is thus imperfect, as most of the co-NP-hard cases are known to be Pspace-complete. Indeed, $x = y \rightarrow y = z$ would be the only open case, if it would to be Pspace-complete [5].

Working Hypothesis. $\mathrm{QCSP}(x = y \rightarrow y = z)$ is Pspace-complete.

Thus the assumption of the working hypothesis would restore the trichotomy to the P, NP-complete or Pspace-complete as stated in [3].

In this paper, we consider the generalisation of the QCSP with counting quantifiers, as pioneered in the recent paper [15]. In [15], the domains of Γ were of finite size n, so the extant quantifiers $\exists^{\geq 1} = \exists$ and $\exists^{\geq n} = \forall$ were augmented with quantifiers of the form $\exists^{\geq j}$, which allow one to assert the existence of at least j elements such that the ensuing property holds. In the world of infinite domains, it makes sense to permit not only quantification above the finite with $\exists^{\geq j}$, but also quantification *below the co-finite* with $\forall^{\geq j}$, whose intended meaning is that the property holds for all but (at most) j elements of the domain. Thus, $\forall = \forall^{\geq 0}$. Counting quantifiers have been extensively studied in finite model theory (see [11,16]), where the focus is on supplementing the descriptive power of various logics. Of more general interest is the majority quantifier $\exists^{\geq n/2}$ (on a structure of domain size n), which sits broadly midway between $\exists$ and $\forall$. Majority quantifiers are studied across diverse fields of logic and have various practical applications, e.g. in cognitive appraisal and voting theory [10,19]. They have also been studied in computational complexity since at least [1] (see also [11]).

We study extensions of $\mathrm{QCSP}(\Gamma)$ in which the input sentence to be evaluated on Γ remains positive conjunctive in its quantifier-free part, but is quantified by various counting quantifiers. For $X \subseteq \{\exists^{\geq 1}, \exists^{\geq 2}, \ldots, \forall^{\geq 0}, \forall^{\geq 1}, \ldots\}$, $X \supseteq \{\exists^{\geq 1}, \forall^{\geq 0}\}$, the X-$\mathrm{CSP}(\Gamma)$ takes as input a sentence given by a conjunction of atoms quantified by quantifiers appearing in X. It then asks whether this sentence is true on Γ. Equality languages admit quantifier elimination of $\forall$ and $\exists$, that is any relation first-order definable in equality is already quantifier-free definable, say as a CNF. An equality language Γ is

- *trivial* if all its relations may be given as a conjunction of equalities,
- *specially negative* if the co-clone $\langle \Gamma \rangle_{\{\exists, \forall, \forall^{\geq 1}\}\text{-pp}}$ does not contain the formula $x \neq y \vee y \neq z$,

- *negative* if all its relations may be given as a conjunction of equalities and disjunctions of disequalities, and
- *positive* if all its relations may be given as a conjunction of disjunctions of equalities.

Similarly, we might use these adjectives on the relations within the equality language. We observe the containments of trivial languages within specially negative languages within negative languages. Further, it is proved in [4] (Proposition 7.3) that the positive languages that are not trivial are precisely the positive languages that are not negative. Our main results are a complete panoply of classifications for $X \supseteq \{\exists^{\geq 1}, \forall^{\geq 0}\}$. It will be seen that the quantifiers $\exists^{\geq 2}, \exists^{\geq 3}, \ldots$ more or less behave as one another and similarly with $\forall^{\geq 2}, \forall^{\geq 3}, \ldots$. However, $\forall^{\geq 1}$ is special and thus our task of classifications for X amounts to choosing subsets of $\{\exists^{\geq 2}, \forall^{\geq 1}, \forall^{\geq 2}\}$ with which to augment $\{\exists^{\geq 1}, \forall^{\geq 0}\}$. A priori there are then eight possibilities, but twice we will see $\forall^{\geq 1}$ being "subsumed" by $\forall^{\geq 2}$. Thus we will give *six distinct classification theorems*: three dichotomies and three trichotomies (one of which is that of [4]). In Fig. 1, these classification theorems are linked to their canonical subsets of $\{\exists^{\geq 2}, \forall^{\geq 1}, \forall^{\geq 2}\}$.

Theorem 1 [4]. *If $X = \{\exists^{\geq 1}, \forall^{\geq 0}\}$, then X-CSP(Γ) displays **trichotomy** on the class of equality languages Γ:*

- *if all relations of Γ are **negative** then X-CSP(Γ) is in L.*
- *if all relations of Γ are **positive** but some relation is not trivial, then X-CSP(Γ) is NP-complete.*
- ***otherwise** X-CSP(Γ) is co-NP-hard.*

Theorem 2. *If $X \subseteq \{\exists^{\geq 1}, \forall^{\geq 0}, \forall^{\geq 1}, \forall^{\geq 2}, \ldots\}$ and contains some $\forall^{\geq j}$ for $j \geq 2$, then X-CSP(Γ) displays **trichotomy** on the class of equality languages Γ:*

- *if all relations of Γ are **trivial**, then X-CSP(Γ) is in L.*
- *if all relations of Γ are **positive**, but some relation is not trivial, then X-CSP(Γ) is NP-complete.*
- ***otherwise** X-CSP(Γ) is Pspace-complete.*

Theorem 3. *If $X = \{\exists^{\geq 1}, \forall^{\geq 0}, \forall^{\geq 1}\}$, then X-CSP(Γ) displays **trichotomy** on the class of equality languages Γ:*

- *if all relations of Γ are **specially negative**, then X-CSP(Γ) is in P.*
- *if all relations of Γ are **positive**, but some relation is not trivial, then X-CSP(Γ) is NP-complete.*
- ***otherwise** X-CSP(Γ) is Pspace-complete.*

Theorem 4. *If $X \subseteq \{\forall^{\geq 0}, \exists^{\geq 1}, \exists^{\geq 2}, \ldots\}$ and contains some $\exists^{\geq j}$ for $j \geq 2$ then X-CSP(Γ) displays **dichotomy** on the class of equality languages Γ:*

- *if all relations of Γ are **negative**, then X-CSP(Γ) is in L.*
- ***otherwise** X-CSP(Γ) is co-NP-hard.*

Theorem 5. *If* $X \subseteq \{\forall^{\geq 0}, \forall^{\geq 1}, \forall^{\geq 2}, \ldots, \exists^{\geq 1}, \exists^{\geq 2}, \ldots\}$ *and contains some* $\exists^{\geq i}$ *and* $\forall^{\geq j}$ *for* $i, j \geq 2$ *then* X-*CSP*(Γ) *displays* **dichotomy** *on the class of equality languages* Γ:

- *if all relations of* Γ *are* **trivial**, *then* X-*CSP*(Γ) *is in* L.
- **otherwise** X-*CSP*(Γ) *is Pspace-complete.*

Theorem 6. *If* $X \subseteq \{\forall^{\geq 0}, \forall^{\geq 1}, \exists^{\geq 1}, \exists^{\geq 2}, \ldots\}$ *and contains* $\forall^{\geq 1}$ *and some* $\exists^{\geq i}$ *for* $i \geq 2$ *then* X-*CSP*(Γ) *displays* **dichotomy** *on the class of equality languages* Γ:

- *if all relations of* Γ *are* **specially negative**, *then* X-*CSP*(Γ) *is in* P.
- **otherwise** X-*CSP*(Γ) *is Pspace-complete.*

Theorem	Subsets
1	$\emptyset$
2	$\{\forall^{\geq 2}\}, \{\forall^{\geq 1}, \forall^{\geq 2}\}$
3	$\{\forall^{\geq 1}\}$
4	$\{\exists^{\geq 2}\}$
5	$\{\forall^{\geq 2}, \exists^{\geq 2}\}, \{\forall^{\geq 1}, \forall^{\geq 2}, \exists^{\geq 2}\}$
6	$\{\forall^{\geq 1}, \exists^{\geq 2}\}$

Fig. 1. Classification theorems linked to canonical subsets of $\{\exists^{\geq 2}, \forall^{\geq 1}, \forall^{\geq 2}\}$

Four of our five new worlds are somewhat more conducive to analysis than that of [4], in that in them we have no gap across co-NP-hardness and Pspace-completeness. For the remaining world of Theorem 4, we are able to demonstrate that improving co-NP- to Pspace-hardness is likely to be as difficult as in Theorem 1. Indeed, from this it follows that our working hypothesis promotes co-NP-hardness to Pspace-hardness for Theorem 4 as well as Theorem 1.

Some of our results are not especially complicated and stem from simple manipulations rather than deep technical nous. Against this we set the nice aesthetic of our results and the way in which they complement [4,6]. For example, the specially negative languages, play an important role in our classifications, but where do they sit in the context of [6]. Do they even form a co-clone? We note that equality languages have been also studied for the abduction problem [18], in the context of which we find a trichotomy among Σ_2^P-complete, NP-complete and problems decidable in P.

The paper is organised as follows. After the preliminaries, we address in Sect. 3 basic upper and lower bounds that play a role in our classification. In Sect. 4 we describe the crucial specially negative languages. Finally, we ponder the difficulty of improving, in Theorem 4, co-NP- to Pspace-hardness. We do the latter by showing that QCSP$(x = y \lor u = v, \neq)$ becomes no more complex when one augments with $\exists^{\geq k}$.

2   Preliminaries

Let $[j] := \{1, \ldots, j\}$. For $i \in \mathbb{N}$, $i \geq 1$, let $\exists^{\geq i}x$ quantify that there exist at least i elements satisfying some property. For $i \in \mathbb{N}$, $i \geq 0$, let $\forall^{\geq i}x$ quantify that for all but at most i elements does some property hold. Thus $\exists$ is $\exists^{\geq 1}$ and $\forall$ is $\forall^{\geq 0}$ – *the base cases being different for these quantifiers.* In this paper we consider languages with first-order (fo) definitions in equality, that is on structures which admit all permutations as automorphisms. Such an *equality language* may be considered a structure of the form $(\mathbb{N}; R_1, \ldots)$ where each R_i is a CNF formula whose atoms are equalities or disequalities (owing to quantifier elimination of $\forall$ and $\exists$ this is equivalent to our saying fo-definable in equality). We typically drop the "$\mathbb{N}$" in referring to an equality language.

Primitive positive (pp) logic is the restriction of fo-logic to the symbols $\{\exists, \wedge, =\}$ and *positive Horn* (pH) likewise to the symbols $\{\forall, \exists, \wedge, =\}$. For $X \subseteq \{\forall^{\geq 0}, \forall^{\geq 1}, \ldots, \exists^{\geq 1}, \exists^{\geq 2}, \ldots\}$, let X-pp denote the logic of prenex sentences whose symbols are among $X \cup \{\wedge, =\}$. We will use small letters such as ϕ to refer to sentences whose quantifier-free part will be denoted Φ. Let Γ be a set of relations on a domain, i.e. a structure, and let $\mathscr{L}$ be a logic. Then the *evaluation problem for $\mathscr{L}$ on Γ* has as input a sentence $\phi \in \mathscr{L}$ and asks whether $\Gamma \models \phi$? The evaluation problem for primitive positive (resp., positive Horn) logic on Γ is better known as CSP(Γ) (resp., QCSP(Γ)). The evaluation problem for X-pp on Γ will henceforth be known as X-CSP(Γ). In this paper we consider only $X \supset \{\exists^{\geq 1}, \forall^{\geq 0}\}$, i.e. extensions of the QCSP.

Let Γ be an equality language, and let $\langle \Gamma \rangle_{\mathrm{pp}}$ be the set of relations pp-definable over Γ. Such a set is termed a *co-clone*. We may abuse notation and write $\langle R \rangle_{\mathrm{pp}}$ when properly we mean $\langle \{R\} \rangle_{\mathrm{pp}}$. A great project was launched in [6] to identify sets of the form $\langle \Gamma \rangle_{\mathrm{pp}}$, charting their inclusion relations in a lattice. The lattice is mostly identified through a dually-isomorphic algebraic lattice of *local clones*, but here we will only be interested in the co-clones. In line with [4,6], we will consider the co-clone $\langle \emptyset \rangle_{\mathrm{pp}} = \langle = \rangle_{\mathrm{pp}}$ to be at the top of the lattice, with the co-clone of all equality-definable relations at the bottom.

In a problem X-CSP(Γ) it is desirable that Γ involve only a finite number of relations lest there arise the question as to how they are encoded. Yet we will enjoy referring to co-clones that contain an infinite number of relations. We resolve this by considering all references to Γ as a language inside X-CSP(Γ) to be restricting of Γ to the relationally finite.[1]

We now recall some basic results from [4,6]. We typically write $\neq$ to indicate the binary relation of disequality (in terms of our canonical notation this abbreviates $x \neq y$). Let $\langle \Gamma \rangle_{\mathrm{pH}}$ be the set of relations pH-definable over Γ.

---

[1] Another typical solution is to give definition to complexity of relationally-infinite Γ along the lines of "easy", if it is easy for all finite subsets, and "hard", if it is hard for some finite subset.

Lemma 1.

(1) $x = y \lor u = v$ is in both $\langle x = y \lor y = z \rangle_{pp}$ and $\langle x = y \lor y = z \lor x = z \rangle_{pp}$.
(2) $x \neq y \lor u \neq v \in \langle x = y \to y = z, \neq \rangle_{pp}$.
(3) $(x_1 = y \land \ldots \land x_m = y) \to y = z \in \langle x = y \to y = z \rangle_{pp}$.
(4) $\langle \{x = y \lor u = v, \neq\} \rangle_{pp}$ contains all equality-definable relations.
(5) $\langle x = y \lor u = v \rangle_{pp}$ contains all positive relations.
(6) If Γ is not positive then $\neq \in \langle \Gamma \rangle_{pp}$.
(7) If Γ is positive but not negative then $x = y \lor u = v \in \langle \Gamma \rangle_{pH}$.

Note that Part 7 does not hold for pp-definability. While $x = y \lor y = z$ is a pp-basis for the positive languages, there is an infinite chain of positive languages up to pp-closure [6]. $\langle x = y \lor y = z \rangle_{pp}$ is at the bottom, being the most expressive, and the trivial $\langle = \rangle_{pp}$ is at the top. Inbetween, is an infinite chain without top element. However, this chain collapses for pH-definability, leaving only the two co-clones up to pH-closure ($\langle = \rangle_{pp}$ and $\langle x = y \lor y = z \rangle_{pp}$).

We contrast Part 1 of Lemma 1 with the knowledge that $x = y \to u = v$ is in neither $\langle x = y \to y = z \rangle_{pp}$ [6] nor $\langle x = y \to y = z \rangle_{pH}$.
Owing to the disparity between the conference and journal versions of [4], we give the following as a specific proposition. Its proof can be derived from [4] by the assiduous reader.

Proposition 1. *Both $QCSP(x = y \lor u = v, \neq)$ and $QCSP(w = z_1 \lor w = z_2 \lor w = z_3, \neq)$ are Pspace-complete.*

It seems that $QCSP(x = y \to u = v)$ is also Pspace-complete [5], but this is harder work. The backbone of a proof appeared in an early unpublished version of [4]. The author has verified this proof but reproducing it here is beyond our scope.

3   Upper and Lower Bounds

Upper bounds. The following lemma is trivial but we state it because we will wish to appeal to it in the future.

Lemma 2 (Substitution of equalities). *Let $X := \{\forall^{\geq 0}, \forall^{\geq 1}, \exists^{\geq 1}, \exists^{\geq 2}, \ldots\}$ and let Φ be an instance of some X-CSP containing an equality $x = y$ in which y appears later in the quantifier order of Φ than x. Then Φ is false if y is quantified by anything other than $\exists^{\geq 1}$. Otherwise, Φ is equivalent to Φ' obtained by substituting all instances of y by x and removing the quantifier $\exists^{\geq 1} y$.*

Lemma 3. *For any $X \subseteq \{\forall^{\geq 0}, \forall^{\geq 1}, \ldots, \exists^{\geq 1}, \exists^{\geq 2}, \ldots\}$, X-CSP(Γ) is in Pspace.*

The following lemma relates to $\{\exists^{\geq 1}, \forall^{\geq i} : i \geq 0\}$ on positive languages.

Lemma 4. *Let $X := \{\exists^{\geq 1}, \forall^{\geq i} : i \geq 0\}$ and Γ be a positive equality language. Then X-CSP(Γ) is in NP.*

Lower bounds. The following proposition relates to $\{\forall^{\geq i}, \forall^{\geq 0}, \exists^{\geq 1}\}$, for $i \geq 2$, on non-positive languages.

Proposition 2. *Let* $X := \{\forall^{\geq i}, \forall^{\geq 0}, \exists^{\geq 1}\}$, *for any* $i \geq 2$ *and* Γ *a non-positive equality language. Then* X-$CSP(\Gamma)$ *is Pspace-complete.*

The following propositions relate to $\{\forall^{\geq 1}, \forall^{\geq 0}, \exists^{\geq 1}\}$ on non-positive languages.

Proposition 3. $\{\forall^{\geq 1}, \forall^{\geq 0}, \exists^{\geq 1}\}$-$CSP(x \neq y \vee u \neq v)$ *is Pspace-complete.*

Proposition 4. $(x \neq y \vee u \neq v)$ *is definable from* $(x \neq y \vee y \neq z)$ *using* $\forall^{\geq 1}$.

Proposition 5. $\{\forall^{\geq 1}, \forall^{\geq 0}, \exists^{\geq 1}\}$-$CSP(x \neq y \vee y \neq z)$ *is Pspace-complete.*

Proposition 6. *Let* $i \geq 2$ *and* Γ *be a positive non-trivial equality language. Then* $\neq$ *is definable from* Γ *using* $\exists^{\geq i}$.

4   Isolating the Specially Negative Languages

We call a CNF Φ *reduced* if it is not logically equivalent to itself with either a clause or a literal in a clause removed. A CNF *depends* on one of its variables v if its truth value can not be given as a propositional function from (the equality type of) only its other variables.

Suppose R is a negative relation which might be given by various reduced CNFs, at least one of which, Φ, is negative. Then Φ may enforce some equalities on its variables, which it plainly does syntactically. Sometimes in this section we will wish to assume that Φ has these equalities factored out by substitution, thus we could assume negative CNFs do not have any positive clauses. The point is that this is an innocuous assumption. Having said that, the process does remove variables and so can't be used strictly within a quantifier elimination procedure. Now, suppose R is also given by another reduced CNF Φ' which is not negative (we will later prove this is not possible). Then we could similarly factor out the implied equalities of Φ', but these might not be obvious unless we have negative Φ where the equalities are syntactically explicit and not semantically implied.

4.1   Non-positive Cases Involving $\forall^{\geq 1}$

Definition 1. *A negative, reduced CNF Φ without equalities is flat if it consists of clauses with no free variable occurring twice in them. Φ is rich if every free variable of Φ occurs in a singleton clause of Φ.*

We will assume that CNFs do not possess dummy variables that do not appear explicitly. Thus, we may consider a CNF to be rich if it is rich once we have discounted such dummy variables.

Theorem 7. *We have the following dichotomy for non-positive equality languages* Γ.

1. Either $\langle \Gamma \rangle_{\{\exists,\forall,\forall^{\geq 1}\}\text{-pp}}$ contains the relation $x \neq y \vee y \neq z$, and then we find $\{\exists,\forall,\forall^{\geq 1}\}\text{-CSP}(\Gamma)$ is Pspace-hard,
2. or $\langle \Gamma \rangle_{\{\exists,\exists^{\geq 2},\ldots,\forall,\forall^{\geq 1}\}\text{-pp}}$ contains only relations whose reduced CNFs are negative, flat and rich; and $\{\exists,\exists^{\geq 2},\ldots,\forall,\forall^{\geq 1}\}\text{-CSP}(\Gamma)$ is in P.

Definition 2. Let Φ be a CNF with variables in V. Let $P_1 \cup \cdots \cup P_k$ be a partition of V. We say that "we weaken Φ around the given partition by keeping $P_1, \ldots, P_j$" if we produce a formula from Φ by the following definition. First we take the conjunction Ψ of all disequalities that are transversal to the partition, i.e., $x \neq y$ with x and y not in the same set of the partition, and produce $\Phi \wedge \Psi$. Then for all $i > j$ we identify the variables in P_i by a new variable w_i. Then we existentially quantify over these new variables.

Note that if Φ pp-defines $\neq$, then this procedure is also a pp-definition from Φ.

Lemma 5. Let Φ be a reduced negative CNF that has a non-flat clause. Then Φ pp-defines $x \neq y \vee y \neq z$.

Lemma 6. From $S(x,y,u,v) := (x \neq y \vee u \neq v) \wedge y \neq u \wedge x \neq u$ we may define with $\forall^{\geq 1}$ and $\exists$ the relation $p \neq q \vee q \neq r$.

Lemma 7. Let Φ be a flat negative CNF with exactly four variables. Assume that Φ depends on each of its variables, and that Φ is not rich. Then $x \neq y \vee y \neq z$ has an $\{\exists,\forall,\forall^{\geq 1}\}$-pp definition in Φ.

Lemma 8. Assume that $\langle \Gamma \rangle_{\{\exists,\forall,\forall^{\geq 1}\}\text{-pp}}$ contains a negative CNF Φ that is not rich. Then $\langle \Gamma \rangle_{\{\exists,\forall,\forall^{\geq 1}\}\text{-pp}}$ contains the formula $x \neq y \vee y \neq z$.

4.2 Quantifier Elimination and Reduction for Negative, Flat and Rich CNF Formulas

We wish to argue that in a certain case we can effect quantifier elimination for negative, flat and rich CNF formulas in polynomial time. Eliminating a $\forall^{\geq 1}$ quantifier over negative, flat and rich formulas can throw up Horn CNFs, so our first task will be to argue that we can compute a reduced form from a Horn CNF Φ in polynomial time. Our task is to determine, whether there are any redundant clauses (and if so remove them) and then whether there are any redundant literals in the remaining clauses (which also must be removed). To test for redundant clauses C it is sufficient to test whether $(\Phi \setminus \{C\})$ implies C, i.e. whether $(\Phi \setminus \{C\}) \wedge \neg C$ is a contradiction. But this is itself the complement of a Horn CSP, itself uniformly tractable by unit propagation (see [7]). Similarly, to determine if a literal ℓ is required in a clause C, we may consider whether $(\Phi \setminus \{C\}) \wedge \ell \wedge \neg(C \setminus \{\ell\})$ is a contradiction. Thus, given a Horn CNF Φ we can compute in polynomial time a reduced CNF (itself Horn) Φ' that is equivalent to Φ.

Consider each $\forall^{\geq 1} x_i \, \Phi(x_1, \ldots, x_k)$, where Φ is negative, flat and rich, without any equalities, where the singletons involving x_i are precisely $x_i \neq x'_{\lambda_1}, \ldots, x_i \neq$

x'_{λ_t}. We can effect quantifier elimination in the following fashion. Substitute each non-singleton clause involving x_i, itself of the form, $(x_i \neq x'_{\mu_1} \vee x_{\mu_2} \neq x'_{\mu_2} \vee \ldots \vee x_{\mu_s} \neq x'_{\mu_s})$, where no variable is repeated, by $(x'_{\mu_1} = x'_{\lambda_1} = \cdots = x'_{\lambda_t} \vee x_{\mu_2} \neq x'_{\mu_2} \vee \ldots \vee x_{\mu_s} \neq x'_{\mu_s})$. Note that we will allow conjuncts of equalities as single relations in our clauses, to avoid having to break the clauses up. If we view this as a notational shorthand then we do not break the condition of Hornness. If x_i appeared in multiple singleton clauses (i.e. $t > 1$), then we need to add the equality $x_{\lambda_1} = \cdots = x_{\lambda_t}$ to the system. If x_i appeared in a single singleton clause then we can now simply remove it. Call the new CNF finally obtained $\tilde{\Phi}$. Plainly, $\tilde{\Phi}$ is logically equivalent to $\forall^{\geq 1} x_i \, \Phi$ and $\tilde{\Phi}$ is Horn. We now apply our effective procedure to establish whether $\tilde{\Phi}$ is reduced and if not reduce it.

The question now naturally arises as to whether $\tilde{\Phi}$ is negative. We argue by the following lemma that it is enough to see whether our reduced CNF form has a non-singleton clause with an equality in it.

Lemma 9. *Let Φ be a reduced CNF representing a relation R in which there is a non-singleton clause that contains an equality. It is not possible that R is negative (i.e. has another reduced CNF that is negative).*

Note that reduced negative CNFs of a relation R are not in general unique, and the method of transitive closure from the first paragraph of the above proof hints at an example: $(x \neq y \vee y \neq z)$ is equivalent to $(x \neq y \vee x \neq z)$.

We will now consider the quantifier elimination of $\exists$, $\exists^{\geq 2}$, $\ldots$ etc. (note that quantification by $\forall$ on a negative, flat and rich formula will always leave it false). On negative formulas, $\exists^{\geq 2}$, $\ldots$ etc. have the same power as $\exists$ and allow us to remove any clause in which the corresponding variable appears. This leaves the formula negative and we again have an effective method for reduction.

We now continue in this vein eliminating quantifiers and reducing CNFs. If we at any point produce a CNF whose reduced form is not negative, flat and rich then we know that Γ $\{\exists, \forall, \forall^{\geq 1}\}$-pp defines $x \neq y \vee y \neq z$. We are now in a position to address Theorem 7.

By way of example for Theorem 7, we note that $\langle \{(u \neq v \vee x \neq y) \wedge u \neq y \wedge v \neq x\}\rangle_{\{\exists, \forall, \forall^{\geq 1}\}\text{-}pp}$ contains only negative, flat and rich formulas.

5 Ennui of co-NP- to Pspace-Completeness

Proposition 7. *For all $1 \leq k \in \mathbb{N}$, $\{\exists^{\geq k}, \forall^{\geq 0}, \exists^{\geq 1}\}$-CSP$(x = y \rightarrow y = z)$ and $\{\forall^{\geq 0}, \exists^{\geq 1}\}$-CSP$(x = y \rightarrow y = z)$ (i.e. QCSP$(x = y \rightarrow y = z)$) are logspace equivalent.*

It follows from Proposition 7 and [4] that our working conjecture that QCSP $(x = y \rightarrow y = z)$ would be Pspace-complete would elevate the co-NP-hardness cases of Theorems 1 and 4 to Pspace-hardness.

References

1. Barrington, D.A.M., Immerman, N., Straubing, H.: On uniformity within NC^1. J. Comput. Syst. Sci. **41**(3), 274–306 (1990)
2. Barto, L., Kozik, M., Niven, T.: The CSP dichotomy holds for digraphs with no sources and no sinks (a positive answer to a conjecture of Bang-Jensen and Hell). SIAM J. Comput. **38**(5), 1782–1802 (2009)
3. Bodirsky, M., Chen, H.: Quantified equality constraints. In: Proceedings of LICS 2007, pp. 203–212 (2007)
4. Bodirsky, M., Chen, H.: Quantified equality constraints. SIAM J. Comput. **39**(8), 3682–3699 (2010)
5. Bodirsky, M., Chen, H.: Personal communication (2012)
6. Bodirsky, M., Chen, H., Pinsker, M.: The reducts of equality up to primitive positive interdefinability. J. Symb. Log. **75**(4), 1249–1292 (2010)
7. Bodirsky, M., Kára, J.: The complexity of equality constraint languages. Theor. Comput. Syst. **43**(2), 136–158 (2008). A preliminary version appeared in the proceedings of CSR 2006
8. Bulatov, A.: A dichotomy theorem for constraint satisfaction problems on a 3-element set. J. ACM **53**(1), 66–120 (2006)
9. Bulatov, A.A., Jeavons, P., Krokhin, A.A.: Classifying the complexity of constraints using finite algebras. SIAM J. Comput. **34**(3), 720–742 (2005)
10. Clark, R., Grossman, M.: Number sense and quantifier interpretation. Topoi **26**(1), 51–62 (2007)
11. Ebbinghaus, H.-D., Flum, J.: Finite Model Theory, 2nd edn. Springer, Heidelberg (1999)
12. Feder, T., Vardi, M.Y.: The computational structure of monotone monadic SNP, constraint satisfaction: a study through Datalog and group theory. SIAM J. Comput. **28**, 57–104 (1999). A preliminary version appeared in the proceedings of STOC 1993
13. Hell, P., Nešetřil, J.: On the complexity of H-coloring. J. Comb. Theor. Ser. B **48**(1), 92–110 (1990)
14. Kolaitis, P.G., Vardi, M.Y.: A logical Approach to Constraint Satisfaction. In: Creignou, N., Kolaitis, P.G., Vollmer, H. (eds.) Finite Model Theory and Its Applications. Texts in Theoretical Computer Science. An EATCS Series. Springer, Heidelberg (2005)
15. Martin, B., Madelaine, F.R., Stacho, J.: Constraint satisfaction with counting quantifiers. SIAM J. Discrete Math. **29**(2), 1065–1113 (2015). Extended abstracts appeared at CSR 2012 and CSR 2014
16. Otto, M.: Bounded Variable Logics and Counting - A Study in Finite Models, vol. 9. Springer, Heidelberg (1997). IX+183pp
17. Schaefer, T.J.: The complexity of satisfiability problems. In: Proceedings of STOC 1978, pp. 216–226 (1978)
18. Schmidt, J., Wrona, M.: The complexity of abduction for equality constraint languages. In: Computer Science Logic (CSL 2013), pp. 615–633 (2013)
19. Szymanik, J.: Quantifiers and Cognition: Logical and Computational Perspectives. Studies in Linguistics and Philosophy. Springer, Cham (2016)

Baire Category Theory and Hilbert's Tenth Problem Inside $\mathbb{Q}$

Russell Miller[1,2]($\boxtimes$)

[1] Queens College – C.U.N.Y., 65–30 Kissena Blvd., Queens, NY 11367, USA
Russell.Miller@qc.cuny.edu
[2] Graduate Center of C.U.N.Y., 365 Fifth Avenue, New York, NY 10016, USA
http://qcpages.qc.cuny.edu/~rmiller

Abstract. For a ring R, Hilbert's Tenth Problem HTP(R) is the set of polynomial equations over R, in several variables, with solutions in R. We consider computability of this set for subrings R of the rationals. Applying Baire category theory to these subrings, which naturally form a topological space, relates their sets HTP(R) to the set HTP($\mathbb{Q}$), whose decidability remains an open question. The main result is that, for an arbitrary set C, HTP($\mathbb{Q}$) computes C if and only if the subrings R for which HTP(R) computes C form a nonmeager class. Similar results hold for 1-reducibility, for admitting a Diophantine model of $\mathbb{Z}$, and for existential definability of $\mathbb{Z}$.

1 Introduction

The original version of Hilbert's Tenth Problem demanded an algorithm deciding which polynomial equations from $\mathbb{Z}[X_1, X_2, \ldots]$ have solutions in the integers. In 1970, Matiyasevic [4] completed work by Davis, Putnam and Robinson [1], showing that no such algorithm exists. In particular, these authors showed that there exists a 1-reduction from the Halting Problem $\emptyset'$ to the set of such equations with solutions, by proving the existence of a single polynomial $h \in \mathbb{Z}[Y, \boldsymbol{X}]$ such that, for each n from the set ω of nonnegative integers, the polynomial $h(n, \boldsymbol{X}) = 0$ has a solution in $\mathbb{Z}$ if and only if n lies in $\emptyset'$. Since the membership in the Halting Problem was known to be undecidable, it followed that Hilbert's Tenth Problem was also undecidable.

One naturally generalizes this problem to all rings R, defining Hilbert's Tenth Problem for R to be the set

$$\mathrm{HTP}(R) = \{f \in R[\boldsymbol{X}] : (\exists r_1, \ldots, r_n \in R^{<\omega})\, f(r_1, \ldots, r_n) = 0\}.$$

R. Miller—The author was supported by Grant # DMS – 1362206 from the N.S.F., and by several grants from the PSC-CUNY Research Award Program. This work grew out of research initiated at a workshop at the American Institute of Mathematics and continued at a workshop at the Institute for Mathematical Sciences of the National University of Singapore. Conversations with Bjorn Poonen and Alexandra Shlapentokh have been very helpful in the creation of this article.

A. Beckmann et al. (Eds.): CiE 2016, LNCS 9709, pp. 343–352, 2016.
DOI: 10.1007/978-3-319-40189-8_35

Here we will examine this problem for one particular class: the subrings R of the field $\mathbb{Q}$ of rational numbers. Notice that in this situation, deciding membership in HTP(R) reduces to the question of deciding this membership just for polynomials from $\mathbb{Z}[\boldsymbol{X}]$, since one readily eliminates denominators from the coefficients of a polynomial. So, for us, HTP(R) will always be a subset of $\mathbb{Z}[X_1, X_2, \ldots]$.

Subrings R of $\mathbb{Q}$ correspond bijectively to subsets W of the set $\mathbb{P}$ of all primes, via the map $W \mapsto \mathbb{Z}[\frac{1}{p} : p \in W]$. We write R_W for the subring $\mathbb{Z}[\frac{1}{p} : p \in W]$. In this article, we will move interchangeably between subsets of ω and subsets of $\mathbb{P}$, using the bijection mapping $n \in \omega$ to the n-th prime p_n, starting with $p_0 = 2$. For the most part, our sets will be subsets of $\mathbb{P}$, but Turing reductions and jump operators and the like will all be applied to them in the standard way. Likewise, sets of polynomials, such as HTP(R), will be viewed as subsets of ω, using a fixed computable bijection from ω onto $\mathbb{Z}[\boldsymbol{X}] = \mathbb{Z}[X_0, X_1, \ldots]$.

We usually view subsets of $\mathbb{P}$ as paths through the tree $2^{<\mathbb{P}}$, a complete binary tree whose nodes are the functions from initial segments of the set $\mathbb{P}$ into the set $\{0, 1\}$. This allows us to introduce a topology on the space $2^{\mathbb{P}}$ of paths through $2^{<\mathbb{P}}$, and thus on the space of all subrings of $\mathbb{Q}$. Each basic open set $\mathcal{U}_\sigma$ in this topology is given by a node σ on the tree: $\mathcal{U}_\sigma = \{W \subseteq \mathbb{P} : \sigma \subset W\}$, where $\sigma \subset W$ denotes that when W is viewed as a function from $\mathbb{P}$ into the set $2 = \{0, 1\}$ (i.e., as an infinite binary sequence), σ is an initial segment of that sequence. Also, we put a natural measure μ on the class $\mathbf{Sub}(\mathbb{Q})$ of all subrings of $\mathbb{Q}$: just transfer to $\mathbf{Sub}(\mathbb{Q})$ the obvious Lebesgue measure on the power set $2^{\mathbb{P}}$ of $\mathbb{P}$. Thus, if we imagine choosing a subring R by flipping a fair coin (independently for each prime p) to decide whether $\frac{1}{p} \in R$, the *measure* of a subclass $\mathcal{S}$ of $\mathbf{Sub}(\mathbb{Q})$ is the probability that the resulting subring will lie in $\mathcal{S}$. Here we will focus on Baire category theory rather than on measure theory, however, as the former yields more useful results. For questions and results regarding measure theory, we refer the reader to Sect. 3 and to the forthcoming [5].

For all $W \subseteq \mathbb{P}$, we have Turing reductions, which in fact are 1-reductions:

$$W \oplus \mathrm{HTP}(\mathbb{Q}) \leq_1 \mathrm{HTP}(R_W) \leq_1 W'.$$

For instance, the Turing reduction from $HTP(R_W)$ to W' can be described by a computable injection which maps each $f \in \mathbb{Z}[\boldsymbol{X}]$ to the code number $h(f)$ of an oracle Turing program which, on every input, searches for a solution $\boldsymbol{x}$ to $f = 0$ in $\mathbb{Q}$ for which the primes dividing the denominators of the coordinates in $\boldsymbol{x}$ all lie in the oracle set W. The reduction from HTP($\mathbb{Q}$) to HTP(R_W) uses the fact that every element of $\mathbb{Q}$ is a quotient of elements of R_W, so that $f(\boldsymbol{X})$ has a solution in $\mathbb{Q}$ if and only if $Y^d \cdot f(\frac{X_1}{Y}, \ldots, \frac{X_n}{Y})$ has a solution in R_W with $Y > 0$. The condition $Y > 0$ is readily expressed using the Four Squares Theorem. Finally, $W \leq_1 \mathrm{HTP}(R_W)$ by mapping p to $(pX - 1)$.

The topological space $2^{\mathbb{P}}$ of all paths through $2^{<\mathbb{P}}$, which we treat as the space of all subrings of $\mathbb{Q}$, is obviously homeomorphic to *Cantor space*, the space 2^ω of all paths through the complete binary tree $2^{<\omega}$. Hence this space satisfies the property of Baire, that no nonempty open set is meager. We recall the relevant definitions. Here as before, $\overline{\mathcal{A}}$ represents the complement of a subset $\mathcal{A} \subseteq 2^{\mathbb{P}}$, and we will write cl($\mathcal{A}$) for the topological closure of $\mathcal{A}$ and Int($\mathcal{A}$) for its interior.

Definition 1. *A subset* $\mathcal{B} \subseteq 2^{\mathbb{P}}$ *is said to be* nowhere dense *if its closure* $cl(\mathcal{B})$ *contains no nonempty open subset of* $2^{\mathbb{P}}$. *In particular, every set* $\mathcal{U}_{\sigma}$ *with* $\sigma \in 2^{<\mathbb{P}}$ *must intersect* $Int(\overline{\mathcal{B}})$, *the interior of the complement of* $\mathcal{B}$.

The union of countably many nowhere dense subsets of 2^{ω} *is called a* meager *set, or a set of first category. Its complement is said to be* comeager.

All sets $W \subseteq \omega$ satisfy $W \oplus \emptyset' \leq_T W'$, and for certain W, Turing-equivalence holds here. Indeed, it is known that the class

$$\mathbf{GL}_1 = \{W \in 2^{\omega} : W' \equiv_T W \oplus \emptyset'\}$$

is comeager, although its complement is nonempty. In computability theory, elements of $\mathbf{GL}_1$ are called *generalized-low$_1$* sets. The low sets – i.e., those W with $W' \leq_T \emptyset'$ – clearly lie in $\mathbf{GL}_1$.

Lemma 1 (Folklore). *There exists a Turing functional Ψ such that* $\{W \subseteq \omega : \Psi^{W \oplus \emptyset'} = \chi_{W'}\}$ *is comeager. It follows that* $\mathbf{GL}_1$ *is comeager.*

Proof. Consider the following oracle program Ψ for computing W' from $W \oplus \emptyset'$. With this oracle, on input e, the program searches for a string $\sigma \subseteq W$ such that either (1) $(\exists s)\ \Phi_{e,s}^{\sigma}(e) \downarrow$, or (2) $(\forall \tau \supseteq \sigma)(\forall s)\ \Phi_{e,s}^{\tau}(e) \uparrow$. The program uses its $\emptyset'$ oracle to check the truth of these two statements for each $\sigma \subseteq W$. If it ever finds that (1) holds, it concludes that $e \in W'$; while if it ever finds that (2) holds, it concludes that $e \notin W'$. Thus, $\Psi^{W \oplus \emptyset'}$ can only fail to compute W' if there exists some $e \notin W'$ such that, for every n, some $\tau \supseteq W \restriction n$ has $\Phi_e^{\tau}(e) \downarrow$. This can happen, but for each single e, the set of those W for which this happens constitutes the boundary of the open set $\{W : e \in W'\}$. This boundary is nowhere dense (cf. Lemma 3 below), so the union of these sets (over all e) is meager, and $\Psi^{W \oplus \emptyset'} = \chi_{W'}$ for every W outside this meager set. $\square$

$\mathbf{GL}_1$ also has measure 1, but no single Turing functional computes W' from $W \oplus \emptyset'$ uniformly on a set of measure 1.

Lemma 2 (Folklore). *If* $A \not\geq_T B$, *then* $\mathcal{C} = \{W : A \oplus W \geq_T B\}$ *is meager.*

Proof. To show that $\mathcal{C}$ is meager, define $\mathcal{C}_e = \{W \subseteq \mathbb{P} : \Phi_e^{A \oplus W} = \chi_B\}$, so $\mathcal{C} = \cup_e \mathcal{C}_e$. We claim that, if $\sigma \in 2^{\mathbb{P}}$ and $\mathcal{U}_{\sigma} \subseteq cl(\mathcal{C}_e)$, the following hold.

1. $\forall x \forall \tau \supseteq \sigma\ [\Phi_e^{A \oplus \tau}(x) \uparrow$ or $\Phi_e^{A \oplus \tau}(x) \downarrow = \chi_B(x)]$.
2. $\forall x \exists \tau \supseteq \sigma\ [\Phi_e^{A \oplus \tau}(x) \downarrow]$.

To see that (1) holds, suppose $\Phi_e^{A \oplus \tau}(x) \downarrow$. With $\mathcal{U}_{\tau} \subseteq \mathcal{U}_{\sigma} \subseteq cl(\mathcal{C}_e)$, some $W \in \mathcal{C}_e$ must have $\tau \subseteq W$. But then $\chi_B(x) = \Phi_e^{A \oplus W}(x) \downarrow = \Phi_e^{A \oplus \tau}(x)$.

To see (2), fix any $W \in \mathcal{C}_e$ with $\sigma \subseteq W$: such a W must exist, since $\mathcal{U}_{\sigma} \subseteq cl(\mathcal{C}_e)$. Then we can take τ to be the restriction of this W to the use of the computation $\Phi_e^{A \oplus W}(x)$ (or $\tau = \sigma$ if the use is $< |\sigma|$).

But now every $\mathcal{C}_e$ must be nowhere dense, since any σ satisfying (1) and (2) would let us compute B from A: given x, just search for some $\tau \supseteq \sigma$ and some s for which $\Phi_{e,s}^{A \oplus \tau}(x) \downarrow$. By (2), our search would discover such a τ eventually, and by (1) we would know $\chi_B(x) = \Phi_{e,s}^{A \oplus \tau}(x)$. Since $A \not\geq_T B$, this is impossible. $\square$

Finally, on a separate topic, it will be important for us to know that whenever R is a semilocal subring of $\mathbb{Q}$, we have $\mathrm{HTP}(R) \leq_1 \mathrm{HTP}(\mathbb{Q})$. Indeed, both the Turing reduction and the 1-reduction are uniform in the complement. (The result essentially follows from work of Julia Robinson in [8]. For a proof by Eisenträger, Park, Shlapentokh, and the author, see [2].) Recall that the *semilocal* subrings of $\mathbb{Q}$ are precisely those of the form R_W where the set W is cofinite in $\mathbb{P}$, containing all but finitely many primes.

Proposition 1 (see Proposition 5.4 in [2]). *There exists a computable function G such that for every n, every finite set $A_0 = \{p_1, \ldots, p_n\} \subset \mathbb{P}$ and every $f \in \mathbb{Z}[\mathbf{X}]$,*

$$f \in HTP(R_{\mathbb{P}-A_0}) \iff G(f, \langle p_1, \ldots, p_n \rangle) \in HTP(\mathbb{Q}).$$

That is, $HTP(R_{\mathbb{P}-A_0})$ is 1-reducible to $HTP(\mathbb{Q})$ for all semilocal $R_{\mathbb{P}-A_0}$, uniformly in A_0.   □

The proof in [2], using work from [3], actually shows how to compute, for every prime p, a polynomial $f_p(Z, X_1, X_2, X_3)$ such that for all rationals q, we have

$$q \in R_{\mathbb{P}-\{p\}} \iff f_p(q, \mathbf{X}) \in \mathrm{HTP}(\mathbb{Q}).$$

2   Baire Category and HTP($\mathbb{Q}$)

For a polynomial $f \in \mathbb{Z}[\mathbf{X}]$ and a subring $R_W \subseteq \mathbb{Q}$, there are three possibilities. First, f may lie in $\mathrm{HTP}(R_W)$. If this holds for R_W, the reason is finitary: W contains a certain finite (possibly empty) subset of primes generating the denominators of a solution. Second, there may be a finitary reason why $f \notin \mathrm{HTP}(R_W)$: there may exist a finite subset A_0 of the complement $\overline{W}$ such that f has no solution in $R_{\mathbb{P}-A_0}$. For each finite $A_0 \subset \mathbb{P}$, the set $\mathrm{HTP}(R_{\mathbb{P}-A_0})$ is 1-reducible to $\mathrm{HTP}(\mathbb{Q})$, by Proposition 1; indeed the two sets are computably isomorphic, with a computable permutation of $\mathbb{Z}[\mathbf{X}]$ mapping one onto the other. Therefore, the existence of such a set A_0 (still for one fixed f) is a $\Sigma_1^{\mathrm{HTP}(\mathbb{Q})}$ problem.

The third possibility is that neither of the first two holds. An example is given in [5], where it is shown that a particular polynomial f fails to lie in $\mathrm{HTP}(R_{W_3})$, where W_3 is the set of all primes congruent to 3 modulo 4, yet that, for every finite set V_0 of primes, there exists some W disjoint from V_0 with $f \in \mathrm{HTP}(R_W)$. We consider sets such as this W_3 to be on the *boundary* of f, in consideration of the topology of the situation. The set $\mathcal{A}(f) = \{W : f \in \mathrm{HTP}(R_W)\}$ is open in the usual topology on $2^{\mathbb{P}}$, since, for any solution of f in R_W and any $\sigma \subseteq W$ long enough to include all primes dividing the denominators in that solution, every other $V \supseteq \sigma$ will also contain that solution. Moreover, one can computably enumerate the collection of those σ such that the basic open set $\mathcal{U}_\sigma = \{W : \sigma \subseteq W\}$ is contained within $\mathcal{A}(f)$. The set $\mathrm{Int}(\overline{\mathcal{A}(f)})$ is similarly a union of basic open sets, and these can be enumerated by an $\mathrm{HTP}(\mathbb{Q})$-oracle, since $\mathrm{HTP}(\mathbb{Q})$

decides HTP(R) uniformly for every semilocal ring R. The *boundary* $\mathcal{B}(f)$ of f remains: it contains those W which lie neither in $\mathcal{A}(f)$ nor in $\text{Int}(\mathcal{A}(f))$. The boundary can be empty, but need not be, as seen in the example mentioned above.

It follows quickly from Baire category theory that the boundary set for a polynomial $f \in \mathbb{Z}[\boldsymbol{X}]$ must be nowhere dense. In general the boundary set $\partial\mathcal{A}$ of a set $\mathcal{A}$ within a space $\mathcal{S}$ is defined to equal $(\mathcal{S} - \text{Int}(\mathcal{A}) - \text{Int}(\overline{\mathcal{A}}))$, and thus is always closed.

Lemma 3. *For every open set $\mathcal{A}$ in a Baire space $\mathcal{S}$, the boundary set $\partial\mathcal{A}$ is nowhere dense. In particular, for each $f \in \mathbb{Z}[\boldsymbol{X}]$, the boundary set $\mathcal{B}(f) = \partial(\mathcal{A}(f))$ must be nowhere dense. Hence the* entire boundary set

$$\mathcal{B} = \{W \subseteq \mathbb{P} : (\exists f \in \mathbb{Z}[\boldsymbol{X}]) \; W \in \mathcal{B}(f)\} = \cup_{f \in \mathbb{Z}[\boldsymbol{X}]} \mathcal{B}(f)$$

is meager.

Proof. Since $\mathcal{A}$ is open, every open subset $\mathcal{V}$ of the closure of $\partial\mathcal{A}$ (namely $\partial\mathcal{A}$ itself) lies within the complement $\overline{\mathcal{A}}$, hence within $\text{Int}(\overline{\mathcal{A}})$, which is also disjoint from $\partial\mathcal{A}$. This proves that $\partial\mathcal{A}$ is nowhere dense. Hence $\mathcal{B}$, the countable union of such sets, is meager.   □

For a set W to fail to lie in $\mathcal{B}$, it must be the case that for every polynomial f, either $f \in \text{HTP}(R_W)$ or else some finite initial segment of W rules out all solutions to f. This is an example of the concept of *genericity*, common in both computability and set theory, so we adopt the term here. With this notion, we can show not only that $\text{HTP}(R_W) \leq W \oplus \text{HTP}(\mathbb{Q})$ for all W in the comeager set $\overline{\mathcal{B}}$, but indeed that the reduction is uniform on $\overline{\mathcal{B}}$.

Definition 2. *A set $W \subseteq \mathbb{P}$ is* HTP-generic *if $W \notin \mathcal{B}$. In this case we will also call the corresponding subring R_W* HTP-generic. *By Lemma 3, HTP-genericity is comeager.*

Proposition 2. *For every HTP-generic set W, $HTP(R_W) \equiv_T W \oplus HTP(\mathbb{Q})$, via uniform Turing reductions. Hence there is a single Turing reduction Φ such that the following set is comeager:*

$$\{W \subseteq \mathbb{P} : \Phi^{W \oplus HTP(\mathbb{Q})} = \chi_{HTP(R_W)}\}.$$

Proof. Given $f \in \mathbb{Z}[\boldsymbol{X}]$ as input, the program for Φ simply searches for either a solution $\boldsymbol{x}$ to $f = 0$ in $\mathbb{Q}$ for which all primes dividing the denominators lie in the oracle set W, or else a finite set $A_0 \subseteq \overline{W}$ such that the HTP($\mathbb{Q}$) oracle, using Proposition 1, confirms that $f \notin \text{HTP}(R_{\mathbb{P}-A_0})$. When it finds either of these, it outputs the corresponding answer about membership of f in HTP(R_W). If it never finds either, then $W \in \mathcal{B}(f)$, and so this process succeeds for every W except those in the meager set $\mathcal{B}$. (The reduction $W \oplus \text{HTP}(\mathbb{Q}) \leq_T \text{HTP}(R_W)$ was described in Sect. 1.)   □

Corollary 1. *For every set $C \subseteq \omega$, the following are equivalent*

1. $C \leq_T HTP(\mathbb{Q})$.
2. $\{W \subseteq \mathbb{P} : C \leq_T HTP(R_W)\} = 2^{\mathbb{P}}$.
3. $\{W \subseteq \mathbb{P} : C \leq_T HTP(R_W)\}$ *is not meager.*

This opens a new possible avenue to a proof of undecidability of $HTP(\mathbb{Q})$: one need not address $\mathbb{Q}$ itself, but only show that for most subrings R_W, $HTP(R_W)$ can decide the halting problem (or some other fixed undecidable set C). Constructions in the style of [6, Theorem 1.3] offer an approach to the problem along these lines: that theorem, proven by Poonen, shows that the set of subrings R with $\emptyset' \leq_T HTP(R)$ has size continuum and is large in certain other senses. Poonen constructs decidable subsets $T_0, T_1 \subseteq \mathbb{P}$, both of asymptotic density 0 within $\mathbb{P}$, such that for every $W \subseteq \mathbb{P}$ with $T_0 \subseteq W$ and $T_1 \cap W = \emptyset$, the subring R_W has $\emptyset' \leq_T HTP(R_W)$. This feels like a substantial collection of subrings, but the conditions $T_0 \subseteq W$ and $T_1 \cap W = \emptyset$ each imply that this set of subrings is nowhere dense, and therefore this set does not by itself enable us to apply Corollary 1. Moreover, it is not clear that any of Poonen's subrings need be HTP-generic.

Proof. Trivially ($1 \implies 2 \implies 3$), since all W satisfy $HTP(\mathbb{Q}) \leq_T HTP(R_W)$. So assume (3). Then by Proposition 2, $C \leq_T W \oplus HTP(\mathbb{Q})$ holds on a non-meager set, as the intersection of a non-meager set with a comeager set cannot be meager. So by Lemma 2, $C \leq_T HTP(\mathbb{Q})$. $\qquad\square$

Having examined classes of subsets of $\mathbb{P}$ defined by Turing reductions involving $HTP(R_W)$, we now replace Turing reducibility by 1-reducibility and ask similar questions about classes so defined. It is not known whether there exists a subring $R \subseteq \mathbb{Q}$ for which $\emptyset' \leq_T HTP(R_W)$ but $\emptyset' \not\leq_1 HTP(R_W)$, and we have no good candidates for such a subring. Ever since the original proof of undecidability of Hilbert's Tenth Problem in [1,4], every Turing reduction ever given from the Halting Problem to any $HTP(R)$ with $R \subseteq \mathbb{Q}$ has in fact been a 1-reduction. Of course, if $\emptyset' \leq_1 HTP(\mathbb{Q})$, then $\emptyset' \leq_1 HTP(R)$ for all subrings R, so in some sense $\mathbb{Q}$ itself is the "only" candidate.

We have a result for 1-reducibility analogous to Corollary 1, but the proof is somewhat different.

Theorem 1. *For every set $C \subseteq \omega$ with $C \not\leq_1 HTP(\mathbb{Q})$, the following class is meager:*

$$\mathcal{O} = \{W \subseteq \mathbb{P} : C \leq_1 HTP(R_W)\}.$$

Proof. One naturally views $\mathcal{O}$ as the union of countably many subclasses $\mathcal{O}_e = \{W \subseteq \mathbb{P} : C \leq_1 HTP(R_W) \text{ via } \varphi_e\}$. Of course, for those e for which the e-th Turing function φ_e is not total, this class is empty. We claim that if any one of these $\mathcal{O}_e$ fails to be nowhere dense, then $C \leq_1 HTP(\mathbb{Q})$, contrary to the assumption of the theorem.

Suppose that indeed $\mathcal{O}_e$ fails to be nowhere dense, and fix a σ for which $\mathcal{U}_\sigma \subseteq \text{cl}(\mathcal{O}_e)$. Let $A_0 = \sigma^{-1}(0)$ contain those primes excluded from all $W \in \mathcal{U}_\sigma$, and set $R = R_{(\mathbb{P}-A_0)}$. Now whenever $n \in C$ and $W \in \mathcal{O}_e$, the polynomial $\varphi_e(n)$ must lie in $\mathrm{HTP}(R_W)$. Since some $W \in \mathcal{O}_e$ lies in $\mathcal{U}_\sigma$, we must have $\varphi_e(n) \in \mathrm{HTP}(R)$, because $R_W \subseteq R$ whenever $W \in \mathcal{U}_\sigma$. On the other hand, suppose $n \notin C$. If R contained a solution to the polynomial $\varphi_e(n)$, then some $\tau \supseteq \sigma$ would by itself invert the finitely many primes required to generate this solution, and thus we would have $\mathcal{U}_\tau \cap \mathcal{O}_e = \emptyset$. With $\mathcal{U}_\sigma \subseteq \mathrm{cl}(\mathcal{O}_e)$, this is impossible, and so, whenever $n \notin C$, we have $\varphi_e(n) \notin \mathrm{HTP}(R)$.

Thus R itself lies in $\mathcal{O}_e$, as φ_e is a 1-reduction from C to $\mathrm{HTP}(R)$. But R is semilocal, inverting all primes p except those with $\sigma(p) = 0$. By Proposition 1, we have $\mathrm{HTP}(R) \leq_1 \mathrm{HTP}(\mathbb{Q})$, and so $C \leq_1 \mathrm{HTP}(\mathbb{Q})$. $\qquad\square$

Now we prove two similar results, one about subrings of $\mathbb{Q}$ which admit diophantine models and one about subrings which admit existential definitions of the integers within the subring. In both cases, the result is a sort of zero-one law: that the given phenomenon must either hold almost everywhere (i.e., on a comeager set of subrings) or almost nowhere (i.e., on a meager set). We begin with the diophantine models.

Definition 3. *In a ring R, a* diophantine model *of $\mathbb{Z}$ consists of three polynomials h, h_+, and $h_\times$, with $h \in R[X_1, \ldots, X_n, \boldsymbol{Y}]$ and $h_+, h_\times \in R[X_1, \ldots, X_{3n}, \boldsymbol{Y}]$ (for some n), such that the set*

$$\{\boldsymbol{x} \in R^n : (\exists \boldsymbol{y} \in R^{<\omega})\, h(\boldsymbol{x}, \boldsymbol{y}) = 0\}$$

(equivalently, $\{\boldsymbol{x} \in R^n : h(\boldsymbol{x}, \boldsymbol{Y}) \in \mathrm{HTP}(R)\}$) is isomorphic to the structure $(\mathbb{Z}, +, \cdot)$ under the binary operations whose graphs are defined by

$$\{(\boldsymbol{x}_1, \boldsymbol{x}_2, \boldsymbol{x}_3) \in R^{3n} : h_+(\boldsymbol{x}_1, \boldsymbol{x}_2, \boldsymbol{x}_3, \boldsymbol{Y}) \in \mathrm{HTP}(R)\}$$

for addition and the corresponding set with $h_\times$ for multiplication.

If a computable ring R admits a diophantine model of $\mathbb{Z}$, then $\mathrm{HTP}(\mathbb{Z})$ can be coded into $\mathrm{HTP}(R)$, and so $\emptyset' \equiv_1 \mathrm{HTP}(\mathbb{Z}) \leq_1 \mathrm{HTP}(R)$. For subrings R_W of $\mathbb{Q}$ for which $\emptyset' \not\leq_T W$, this is the only known method of showing that $\emptyset' \leq_T \mathrm{HTP}(R_W)$ (apart from the original proof by Matiyasevich, Davis, Putnam, and Robinson for the case $W = \emptyset$, of course, which is what allows this method to succeed).

Definition 4. $\mathcal{D} = \{W \subseteq \mathbb{P} : R_W \text{ admits a diophantine model of } \mathbb{Z}\}$.

In this section we address the question of the size of the class $\mathcal{D}$. The main result fails to resolve this question, but shows it to have an "all-or-nothing" character.

Theorem 2. *The class $\mathcal{D}$ is non-meager if and only if there exists a particular triple $(h, h_+, h_\times)$ of polynomials over $\mathbb{Z}$ and a finite binary string $\sigma \in 2^{<\mathbb{P}}$ such that, for every HTP-generic $V \in \mathcal{U}_\sigma$, R_V admits a diophantine model of $\mathbb{Z}$ via these three polynomials.*

Moreover, if $\mathcal{D}$ is non-meager, then $\mathbb{P} \in \mathcal{D}$ (i.e., $\mathbb{Q}$ admits a diophantine model of $\mathbb{Z}$).

Proof. For each triple $\boldsymbol{h} = (h, h_+, h_\times)$ of polynomials of appropriate lengths over $\mathbb{Z}$, we set $\mathcal{D}_{\boldsymbol{h}}$ to contain those W for which $\boldsymbol{h}$ defines a diophantine model of $\mathbb{Z}$ within R_W. If each $\mathcal{D}_{\boldsymbol{h}}$ is nowhere dense, their countable union $\mathcal{D}$ is meager.

Now suppose that $\mathcal{D}$ is non-meager, so some class $\mathcal{D}_{\boldsymbol{h}}$ fails to be nowhere dense. Then there must be a string σ such that $\mathcal{U}_\sigma \subseteq \mathrm{cl}(\mathcal{D}_{\boldsymbol{h}})$. Using this σ and this $\boldsymbol{h}$, we now prove the main claim: all $W \in \mathcal{U}_\sigma$ with HTP-generic R_W lie in $\mathcal{D}_{\boldsymbol{h}}$. Let $R_0 = R_{\sigma^{-1}(1)}$ and $R_1 = R_{\mathbb{P}-\sigma^{-1}(0)}$ be the smallest and largest subrings (under $\subseteq$) in $\mathcal{U}_\sigma$, so R_0 is finitely generated and R_1 is semilocal.

Fix a single $W \supset \sigma$ with $W \in \mathcal{D}_{\boldsymbol{h}}$, and fix the tuples $\boldsymbol{x}_0$ and $\boldsymbol{x}_1$ from R_W which represent the elements 0 and 1 in the diophantine model defined in R_W by $\boldsymbol{h}$. It follows that $h_\times(\boldsymbol{x}_0, \boldsymbol{x}_0, \boldsymbol{x}_0, \boldsymbol{Y}) \in \mathrm{HTP}(R_W)$ and $h_\times(\boldsymbol{x}_1, \boldsymbol{x}_1, \boldsymbol{x}_1, \boldsymbol{Y}) \in \mathrm{HTP}(R_W)$. Now if any other tuple $\boldsymbol{x}$ from R_1 had $h(\boldsymbol{x}, \boldsymbol{Y}) \in \mathrm{HTP}(R_1)$ and $h_\times(\boldsymbol{x}, \boldsymbol{x}, \boldsymbol{x}, \boldsymbol{Y}) \in \mathrm{HTP}(R_1)$, then we could set $\tau = \sigma^{\smallfrown}111\cdots 1$ to contain enough primes that $R_{\tau^{-1}(1)}$ would contain $\boldsymbol{x}$, $\boldsymbol{x}_0$, and $\boldsymbol{x}_1$. This would mean that $\boldsymbol{h}$ could not define a diophantine model of $\mathbb{Z}$ in any R_V with $V \in \mathcal{U}_\tau$, contrary to hypothesis. Therefore, no other $\boldsymbol{x}$ from R_1 can do this. Now suppose that $\boldsymbol{x}_0$ does *not* lie within R_0. In this case, some extension $\rho = \sigma^{\smallfrown}000\cdots 0$ would exclude enough primes to ensure that $\boldsymbol{x}_0$ does not lie in $R_{\mathbb{P}-\rho^{-1}(0)}$, and then no $\tau \supset \rho$ would admit a diophantine model via $\boldsymbol{h}$, since no other tuple with the right properties lies in R_1. Again, this contradicts our hypothesis that $\mathcal{U}_\sigma \subseteq \mathrm{cl}(\mathcal{D}_{\boldsymbol{h}})$, since $\mathcal{D}_{\boldsymbol{h}} \cap \mathcal{U}_\rho$ would be empty, and so $\boldsymbol{x}_0$ lies in R_0. Similarly so does $\boldsymbol{x}_1$.

Now one proceeds by induction on the subsequent elements of the diophantine model in R_1. Some tuple $\boldsymbol{x}_2$ from R_W must satisfy $h(\boldsymbol{x}_2, \boldsymbol{Y}) \in \mathrm{HTP}(R_W)$ and $h_+(\boldsymbol{x}_1, \boldsymbol{x}_1, \boldsymbol{x}_2, \boldsymbol{Y}) \in \mathrm{HTP}(R_W)$, and by the same arguments as above, we see that $\boldsymbol{x}_2$ is the only tuple in R_1 with this property, and then that $\boldsymbol{x}_2$ actually lies in R_0. Likewise, $\boldsymbol{x}_{-1}$ must satisfy $h(\boldsymbol{x}_{-1}, \boldsymbol{Y}) \in \mathrm{HTP}(R_W)$ and $h_+(\boldsymbol{x}_1, \boldsymbol{x}_{-1}, \boldsymbol{x}_0, \boldsymbol{Y}) \in \mathrm{HTP}(R_W)$, and again this forces $\boldsymbol{x}_{-1}$ to lie in R_0 and to be the unique tuple with these properties in R_1.

Continuing this induction, we see that every tuple in the domain of the diophantine model of $\mathbb{Z}$ in R_W actually lies in R_0, and hence in every R_W with $W \in \mathcal{U}_\sigma$; and moreover that these are the only tuples $\boldsymbol{x}$ in R_1 for which $h(\boldsymbol{x}, \boldsymbol{Y}) \in \mathrm{HTP}(R_1)$. Likewise, if some $\boldsymbol{x}_m$, $\boldsymbol{x}_n$ and $\boldsymbol{x}_p$ (representing m, n, and p in the diophantine model) satisfy $h_+(\boldsymbol{x}_m, \boldsymbol{x}_n, \boldsymbol{x}_p, \boldsymbol{Y}) \in \mathrm{HTP}(R_1)$, then for some k, $\tau = \sigma^{\smallfrown}1^k$ is long enough to ensure that every W extending τ must have $h_+(\boldsymbol{x}_m, \boldsymbol{x}_n, \boldsymbol{x}_p, \boldsymbol{Y}) \in \mathrm{HTP}(R_W)$. But some such W lies in $\mathcal{D}_{\boldsymbol{h}}$, so we must have $m + n = p$. The same works for $h_\times$, so $\boldsymbol{h}$ defines a diophantine model of $\mathbb{Z}$ in R_1.

It is not clear whether $\boldsymbol{h}$ defines a diophantine model in the subring R_0 (which, being finitely generated, lies in $\mathcal{B}$). The domain elements of the model in R_1 all lie in R_0, but the witnesses might not. However, suppose that $V \in \mathcal{U}_\sigma$ is HTP-generic, and fix any domain element $\boldsymbol{x}$. Let $\tau = V \restriction m$, for any $m \geq |\sigma|$. Then some $U \supseteq \tau$ lies in $\mathcal{D}_{\boldsymbol{h}}$, and so some extension of τ yields a solution to $h(\boldsymbol{x}, \boldsymbol{Y})$. Since V is HTP-generic (that is, $V \notin \mathcal{B}$), this forces $h(\boldsymbol{x}, \boldsymbol{Y}) \in \mathrm{HTP}(R_V)$. Likewise, for each fact coded by h_+ or $h_\times$ about domain elements of the model, some extension of $V \restriction m$ must yield a witness to that fact, and therefore

R_V itself contains such a witness. So h defines this same diophantine model in *every* HTP-generic subring R_V with $V \in \mathcal{U}_\sigma$, as required by the theorem.

Cases (1) and (2) of the theorem cannot both hold, because under (2), $\mathcal{U}_\sigma \cap \overline{\mathcal{B}}$ would be a nonmeager subset of $\mathcal{D}$. Moreover, the 1-reduction $HTP(R_1) \leq_1 HTP(\mathbb{Q})$ given in [2, Proposition 5.4] has sufficient uniformity that the images of h, h_+, and $h_\times$ under this reduction define a diophantine model of $\mathbb{Z}$ inside $\mathbb{Q}$. (Specifically, $h(X, Y)$ maps to the sum of h^2 with several other squares of polynomials in such a way as to guarantee that all solutions use values from R_1 for the variables X and Y; likewise with h_+ and $h_\times$.) This proves the final statement of the theorem. □

Now we continue with the question of existential definability of the integers.

Definition 5. *In a ring R, a polynomial $g \in \mathbb{Z}[X, Y]$ existentially defines $\mathbb{Z}$ if, for every $q \in R$,*

$$q \in \mathbb{Z} \iff g(q, Y) \in HTP(R).$$

$\mathbb{Z}$ *is* existentially definable in R *if such a polynomial g exists.*

A ring in which $\mathbb{Z}$ is existentially definable must admit a very simple diophantine model of $\mathbb{Z}$, given by the polynomial g along with $h_+ = X_1 + X_2 - X_3$ and $h_\times = X_1 X_2 - X_3$. The question of definability of $\mathbb{Z}$ in the field $\mathbb{Q}$ was originally answered by Julia Robinson (see [8]), who gave a Π_4 definition. Subsequent work by Poonen [7] and then Koenigsmann [3] has resulted in a Π_1 definition of $\mathbb{Z}$ in $\mathbb{Q}$, but it remains unknown whether any existential formula defines $\mathbb{Z}$ there.

Definition 6. $\mathcal{E}$ *is the class of subrings of $\mathbb{Q}$ where $\mathbb{Z}$ is existentially definable:*

$$\mathcal{E} = \{W \subseteq \mathbb{P} : \mathbb{Z} \text{ is existentially definable in } R_W\}.$$

We now address the question of the size of the class $\mathcal{E}$. As with $\mathcal{D}$, we show $\mathcal{E}$ to be either very large or very small, in the sense of Baire category.

Theorem 3. *The following are equivalent.*

1. *The class $\mathcal{E}$ is not meager.*
2. *There is a $\sigma \in 2^{<\mathbb{P}}$, and a single polynomial g which existentially defines $\mathbb{Z}$ in all HTP-generic subrings R_V with $V \in \mathcal{U}_\sigma$.*
3. $\mathbb{P} \in \mathcal{E}$; *that is, $\mathbb{Z}$ is existentially definable in $\mathbb{Q}$.*
4. *There is a single existential formula which defines $\mathbb{Z}$ in every subring of $\mathbb{Q}$.*

Proof. The proof that (1) $\implies$ (2) $\implies$ (3) proceeds along the same lines as that of Theorem 2, with $\mathcal{E}_g$ as the class of those W for which the polynomial g existentially defines $\mathbb{Z}$ within R_W. If every one of these classes is nowhere dense, then their countable union $\mathcal{E}$ is meager. Otherwise one proves (2), and from that (3), by a simplification of the same method as before, with no induction required. To see that (3) implies (4), notice that if $\mathbb{Z}$ is defined in $\mathbb{Q}$ by the formula $\exists Y\, f(X, Y) = 0$, and d is the total degree of f, then the formula

$$\exists Y \exists Z\, [Z^d \cdot f\left(X, \frac{Y_1}{Z}, \ldots, \frac{Y_n}{Z}\right) = 0\ \&\ Z > 0]$$

defines $\mathbb{Z}$ in R_W. □

It is possible to turn Theorem 2 into an equivalence analogous to that in Theorem 3, with the third condition stating that $\mathbb{P} \in \mathcal{D}$. As far as we know, however, it is necessary to consider diophantine *interpretations* in subrings R_W, rather than diophantine models, in order to accomplish this.

3   Measure Theory

Normally there is a strong connection between measure theory and Baire category theory. Each defines a certain Σ-ideal of sets to be "small": the sets of measure 0, and the meager sets. In Cantor space, neither of these properties implies the other, but empirically they appear closely connected, especially when the sets are given by natural definitions: sets of measure 0 are often meager, and vice versa. (Exceptions to this principle do exist, however, and another difference was mentioned in the context of Lemma 1.)

Our results here rely heavily on the simple Lemma 3, stating that the boundary set $\mathcal{B}(f)$ of a polynomial f is nowhere dense. Most of our subsequent results have measure-theoretic analogues which would go through fairly easily, provided that these sets $\mathcal{B}(f)$ also have measure 0. However, determining the measure of the boundary set of a polynomial appears to be a nontrivial problem. It is unknown whether there exists any polynomial f for which $\mu(\mathcal{B}(f)) > 0$. Indeed, in work to appear elsewhere, the author has shown that if $\mu(\mathcal{B}(f)) = 0$ for all $f \in \mathbb{Z}[\boldsymbol{X}]$, then there is no existential definition of the set $\mathbb{Z}$ within the field $\mathbb{Q}$.

Moreover, if an f exists with $\mu(\mathcal{B}(f)) > 0$, it is unclear what other constraints on the real number $\mu(\mathcal{B}(f))$ exist, apart from the computability-theoretic upper bound given by its definition as $\mu(\mathcal{B}(f))$. Could such a number be transcendental? Or noncomputable? If not, is there an algorithm computing $\mu(\mathcal{B}(f))$ uniformly in f? These appear to be challenging questions, often with a more number-theoretic flavor than most of this article. Resolving them might make it possible to determine whether Hilbert's Tenth Problem on subrings of $\mathbb{Q}$ has measure-theoretic zero-one laws similar to those proven here for Baire category.

References

1. Davis, M., Putnam, H., Robinson, J.: The decision problem for exponential diophantine equations. Ann. Math. **2**(74), 425–436 (1961)
2. Eisenträger, K., Miller, R., Park, J., Shlapentokh, A.: As easy as $\mathbb{Q}$: Hilbert's Tenth Problem for subrings of the rationals (submitted for publication)
3. Koenigsmann, J.: Defining $\mathbb{Z}$ in $\mathbb{Q}$. Ann. Math. **183**(1), 73–93 (2016)
4. Matijasevič, Y.: The diophantineness of enumerable sets. Dokl. Akad. Nauk SSSR **191**, 279–282 (1970)
5. Miller, R.: Measure theory and Hilbert's Tenth Problem inside $\mathbb{Q}$ (submitted)
6. Poonen, B.: Hilbert's tenth problem and mazur's conjecture forlarge subrings of $\mathbb{Q}$. J. AMS **16**(4), 981–990 (2003)
7. Poonen, B.: Characterizing integers among rational numbers with a universal-existential formula. Am. J. Math. **131**(3), 675–682 (2009)
8. Robinson, J.: Definability and decision problems in arithmetic. J. Symbolic Logic **14**, 98–114 (1949)

Partial Orders and Immunity in Reverse Mathematics

Ludovic Patey[✉]

Laboratoire PPS, Université Paris Diderot, Paris, France
ludovic.patey@computability.fr

Abstract. We identify computability-theoretic properties enabling us to separate various statements about partial orders in reverse mathematics. We obtain simpler proofs of existing separations, and deduce new compound ones. This work is part of a larger program of unification of the separation proofs of various Ramsey-type theorems in reverse mathematics in order to obtain a better understanding of the combinatorics of Ramsey's theorem and its consequences. We also answer a question of Murakami, Yamazaki and Yokoyama about pseudo Ramsey's theorem for pairs.

1 Introduction

Many theorems of "ordinary" mathematics are of the form

$$(\forall X)[\Phi(X) \to (\exists Y)\Psi(X, Y)]$$

where Φ and Ψ are arithmetic formulas. They can be seen as *mathematical problems*, whose *instances* are sets X such that $\Phi(X)$ holds, and whose *solutions* to X are sets Y such that $\Psi(X, Y)$ holds. For example, König's lemma asserts that every infinite, finitely branching tree admits an infinite path through it.

There exist many ways to calibrate the strength of a mathematical problem. Among them, *reverse mathematics* is a vast foundational program that seeks to determine the weakest axioms necessary to prove ordinary theorems. It uses the framework of subsystems of second-order arithmetic, within the base theory $\mathsf{RCA_0}$, which can be thought of as capturing *computable mathematics*. An ω-*structure* is a structure whose first-order part consists of the standard integers. The ω-models of $\mathsf{RCA_0}$ are those whose second-order part is a *Turing ideal*, that is, a collection of sets $\mathcal{S}$ downward-closed under the Turing reduction and closed under the effective join.

In this setting, a ω-model $\mathcal{M}$ satisfies a mathematical problem P if every P-instance in $\mathcal{M}$ has a solution in $\mathcal{M}$. A standard way of proving that a problem P does not imply another problem Q consists of creating an ω-model $\mathcal{M}$ satisfying P but not Q. Such a model is usually constructed by taking a ground Turing ideal, and extending it by iteratively adding solutions to its P-instances. However, while taking the closure of the collection $\mathcal{M} \cup \{Y\}$ to obtain a Turing ideal, one may

© Springer International Publishing Switzerland 2016
A. Beckmann et al. (Eds.): CiE 2016, LNCS 9709, pp. 353–363, 2016.
DOI: 10.1007/978-3-319-40189-8_36

add solutions to Q-instances as well. The whole difficulty of this construction consists of finding the right computability-theoretic notion preserved by P but not by Q.

We conduct a program of identification of the computability-theoretic properties enabling us to distinguish various Ramsey-type theorems in reverse mathematics, but also under computable and Weihrauch reducibilities. This program puts emphasis on the interplay between computability theory and reverse mathematics, the former providing tools to separate theorems in reverse mathematics over standard models, and the latter exhibiting new computability-theoretic properties.

Among the theorems studied in reverse mathematics, the ones coming from Ramsey's theory play a central role. Their strength are notoriously hard to gauge, and required the development of involved computability-theoretic frameworks. Perhaps the most well-known example is Ramsey's theorem.

Definition 1 (Ramsey's theorem). *A subset H of ω is* homogeneous *for a coloring $f : [\omega]^n \to k$ (or f-homogeneous) if each n-tuples over H are given the same color by f. RT^n_k is the statement "Every coloring $f : [\omega]^n \to k$ has an infinite f-homogeneous set".*

Jockusch [11] conducted a computational analysis of Ramsey's theorem. He proved in particular that RT^n_k implied the existence of the halting set whenever $n \geq 3$. There has been a lot of literature around the strength of Ramsey's theorem for pairs [4,6,9,19] and its consequences [3,5,10]. We focus on some mathematical statements about partial orders which are consequences of Ramsey's theorem for pairs.

Definition 2 (Chain-antichain). *A chain in a partial order $(P, \leq_P)$ is a set $S \subseteq P$ such that $(\forall x, y \in S)(x \leq_P y \vee y \leq_P x)$. An antichain in P is a set $S \subseteq P$ such that $(\forall x, y \in S)(x \neq y \to x |_P y)$ (where $x |_P y$ means that $x \not\leq_P y \wedge y \not\leq_P x$). CAC is the statement "every infinite partial order has an infinite chain or an infinite antichain."*

The chain-antichain principle was introduced by Hirschfeldt and Shore [10] together with the ascending descending sequence (ADS). They studied extensively cohesive and stable versions of the statements, and proved that CAC is computationally weak, in that it does not even imply the existence of a diagonally non-computable function. However, their proof has an ad-hoc flavor, in that it is a direct separation involving the two statements. Later, Lerman et al. [13] separated ADS from CAC over ω-models by using an involved iterated forcing argument.

In this paper, we revisit the two proofs and emphasis on the combinatorial nature of the principles by identifying the computability-theoretic properties separating them. Those properties happen to be very natural and coincide on co-c.e. sets to some well-known computability-theoretic notions, namely, immunity and hyperimmunity. The proof of the separation of ADS from CAC is significantly simpler and more modular, as advocated by the author in [16].

1.1 Notation and Definitions

Given two sets A and B, we denote by $A < B$ the formula $(\forall x \in A)(\forall y \in B)[x < y]$ and by $A \subseteq^* B$ the formula $(\forall^\infty x \in A)[x \in B]$, meaning that A is included in B *up to finitely many elements*. A *Mathias condition* is a pair (F, X) where F is a finite set, X is an infinite set and $F < X$. A condition (F_1, X_1) *extends* (F, X) (written $(F_1, X_1) \leq (F, X)$) if $F \subseteq F_1$, $X_1 \subseteq X$ and $F_1 \setminus F \subset X$. A set G *satisfies* a Mathias condition (F, X) if $F \subset G$ and $G \setminus F \subseteq X$.

2 Preservation of Properties for Co-c.e. Sets

Ramsey's theorem for k colors has a deeply disjunctive nature. One cannot know in a finite amount of time whether a coloring will admit an infinite homogeneous set for a fixed color, and one must therefore build multiple homogeneous sets simultaneously, namely, one for each color. This disjunction was exploited by the author to show for example that ADS does not preserve 2 hyperimmunities simultaneously, whereas the Erdős-Moser theorem does [16]. This idea was also used in the context of computable reducibility to show that RT^2_{k+1} does not computably reduce to RT^2_k whenever $k \geq 1$, by showing that RT^2_k preserves 2 among $k + 1$ hyperimmunities simultaneously whereas RT^2_{k+1} does not [18]. In this section, we shall see that this disjunctive flavor disappears whenever considering co-c.e. sets. In particular, RT^2_2 admits preservation of countably many hyperimmune co-c.e. sets simultaneously.

Definition 3 (Hyperimmunity). *An* array *is a sequence of mutually disjoint finitely coded sets. A set A is X-hyperimmune if for every X-c.e. array $F_0, F_1, \ldots$, there is some i such that $F_i \cap A = \emptyset$.*

Equivalently, a set is X-hyperimmune if its principal function is not dominated by any X-computable function, where the *principal function* p_A of a set $A = \{x_0 < x_1 < \ldots\}$ is defined by $p_A(i) = x_i$.

Definition 4 (Preservation of hyperimmunity for co-c.e. sets). *A Π^1_2 statement* P *admits* preservation of hyperimmunity for co-c.e. sets *if for every set Z, every sequence of Z-co-c.e. Z-hyperimmune sets $A_0, A_1, \ldots$ and every* P-*instance $X \leq_T Z$, there is a solution Y to X such that the A's are $Y \oplus Z$-hyperimmune.*

Hirschfeldt and Shore [10] proved that CAC is equivalent to the existence of homogeneous sets for semi-transitive colorings. A coloring $f : [\mathbb{N}]^2 \to 2$ is *semi-transitive* if whenever $f(x, y) = 1$ and $f(y, z) = 1$, then $f(x, z) = 1$ for $x < y < z$.

Theorem 5. CAC *admits preservation of hyperimmunity for co-c.e. sets.*

Proof. Fix a set Z and a countable sequence of Z-co-c.e. Z-hyperimmune sets $A_0, A_1, \ldots$ Let $f : [\omega]^2 \to 2$ be a Z-computable semi-transitive coloring. We shall assume that there is no infinite Z-computable f-homogeneous set for color 0, otherwise we are done. We will build an infinite set G f-homogeneous for color 1 such that the A's are $G \oplus Z$-hyperimmune. The construction is done by a Mathias forcing (F, X), where F is a finite set, X is an infinite Z-computable set such that $max(F) < min(X)$, and for every $x \in X$, $F \cup \{x\}$ is f-homogeneous for color 1. The condition extension is the usual Mathias extension. A set G *satisfies* (F, X) if it satisfies the Mathias condition (F, X) and is f-homogeneous for color 1. Lemma 6 shows that every sufficiently generic filter for this notion of forcing yields an infinite set.

Lemma 6. *Every condition* $c = (F, X)$ *has an extension* (E, Y) *such that* $|E| > |F|$.

In what follows, we say that a condition c *forces* a formula property $\varphi(G)$ if $\varphi(G)$ holds for every set G satisfying c.

Lemma 7. *For every condition* $c = (F, X)$ *and every pair of indices* e, i, *there is an extension forcing* $\Phi_e^{G \oplus Z}$ *not to dominate* p_{A_i}.

Proof. Define the Z-partial computable function h which on input x, searches for a finite set $E_x \subseteq X$ f-homogeneous for color 1 such that $\Phi_e^{(F \cup E_x) \oplus Z}(x) \downarrow$. If found, $h(x) = \Phi_e^{(F \cup E_x) \oplus Z}(x)$, otherwise $h(x) \uparrow$. We have two cases.

- Case 1: h is total. By Z-hyperimmunity of p_{A_i}, there are infinitely many x such that $h(x) < p_{A_i}(x)$. If there is such an x such that the set $Y = \{y \in X : (\forall z \in E_x) f(z, y) = 1\}$ is infinite, then the condition $(F \cup E_x, Y)$ is an extension of c forcing $\Phi_e^{G \oplus Z}(x) < p_{A_i}(x)$. If there is no such x, then by semi-transitivity of f, for every x such that $h(x) < p_{A_i}(x)$, for almost every $y \in X$, $f(max(E_x), y) = 0$. Since A_i is co-c.e., one can find a Z-computable infinite subset Y of $\{max(E_x) : h(x) < p_{A_i}(x)\}$. The set Y is Z-computable and limit-homogeneous for color 0, and therefore computes an infinite f-homogeneous set for color 0, contradicting our assumption.
- Case 2: there is some x such that $h(x) \uparrow$. By definition of h, the condition c already forces $\Phi_e^{G \oplus Z}(x) \uparrow$. □

Corollary 8. RT_2^2 *admits preservation of hyperimmunity for co-c.e. sets.*

Proof. Bovykin and Weiermann [2] studied the reverse mathematics of the Erdős-Moser theorem (EM) and proved that $\mathsf{RCA}_0 \vdash \mathsf{RT}_2^2 \leftrightarrow [\mathsf{CAC} \wedge \mathsf{EM}]$. The author proved in [16] that EM admits preservation of hyperimmunity. Together with Theorem 5, we deduce that $\mathsf{CAC} \wedge \mathsf{EM}$, hence RT_2^2, admits preservation of hyperimmunity for co-c.e. sets.

3   CAC and Constant-Bound Immunity

Hirschfeldt and Shore [10] separated CAC from DNC in reverse mathematics by a direct construction. DNC is the statement asserting, for every set X, the existence of a function f such that $f(e) \neq \Phi_e^X(e)$ for every e. In this section, we extract the core of the combinatorics of their forcing argument to exhibit a computability-theoretic property separating the two notions, namely, constant-bound immunity.

Definition 9 (Constant-bound immunity). *A k-enumeration (k-enum) of a set A is an infinite sequence of k-sets $F_0 < F_1 < \ldots$ such that for every $i \in \omega$, $F_i \cap A \neq \emptyset$. A* constant-bound enumeration *(c.b-enum) of a set A is a k-enumeration of A for some $k \in \omega$. A set A is k-immune (c.b-immune) relative to X if it admits no X-computable k-enumeration (c.b-enumeration).*

In particular, 1-immunity coincides with the standard notion of immunity. Also note that one can easily create a c.b-immune set computing no effectively immune set. The following lemma shows that c.b-immunity and immunity coincide for co-c.e. sets.

Lemma 10. *An X-co-c.e. set A is c.b-immune relative to X iff it is X-immune.*

Definition 11 (Preservation of c.b-immunity). *A Π_2^1 statement P admits preservation of c.b-immunity if for every set Z, every set A which is c.b-immune relative to X, and every P-instance $X \leq_T Z$, there is a solution Y to X such that A is c.b-immune relative to $Y \oplus Z$.*

We can easily relate the notion of preservation of c.b-immunity with the existing notion of constant-bound enumeration avoidance defined by Liu [14] to separate RT_2^2 from WWKL over RCA_0.

Lemma 12. *If P admits preservation of c.b-immunity, then it admits constant-bound enumeration avoidance.*

Theorem 13. CAC *admits preservation of c.b-immunity.*

Proof. Let A be a set c.b-immune relative to some set Z, and let $f : [\omega]^2 \to 2$ be a Z-computable semi-transitive coloring. Assume that there is no infinite f-homogeneous set H such that A is c.b-immune relative to $H \oplus Z$, otherwise we are done. We will build two infinite sets G_0 and G_1, such that G_i is f-homogeneous for color i for each $i < 2$, and such that A is c.b-immune relative to $G_i \oplus Z$ for some $i < 2$.

The construction is done by a variant of Mathias forcing (F_0, F_1, X), where F_0 and F_1 are finite sets, X is infinite set such that $max(F_0, F_1) < min(X)$, and A is c.b-immune relative to $X \oplus Z$. Moreover, we require that for every $i < 2$ and every $x \in X$, $F_i \cup \{x\}$ is f-homogeneous for color i. A condition (E_0, E_1, Y) *extends* (F_0, F_1, X) if (E_i, Y) Mathias extends (F_i, X) for each $i < 2$. A pair of sets G_0, G_1 *satisfies* a condition $c = (F_0, F_1, X)$ if G_i is f-homogeneous for color i and satisfies the Mathias condition (F_i, X) for each $i < 2$.

Lemma 14. *For every condition* $c = (F_0, F_1, X)$ *and every* $i < 2$, *there is an extension* (E_0, E_1, Y) *of* c *such that* $|E_i| > |F_i|$.

In what follows, we interpret $\Phi_0, \Phi_1, \ldots$ as Turing functionals outputting non-empty finite sets such that if $\Phi_e^X(x)$ and $\Phi_e^X(x+1)$ both halt, $max(\Phi_e^X(x)) < min(\Phi_e^X(x+1))$. We want to satisfy the following requirements for each $e_0, k_0, e_1, k_1 \in \omega$:

$$\mathcal{R}_{e_0, k_0, e_1, k_1} : \quad \mathcal{R}_{e_0, k_0}^{G_0} \quad \vee \quad \mathcal{R}_{e_1, k_1}^{G_1}$$

where $\mathcal{R}_{e,k}^G$ is the requirement

$$(\exists x)\left(\Phi_e^{G \oplus Z}(x) \uparrow \vee |\Phi_e^{G \oplus Z}(x)| > k \vee \Phi_e^{G \oplus Z}(x) \cap A = \emptyset\right)$$

In other words, $\mathcal{R}_{e,k}^G$ asserts that $\Phi_e^{G \oplus Z}$ is not a k-enumeration of A. A condition c *forces* a formula $\varphi(G_0, G_1)$ if $\varphi(G_0, G_1)$ holds for every pair of *infinite* sets G_0, G_1 satisfying c.

Lemma 15. *For every condition* c *and every vector of indices* $e_0, k_0, e_1, k_1 \in \omega$, *there is an extension* d *of* c *forcing* $\mathcal{R}_{e_0, k_0, e_1, k_1}$.

Proof. Fix a condition $c = (F_0, F_1, X)$, and let $P_0, P_1, \ldots$ be an $X \oplus Z$-computable sequence of sets where $P_n = \Phi_{e_0}^{(F_0 \cup E_0) \oplus Z}(x_0) \cup \Phi_{e_1}^{(F_1 \cup E_1) \oplus Z}(x_1)$ for a pair of sets $E_1 < E_0 \subseteq X$ and some $x_0, x_1 \in \omega$ such that E_0 is f-homogeneous for color 0, $E_1 \cup \{y\}$ is f-homogeneous for color 1 for each $y \in E_0$, and for each $i < 2$, $max(P_{n-1}) < min(\Phi_{e_i}^{(F_i \cup E_i) \oplus Z}(x_i))$ and $|\Phi_{e_i}^{(F_i \cup E_i) \oplus Z}(x_i)| \leq k_i$. We have two cases.

- Case 1: the sequence of the P's is finite and is defined, say to level $n - 1$. If there is a pair of infinite sets G_0, G_1 satisfying c and some $x_1 \in \omega$ such that $\Phi_{e_1}^{G_1 \oplus Z}(x_1) \downarrow$, $max(P_{n-1}) < min(\Phi_{e_1}^{G_1 \oplus Z}(x_1))$, and $|\Phi_{e_1}^{G_1 \oplus Z}(x_1)| \leq k_1$, then let $E_1 \subseteq G_1$ be such that $F_1 \cup E_1$ is an initial segment of G_1 for which $\Phi_{e_1}^{(F_1 \cup E_1) \oplus Z}(x_1) \downarrow$. The set $Y = \{y \in X : E_1 \cup \{y\} \text{is} f\text{-homogeneous for color 1}\}$ is a superset of G_1, hence is infinite. The condition $d = (F_0, F_1 \cup E_1, Y)$ is an extension of c forcing $\mathcal{R}_{e_0, k_0}^{G_0}$, hence forcing $\mathcal{R}_{e_0, k_0, e_1, k_1}$. If there is no such pair of infinite sets G_0, G_1, then the condition c already forces $\mathcal{R}_{e_1, k_1}^{G_1}$, hence $\mathcal{R}_{e_0, k_0, e_1, k_1}$.
- Case 2: the sequence of the P's is infinite. By c.b-immunity of A relative to $X \oplus Z$, $P_n \cap A = \emptyset$ for some $n \in \omega$. Let $E_1 < E_0 \subseteq X$ and $x_0, x_1 \in \omega$ witness the existence of P_n. If $Y_0 = \{y \in X : E_0 \cup \{y\} \text{is} f\text{-homogeneous for color 1}\}$ is infinite, then the condition $(F_0 \cup E_0, F_1, Y_0)$ is an extension of c forcing $\mathcal{R}_{e_0, k_0}^{G_0}$. If Y_0 is finite, then for almost every $y \in X$, there is some $x_y \in E_0$ such that $f(x_y, y) = 1$, and by transitivity of f for color 1, $E_1 \cup \{y\}$ is f-homogeneous for color 1. Indeed, E_1 is f-homogeneous for color 1 and for each $x \in E_1$, $f(x, x_y) = f(x_y, y) = 1$. In this case, $(F_0, F_1 \cup E_1, Y_1)$ is an extension of c forcing $\mathcal{R}_{e_1, k_1}^{G_1}$, for some $Y_1 =^* X$. In both cases, there is an extension of c forcing $\mathcal{R}_{e_0, k_0, e_1, k_1}$.

This completes the proof of Theorem 13. □

Theorem 16. DNC *does not admit preservation of c.b-immunity.*

Proof (Proof sketch). Let $\mu_{\emptyset'}$ be the modulus function of $\emptyset'$, that is, such that $\mu_{\emptyset'}(x)$ is the minimum stage s at which $\emptyset'_s {\restriction} x = \emptyset' {\restriction} x$.

Computably split ω into countably many columns $X_0, X_1, \ldots$ of infinite size. For example, set $X_i = \{\langle i, n\rangle : n \in \omega\}$ where $\langle \cdot, \cdot\rangle$ is a bijective function from ω^2 to ω. For each i, let F_i be the set of the $\mu_{\emptyset'}(i)$ first elements of X_i. The sequence $F_0, F_1, \ldots$ is $\emptyset'$-computable. By a simple finite injury priority argument (see appendix), one can construct a c.e. set W such that the Δ_2^0 set $A = \bigcup_i F_i \smallsetminus W$ is c.b-immune, and such that $|X_i \cap W| \leq i$. We claim that every DNC function computes an infinite subset of A.

Let f be any DNC function. By a classical theorem about DNC functions (see Bienvenu et al. [1] for a proof), f computes a function $g(\cdot, \cdot, \cdot)$ such that whenever $|W_e| \leq n$, then $g(e, n, i) \in X_i \smallsetminus W_e$. For each i, let e_i be the index of the c.e. set $W_{e_i} = W \cap X_i$, and let $n_i = g(e_i, i, i)$. Since $|X_i \cap W| \leq i$, $|W_{e_i}| \leq i$, hence $n_i = g(e_i, i, i) \in X_i \smallsetminus W_{e_i} = X_i \smallsetminus W$. We then have two cases.

- Case 1: $n_i \in F_i$ for infinitely many i's. One can f-computably find infinitely many of them since $\mu_{\emptyset'}$ is left-c.e. and the sequence of the n's is f-computable. Therefore, one can f-computably find an infinite subset of $\bigcup_i F_i \smallsetminus W = A$.
- Case 2: $n_i \in F_i$ for only finitely many i's. Then the sequence of the n_i's dominates the modulus function $\mu_{\emptyset'}$, and therefore computes the halting set. Since the set A is Δ_2^0, f computes an infinite subset of A. □

Corollary 17 (Hirschfeldt and Shore [10]). $\mathsf{RCA}_0 \wedge \mathsf{CAC} \nvdash \mathsf{DNC}$.

4   ADS and Dependent Hyperimmunity

Lerman et al. [13] separated the ascending descending sequence principle from a stable version of CAC by using a very involved iterated forcing argument. According to our previous simplification of their general framework [16], we reformulate their proof in terms of preservation of dependent hyperimmunity, and extend it to pseudo Ramsey's theorem for pairs.

Definition 18 (Ascending descending sequence). *Given a linear order* $(L, <_L)$, *an* ascending *(descending) sequence is a set S such that for every $x <_{\mathbb{N}} y \in S$, $x <_L y$ ($x >_L y$).* ADS *is the statement "Every infinite linear order admits an infinite ascending or descending sequence".*

Pseudo Ramsey's theorem for pairs was introduced by Friedman [7] and later studied by Friedman and Pelupessy [8], and Murakami et al. in [15] who proved that it is between the chain antichain principle and the ascending descending sequence principle over RCA_0. Steila [20] and the author [17] independently proved that it is actually equivalent to ADS.

Definition 19 (Pseudo Ramsey's theorem). *A set* $H = \{x_0 < x_1 < \ldots\}$ *is* pseudo-homogeneous *for a coloring* $f : [\mathbb{N}]^n \to k$ *if* $f(x_i, \ldots, x_{i+n-1}) = f(x_j, \ldots, x_{j+n-1})$ *for every* $i, j \in \mathbb{N}$. psRT_k^n *is the statement "Every coloring* $f : [\mathbb{N}]^n \to k$ *has an infinite pseudo-homogeneous set".*

Definition 20 (Dependent hyperimmunity). *A formula* $\varphi(U, V)$ *is essential if for every* $x \in \omega$, *there is a finite set* $R > x$ *such that for every* $y \in \omega$, *there is a finite set* $S > y$ *such that* $\varphi(R, S)$ *holds. A pair of sets* $A_0, A_1 \subseteq \omega$ *is* dependently X-hyperimmune *if for every essential* $\Sigma_1^{0,X}$ *formula* $\varphi(U, V)$, $\varphi(R, S)$ *holds for some* $R \subseteq \overline{A}_0$ *and* $S \subseteq \overline{A}_1$.

In particular, if the pair A_0, A_1 is dependently hyperimmune, then A_0 and A_1 are both hyperimmune.

Definition 21 (Preservation of dependent hyperimmunity). *A* Π_2^1 *statement* P *admits* preservation of dependent hyperimmunity *if for every set* Z, *every pair of dependently* Z-hyperimmune *sets* $A_0, A_1 \subseteq \omega$ *and every* P-*instance* $X \leq_T Z$, *there is a solution* Y *to* X *such that* A_0, A_1 *are dependently* $Y \oplus Z$-hyperimmune.

A partial order $(P, \leq_P)$ is *stable* if either $(\forall i \in P)(\exists s)[(\forall j > s)(j \in P \to i \leq_P j) \vee (\forall j > s)(j \in P \to i \mid_P j)]$ or $(\forall i \in P)(\exists s)[(\forall j > s)(j \in P \to i \geq_P j) \vee (\forall j > s)(j \in P \to i \mid_P j)]$. SCAC is the restriction of CAC to stable partial orders. A simple finite injury priority argument shows that SCAC does not admit preservation of dependent hyperimmunity.

Theorem 22. *There exists a computable, stable semi-transitive coloring* $f : [\omega]^2 \to 2$ *such that the pair* $\overline{A}_0, \overline{A}_1$ *is dep. hyperimmune, where* $A_i = \{x : \lim_s f(x, s) = i\}$.

Corollary 23. SCAC *does not admit preservation of dependent hyperimmunity.*

Proof. Let $f : [\omega]^2 \to 2$ be the coloring of Theorem 22. By construction, the pair $\overline{A}_0, \overline{A}_1$ is dependently hyperimmune, where $A_i = \{x : \lim_s f(x, s) = i\}$. Let H be an infinite f-homogeneous set. In particular, $H \subseteq A_0$ or $H \subseteq A_1$. We claim that the pair $\overline{A}_0, \overline{A}_1$ is not dependently H-hyperimmune. The $\Sigma_1^{0,H}$ formula $\varphi(U, V)$ defined by $U \neq \emptyset \wedge V \neq \emptyset \wedge U \cup V \subseteq H$ is essential since H is infinite. However, if there is some $R \subseteq A_1$ and $S \subseteq A_0$ such that $\varphi(R, S)$ holds, then $H \cap A_0 \neq \emptyset$ and $H \cap A_1 \neq \emptyset$, contradicting the choice of H. Therefore $\overline{A}_0, \overline{A}_1$ is not dependently H-hyperimmune. Hirschfeldt and Shore [10] proved that SCAC is equivalent to stable semi-transitive Ramsey's theorem for pairs over RCA$_0$. Therefore SCAC does not admit preservation of dependent hyperimmunity. $\square$

We will now prove the positive preservation result.

Theorem 24. *For every $k \geq 2$, psRT_k^2 admits preservation of dep. hyperimmunity.*

Proof. The proof is done by induction over $k \geq 2$. Fix a pair of sets $A_0, A_1 \subseteq \omega$ dependently Z-hyperimmune for some set Z. Let $f : [\omega]^2 \to k$ be a Z-computable coloring and suppose that there is no infinite set H over which f avoids at least one color, and such that the pair A_0, A_1 is dependently $H \oplus Z$-hyperimmune, as otherwise, we are done by induction hypothesis. We will build k infinite sets $G_0, \dots, G_{k-1}$ such that G_i is pseudo-homogeneous for f with color i for each $i < k$ and such that A_0, A_1 is dependently $G_i \oplus Z$-hyperimmune for some $i < k$. The sets $G_0, \dots, G_{k-1}$ are built by a variant of Mathias forcing $(F_0, \dots, F_{k-1}, X)$ such that

(i) $F_i \cup \{x\}$ is pseudo-homogeneous for f with color i for each $x \in X$
(ii) X is an infinite set such that A_0, A_1 is dependently $X \oplus Z$-hyperimmune

A condition $d = (H_0, \dots, H_{k-1}, Y)$ *extends* $c = (F_0, \dots, F_{k-1}, X)$ (written $d \leq c$) if (H_i, Y) Mathias extends (F_i, X) for each $i < k$. A tuple of sets $G_0, \dots, G_{k-1}$ *satisfies* c if for every $n \in \omega$, there is an extension $d = (H_0, \dots, H_{k-1}, Y)$ of c such that $G_i{\restriction}n \subseteq H_i$ for each $i < k$. Informally, $G_0, \dots, G_{k-1}$ satisfy c if the sets are generated by a decreasing sequence of conditions extending c. In particular, G_i is pseudo-homogeneous for f with color i and satisfies the Mathias condition (F_i, X). The first lemma shows that every sufficiently generic filter yields a k-tuple of infinite sets.

Lemma 25. *For every condition $c = (F_0, \dots, F_{k-1}, X)$ and every $i < k$, there is an extension $d = (H_0, \dots, H_{k-1}, Y)$ of c such that $|H_i| > |F_i|$.*

Fix an enumeration $\varphi_0(G, U, V), \varphi_1(G, U, V), \dots$ of all $\Sigma_1^{0,Z}$ formulas. We want to satisfy the following requirements for each $e_0, \dots, e_{k-1} \in \omega$:

$$\mathcal{R}_{\vec{e}}: \qquad \mathcal{R}_{e_0}^{G_0} \quad \vee \quad \dots \quad \vee \quad \mathcal{R}_{e_{k-1}}^{G_{k-1}}$$

where $\mathcal{R}_e^G$ is the requirement "$\varphi_e(G, U, V)$ essential $\to \varphi_e(G, R, S)$ for some $R \subseteq \overline{A_0}$ and $S \subseteq \overline{A_1}$". We say that a condition c *forces* $\mathcal{R}_{\vec{e}}$ if $\mathcal{R}_{\vec{e}}$ holds for every k-tuple of sets satisfying c. Note that the notion of satisfaction has a precise meaning given above.

Lemma 26. *For every condition c and every k-tuple of indices $e_0, \dots, e_{k-1} \in \omega$, there is an extension d of c forcing $\mathcal{R}_{\vec{e}}$.*

Proof. Fix a condition $c = (F_0, \dots, F_{k-1}, X)$. Let $\psi(U, V)$ be the $\Sigma_1^{0,X \oplus Z}$ formula which holds if there is a k-tuple of sets $E_0, \dots, E_{k-1} \subseteq X$ and a $z \in X$ such that for each $i < k$,

(i) $z > max(E_i)$
(ii) $F_i \cup E_i \cup \{z\}$ is pseudo-homogeneous for color i.
(iii) $\varphi_{e_i}(F_i \cup E_i, U_i, V_i)$ holds for some $U_i \subseteq U$ and $V_i \subseteq V$

Suppose that c does not force $\mathcal{R}_{\bar{e}}$, otherwise we are done.

We claim that ψ is essential. Since c does not force $\mathcal{R}_{\bar{e}}$, there is a k-tuple of infinite sets $G_0, \ldots, G_{k-1}$ satisfying c and such that $\varphi_{e_i}(G_i, U, V)$ is essential for each $i < k$. Fix some $x \in \omega$. By definition of being essential, there are some finite sets $R_0, \ldots, R_{k-1} > x$ such that for every $y \in \omega$, there are finite sets $S_0, \ldots, S_{k-1} > y$ such that $\varphi_{e_i}(G_i, R_i, S_i)$ holds for each $i < k$. Let $R = \bigcup R_i$ and fix some $y \in \omega$. There are finite sets $S_0, \ldots, S_{k-1} > y$ such that $\varphi_{e_i}(G_i, R_i, S_i)$ holds for each $i < k$. Let $S = \bigcup S_i$. By continuity, there are finite sets $E_0, \ldots, E_{k-1}$ such that $G_i \restriction max(E_i) = F_i \cup E_i$ and $\varphi_{e_i}(F_i \cup E_i, R_i, S_i)$ holds for each $i < k$. By our precise definition of satisfaction, we can even assume without loss of generality that $(F_0 \cup E_0, \ldots, F_{k-1} \cup E_{k-1}, Y)$ is a valid extension of c for some infinite set $Y \subseteq X$. Let $z \in Y$. In particular, by the definition of being a condition extending c, $z \in X$, $z > max(E_0, \ldots, E_{k-1})$ and $F_i \cup E_i \cup \{z\}$ is pseudo-homogeneous for color i for each $i < k$. Therefore $\psi(R, S)$ holds, as witnessed by $E_0, \ldots, E_{k-1}$ and z. Thus $\psi(R, S)$ is essential.

Since A_0, A_1 is dependently $X \oplus Z$-hyperimmune, then $\psi(R, S)$ holds for some $R \subseteq \overline{A_0}$ and some $S \subseteq \overline{A_1}$. Let $E_0, \ldots, E_{k-1} \subseteq X$ be the k-tuple of sets and $z \in X$ be the integer witnessing $\psi(R, S)$. Let $i < k$ be such that the set $Y = \{w \in X \setminus [0, max(E_i)] : f(z, w) = i\}$ is infinite. The condition $d = (F_0, \ldots, F_{i-1}, F_i \cup E_i \cup \{z\}, F_{i+1}, \ldots, F_{k-1}, Y)$ is a valid extension of c forcing $\mathcal{R}_{\bar{e}}$. $\square$

Theorem 27. *Fix some set Z and a pair of sets A_0, A_1 dependently Z-hyperimmune. If Y is sufficiently random relative to Z, then the pair A_0, A_1 is dependently $Y \oplus Z$-hyperimmune.*

Corollary 28. WWKL *admits preservation of dependent hyperimmunity.*

Proof. Let Z be a set and A_0, A_1 be a pair of dependently Z-hyperimmune sets. Fix a Z-computable tree of positive measure $T \subseteq 2^{<\omega}$. By Theorem 27, the pair A_0, A_1 is dependently $Y \oplus Z$-hyperimmune for some Martin-Löf random Y relative to Z. By Kučera [12], Y is, up to finite prefix, a path through T. $\square$

Corollary 29. *For every $k \geq 2$,* $\mathsf{RCA}_0 \wedge \mathsf{psRT}_k^2 \wedge \mathsf{WWKL} \nvdash \mathsf{SCAC}$.

Whenever requiring the sets A_0 and A_1 to be co-c.e., we recover the standard notion of hyperimmunity. Therefore, the restriction of the preservation of dependent hyperimmunity to co-c.e. sets is not a good computability-theoretic property to distinguish consequences of Ramsey's theorem for pairs.

Lemma 30. *Fix two sets A_0, A_1 such that A_0 is X-co-c.e. The pair A_0, A_1 is dependently X-hyperimmune iff A_0 and A_1 are X-hyperimmune.*

Corollary 31. RT_2^2 *admits preservation of dependent hyperimmunity for co-c.e. sets.*

References

1. Bienvenu, L., Patey, L., Shafer, P.: On the logical strengths of partial solutions to mathematical problems (2015). http://arxiv.org/abs/1411.5874
2. Bovykin, A., Weiermann, A.: The strength of infinitary Ramseyan principles can be accessed by their densities. Ann. Pure Appl. Logic, 4 (2005 to appear)
3. Cholak, P.A., Giusto, M., Hirst, J.L., Jockusch, Jr., C.G.: Free sets and reverse mathematics. Reverse Math. **21**, 104–119 (2001)
4. Cholak, P.A., Jockusch, C.G., Slaman, T.A.: On the strength of Ramsey's theorem for pairs. J. Symbolic Logic **66**(01), 1–55 (2001)
5. Csima, B.F., Mileti, J.R.: The strength of the rainbow Ramsey theorem. J. Symbolic Logic **74**(04), 1310–1324 (2009)
6. Dzhafarov, D.D., Jockusch, C.G.: Ramsey's theorem and cone avoidance. J. Symbolic Logic **74**(2), 557–578 (2009)
7. Friedman, H.: Adjacent ramsey theory. preprint, 8 (2010). https://u.osu.edu/friedman.8/
8. Friedman, H., Pelupessy, F.: Independence of Ramsey theorem variants using ε_0. Proc. Am. Math. Soc. **144**(2), 853–860 (2016)
9. Hirschfeldt, D.R., Jockusch, C.G., Kjos-Hanssen, B., Lempp, S., Slaman, T.A.: The strength of some combinatorial principles related to Ramsey's theorem for pairs. Comput. Prospects Infinity Part II: Presented Talks **15**, 143–161 (2008). World Scientific Press, Singapore
10. Hirschfeldt, D.R., Shore, R.A.: Combinatorial principles weaker than Ramsey's theorem for pairs. J. Symbolic Logic **72**(1), 171–206 (2007)
11. Jockusch, C.G.: Ramsey's theorem and recursion theory. J. Symbolic Logic **37**(2), 268–280 (1972)
12. Kučera, A.: Measure, Π_1^0 classes, and complete extensions of PA. In: Ebbinghaus, H.-D., Müller, G.H., Sacks, G.E. (eds.) Recursion Theory Week. Lecture Notes in Mathematics, vol. 1141, pp. 245–259. Springer, Heidelberg (1985)
13. Lerman, M., Solomon, R., Towsner, H.: Separating principles below Ramsey's theorem for pairs. J. Math. Logic **13**(02), 1350007 (2013)
14. Liu, L.: Cone avoiding closed sets. Trans. Am. Math. Soc. **367**(3), 1609–1630 (2015). http://dx.org/10.1090/S0002-9947-2014-06049-2
15. Murakami, S., Yamazaki, T., Yokoyama, K.: On the Ramseyan factorization theorem. In: Beckmann, A., Csuhaj-Varjú, E., Meer, K. (eds.) CiE 2014. LNCS, vol. 8493, pp. 324–332. Springer, Heidelberg (2014)
16. Patey, L.: Iterative forcing and hyperimmunity in reverse mathematics. In: Beckmann, A., Mitrana, V., Soskova, M. (eds.) CiE 2015. LNCS, vol. 9136, pp. 291–301. Springer, Heidelberg (2015)
17. Patey, L.: The reverse mathematics of Ramsey-type theorems. Ph.D. thesis. Université Paris Diderot (2016)
18. Patey, L.: The weakness of being cohesive, thin or free in reverse mathematics. Isr. J. Math. (2016 to appear). http://arxiv.org/abs/1502.03709
19. Seetapun, D., Slaman, T.A.: On the strength of Ramsey's theorem. Notre Dame J. Formal Logic **36**(4), 570–582 (1995)
20. Steila, S., Yokoyama, K.: Reverse mathematical bounds for the termination theorem, to appear

A Direct Constructive Proof
of a Stone-Weierstrass Theorem
for Metric Spaces

Iosif Petrakis[(✉)]

University of Munich, Munich, Germany
petrakis@math.lmu.de

Abstract. We present a constructive proof of a Stone-Weierstrass theorem for totally bounded metric spaces (**SWtbms**) which implies Bishop's Stone-Weierstrass theorem for compact metric spaces (**BSWcms**) found in [3]. Our proof has a clear computational content, in contrast to Bishop's highly technical proof of **BSWcms** and his hard to motivate concept of a (Bishop-)separating set of uniformly continuous functions. All corollaries of **BSWcms** in [3] are proved directly by **SWtbms**. We work within Bishop's informal system of constructive mathematics BISH.

1   Introduction

According to the classical Stone-Weierstrass theorem (**SWchts**), if X is a compact Hausdorff topological space and Φ is a separating subalgebra of the continuous real-valued functions $C(X)$ on X that contains a non-zero constant function, then the uniform closure of Φ is $C(X)$ (see [10], p. 282). Recall that Φ is *separating*, if $\forall_{x,y \in X}(x \neq y \rightarrow \exists_{f \in \Phi}(f(x) \neq f(y)))$.

There are some constructive versions of this theorem depending on the notion of space under constructive study. In [1] Banaschewski and Mulvey considered a compact, completely regular locale instead of a compact Hausdorff topological space. In [7] Coquand gave a simple, constructive localic proof of it, replacing the ring structure of $C(X)$ by its lattice structure, while in [8] he studied the usual formulation of the Stone-Weierstrass theorem in this point-free topological framework.

For reasons which we discuss in [15], Bishop did not pursue a constructive reconstruction of abstract topology. Although he introduced two constructive alternatives to the notion of topological space, the notion of *neighborhood space*, see [11,13], and the notion of *function space*, or *Bishop space*, see [4,12,15–17], he never elaborated them, restricting his studies to metric spaces. In [2,3] Bishop formulated a theorem of Stone-Weierstrass type for *compact* metric spaces (i.e., complete and totally bounded metric spaces) using the notion of a Bishop-separating set of uniformly continuous functions[1]. Since Bishop's results, as well as ours, hold

---

[1] Bishop's original term is that of a separating set, which we avoid here in the presence of the standard classical notion of a separating subset of $C(X)$.

© Springer International Publishing Switzerland 2016
A. Beckmann et al. (Eds.): CiE 2016, LNCS 9709, pp. 364–374, 2016.
DOI: 10.1007/978-3-319-40189-8_37

for totally bounded metric spaces, we formulate all related concepts and results for them without restricting to compact metric spaces. Recall that a metric space (X, d) is *totally bounded*, if for every $\epsilon > 0$ there exists a finite ϵ-approximation of X, and a set A is *finite* if there exists a one-one mapping of $\{1, \ldots, n\}$ onto A, for some $n > 0$ (see [6], p. 29). Hence, a totally bounded metric space is always inhabited.

Throughout this paper (X, d) is a totally bounded metric space, $C_u(X)$ denotes the uniformly continuous real-valued functions on X, and $\Phi \subseteq C_u(X)$.

Definition 1. Φ *is called Bishop-separating, if there is* $\delta : \mathbb{R}^+ \to \mathbb{R}^+$ *such that:*
(Bsep$_1$) *For all* $\epsilon > 0$ *and* $x_0, y_0 \in X$, *if* $d(x_0, y_0) \geq \epsilon$, *there exists* $g_{\epsilon, x_0, y_0} \in \Phi$ *such that*

$$\forall_{z \in X}(d_{x_0}(z) \leq \delta(\epsilon) \to |g_{\epsilon, x_0, y_0}(z)| \leq \epsilon) \text{ and}$$

$$\forall_{z \in X}(d_{y_0}(z) \leq \delta(\epsilon) \to |g_{\epsilon, x_0, y_0}(z) - 1| \leq \epsilon).$$

(Bsep$_2$) *For all* $\epsilon > 0$ *and* $x_0 \in X$ *there exists* $g_{\epsilon, x_0} \in \Phi$ *such that*

$$\forall_{z \in X}(d_{x_0}(z) \leq \delta(\epsilon) \to |g_{\epsilon, x_0}(z) - 1| \leq \epsilon).$$

Note that in Definition 1 g_{ϵ, x_0, y_0} and g_{ϵ, x_0} are just notations that do not involve the use of some choice principle. Recall also that for every $x_0 \in X$ the map $d_{x_0} : X \to \mathbb{R}$, defined by $x \mapsto d(x_0, x)$, is in $C_u(X)$ with modulus of continuity $\omega_{d_{x_0}} = \mathrm{id}_{\mathbb{R}^+}$. If $a \in \mathbb{R}$, we denote by $\bar{a}$ the constant map on X with value a, and their set by $\mathrm{Const}(X)$. We define

$$U_0(X) := \{d_{x_0} \mid x_0 \in X\}.$$

$$U_0^*(X) := U_0(X) \cup \{\bar{1}\}.$$

We call Φ *positively separating*, if $\forall_{x,y \in X}(x \bowtie_d y \to \exists_{g \in \Phi}(g(x) \bowtie_{\mathbb{R}} g(y)))$, where $x \bowtie_d y \leftrightarrow d(x, y) > 0$, for every $x, y \in X$, and $a \bowtie_{\mathbb{R}} b \leftrightarrow |a - b| > 0 \leftrightarrow a < b \vee b < a$, for every $a, b \in \mathbb{R}$, are the canonical point-point apartness relations on X and $\mathbb{R}$, respectively. The notion of a positively separating set Φ is the positive version of the classical notion of a separating subset of $C(X)$ for metric spaces. Clearly, $U_0(X)$ is positively separating.

Remark 1. If Φ is Bishop-separating, then Φ is positively separating.

Proof. By the Archimedean property of $\mathbb{R}$ (see [5], p. 57), if $x_0, y_0 \in X$ such that $d(x_0, y_0) > 0$, there is some natural number $N > 2$ such that $d(x_0, y_0) > \frac{1}{N}$. By Bsep$_1$ we have that $|g_{\frac{1}{N}, x_0, y_0}(x_0)| \leq \frac{1}{N}$ and $|g_{\frac{1}{N}, x_0, y_0}(y_0) - 1| \leq \frac{1}{N}$, for some $g_{\frac{1}{N}, x_0, y_0} \in \Phi$, therefore $g_{\frac{1}{N}, x_0, y_0}(x_0) \bowtie_{\mathbb{R}} g_{\frac{1}{N}, x_0, y_0}(y_0)$.

In [3], p. 106, Bishop formulated a theorem of Stone-Weierstrass type for compact metric spaces using the notion of a Bishop-separating set as the property that corresponds to the classical notion of a separating set in the formulation of **SWchts**. Bishop's proof of this theorem is non-trivial and does not involve the completeness property of X. Following Bishop, we denote by $\mathcal{A}(\Phi)$ the least subset of $C_u(X)$ that includes Φ and it is closed with respect to addition, multiplication, and multiplication by reals. Bishop didn't define $\mathcal{A}(\Phi)$ inductively but explicitly, as the set of compositions of strict real polynomials in several variables with vectors of elements of Φ (see [3], p. 105).

Theorem 1 (Bishop's Stone-Weierstrass theorem for totally bounded metric spaces (BSWtbms)). *If Φ is Bishop-separating, then $\mathcal{A}(\Phi)$ is dense in $C_u(X)$.*

The condition of Φ being Bishop-separating implies that the constant map $\bar{1}$ is in the closure of $\mathcal{A}(\Phi)$ (see [3], p. 106). Bishop's formulation of **BSWtbms** represents a non-trivial technical achievement, namely to find a formulation of a theorem of Stone-Weierstrass type in the constructive theory of metric spaces that resembles the formulation of the classical **SWchts**. As Coquand and Spitters mention in [9], pp. 339–340, constructive proofs using a concrete presentation of topological notions (e.g., the Gelfand spectrum as a lattice) are "more direct than proofs via an encoding of topology in metric spaces, as is common in Bishop's constructive mathematics".

In the next two sections we present a Stone-Weierstrass theorem for metric spaces which avoids the concept of a Bishop-separating set, it has an informative and direct proof, it implies **BSWtbms**, and it proves directly all corollaries of **BSWtbms**.

2   A Stone-Weierstrass Theorem for Totally Bounded Metric Spaces

Definition 2. *If $f, g \in C_u(X)$ and $\epsilon > 0$, then $f \wedge g := \min\{f, g\}, f \vee g := \max\{f, g\}$, and the uniform closure $\mathcal{U}(\Phi)$ of Φ is defined by*

$$U(g, f, \epsilon) :\leftrightarrow \forall_{x \in X}(|g(x) - f(x)| \leq \epsilon),$$

$$U(\Phi, f) :\leftrightarrow \forall_{\epsilon > 0} \exists_{g \in \Phi}(U(g, f, \epsilon)),$$

$$\mathcal{U}(\Phi) := \{f \in C_u(X) \mid U(\Phi, f)\}.$$

The following remark is immediate to show.

Remark 2. If Φ is closed under addition, multiplication by reals and multiplication, then $\mathcal{U}(\Phi)$ is closed under addition, multiplication by reals and multiplication. Moreover, if Φ is closed under $|.|$, then $\mathcal{U}(\Phi)$ is closed under $|.|$.

The next two lemmas are proved in [3], pp. 105–106 (Lemmas 5.11 and 5.12).

Lemma 1. *If $\mathrm{Const}(X) \subseteq \Phi$, and Φ is closed under addition and multiplication (or if Φ is closed under addition, multiplication by reals, and multiplication), then $\mathcal{U}(\Phi)$ is closed under $|.|, \vee$ and $\wedge$.*

Lemma 2. *If Φ is closed under addition, multiplication by reals, and multiplication, and $f \in \mathcal{U}(\Phi)$ such that $\forall_{x \in X}(|f(x)| \geq c)$, for some $c > 0$, then $\frac{1}{f} \in \mathcal{U}(\Phi)$.*

Corollary 1. *If $x_0, y_0 \in X$ such that $d(x_0, y_0) > 0$, then $\bar{1} \in \mathcal{U}(\mathcal{A}(U_0(X)))$.*

Proof. If $x \in X$, then $0 < d(x_0, y_0) \le d(x_0, x) + d(x, y_0) = d_{x_0}(x) + d_{y_0}(x)$ i.e., $d(x_0, y_0) \le d_{x_0} + d_{y_0} \in \mathcal{A}(U_0(X))$. By Lemma 2 we get that $\frac{1}{d_{x_0} + d_{y_0}} \in \mathcal{U}(\mathcal{A}(U_0(X)))$, therefore $\bar{1} \in \mathcal{U}(\mathcal{A}(U_0(X)))$.

The existence of $x_0, y_0 \in X$ such that $d(x_0, y_0) > 0$ is equivalent to the positivity of the diameter of (X, d) (see the footnote in the proof of Lemma 3).

Definition 3. *If $\mathbb{F}(X)$ denotes the set of real-valued functions on X, the set of Lipschitz functions $\mathrm{Lip}(X)$ on (X, d) is defined by*

$$\mathrm{Lip}(X, k) := \{ f \in \mathbb{F}(X) \mid \forall_{x,y \in X}(|f(x) - f(y)| \le kd(x, y)) \},$$

$$\mathrm{Lip}(X) := \bigcup_{k \ge 0} \mathrm{Lip}(X, k).$$

Remark 3. The set $\mathrm{Lip}(X) \subseteq C_u(X)$ includes $U_0(X)$, $\mathrm{Const}(X)$ and it is closed under addition, multiplication by reals, and multiplication.

Proof. If $x_0 \in X$, then $|d(x_0, x) - d(x_0, y)| \le d(x, y)$, for every $x, y \in X$, therefore $U_0(X) \subseteq \mathrm{Lip}(X, 1)$. Clearly, $\mathrm{Const}(X) \subseteq \mathrm{Lip}(X, k)$, for every $k \ge 0$. Recall that $f \cdot g = \frac{1}{2}((f + g)^2 - f^2 - g^2)$, and if $M_f > 0$ is a bound of f, it is immediate to see that

$$f \in \mathrm{Lip}(X, k_1) \to g \in \mathrm{Lip}(X, k_2) \to f + g \in \mathrm{Lip}(X, k_1 + k_2),$$

$$f \in \mathrm{Lip}(X, k) \to \lambda \in \mathbb{R} \to \lambda f \in \mathrm{Lip}(X, |\lambda| k),$$

$$f \in \mathrm{Lip}(X, k) \to f^2 \in \mathrm{Lip}(X, 2M_f k).$$

Lemma 3. *If $\Phi = \mathcal{A}(U_0^*(X))$, then $\mathrm{Lip}(X) \subseteq \mathcal{U}(\Phi)$.*

Proof. It suffices to show that $\mathrm{Lip}(X, 1) \subseteq \mathcal{U}(\Phi)$, since if $f \in \mathrm{Lip}(X, k)$, for some $k > 0$, then $\frac{1}{k} f \in \mathrm{Lip}(X, 1)$ and we have, for every $\epsilon > 0$ and $\theta \in \Phi$,

$$U(\theta, \frac{1}{k} f, \frac{\epsilon}{k}) \to U(k\theta, f, \epsilon).$$

Suppose next that $f \in \mathrm{Lip}(X, 1)$ and $\epsilon > 0$. We find $g \in \mathcal{U}(\Phi)$ such that $U(g, f, \epsilon)$, therefore $f \in \mathcal{U}(\mathcal{U}(\Phi)) = \mathcal{U}(\Phi)$. More specifically, if $\{z_1, \ldots, z_m\}$ is an $\frac{\epsilon}{2}$-approximation of X, we find $g \in \mathcal{U}(\Phi)$ such that $g(z_i) = f(z_i)$, for every $i \in \{1, \ldots, m\}$, and $|g(x) - g(z_i)| = |g(x) - f(z_i)| \le \frac{\epsilon}{2}$, for every $x \in X$ and z_i such that $d(x, z_i) \le \frac{\epsilon}{2}$. Consequently,

$$\begin{aligned} |g(x) - f(x)| &\le |g(x) - g(z_i)| + |g(z_i) - f(z_i)| + |f(z_i) - f(x)| \\ &\le \frac{\epsilon}{2} + 0 + d(z_i, x) \\ &\le \frac{\epsilon}{2} + \frac{\epsilon}{2} \\ &= \epsilon. \end{aligned}$$

We define

$$g := \bigwedge_{k=1}^{m} (\overline{f(z_k)} + d_{z_k}).$$

Since $\overline{f(z_k)} + d_{z_k} \in \Phi$ and since by Lemma 1 $\mathcal{U}(\Phi)$ is closed under $\wedge$ we get $g \in \mathcal{U}(\Phi)$. Moreover,

$$g(z_i) = \bigwedge_{k=1}^{m} (f(z_k) + d_{z_k}(z_i)) \leq f(z_i) + d_{z_i}(z_i) = f(z_i).$$

For the converse inequality we suppose that $g(z_i) < f(z_i)$ and reach a contradiction (here we use the fact that $\neg(a < b) \to a \geq b$, for every $a, b \in \mathbb{R}$ (see [3], p. 26)). If $a, b, c \in \mathbb{R}$, then one shows[2] that $a \wedge b < c \to a < c \vee b < c$. Hence

$$\bigwedge_{k=1}^{m} (f(z_k) + d_{z_k}(z_i)) < f(z_i) \to \exists_{j \in \{1,\dots,m\}} (f(z_j) + d(z_j, z_i) < f(z_i))$$

$$\to d(z_j, z_i) < f(z_i) - f(z_j) \leq |f(z_i) - f(z_j)| \leq d(z_j, z_i),$$

which is a contradiction. Using the equality $g(z_i) = f(z_i)$ we have that

$$g(x) = \bigwedge_{k=1}^{m} (f(z_k) + d_{z_k}(x)) \leq f(z_i) + d_{z_i}(x) \to$$

$$g(x) - g(z_i) \leq f(z_i) + d_{z_i}(x) - g(z_i) = f(z_i) + d_{z_i}(x) - f(z_i) = d(x, z_i) \leq \frac{\epsilon}{2}.$$

If $k \in \{1,\dots,m\}$, then $f(z_i) - f(z_k) \leq |f(z_i) - f(z_k)| \leq d(z_i, z_k) \leq d(z_i, x) + d(x, z_k)$, therefore

$$\forall_{k \in \{1,\dots,m\}} (f(z_i) - d(z_i, x) \leq f(z_k) + d(z_k, x)) \to$$

$$f(z_i) - d(z_i, x) \leq \bigwedge_{k=1}^{m} (f(z_k) + d(z_k, x)) \leftrightarrow$$

$$f(z_i) - \bigwedge_{k=1}^{m} (f(z_k) + d(z_k, x)) \leq d(z_i, x) \to$$

$$g(z_i) - g(x) \leq d(z_i, x) \to$$

$$g(z_i) - g(x) \leq \frac{\epsilon}{2}.$$

From $g(x) - g(z_i) \leq \frac{\epsilon}{2}$ and $g(z_i) - g(x) \leq \frac{\epsilon}{2}$ we get $|g(x) - g(z_i)| \leq \frac{\epsilon}{2}$.

---

[2] The proof goes as follows. By the constructive trichotomy property (see [3], p. 26) either $a < c$ or $a \wedge b < a$. In the first case we get immediately what we want to show. In the second case we get that $b \leq a$, since if $b > a$, we have that $a = a \wedge b < a$, which is a contradiction. Thus $a \wedge b = b$ and the hypothesis $a \wedge b < c$ becomes $b < c$.

Lemma 4. *If $f \in C_u(X)$ and $\epsilon > 0$, there exist $\sigma > 0$ and $g, g^* \in \mathrm{Lip}(X, \sigma)$ such that*
(i) $\forall_{x \in X}(f(x) - \epsilon \leq g(x) \leq f(x) \leq g^*(x) \leq f(x) + \epsilon)$.
(ii) For every $e \in \mathrm{Lip}(X, \sigma)$, if $e \leq f$, then $e \leq g$.
(iii) For every $e^* \in \mathrm{Lip}(X, \sigma)$, if $f \leq e^*$, then $g^* \leq e^*$.

Proof. (i) Let ω_f be a modulus of continuity of f and $M_f > 0$ a bound of f. We define the functions $h_x : X \to \mathbb{R}$ and $g : X \to \mathbb{R}$ by

$$h_x := f + \sigma d_x,$$

$$\sigma := \frac{2M_f}{\omega_f(\epsilon)} > 0,$$

$$g(x) := \inf\{h_x(y) \mid y \in X\} = \inf\{f(y) + \sigma d(x, y) \mid y \in X\},$$

for every $x \in X$. Note that $g(x)$ is well-defined, since $h_x \in C_u(X)$ and the infimum of h_x exists (see [3], p. 38, 94). First we show that $g \in \mathrm{Lip}(X, \sigma)$. If $x_1, x_2, y \in X$ the inequality $d(x_1, y) \leq d(x_2, y) + d(x_1, x_2)$ implies that $f(y) + \sigma d(x_1, y) \leq (f(y) + \sigma d(x_2, y)) + \sigma d(x_1, x_2)$, hence $g(x_1) \leq (f(y) + \sigma d(x_2, y)) + \sigma d(x_1, x_2)$, therefore $g(x_1) \leq g(x_2) + \sigma d(x_1, x_2)$, or $g(x_1) - g(x_2) \leq \sigma d(x_1, x_2)$. Starting with the inequality $d(x_2, y) \leq d(x_1, y) + d(x_1, x_2)$ and working similarly we get that $g(x_2) - g(x_1) \leq \sigma d(x_1, x_2)$, therefore $|g(x_1) - g(x_2)| \leq \sigma d(x_1, x_2)$. Next we show that

$$\forall_{x \in X}(f(x) - \epsilon \leq g(x) \leq f(x)).$$

Since $f(x) = f(x) + \sigma d(x, x) = h_x(x) \geq \inf\{h_x(y) \mid y \in X\} = g(x)$, for every $x \in X$, we have that $g \leq f$. Next we show that $\forall_{x \in X}(f(x) - \epsilon \leq g(x))$. For that we fix $x \in X$ and we show that $\neg(f(x) - \epsilon > g(x))$. Note that if $A \subseteq \mathbb{R}, b \in \mathbb{R}$, then[3] $b > \inf A \to \exists_{a \in A}(a < b)$. Therefore,

$$f(x) - \epsilon > g(x) \leftrightarrow$$
$$f(x) - \epsilon > \inf\{f(y) + \sigma d(x, y) \mid y \in X\} \to$$
$$\exists_{y \in X}(f(x) - \epsilon > f(y) + \sigma d(x, y)) \leftrightarrow$$
$$\exists_{y \in X}(f(x) - f(y) > \epsilon + \sigma d(x, y)).$$

For this y we show that $d(x, y) \leq \omega_f(\epsilon)$. If $d(x, y) > \omega_f(\epsilon)$, we have that

$$2M_f \geq f(x) + M_f \geq f(x) - f(y) > \epsilon + 2M_f \frac{d(x, y)}{\omega_f(\epsilon)} > \epsilon + 2M_f > 2M_f,$$

which is a contradiction. Hence, by the uniform continuity of f we get that $|f(x) - f(y)| \leq \epsilon$, therefore the contradiction $\epsilon > \epsilon$ is reached, since

$$\epsilon \geq |f(x) - f(y)| \geq f(x) - f(y) > \epsilon + \sigma d(x, y) \geq \epsilon.$$

---

[3] By the definition of $\inf A$ in [3], p. 37, we have that $\forall_{\epsilon > 0} \exists_{a \in A}(a < \inf A + \epsilon)$, therefore if $b > \inf A$ and $\epsilon = b - \inf A > 0$ we get that $\exists_{a \in A}(a < \inf A + (b - \inf A) = b)$.

Next we define the functions $h_x^* : X \to \mathbb{R}$ and $g^* : X \to \mathbb{R}$ by

$$h_x^* := f - \sigma d_x,$$

$$g^*(x) := \sup\{h_x^*(y) \mid y \in X\} = \sup\{f(y) - \sigma d(x,y) \mid y \in X\},$$

for every $x \in X$, and $\sigma = \frac{2M_f}{\omega_f(\epsilon)}$. Similarly[4] to g we get that $g^* \in \mathrm{Lip}(X, \sigma)$ and

$$\forall_{x \in X}(f(x) \leq g^*(x) \leq f(x) + \epsilon).$$

(ii) Let $e \in \mathrm{Lip}(X, \sigma)$ such that $e \leq f$. If we fix some $x \in X$, then for every $y \in X$ we have that $e(x) - e(y) \leq |e(x) - e(y)| \leq \sigma d(x,y)$, hence $e(x) \leq e(y) + \sigma d(x,y) \leq f(y) + \sigma d(x,y)$, therefore $e(x) \leq g(x)$.

(iii) Let $e^* \in \mathrm{Lip}(X, \sigma)$ such that $f \leq e^*$. If we fix some $x \in X$, then for every $y \in X$ we have that $e^*(y) - e^*(x) \leq |e^*(y) - e^*(x)| \leq \sigma d(x,y)$, hence $f(y) - \sigma d(x,y) \leq e^*(y) - \sigma d(x,y) \leq e^*(x)$, therefore $g^*(x) \leq e^*(x)$.

Hence g is the largest function in $\mathrm{Lip}(X, \sigma)$ which is smaller than f, and g^* is the smallest function in $\mathrm{Lip}(X, \sigma)$ which is larger than f. So, if there is some $e' \in \mathrm{Lip}(X)$ such that $e' \leq f$ and $g(x) < e'(x)$, for some $x \in X$, then $e' \in \mathrm{Lip}(X, \sigma')$, for some $\sigma' > \sigma$. It is interesting that Lemma 4 is in complete analogy to the McShane-Kirszbraun theorem. To make this clear we include a constructive version of this theorem (for a classical presentation see [18], p. 6). Recall that $A \subseteq X$ is *located*, if the distance $d(x, A) := \inf\{d(x,y) \mid y \in Y\}$ exists for every $x \in X$, and that a located subset of a totally bounded metric space is totally bounded (see [3], p. 95).

Proposition 1 (McShane-Kirszbraun theorem for totally bounded metric spaces). *If $\sigma > 0$, $A \subseteq X$ is located, and $f : A \to \mathbb{R} \in \mathrm{Lip}(A, \sigma)$, then there exist $g, g^* \in \mathrm{Lip}(X, \sigma)$ such that $g_{|A} = g^*_{|A} = f$ and for every $e \in \mathrm{Lip}(X, \sigma)$ such that $e_{|A} = f$ we have that $g^* \leq e \leq g$.*

Proof. The functions g, g^* defined by $g(x) := \inf\{f(a) + \sigma d(x,a) \mid a \in A\}$, and $g^*(x) := \sup\{f(a) - \sigma d(x,a) \mid a \in A\}$, for every $x \in X$, are well-defined and satisfy the required properties.

Corollary 2. $\mathcal{U}(\mathrm{Lip}(X)) = C_u(X)$.

Proof. If $\epsilon > 0$, then the functions $g, g^* \in \mathrm{Lip}(X, \sigma)$ of Lemma 4 satisfy $U(g, f, \epsilon)$, $U(g^*, f, \epsilon)$, respectively.

Next follows our Stone-Weierstrass theorem for totally bounded metric spaces.

Theorem 2 (Stone-Weierstrass theorem for totally bounded metric spaces (SWtbms)). *If $\Phi = \mathcal{A}(U_0^*(X))$, then $C_u(X) = \mathcal{U}(\Phi)$.*

Proof. First we show that $C_u(X) \subseteq \mathcal{U}(\Phi)$. If $f \in C_u(X)$ and $\epsilon > 0$, then by Corollary 2 there exists $h \in \mathrm{Lip}(X)$ such that $U(h, f, \frac{\epsilon}{2})$, while by Lemma 3 there exists $g \in \Phi$ such that $U(g, h, \frac{\epsilon}{2})$. Consequently, $U(g, f, \epsilon)$. The converse inclusion follows from the immediate fact that all elements of $\mathcal{U}(\Phi)$ are in $C_u(X)$.

---

[4] To show that $\neg(g^*(x) > f(x) + \epsilon)$ we just use the fact that if $A \subseteq \mathbb{R}, b \in \mathbb{R}$, then $\sup A > b \to \exists_{a \in A}(a > b)$. The function g^* is mentioned in [19], where non-constructive properties of the classical $(\mathbb{R}, <)$ are used.

3   Corollaries of SWtbms

Proposition 2. SWtbms *implies* **BSWtbms**

Proof. The proof follows immediately by inspection of the proof of **BSWtbms** in [3], pp. 106–108. Bishop shows there that if Φ is Bishop-separating, then $\overline{1} \in \mathcal{U}(\mathcal{A}(\Phi))$, and by his Lemma 5.14.1 one shows that $U_0(X) \subseteq \mathcal{U}(\mathcal{A}(\Phi))$ - this is a slight simplification of the final part of Bishop's proof that $C_u(X) \subseteq \mathcal{U}(\mathcal{A}(\Phi))$. Since $U_0^*(X) \subseteq \mathcal{U}(\mathcal{A}(\Phi))$, then by Remark 2 $\mathcal{A}(U_0^*(X)) \subseteq \mathcal{U}(\mathcal{A}(\Phi))$, therefore $C_u(X) = \mathcal{U}(\mathcal{A}(U_0^*(X))) \subseteq \mathcal{U}(\mathcal{U}(\mathcal{A}(\Phi))) = \mathcal{U}(\mathcal{A}(\Phi))$.

In the proof of Corollary 5.16 in [3], pp. 108–109, it is shown that if (X, d) has positive diameter, then $\mathcal{A}(U_0(X))$ is a Bishop-separating set, therefore by **BSWtbms** we get that $\mathcal{U}(\mathcal{A}(U_0(X))) = C_u(X)$. Hence **SWtbms** is only "slightly" stronger than **BSWtbms**. If we use **SWtbms**, we get immediately the same result.

Corollary 3. *If (X, d) has positive diameter, then $\mathcal{U}(\mathcal{A}(U_0(X))) = C_u(X)$.*

Proof. The hypothesis of positive diameter implies the hypothesis of Corollary 1, therefore $\overline{1} \in \mathcal{U}(\mathcal{A}(U_0(X))) \subseteq C_u(X)$. Hence $U_0^*(X) \subseteq \mathcal{U}(\mathcal{A}(U_0(X)))$, and by Remark 2 we get that $\mathcal{A}(U_0^*(X)) \subseteq \mathcal{A}(\mathcal{U}(\mathcal{A}(U_0(X)))) = \mathcal{U}(\mathcal{A}(U_0(X)))$. Thus $C_u(X) = \mathcal{U}(\mathcal{A}(U_0^*(X))) \subseteq \mathcal{U}(\mathcal{U}(\mathcal{A}(U_0(X)))) = \mathcal{U}(\mathcal{A}(U_0(X)))$.

Next we prove Corollary 5.15 in [3], p. 108 and its finite version using **SWtbms**. If $(X, d), (Y, \rho)$ are totally bounded, then $(X \times Y, \sigma)$ is totally bounded, where $\sigma((x_1, y_1), (x_2, y_2)) := d(x_1, x_2) + \rho(y_1, y_2)$, for every $x_1, x_2 \in X$ and $y_1, y_2 \in Y$; if $A = \{x_1, \ldots, x_n\}$ is an $\frac{\epsilon}{2}$-approximation of X and $B = \{y_1, \ldots, y_m\}$ is an $\frac{\epsilon}{2}$-approximation of Y, then $A \times B$ is an ϵ-approximation of $X \times Y$. We denote by π_1 the projection of $X \times Y$ onto X and by π_2 its projection onto Y.

Corollary 4. *If $(X, d), (Y, \rho)$ are totally bounded metric spaces and*

$$\Phi := \{\sum_{i=1}^{n}(f_i \circ \pi_1)(g_i \circ \pi_2) \mid f_i \in C_u(X), g_i \in C_u(Y), 1 \leq i \leq n, n \in \mathbb{N}\},$$

then $\mathcal{U}(\Phi) = C_u(X \times Y)$.

Proof. Clearly, $\Phi \subseteq C_u(X \times Y)$, Φ is an algebra (actually, $\Phi = \mathcal{A}((C_u(X) \circ \pi_1) \cup (C_u(Y) \circ \pi_2))$, where $C_u(X) \circ \pi_1 = \{f \circ \pi_1 \mid f \in C_u(X)\}$ and $C_u(Y) \circ \pi_2 = \{g \circ \pi_2 \mid g \in C_u(Y)\}$), and $\mathcal{U}(\Phi) \subseteq C_u(X \times Y)$. The constant $\overline{1}$ on $X \times Y$ is equal to $(\overline{1} \circ \pi_1)(\overline{1} \circ \pi_2)$. If $x_0, x \in X$ and $y_0, y \in Y$, then $\sigma_{(x_0, y_0)}((x, y)) = \sigma((x_0, y_0), (x, y)) = d(x_0, x) + \rho(y_0, y) = d_{x_0}(x) + \rho_{y_0}(y) = (d_{x_0} \circ \pi_1)((x, y)) + (\rho_{y_0} \circ \pi_2)((x, y))$, therefore $\sigma_{(x_0, y_0)} = (d_{x_0} \circ \pi_1) + (\rho_{y_0} \circ \pi_2) = (d_{x_0} \circ \pi_1)(\overline{1} \circ \pi_2) + (\overline{1} \circ \pi_1)(\rho_{y_0} \circ \pi_2) \in \Phi$. Since $U_0^*(X \times Y) \subseteq \mathcal{U}(\Phi)$, by **SWtbms** we get that $C_u(X \times Y) \subseteq \mathcal{U}(\Phi)$.

If (X_n, d_n) is totally bounded, where without loss of generality $d_n \leq \bar{1}$, for every $n \in \mathbb{N}$, then (X, σ_∞), where $X = \prod_{n=1}^\infty X_n$ and $\sigma_\infty((x_n)_{n=1}^\infty, (y_n)_{n=1}^\infty) :=$ $\sum_{n=1}^\infty \frac{d_n(x_n, y_n)}{2^n}$, is totally bounded; if $A(X_n, \epsilon)$ is an ϵ-approximation of X_n and $x_{0,n}$ inhabits X_n, then $A(X, \epsilon) = \prod_{k=1}^{n_0} A(X_k, \frac{2^{k-1}\epsilon}{n_0}) \times \prod_{k=n_0+1}^\infty \{x_{0,k}\}$ is an ϵ-approximation of X, where $n_0 \in \mathbb{N}$ such that $\sum_{k=n_0+1}^\infty \frac{1}{2^k} \leq \frac{\epsilon}{2}$.

Corollary 5. *If (X, σ_∞) is the product of a sequence $(X_n, d_n)_{n=1}^\infty$ of totally bounded metric spaces, then $\mathcal{U}(\Phi) = C_u(X)$, where*

$$\Phi_0 := \{\prod_{i=1}^n (f_i \circ \pi_i) \mid f_i \in C_u(X_i), 1 \leq i \leq n, n \in \mathbb{N}\},$$

$$\Phi := \{\sum_{k=1}^n h_k \mid h_k \in \Phi_0, 1 \leq k \leq n, n \in \mathbb{N}\}.$$

Proof. Without loss of generality let $d_n \leq \bar{1}$, for every $n \in \mathbb{N}$. The only difference with the proof of Corollary 4 is treated as follows. If $(x_n^0)_{n=1}^\infty \in X$ and $\epsilon > 0$, let

$$g := \sum_{k=1}^{n_0} \frac{d_{k, x_k^0} \circ \pi_k}{2^k} = \sum_{k=1}^{n_0} (\frac{d_{k, x_k^0}}{2^k}) \circ \pi_k \in \Phi,$$

where $n_0 \in \mathbb{N}$ such that $\sum_{k=n_0+1}^\infty \frac{1}{2^k} \leq \epsilon$. We get $U(g, \sigma_{\infty, (x_n^0)_{n=1}^\infty}, \epsilon)$, since

$$|g((y_n)_{n=1}^\infty) - \sigma_{\infty, (x_n^0)_{n=1}^\infty}((y_n)_{n=1}^\infty)| = |\sum_{k=n_0+1}^\infty \frac{d_{k, x_k^0}(y_k)}{2^k}| \leq \sum_{k=n_0+1}^\infty |\frac{d_k(x_k^0, y_k)}{2^k}| \leq \epsilon.$$

Recall that a totally bounded metric space is separable (see [3], p. 94). The separability of $C_u(X)$ follows by the next corollary.

Corollary 6. *If $Q = \{q_n \mid n \in \mathbb{N}\}$ is dense in (X, d), $U_0(Q) := \{d_{q_n} \mid n \in \mathbb{N}\}$, and $\Phi_0^* = \mathcal{A}(U_0(Q) \cup \{\bar{1}\})$, then $\mathcal{U}(\Phi_0^*) = C_u(X)$.*

Proof. If $(x_n)_{n=1}^\infty \in X^\mathbb{N}$ converges pointwise to x, then $(d_{x_n})_{n=1}^\infty$ converges uniformly to d_x [i.e., if $\forall_{\epsilon>0}\exists_{n_0}\forall_{n \geq n_0}(d(x_n, x) \leq \epsilon)$, then $\forall_{\epsilon>0}\exists_{n_0}\forall_{n \geq n_0}\forall_{y \in X}(|d(x_n, y) - d(x, y)| \leq \epsilon)$]. If $\epsilon > 0$ and $n \geq n_0$, then $d(x_n, y) \leq d(x_n, x) + d(x, y) \to d(x_n, y) - d(x, y) \leq d(x_n, x) \leq \epsilon$, and similarly $d(x, y) - d(x_n, y) \leq d(x_n, x) \leq \epsilon$. By **SWtbms** it suffices to show that $U_0(X) \subseteq \mathcal{U}(\mathcal{A}(U_0(Q)))$. If $d_x \in U_0(X)$, for some $x \in X$, and $(q_{k_n})_{n=1}^\infty$ is a subsequence of Q that converges pointwise to x, then $(d_{q_{k_n}})_{n=1}^\infty$ converges uniformly to d_x, therefore $d_x \in \mathcal{U}(\mathcal{A}(U_0(Q)))$.

4   Concluding Comments

We presented a direct constructive proof of **SWtbms** with a clear computational content. Its translation to Type Theory and its implementation to a proof assistant like Coq are expected to be straightforward. Although **SWtbms** does not look like a theorem of Stone-Weierstrass type, as **BSWtbms** does, it has certain advantages over it. Its proof is "natural", in comparison to Bishop's technical proof and his difficult to motivate concept of a Bishop-separating set. As we explained, **SWtbms** implies **BSWtbms**, and all applications of **BSWtbms** in [3] are proved directly by **SWtbms**. We know of no application of **BSWtbms** which cannot be derived by **SWtbms** (in [3] we found only one application of Corollary 4 and one of the Weierstrass approximation theorem[5]).

An interesting question related to Corollary 2 is if for (X, d) totally bounded and (Y, ρ) complete metric space, the set of Lipschitz functions $\mathrm{Lip}(X, Y)$ between them is a dense subset of the uniformly continuous functions $C_u(X, Y)$ between them. A similar classical result, see [14], requires a Lipschitz extension property, which indicates that the correlation of Lemma 4 to the McShane-Kirszbraun theorem may not be accidental.

References

1. Banaschewski, B., Mulvey, C.J.: A constructive proof of the Stone-Weierstrass theorem. J. Pure Appl. Algebra **116**, 25–40 (1997)
2. Bishop, E.: Foundations of Constructive Analysis. McGraw-Hill, New York (1967)
3. Bishop, E., Bridges, D.: Constructive Analysis. Grundlehren der mathematischen Wissenschaften, vol. 279. Springer, New York (1985)
4. Bridges, D.S.: Reflections on function spaces. Ann. Pure Appl. Logic **103**(12), 101 110 (2012)
5. Bridges, D.S., Vîţă, L.S.: Techniques of Constructive Analysis: Universitext. Springer, New York (2006)
6. Bridges, D.S., Richman, F.: Varieties of Constructive Mathematics. Cambridge University Press, New York (1987)
7. Coquand, T.: A Constructive Analysis of the Stone-Weierstrass Theorem, Manuscript (2001)
8. Coquand, T.: About Stone's notion of spectrum. J. Pure Appl. Algebra **197**, 141–158 (2005)
9. Coquand, T., Spitters, B.A.W.: Constructive Gelfand duality for C^*-algebras. Math. Proc. Camb. Philos. Soc. **147**(2), 339–344 (2009)
10. Dugundji, J.: Topology. Brown Publishers Wm. C, Dubuque (1989)
11. Ishihara, H.: Two subcategories of apartness spaces. Ann. Pure Appl. Logic **163**, 132–139 (2013)
12. Ishihara, H.: Relating Bishop's function spaces to neighborhood spaces. Ann. Pure Appl. Logic **164**, 482–490 (2013)
13. Ishihara, H., Mines, R., Schuster, P., Vîţă, L.S.: Quasi-apartness and neighborhood spaces. Ann. Pure Appl. Logic **141**, 296–306 (2006)
14. Miculescu, R.: Approximations by Lipschitz functions generated by extensions. Real Anal. Exch. **28**(1), 33–40 (2002)

---

[5] The first, in p. 414, is the uniform approximation of a test function $f(x, y)$ on $G \times G$, where G is a locally compact group, by finite sums of the form $\sum_i f_i(x) g_i(y)$, and the second, in p. 375, is a density theorem in the theory of Hilbert spaces.

15. Petrakis, I.: Constructive Topology of Bishop Spaces, Ph.D. Thesis. Ludwig-Maximilians-Universität, München (2015)
16. Petrakis, I.: Completely regular Bishop spaces. In: Beckmann, A., Mitrana, V., Soskova, M. (eds.) CiE 2015. LNCS, vol. 9136, pp. 302–312. Springer, Heidelberg (2015)
17. Petrakis, I.: The Urysohn extension theorem for Bishop spaces. In: Artemov, S., et al. (eds.) LFCS 2016. LNCS, vol. 9537, pp. 299–316. Springer, Heidelberg (2016). doi:10.1007/978-3-319-27683-0_21
18. Tuominen, H.: Analysis in Metric Spaces, Lecture notes (2014)
19. http://math.stackexchange.com/questions/665587/

Author Index

Printed in the United States
By Bookmasters